CONSERVATION BIOLOGY AND BIODIVERSITY

ECOLOGICAL RECIPROCITY

A TREATISE ON KINDNESS

CONSERVATION BIOLOGY AND BIODIVERSITY

Additional books and e-books in this series can be found on Nova's website under the Series tab.

CONSERVATION BIOLOGY AND BIODIVERSITY

ECOLOGICAL RECIPROCITY

A TREATISE ON KINDNESS

MICHAEL CHARLES TOBIAS
AND
JANE GRAY MORRISON

(TRANSLATION INTO GREEK
BY NIKI STAVROU)

nova
science publishers
New York

https://doi.org/10.52305/TQZH6902

Library of Congress Cataloging-in-Publication Data

ISBN: 978-1-53619-957-4

Published by Nova Science Publishers, Inc. † New York

IN ENGLISH

Milford Sound, Fiordland National Park, Southern New Zealand. Photo © M. C. Tobias.

CONTENTS

IN ENGLISH

Acknowledgments

The Authors thank Ms. Nadya Columbus, President of Nova Science Publishers, for her friendship and her belief in this work. We are incredibly grateful to Ms. Niki Stavrou for her extraordinary efforts in translating this text into Greek. And we are also incredibly thankful for Dimitris Moutafis for proofreading and integrating the scientific terminology in Greek.

Michael Charles Tobias
Jane Gray Morrison

INTRODUCTION

From our earliest ethical injunctions we have been advised to "love thy neighbor as thyself" (Leviticus 19:18), a challenge reiterated throughout the Old and New Testaments. A message alerting us to what, in fact, is a biological rubric: to engage in fellowship and kindness without expectation or hope of anything in reward. It has been phrased and re-phrased as a universal Golden Rule, among many, that exhorts the Oaths of Hippocrates, as well as what biologists have noticed throughout most of the plant kingdom, namely relations that can be characterized as mutualistic. Similar relations have been examined in nearly every domain of life, including species of amphipods (*Podocerus jinbe*) in the mouths of whale sharks, amongst tarantulas and frogs, sloths and algae, some insects (lacewing larvae) and lichens, the mongoose and the hornbill, cattle egrets and their numerous hosts, from bovines to hippos, leafcutter ants and leaves, and amongst every plant-pollinator from flies to hummingbirds.

Indeed, the whole biological world functions in one symphonic unison.

We seldom consider the love of nature in the same realm as the love of a specific mouse, or the bacteria teeming in our own bodies for whom we are most generous hosts. But, in fact, natural selection selects for that which is loved, not just by our species. Selection is both mystical and loving, pure, practical and idealistic, all of these things. But it is not violent, vindictive, or premeditated. While we may think ourselves somehow marooned outside the tenacious blueprints of evolution, it is our very predilection for empathy that must remind us that we, like every other organism, have a noble calling that is, generally speaking, in tune with all others. Whatever ethical differences we are making, the hurdles we surmount, the world is noticing. There is a momentum that enacts change in ecological terms, whether by the very prosociality that grants us basic friendly exchanges, a proliferation of handshakes, as described by developmental psychologists, or by the 560+ multilateral treaties (500 of which are environmental treaties), in addition to over 3,716 other environmental agreements from nation to nation currently in force.

These agreements to honor biodiversity may be vastly imperfect, but they nonetheless represent growth towards an ideal which all of humanity has shared for the more than 330,000 years of our antiquity.

While countering numbers are harsh, measuring up to what we term the Anthropocene – the sixth extinction spasm ever recorded in the entire history of life on earth; a cascade of ecological woes unleashed solely by our own species – there is also a new shape and psychology mentoring our 11th hour endeavors to resurrect environmental sanity and a fundamental generosity that is motivated by a love of our fellow co-habitants here on Earth.

Yes, we have destroyed something like 60% of all wildlife, whilst overpowering with our selfish regard nearly half of the terrestrial planet, fouling in any number of ways the other half, not least the marine environs.

Yet, having set aside nearly 265,000 protected areas, and mobilizing in coalitions throughout the world to save what's left of the precious biological "hotspots," (those areas hosting the most abundant endemic species and genetic diversity), our newly declared goals encompass a thoroughly re-vitalized compassion. An aspiration shared by hundreds of millions of people that is grounded in intelligent and inspiring desires to see human hunger, poverty, racism, gender oppression and a callous misappropriation of goods and services, all rectified in this and the next generation, while there is still time to do so prior to total biospheric collapse.

Ecological Reciprocity: A Treatise on Kindness succinctly examines a myriad of components of what evolution tends to view as true conscience, manifested in behavioral norms that have nurtured and stabilized biological connections over hundreds-of-millions of years. The Authors have worked with Ms. Niki Stavrou in Athens who translated this book into Greek for the first bilingual Nova Science Publishers edition; Greek for a most prominent reason. It was in Greece that the first Western impulses towards physiolatry, the love of nature, were first enunciated in great tomes of ecological philosophy, animal rights tractates, and dramatic form. In works by the likes of Thales, Aesop, Heraclitus, Pythagoras, and Parmenides; Aristotle, Plotinus, Plutarch and Porphyry. From the monks of Mount Athos to the great 20th century philosopher/author from Crete, Nikos Kazantzakis, the Greek poetic voice has inhabited the annals of cosmology and biology with a spiritual search that is all but unmatched in world literature. Ms. Stavrou is Eleni Kazantzakis' goddaughter.

Ecological Reciprocity focuses primarily upon the very metaphysics inherent to our individual and collective will to protect the world. It is a mystery that we strive to do so, until we recognize that those same urgings compel all biological organisms. How many have set aside serious time and effort to save a fly, *Musca domestica*, weighing approximately 0.00035th of an ounce? With brilliant eyesight, unmatchable aerodynamics and its own great passion to live, to survive, even if for only 2-4 weeks in human time, every fly is precious. Try to build one! A single fly, or mouse, or spider, or whale, or rose. We love them all and such concern and fidelity constitute nothing less than the underlying rule of life, those life forces whose loyalties are not divided, not on this planet. Indeed, it is our very veneration and longing to be connected to wildness that guarantees that *Homo sapiens* will either survive as a species, or disappear with the shortest lived of any known large vertebrate epitaph on earth.

This work tracks numerous poets, scientists, engineers, mathematicians, artists and philosophers who all construed different pathways towards a unison of natural positivism and embrace. Immanuel Kant, Philip Gilbert Hamerton, Samuel Von Baron Pufendorf, George Orwell, the great Quaker painter of some 62 versions of his "Peaceable Kingdom," Edward Hicks, and the genius of the Northern Renaissance, Albrecht Dürer whose protagonist, the Roman general Placidus would be martyred and canonized as St. Eustace, the hunter who became a devotee of the mighty stag, in whose brace Placidus had a vision of Christ speaking to him of love. The love with which Buddha divined his Noble Eightfold Path, that biocalculus for attaining a place in which all suffering ceased.

Not unlike Albert Camus' childhood memories of his grandmother killing a family hen, the ethological revelations noted by famed ornithologist Alexander Skutch, or the botanical insights of Nobel Laureate Maurice Maeterlinck's "The Intelligence of Flowers," this work looks at animal and plant behavior through the lens of that indigenous critical mass that would give us to understand neural and biocultural thresholds that can facilitate our own communions with the natural world. That is the only goal here. How to transmute the promiscuity of our violence into an entirely other shape-shifting mentality aligned with tenderness and generosity to all. It is a disposition at once perfectly fitted to the ethic of our time, and to which all the sciences must ultimately convene, if they are to meaningfully illuminate, and spare Nature.

The countless transitional states to reach beyond the ecological dialectics of human behavior will necessarily result – if we remain steadfast – in such proclamations as that of a recent French Agriculture Minister who called for an end in 2021 to all killing of male chicks across France (the literal grinding up of live birds in shredding machines because the male chicks are unwanted). Even its description evokes in this generation a horror that was but a blind-spot in our species lens just a few generations ago. We are improving as a capably empathic species, and such changes must be accelerated if we are to effectively prevent and ameliorate the wanton ruination that is our present ecological legacy.

To ensure that great escalation of kindness, we find ourselves more indebted than ever – emotionally, psychologically, spiritually – to all those biomes our predecessors thought nothing of overrunning in the great geopolitical games to acquire territory and stake it out as their own. A ferocious battle of egos and nationalities clamoring after resources which were not natural at all. As other, indigenous humans were oppressed and/or slaughtered, so was one ecosystem after another, leaving us in this generation an opportunity unlike any in our history: the opportunity to thoroughly demonstrate unconditional love. Darwin spoke of evolution as obtaining a "general plan" and this is what it was. The great Jain sage, Umaswati (2nd-century to 5th century CE) declared in his famed *Tattvartha Sutra* that "souls render service to one another," echoing the 24th Jain sage, Mahavira, who spoke of biological interdependence, or Parasparopagraho Jīvānām.

This transition to a virtue ethic comports with an essential neural plasticity that has favored ideals over casual ideas; championed truths for all species, not just a few; and engendered the preconditions for a biological scaling that encompasses nothing less than a *new nature*; a testimony to causes-and-consequences that have added up the moral sums and recognized that there is pertinacity, survival and grace in etiquette, courtesy, symbiosis. This is not a philosophy of altruistic exclusiveness, or net gains. Rather, a cause of action that is mutually dependent on another cause; helping others, sharing, giving, without condition.

We call these causes "pairings" and they are the essence of empathy. Between every two, there is a one, and this one is the evolutionary result of shared endeavors that mean something. They are critical to 4 billion+ years of life. The sum-total of all those shared efforts describes a quantum ecodynamic that underlies the so-called "readiness potential" long described by neurophysiologists. A cumulative ingredient in the vast workshop of the human spirit, namely, our ability to forgive, to love, to mentor others in our love. It is as ethical as it is scientific, and that is the reality of the 21st century. A non-violent resolve that cannot simply be quantified in the biochemical annals of research over thousands of years, because there is a quintessential spirituality within it that gains by the insights of every past and living culture.

As the weight of human impact has become only too increasingly clear, we know now that this generation is tasked with getting it right. Finding the balance that will enable us to carry on as a species, but with the epiphanies and spiritual uplift of all those individuals who readily perceive what is now at stake: a wild companionship between humans and all others which is the true basis of a realistic prognosis for, and reckoning with the future. A future of kindness.

Chapter 1

A BASIS FOR PHYSIOLATRY

Search every library and database – hundreds-of-millions of citations, articles, journals, books[1]; and explore every continent for most of a lifetime. Visit multitudes of art and natural history/science museums. Count, and reflect upon, as many trees and flowers, and all the other life forms as humanly possible; or, at least contemplate a vast number of representative organisms across the planet. Others have, in part, done so, in the manner of a Mahavira, Pliny the Elder, Aristotle, Lucretius, Gessner, Linnaeus, Buffon, Humboldt, Audubon, Wallace and Darwin, to name but a few; compiling all the relevant statistics, thereby formulating a method, an opinion, a plan, a conviction. Consolidate all of that data, and your own perceptions of it into a repository that can accommodate up-to-the-minute information. The late philosopher, historian of science, Thomas Kuhn, meditated on the structural impediments and advantages to engaging in something like the equivalent of that, and concluded (among other things) that scientists are eventually "able to see nature in a different way"; but that such novelties can also backfire: that "new paradigms arise with destructive changes in beliefs about nature."[2]

An endlessly tempting, dangling modifier attends upon the history, and current science of eco-dynamics: the life histories of individuals, species and the vast, vulnerable collection of global habitats we think of as the biosphere. That *modifier* is present in both quantities and qualities of how we conceive of our fellow beings. We cherish the child far more than her/his pluralities. Hence, Mao Zedong is said to have once declared words to the effect that every child is precious but too many of them can become a problem. Our daily news cycles reinforce the uncomfortable reality of our abilities to tune in, and with equal dispatch, to tune out of so much intensely disappointing causes and effects across the mental and physical landscapes in which we all cohabit.

In some cases, tuning out is actually tuning it. St. Augustine and John Muir both said as much in reference to the libraries of nature that, experienced firsthand come to mean both outward interaction and internal deliberation. For Muir, between his early days in the

High Sierras in the late 1860s, and his death in 1914, the sheepherder and saw-mill operator, would become a mystic, a poet, mountaineer/explorer, botanist, essayist, political activist, founder of the Sierra Club, as well as a husband and father.[3] He had his own blind spots, as history has recorded. In his time, he watched a pivotal dialectic in action: the destruction of ecosystems by human greed, anger, ego and indifference; as well as the establishment of bonds and alliances by good people who all agreed to come together upon acts of ethical resolution, and outright legislation to protect the world around us. In the mid-nineteenth century, there were fewer than a dozen politically protected ecological parks and reserves in the world. Today, nearly 265,000 of them[4].

Such conservation in protected places mirrors a primordial set of codes enshrined in the evolution of all species. Genetically, multitudinous studies have examined the apparent crossroads that individuals, populations and species encounter in the multi-tiered reactions that keep changing around a set of seemingly solid injunctions that humans have worked into what may be the oldest, universally ascribed-to credo, namely, a fundamentally unidimensional Golden Rule[5]. It would seem that every singular spiritual tradition has enshrined some version of the Leviticus 19:18, "thou shalt love they neighbour as thyself." Reiterated in the first book of the New Testament within the Gospel of Matthew 22:35-40, as well as in Mark and Luke, whereas Christ is said to have considered this declaration to be the most fundamental message of God: intentional reciprocity without rank, expectation or self-interest. Rather, a pantheism that, by no surprise, has come to be the hallmark of both ecology and psychology. Kindness to and of strangers invokes both cynicism and awe; the incredulity that sweeps over all *survival of the fittest* paradigms; elucidating with pained but hopeful realizations our not unbiased logic regarding the Serengeti (the number of carnivore versus herbivores, for example); and explicitly poignant truths worthy of every optimism, in spite of ourselves. An initial search on the World Catalogue database reveals well over 3.1 million books and articles that deploy the word "kindness". That includes a powerful early work on animal rights by the most prolific female author of the 19th century, Charlotte Elizabeth Tonna (1790-1846), *Kindness To Animals: Or, The Sin of Cruelty Exposed and Rebuked* (Philadelphia, 1845). "It is their delicate sense of feeling," she says, speaking of bears, worms, bats, fish, and every other being, that ought to astonish our deepest, most sovereign and loving attitudes to the world.

Beginning in the Old English of the 14th century, kindness invokes nations and gain, but morphed by the time of Middle English to reference noble deeds. Its etymology encompasses all *kinds*, suggesting that no one, nothing be left out. In the Judeo-Christian Bible, and 5th century CE Tamil language *Tirukkural* (திருக்குறள்) by Saint Valluvar, kindness (and in the latter, moral vegetarianism) is at the core of all the teachings. In both the Old and New Testaments, it is the highest virtue, just as ahimsa, non-violence, dominates nearly all Eastern religious texts. In North Africa, the Old Egyptian *Pyramid Texts*, circa 2375 BCE, first mentioned the concept of *maat*, symbolized by the ostrich, the essence of truth and honorable deeds, and one of the oldest known religions in all of Africa.

The geographical expanse of kindness in human cultures is a vastly understudied realm. But in late 2019, UCLA established the multidisciplinary Bedari Kindness Institute at the Luskin School of Public Affairs to enshrine serious research involving genetics, psychiatry, anthropology, and the broadest medical and sociological approaches to grasping the evolutionary importance and future potential of simply being kind.

Given that such texts are invoked as the highest extant of representative gravitas in human nature, it is fitting that one of the most readily accessed and operative synonyms applied to the action of kindness compounds the techniques, motives, and actualizations of *reciprocity*. The word itself, like kindness, is steeped in the annals of an ideal, even as its core mechanisms are propelled by most – though not all – biological nurturance. We see it in reciprocal altruism, from bacteria to birds to humans[6]; numerous forms of "direct," "spatial," "upstream" gratitude akin to rudimentary forms of cooperation, each of which has been explored in terms of natural selection and the evolution of organizationally effective collaboration.[7] This would include all those abundant actions of mutual exchange, sharing and sacrifice.[8] Even "generalized reciprocity" has been shown to operate between Norway rats with a resulting extrapolation suggesting for a call to researchers to study the likelihood that it is, in fact, "a common phenomenon."[9]

Another version of this interactive near constant is mutualism, that is seen in nearly half-to-90% of all plant relations, either at the level of chemical exchange or service-to-service seed dispersal[10]. Coevolution, symbioses, commensalisms that favor at least one species without harming another,[11] all lead to both social and individual embraces of Others, for any number of motives. E. O. Wilson has written that "The biophilic instinct can be counted as one of humanity's fortunate irrationalities…"[12] But let us assume, instead, a most suprarational compass orientation that spans all living realms, from rhizobial nitrogen fixation, as the world's bacteria convert atmospheric nitrogen into ammonia, thereby providing plants the ability to synthesize strings of amino acids – proteins,[13] to the inherent teamwork, positive affiliations, common sense assistance, and various applied Oaths of Hippocrates (460-370 BC): rudimentary ecologies of all healing, and amelioration. One of our grandparents was a cardiologist and he used to juggle for his patients to assist, he said, in their rapid recovery. In other words, they smiled. That optimism is curative, and it stems from the imagination, choice, curiosity and love. Combined, these attitudes and proclivities shed light on what is truly a constant at the heart of physiolatry, the worship of nature. Hence, an iterative window on the ultimate objects of kindness: individuals, and the entire world upon which we depend. If we can view that all-encompassing biosphere without granting any gaps, then our observance is nothing less than honest fidelity and unwavering fellowship.

If physiolatrous behavior is (we think quite obviously) at the root of *all* behavior then evolution and natural selection have tended to favor outcomes that perpetuate life, on the basis of life herself commending and commandeering collaboration. When we first looked at predator-prey relations in East Africa, back in the late 1980s, to determine approximate numbers of each, the ratios of carnivores to herbivores was on the order of 1-to-4,000 individuals. Today, the herbivores appear to have been reduced in that equation by a magnitude of ten. In other words, predator-to-prey is closer to 1-to-400; 400 grazing ungulates, mostly, for every big cat. The stressors of the Anthropocene (climate change, fires and other habitat destruction, most notably from human overpopulation and expansive development, disappearing fresh water and poaching) have each taken their toll in ways that do not complement this physiolatry hypothesis, any more than we can assert that all competition and carnivorous behavior are ever likely to become obsolete, as much as we might wish. Whatever the biological rudiments may be in store for us all in the future, the past has certainly raised doubts as to any universal logic of what, in Jain, Buddhist and Hindu traditions goes by the word, ahimsa, or non-violence in Sanskrit.

Hinsa, violence, does not take anything away from the Golden Rule. But it does challenge 21st century humans with an abundance of considerations that now verge upon the matrix of our very survival, as we see every day the unrelenting, escalating inflictions of our kind on the natural world. The tell-all's are couched in big numbers that are dizzying and impossible to internalize. For example, the 349 billion gallons of groundwater used by Americans every 24 hours. It is a significant volume, even against the context of an estimated 2.78 million trillion perpetually cycling gallons on earth, 30.1 percent of which is fresh water, which, in turn, is less than 1% of all *available* fresh water[14]. The majority of that (and over 40% of all grains) goes to feed animals – comprising more than 60% of all mammalian biomass – which, in turn, we exploit and slaughter. At the same time 2019 saw 43.1 billion tons of CO^2 pumped by humans into the atmosphere.[15] While the numbers are revealing in a vague way, we know that anything we do, in various proportions depending on individual exploitive/extractive social mosaics, will be multiplied by nearly eight billion others. No one has ever seen eight billion humans. We don't know what that looks like. But we can dimly appreciate the vast scale of its reciprocal mass. The enormity slams into the brain which, in turn, rejects the sensation reflexively.

That mental eschewal is not easily fitted into any evolutionary principle. For example, David Sloan Wilson famously suggests that "group-level functional organization evolves primarily by natural selection between groups" not "within groups"[16]. There is no direct scientific evidence for the superorganism hypothesis (natural selection working on groups as well as individuals), other than by semantic deferral to the behavior and success of such vast aggregates and aggregation efficacy witnessed by coral reefs, neotropical forests, Utah's 80,000-year-old "Pando" clonal grove of some "47,000 genetically identical quaking aspen trees[17]," beehives, termitaries and ant colonies, for example. But this conceptualization of biological entities -as population or as individual, as well as between clusters - is an arbitrary class of taxonomies, emphasizing differentiated ideas that blur the borders of subjective and objective classification. Indeed, the differing inflections are as much an impressionable reaction to large numbers as to the very notion of functionality across expansive territories. That debate, between the borders of an individual and of a species, has never ceased, particularly at the cellular level. But also, in terms of large-scale management by humans – in differing types of the exercise of control – of natural resources that happen to be perceived in common, or as commons, i.e., air, water, soil, biochemistry[18]. The point is one of grammar, not necessarily reality (much like the 1972 George "Price Equation" that suspected a mathematical "equivalence thesis" between "kin selection and multilevel selection…").[19]

The numbers, large and small, rational or irrational, represent ideas which, by turns, tease out motives from within the history of biology. That, in turn, connotes the geopolitical and economic rudiments of both partnerships and ruination. Of tyranny and benevolence. These mixed affairs are our story in isolation, but also our epitaph with respect to most biodiversity.

Endnotes

[1]See, for example, *Environmental Resources*, NJ State Library, https://libguides.njstatelib.org/environment/general; and https://libguides.njstatelib.org/ environment/environmental_libraries

[2] See Frank Pajares' "Outline and Study Guide," to Thomas S. Kuhn's *The Structure of Scientific Revolutions*, https://www.uky.edu/~eushe2/Pajares/Kuhn.html. See Thomas Kuhn's The Structure of Scientific Revolutions, University of Chicago Press, Chicago, Ill., 1962.

[3] "Going out, I found I was really going in: John Muir's spiritual and political journey," by Tim Flinders, April 18, 2016, *openDemocracy*, https://www.opendemocracy.net/ en/transformation/going-out-i-found-i-was-really-going-in-john-muir-s-spiritual-and-politi/

[4] See "Protected Areas," *WDPA*, https://www.protectedplanet.net/en/thematic-areas/wdpa; see also, *On the Nature of Ecological Paradox*, by M. C. Tobias and J. G. Morrison, Springer Nature, New York and Switzerland, 2021. See also, *Sanctuary: Global Oases of Innocence*, by M. C. Tobias and J. G. Morrison, A Dancing Star Foundation Book, Council Oaks Books, San Francisco and Tulsa, 2008.

[5] See, for example, The Golden Rule – The Ethics of Reciprocity In *World Religions*, Edited by Jacob Neusner and Bruce Chilton, Continuum Books, New York, 2008.

[6] See Trivers, R.L. (1971). "The evolution of reciprocal altruism". *Quarterly Review of Biology*. 46: 35–57. doi:10.1086/406755. S2CID 19027999.

[7] See "Upstream reciprocity and the evolution of gratitude," by Martin A. Nowak and Sébastien Roch, PMC, *Proc Biol Sci.* 2007 Mar 7; 274(1610): 605–610. Published online 2006 Dec 19. doi: 10.1098/rspb.2006.0125, PMCID: PMC2197219, PMID: 17254983, *The Royal Society Publishing, Proceedings B*, https://www.ncbi.nlm.nih.gov/pmc/articles/PMC2197219/ Accessed October 19, 2020. See also, "Five rules for the evolution of cooperation," by Martin A. Nowak, Science. 2006 Dec 8; 314(5805): 1560–1563. doi: 10.1126/science.1133755, PMCID: PMC3279745 NIHMSID: NIHMS49939 PMID: 17158317, https://www.ncbi.nlm.nih.gov/pmc/articles/PMC3279745/Accessed October 19, 2020.

[8] See "The Natural Model of Reciprocity," by Madronna Holden, *Our Earth/Ourselves,* https://holdenma.wordpress.com/culture-and-environment/reciprocity/Accessed October 19, 2020.

[9] "Assortment and the Evolution of Generalized Reciprocity," by Daniel J. Rankin and Michael Taborsky, doi:10.1111/j.1558-5646.2009.00656.x, January 10, 2009, *Evolution* 63-7: 1913-1922, https://onlinelibrary.wiley.com/doi/pdf/10.1111/j.1558-5646.2009. 00656.x

[10] Bronstein, JL (1994). "Our current understand of mutualism". *Quarterly Review of Biology. 69 (1): 31–51.* doi:10.1086/418432.

[11] Mikula P, Hadrava J, Albrecht T, Tryjanowski P (2018). "Large-scale assessment of commensalistic-mutualistic associations between African birds and herbivorous mammals using internet photos". *Peer J. 6: e4520.* doi:10.7717/peerj.4520. PMC 5863707. PMID 29576981.

[12] See *The Future Of Life*, by Edward O. Wilson, Alfred A. Knopf, New York, 2001, p.140.

[13] See "Nutritional Adaptations of Plants," in *Nitrogen Fixation: Root and Bacteria Interactions*, https://courses.lumenlearning.com/boundless-biology/chapter/nutritional-adaptations-of-plants/; See also, *Biological Nitrogen Fixation*, by Stephen C. Wagner, 2011, Nature Education Knowledge, 3(10):15, https://www.nature.com/scitable/knowledge/ library/biological-nitrogen-fixation-23570419/

[14] "Groundwater Facts," *Groundwater Summit*, December 8-10, 2020, NGWA, https://www.ngwa.org/what-is-groundwater/About-groundwater/groundwater-facts

[15] See *The World Counts – Tons of CO^2 emitted into the atmosphere*, https://www.theworldcounts.com/challenges/climate-change/global-warming/global-co2-emissions/story; Other sources suggest fewer and/or larger sums. See "Feeding the World Vertically," by Geoffrey Holland, *The Planetary Press*, October 19, 2020, https://www.theplanetarypress.com/2020/10/feeding-the-world-vertically/

[16] *Does Altruism Exist? Culture, Genes, and the Welfare of Others*, By David Sloan Wilson, Yale University Press/Templeton Press, New Haven, Conn., 2015, p. 21.

[17] See "Pando, One of the World's Largest Organisms, Is Dying," by Brigit Katz, *Smithsonianmag.com*, October 18, 2018, https://www.smithsonianmag.com/smart-news/pano-one-worlds-largest-organisms-dying-180970579/

[18] See, for example, the concept of "optimal scale" and "optimal governance" elucidated by 2009 Nobel Prize winning economist, Elinor Ostrom, in her book, *Governing the Commons: The Evolution of Institutions for Collective Action*, Cambridge University Press, Cambridge, UK, 1990.

[19] See "The Biology of Being Good to Others," by H. Allen Orr, *The New York Review*, March 191 2015, https://www.nybooks.com/articles/2015/03/19/biology-being-good-others/

Chapter 2

OPTIONS FOR NATURAL SELECTION

Various words have long been our companions, or in our shadow: reciprocity, kindness, altruism, the conscience, and empathy. To each of these concepts we apply an endlessly studious etymology, as well as thousands of books and essays detailing the social, economic, moral and political cores and boundaries of these words and ideals. A word like "reciprocally" in 1632 refers to backwards and forwards, "two high trees, reciprocally waning in the ayre"; or, in 1622, "The Lungs, which diluting themselves as a paire of bellowes reciprocally fetch it (the air) and send it out." In mathematics, "Inversely" or "Conversely". In 1776, Thomas Paine in his *Rights of Man*, stated, "A Declaration of Rights is, by reciprocity, a Declaration of Duties also."[1] Altruism, in French and Italian refers to "somebody else"; conscience = Latin, conscientia, "being privy to"; compassion = compati, to "suffer with". Together, these enunciations elicit the best in humanity, when applied.

Three Enlightenment and Renaissance philosophers, Edmund Burke, Francis Bacon and Sir Thomas Browne, lent instructive, dualistic impressions regarding such terminologies and their applicability to human affairs. The words themselves manifest unbridled and peculiar interpretations and inventions. Take "sympathy," (Greek + Latin = feeling with) which might seem to unify all four of the above specific categories of language and thought. In fact, much has been written about the historically nuanced differences between sympathy, pity, and "the mechanism of changing places with the object of our observation," as in the writings of Plato and Aristotle, Adam Smith and David Hume.[2] But the sense of it was taken to mean something utterly unexpected by Edmund Burke, when he writes of "the distresses of others" and of "the delight we have in such things, [which] hinders us from shunning scenes of misery; and the pain we feel, prompts us to relieve ourselves in relieving those who suffer; and all this antecedent to any reasoning, by an instinct that works us to its own purposes without our concurrence."[3] One year before Burke wrote *On the Sublime and Beautiful*, he published *A Vindication Of*

Natural Society, in which he supposed that human barbarism on the battlefields of history has resulted in approximately "thirty-six millions" of warriors who had been slaughtered, including "a low estimation effected by the murder of ten millions of the species" by the "conquest of Spanish America"[4]. Indeed, he goes on to point out that "It is an incontestable truth, that there is more havock made in one year by men, of men, than has been made by all the lions, tygers, panthers, ounces [snow leopards], leopards, pards, hyenas, rhinoceroses, elephants, bears, and wolves upon their several species, since the beginning of the world…"[5]

Writing in his *Essays Civil and Moral* (1597), Francis Bacon considered two essentials of human nature. First, that "the inclination to goodness is imprinted deeply in the nature of man; insomuch, that if it issue not towards men, it will take unto other living creatures…"[6] And second, that "Notwithstanding we see that the greatest number of men give themselves over to their imaginations of hope and apprehensions of the mind in such sort, that ungrateful towards things past, and in a manner unmindful of things present, as if they were ever children and beginners, they are still in longing for things to come."[7] And in Sir Thomas Browne's *Religio Medici* (1643), "Charity begins at home, is the voice of the World…It is not the tears of our own eyes only, but of our friends also, that do exhaust the current of our sorrows…[8]

In all such meditations we see a similar struggle with a dialectic that pivots upon the sense that there are natural laws; that humans are caught out in the whorls of evolution, implicating our dispositions and actions without any precise understanding, on our part, or where this is all going.

So let us start with empathy, a word that the psychologist Edward Titchener (1867-1927)[9] coined in 1909, from the German "Einfühlung" ("feeling into"). Earlier, the German philosopher Theodor Lipps (1851-1914) had thought of it as "projecting oneself onto the object of perception," the mirror neurons[10] currently investigated by scientists from many disciplines. Or, as Montag writes, "Lipps's theory of 'inner imitation' may find some reflection in present-day concepts of emotional contagion and perception-action models of empathy."[11] But the philosophy and psychology of empathy have been diminished from the activist stance, to a predominantly academic, and largely off-putting series of didactic analyses that beg a variety of otherwise fundamental questions, most basically: Why should we care? What are the biases of moral agency? What motivations are really guiding our alleged instincts? Are "other minds" actually a "problem" or the basis for solutions? Is there a difference between feeling bad, and feeling bad about something specific, some other person or Other? Is charity borne of selfishness, altruism, or actual belief in the power of empathy in action? Does empathy engender altruism, as well as other sympathetic modes of feeling and thinking? If so, are such affects universal, selective, or narrowly distributed? Are moral norms subject to development aspects of psychology, or is this mere verbiage, underscored by the literature of verbiage? Ultimately, the question persists: why should we even care about any of these aspects of the so-called *moral compass* in humans?[12] Does caring make a difference? And does *a difference* make a difference?

Does natural selection select for such differences?

Our distance from other people and organisms compels less immediacy the further in time and space our own sovereignty is assured of the very protection from want, loss, pain and prolonged suffering about and to which we are somehow privy; in the news, by word of mouth, or more distantly still, in reading a re-count, or history of what has gone on before and/or elsewhere. Unless they are direct predecessors, family, friends, it is more difficult to feel "pain" for all those who perished during the Black Plagues; or for dinosaurs that went extinct. Our sympathies with, or compassion for, everyone who has died closer to our own contemporary realities gain momentum the nearer to ourselves the mortality events occur. There is no equation to measure the closing of that gap between our feelings, sadness contagion, and what has preceded our moment in time. Is this even an issue?

Edmund Burke, and certainly the remorseful lessons of history conveyed by his contemporary, Edward Gibbon in his six-volume *The History of the Decline and Fall of the Roman Empire* (1776) communicate something other than neutrality on this closing of the gap. Gibbon points the finger at the "deluge of Barbarians" and then, century by century, prodigious and selfish excess, waste and inward decay against which, future rationalism should prove victorious. But even by the time of Petrarch marveling at the ancient ruins, writes Gibbon, Roman locals had themselves utterly grown beyond such nostalgia; had forgotten any sense of beauty, abandoned any meaning or majesty of such discarded scraps of the twelve centuries before their time; had forsaken their own history ("supine indifference," as Gibbon put it)[13] or the relevancy of the ruins which visitors to Rome found so compelling, and continue to value as living museums. We flock to situate and photograph ourselves amid these testimonials of grandeur, ultimate futility, suffering and hope. Gibbon knew that posterity will always condemn human folly, even if humans themselves continue on, unabated in our stubborn, repetitive errors. And this gives great pause to any implied rule of lessons learned. Indeed, the gap between observation and memory is not in any way a predictable science of behavior. Slaughterhouse workers are the obvious testimony to this truism. At one of the largest known slaughterhouses in the United States, as many as 5,000 laborers hear, inhale, and view as many as 32,000 pigs executed daily. Many of those workers slosh about in the blood and entrails of the writhing victims. They take an active role in seeing to the annihilation. If we are *all* human, of by and large the same DNA bits, how is this possible?

What is innate, inchoate, instinctive, basic to our genome? Let us argue, for the moment, that the genome = zero, in statistical terms. That no evolutionary bias attends the personification of our requisite physiology, in terms either of individual behavior or that distribution of such behaviors. Rather, that descent, as Darwin theorized, entails every necessary modification, his "mystery of mysteries"; a matter of real-time evolution by which an individual exhibits the power to accelerate, announce, champion and defend some new characteristic in human evolution. This singular safeguard may take 100,000 years to show itself across the boundaries otherwise demarcating populations, hamlets, communities, nation states. This would uphold the notion that natural selection has selected in this post-Holocene set of circumstances a distinctive outcry that utterly contests any and all gradualism between fossilized indicators, rallying instead to an entirely sudden outrage accompanying our knowledge of human-induced mass extinctions, of factory farming, climate change, global destruction at a mere moment in geological time.

How very different Henry David Thoreau's worship of stones, writing, "It would imply the regeneration of mankind, if they were to become elevated enough to truly worship stocks and stones… I would fain improve every opportunity to wonder and worship, as a sunflower welcomes the light…a stone appeals to me and elevates me, tells me how many miles I have come, how many remain to travel…If it did the same service to all, it might well be a matter of public rejoicing."[14]

The pictorial sway of emotions that are human history, as a function of our time/space distance from the events in question should not obfuscate the persistent ideations devolving from meaning, no matter how attenuated the fact of its falling leaves in season, or remote customs. Reaction times may have been blunted in the age of instantaneity, but basic moral reflexes have only darkened, heightening levels of depression, grief and awareness. If, as a species, we have evolved since the varying timeframes of the Neolithic, from continent to continent, our handprints and doorknobs, beds and pillows have not quintessentially changed. And prevailing neuropsychological studies tend towards the premise that "To survive, you have to care about yourself first. It doesn't mean that you don't care for others, but [to reiterate] you have to survive first," writes scientist Jean Decety.[15] This maxim, employed in every seatbelt instruction on a commercial airliner, also comports with the emergence of "functional biodiversity," which cites three particularly salient new taxonomic classification concepts: "Guilds," "Functional Response Groups," and "Functional Effect Groups".[16] Such categories render qualitative analysis of organismic relations as possibly more direct and revealing in terms of contributory components, or basal collateral involvement in the matrices used to divine Anthropocenic fall-out. In other words, across numerous sectors of human experience, reactions and the nature of response can vary; grow; learn. The honing of our pain reflexes and suffering memory mechanisms, is of particular importance.[17]

If cause and effect over spatial and temporal quantification can be expressed as functions of neuroevolutionary constraints, pressures, and changes, then the neural infrastructure itself is complicit by yardsticks yet to be understood in all organisms, with the known exceptions of sponges and Trichoplax (primordial multicellular Placozoa), believed to have *no* neurons; versus, the killer whale with over 43 billion, or twice the number of neurons found in humans, a significant proportion of them in the cerebral cortex. Yet, this is not to lose sight of relational magnitude of the neuron-synapse-free phylum Placozoa whose 11,514 genes are similar to those in many other animals.[18] Moreover, they also communicate from cell to cell by the mechanism of peptides, or groups of up to 20 amino acids. Their favorite food tends to be green algae. One of the Trichoplax, the species *adhaerens* – one of three Placozoa species – has been proposed as a model organism on earth for cryptic complexity, meaning, it is related individually and in aggregate to others like it. In fact, their signaling capacity has been called "huge".[19]

Hence, a splendid confusion that helps to break down barriers in the projection of similarities and differences between organisms; while again raising the specter of that gap of reflexes galvanizing support for one individual, but not another. We hear and recognize at once the valid indicator of these callibrations. In the manner of Shakespeare's character of Berowne attending Lord to King Ferdinand of Navarre in "Love's Labour's Lost," (1595/1596) who declares, "For wisdom's sake, a word that all men love, Or for love's sake, a word that loves all men…"[20] Or, in that total similitude that breaches all difference to insinuate some other entity entire, a species/individual supraorganism whose intent and method of subsistence transcends our feckless differentiations. Instead, a being of sense, metabolism, travail and inner knowledge, without any of the obvious factors of life to calculate; a force of nature we have yet to identify that comes from deeper within than the mere nucleus, or atomist connections, friction, mass and energy. Again, an entirely novel form of Life. As the late poet Charles Tomlinson wrote in one of his earliest works, "You think it for a second/hugely dead, until the ripple soundless on the further corn, is roaring in it."[21]

In all this we are speaking of *relations*. Whether in the means by which human history records our recapitulations, or researchers catalogue fossils. In both cases we start from an idea in our interpretations of nature. These may be personal effects, loose strings from the past that have come to remind us of the present. In his chapter entitled, "Political Autobiography of a Young Man," Italo Calvino writes, "…he who, in the general catastrophe of human affairs when they are left to biological impulses, knows how to choose what needs to be saved and defended and developed and made to bear fruit."[22] Ideas, interpretations, methods, and the epiphanies that help us to orient our lives within the biosphere are constantly changing, despite periodic fixtures of seeming permanence, like the Laws of Nature, of physics, mathematics, whole ecosystems. Yet, only recently – long after Relativity, Game, Information, and Quantum theories, Heliocentrism, or revelations pertaining to the "psychophysiology of vision"[23] – have ecologists understood that there is ecosystem and community succession in ways that go far beyond the evolution of individuals; bypassing the rules that dictate core notions of genes, rather, invoking "ecosystem evolution."[24] For all those same relations that have historically engendered presuppositions in science, it is the philosophical *character of Being* that challenges stasis, on no particular plateau of principles, but sheer momentum, the restlessness inherent to life and its reciprocating impulses.

What is that nomadic turn in attitude, personality, volition? Everything we have thus far learned of biodiversity mentors us in flux, without giving away the actual storyline; only that change is a pre-requisite of evolution. But beyond that, schooled merely in the shortest-term of equations, namely, survival of the individual, we know instinctively that populations proceed us; ancestors are recycled; predecessors, family or not, pre-ordain some form of continuation, even at the risk of associated extinctions, which we know to be the rare norm. But then what? How do we get beyond the awkward, vainglorious, *but what about us*? We know for certain that such concern is relevant in the present, but of likely little or no consequence in the near future.

ENDNOTES

[1] *The Oxford English Dictionary, XIII Volumes, Volume VIII*, Oxford: At the Clarendon Press, UK, 1961, pp. 242-243.

[2] See "Pity and Sympathy: Aristotle Versus Plato And Smith Versus Hume," by Christos Grigoriou, *The Journal of Scottish Philosophy* 16.1 (2018): 63-68. doi: 10.3366/jsp.2018.0183, Edinburgh University Press, https://www.researchgate.net/publication/323491569_Pity_and_Sympathy_Aristotle_versus_Plato_and_Smith_versus_Hume

[3] From A Philosophical Enquiry Into The Origin Of Our Ideas Of The Sublime And Beautiful, in *The Works Of The Right Honourable Edmund Burke, Collected In Three Volumes, Vol. 1*, J. Dodsley, London, 1792, p. 107. First published in 1757.

[4] ibid., *The Works... in A Vindication Of Natural Society*, p. 23.

[5] ibid, *A Vindication*…pp. 25-26.

[6] Essays Civil And Moral, XIII Of Goodness And Goodness Of Nature, in *The Works Of Francis Bacon, Lord Chancellor of England, A New Edition, With A Life Of The Author*, by Basil Montagu, in Three Volumes, Vol. 1, A. Hart, Philadelphia, 1852, p. 21.

[7] ibid., in *Meditationes Sacrae*, p. 68.

[8] Titchener E.B. (1909/2014) *Introspection and empathy Dialogues in Philosophy, Mental and Neuro Sciences* 2014; 7: 25–30; See also, https://plato.stanford.edu/entries/empathy/#HisInt

[9] Sir Thomas Browne, *Religio Medici, Sixth Edition, With Annotations*, Also *Observations by Sir Kenelm Digby*, London, 1669, pp. 142-144.

[10] Rizzolatti, G. and C. Sinigaglia, 2008. *Mirrors in the Brain: How our Minds Share Actions and Emotions*, Oxford: Oxford University Press.

[11] "Theodor Lipps and the Concept of Empathy: 1851-1914," to Dr. Montag, Images in Psychiatry, *Am J Psychiatry* 165:10, October 2008, p. 1261, ajp.psychiatryonline.org; See Lipps T: Ästhetik: Psychologie des Schönen und der Kunst: Grundlegung der Ästhetik, Erster Teil. Hamburg, Germany, L Voss, 1903 13. Preston SD, DeWaal FBM: Empathy: its ultimate and proximate bases. *Behav Brain Sci* 2002; 25:1–72

[12] Karsten Stueber, Stanford Center for the Study of Language and Information, The Metaphysics Research Lab, "Empathy," *Stanford Encyclopedia of Philosophy*, June 27, 2019, https://plato.stanford.edu/entries/empathy/#HisInt

[13] Edward Gibbon, *The History Of The Decline And Fall Of The Roman Empire, in Twelve Volumes, Vol. XII*, London, 1821, p. 420.

[14] *The Journal Of Henry D. Thoreau, In Fourteen Volumes, Bound as Two, Volumes VIII-XIV* 9 November, 1855-1861), Edited by Bradford Torrey and Francis H. Allen, With a Foreword By Walter Harding, Dover Publications, Inc., New York, 1962, pp. 1063-1064.

[15] See, "How do we balance self-interest with fairness for others?" by Jack Wang, August 25, 2020, *University of Chicago News*, https://news.uchicago.edu/story/how-do-we-balance-self-interest-fairness-others;

[16] See "Functional Biodiversity and the Concept of Ecological Function," by Antoine C. Dussault, in *From Assessing to Conserving Biodiversity-Conceptual and Practical Challenges*, Edited by Elena Casetta, Jorge Marques da Silva and Davide Vecchi, Springer Open Access Book, History, Philosophy and Theory of the Life Sciences book series, Vol. 24, https://link.springer.com/book/10.1007/978-3-030-10991-2

[17] See "Commonalities between pain and memory mechanisms and their meaning for understanding chronic pain," by Theodore J. Price and Kufreobong E. Inyang, *Prog Mol Biol Transl Sci.* Author manuscript; available in PMC 2015 Nov 30. *Prog Mol Biol Transl Sci.* 2015; 131: 409–434. Published online 2015 Jan 30. doi: 10.1016/bs.pmbts.2014.11.010, PMCID: PMC4664460 NIHMSID: NIHMS727455, PMID: 25744681, https://www.ncbi.nlm.nih.gov/pmc/articles/PMC4664460/

[18] Eitel, Michael; Osigus, Hans-Jürgen; DeSalle, Rob; Schierwater, Bernd (2 April 2013). "Global Diversity of the Placozoa". *PLOS ONE.* 8 (4): e57131. Bibcode:2013PLoSO...857131E. doi:10.1371/journal.pone.0057131. PMC 3614897. PMID 23565136. See also, Knowlton N. (February 2000). "Molecular genetic analyses of species boundaries in the sea". *Hydrobiologia.* 420 (1): *73–90.* doi:10.1023/A:1003933603879. ISSN 0018-8158. S2CID 40909997.

[19] See "Before nerves, there were peptides," by Nick Carne, October 18, 2018, *Cosmos,* https://cosmosmagazine.com/biology/before-nerves-there-were-peptides

[20] *The Complete Works Of William Shakespeare*, Edited with a Glossary by W. J. Craig, Oxford University Press, Oxford, UK, 1947, "Love's Labour's Lost," ll. 358-359, p. 159.

[21] Charles Tomlinson, from "An Insufficiency of Earth," in *A Peopled Landscape-Poems*, Oxford University Press, London, 1963, p. 21.

[22] *Hermit In Paris-Autobiographical Writings*, by Italo Calvino, Translated from the Italian by Martin McLaughlin, Pantheon Books, New York, 2003, p. 146.

[23] See *Objectivity*, by Lorraine Daston and Peter Galison, Zone Books, New York, 2007, p. 380.

[24] See "The creation of 'Ecosystem Core' hypothesis to explain ecosystem evolution," by Kun Wang and Xiajie Zhai, *BMC Ecology*, 19, Article number: 33 (2019), BMC Ecology, doi: https://doi.org/10.1186/s12898-019-0251-y, https://bmcecol.biomedcentral.com/articles/10.1186/s12898-019-0251-y

Two Friends at a Rajasthani Temple. Photo © J.G. Morrison.

Chapter 3

PROSOCIALITY

Prosociality in all species – making others happy, kind gestures at any and every juncture of life – has as a discipline of enquiry enjoyed a social scientific renaissance during the last decade.[1] Mental health research shows unambiguously a "positive affect" and the amelioration of various neuroses that can be directly tied to positive social actions by individuals.[2] Child developmental psychology has been particularly galvanized by such heuristic research.[3] From sustainable marriages, support for elders, to general global studies of human happiness, every correlate of social support networks reveals greater cohesion, mental health and general well-being through the agency of prosocial engagement.[4]

There is an existing bias of investigations fixated on alleged higher evolutionary prosociality in *H. sapiens*, than in all other primates, and, to paraphrase Martin Luther King, Jr., that *yes, hate is too great of a burden to bear.*[5] It is true that there is evidence of murder among chimpanzees and the occasional fracas amongst other primate species.[6] Or that a barn swallow "can consume 60 insects per hour or a whopping 850 per day."[7] But any comparisons surrounding the propensity for violence in our species versus others ends before the argument is even begun.

Yet, the overwhelming universality of various levels of cooperation, as perceived through our specifically biased lens, gives incontestable loft to the very centrality of both positive affect and effect, through affiliations that are singularly helpful, whether among cells, or throughout the entire plant and animal kingdoms. And certainly amongst we humans who, long ago, were eloquently admonished in *Exodus* 23:2, to *not follow a multitude to do evil.*

An obvious empirical case study *context* amongst northern New Mexico avifauna in late October 2020 challenges easy generalizations regarding the estimated 500 million tons (quadrillions) of insects that the more than 18,000 species of birds collectively devour every year throughout the world. This is no radical equation that would seek false biological equivalencies between avian insectivores and human behavior, which other data sets resoundingly point to the killing of between two-and-five trillion vertebrates annually. And, additionally, at least two billion people also kill hundreds-of-trillions of invertebrates, including cephalopoda (squid, octopus, etc.), marine crustacea (e.g., lobsters) and arthropoda in general, a phylum comprising the largest groups of organisms on the planet (other than plant-like phytoplankton), namely, insects and spiders. On this given Autumn mid-day, Jane and I are watching 50+ individual birds of at least 16 species: Robins, Black-billed Magpies, Mourning Doves, House and Rock Wrens, Dark-Eyed Juncos, Cassin's Finches, Pine Siskin, Mountain Chickadees, Juniper Titmice, Fox, Song, House, Lincoln, Chipping and White-crowned Sparrows. Along with rock, red, Abert's and gray hybrid squirrels and quietly working Alder bark beetles, all sharing in a variety of foodstuff, zero competition, zero conflict. A herd of deer wander through the scene, browsing on the twigs of juniper, acorns, wild raspberry, cedar, willow and dogwood. The paradise-like scene evokes the wonderment of comparisons: can a human idea, transfixed into an ideal, be vulnerable in the best sense to natural selection?

Can selection move rapidly enough to re-evolve in real time, moving from the idea to its applied engineering? The examples of the Wright Brothers or Thomas Savery's steam engine would suggest, yes. But what about that transition between hope and physical manifestation? Hunger and the delivery of food? Illness and healing? Non-violence and real non-violence? A point transformed into a pattern, from the one to the many?

For these neighborhood wild creatures, there appears to be, at least for now, a 100% survival rate during the sudden onset of 15-degree F nighttime temperatures, and a foot of snow, after three months of Summer drought. This seemingly low-stress, cooperative milieu has occurred at the same time that hundreds-of-thousands, possibly millions of migratory birds – particularly, "Warblers, sparrows, and western wood pewees" that normally fly South from New Mexico in Winter, have died off, possibly as a result of damaged lungs from the smoke coming from the raging fires up and down the West Coast.[8] That being the landscape-sized context, one can only interpolate data that underscores just one of countless consequences of the Anthropocene. Yet, it is without fanfare or any show of territorial distress that these many species are commingling on this day, with such ease and emphatic friendship; sharing food with full-on collaborative indulgences. No insects are being eaten, worms deep down ignored, which might argue, theoretically, for some ideal of nutritional deliverance; widespread distribution of native food types otherwise in scarce supply, around the world, in pursuit of an outcome that would alter the biological mosaic, giving weight to the promise of an ideal human redemptive passage from non-involvement to involvement, by way of observational collaboration. Cooperative assistance not unlike any other food aid delivery in time of emergencies. Species throughout the world are starving to death. Is it not incumbent upon us to help? Most state laws forbid the feeding of wildlife.[9]

We are as quick to point out the ambiguous frailties of all such ecological regime intervention (controlled burns versus natural wildfires being another storied case). But there is more force of logic when the heart refuses to collaborate by practical indifference. In many cities (Phoenix and Las Vegas among them) it is illegal to feed the pigeons. As well as many parts of London (e.g., Trafalgar Square). In Paris, a 450 Euro fine for doing so. In the state of Utah, the theft of any animal being raised for commercial purposes, whether a sparrow or a pig that will be slaughtered, is considered a felony. Animal rights/animal liberation felons languish behind bars in state and country after country.[10]

But here, today in northern New Mexico, interspecies conviviality underscores the anxiety of being human and watching this feast within a hostile bubble. Ambient air temperatures to which all these birds must physiologically combat persistent frigid temperatures that have hit them suddenly. High metabolisms, fat storage, oil glands, preening methods, air storage insulation techniques, large food consumption per day, other anatomical features that all work to stave off enormous morality rates. The risks of migration for some 75% of all North American birds, versus the dangers of staying through northern Winter months[11]. Reproductive and vagrancy styles and compulsions; the combination of human-induced risks.[12] It is too much to take in.

Yet, throughout all of these travails towards which centuries and millions of years have given birds their tactical inner traditions and grace, ideas and concepts, the very wisdom born of mechanisms of anatomy implicating every morphological ingenuity with which to cope, there stands out not just the air within feathers which are fluffed, but this other green light of unconditional partnerships, notwithstanding erratic, reactionary lawmakers. The pathways of kindness in sync with human perception finds that commences with the premise, "Cooperation and prosciality are hallmarks of human behavior, and it is generally believed that our extra-ordinary levels of cooperation have contributed to the success of our species."[13] The goal of such studies typically rests upon the desire to find stable, or phylogenetic specific traits that can be said to converge across species boundaries throughout evolution, particularly among primates as well as avians. In 2016, researchers at the University of Vienna published data suggesting that Azure-winged magpies "show human-like generosity."[14] At the Loro Parque Fundación, in Puerto de la Cruz, Tenerife, an African grey parrot named Bella was shown to "help others to obtain food," and "no jealousy if others get better treatments," findings hailed as equivalent to the "level of intelligence [in African grey parrots] comparable to that of great apes, crows and dolphins."[15]

In trying to narrow down one possible mechanism driving prosocial aptitude and expression, one group of researchers sought the equivalent for humans of the "neuropeptide oxytocin hormonal mechanism" for which the avian homologue appears to be "mesotocin". It certainly appears to prompt similar mammalian-like prosociality in pinyon jays, "suggesting an evolutionarily conserved hormonal mechanism for prosociality."[16] Such behavior may be construed as a constellation of honest signaling, and, hence, stable, evolutionary traits. And because it is – as thus far construed – a universal form of signaling inherent to all kinship and nurturance, the more interesting facet of such behavior is its occurrence in heterospecific landscapes – *between species*.

ENDNOTES

[1] See *In the Light of Evolution V: Cooperation*, Organized by Joan E. Strassmann, David C. Queller, John C. Avise, and Francisco J. Ayala, January 7-8, 2011, Irvine, California, National Academy of Sciences Colloquia, www.nasonline.org/SACKLER_cooperation.

[2] Snippe, E.; et al. (2017). "The Reciprocity of Prosocial Behavior and Positive Affect in Daily Life". Journal of Personality. 86 (2): 139–146. doi:10.1111/jopy.12299. PMID 28093772. See also, Schwartz, C. E.; Meisenhelder, J. B.; Yusheng, A.; Reed, G. (2003). "Altruistic social interest behaviors are associated with better mental health". *Psychosomatic Medicine*. 65(5): 778–785. doi:10.1097/01.psy.0000079378.39062.d4. PMID 14508020.

[3] For hundreds of research studies on the subject, see: https://psycnet.apa.org/ search/results?id=51e158b8-97f2-40db-f052-e436f82f393d; See Eisenberg, Nancy; Paul Henry Mussen (1989-08-25). The Roots of Prosocial Behavior in Children. Cambridge University Press. See also, Eisenberg, Nancy; Fabes, Richard A; Spinrad, Tracy L (2007). "Prosocial Development". *Handbook of Child Psychology*. doi:10.1002/ 9780470147658.chpsy0311.

[4] Baumeister & Bushman (2007). Social Psychology and Human Nature. Cengage Learning. p. 254. See also, Helliwell, J. F.; Putnam, R. D. (2004). "The social context of well-being". *Philosophical Transactions of the Royal Society B: Biological Sciences*. 359 (1449): 1435–1446. doi:10.1098/rstb.2004.1522. PMC 1693420. PMID 15347534. See also, Silk, J & House, B (2011). "Evolutionary foundations of human prosocial sentiments". *PNAS*. 108 Suppl 2: 10910–7. Bibcode:2011PNAS..10810910S. doi:10.1073/pnas.1100305108. PMC 3131813. PMID 21690372. See also, Alden; Trew (2013). "If it makes you happy: Engaging in kind acts increases positive affect in socially anxious individuals". *Emotion*. 13 (1): 64–75. doi:10.1037/a0027761. PMID 22642341. See also, Penner, Louis A; Dovidio, John F; Piliavin, Jane A; Schroeder, David A (2005). "Prosocial Behavior: Multilevel Perspectives". *Annual Review of Psychology*. 56: 365–92. doi:10.1146/annurev.psych.56.091103.070141. PMID 15709940. *See I've Decided To Stick with Love*, by Javon Collines, Covelli Enterprises, https://www.covelli.com/wp-content/uploads/2015/02/Covelli-MLK-1st-place-essay_w-covelli_logo.pdf

[5] *Philos Trans R Soc Lond B Biol Sci.* 2010 Sep 12; 365(1553): 2723–2735. doi: 10.1098/rstb.2010.0118, PMCID: PMC2936168, PMID: 20679115 "On the psychology of cooperation in humans and other primates: combining the natural history and experimental evidence of prosociality," Adrian V. Jaeggi, Judith M. Burkart,and Carel P. Van Schaik, The Royal Society Publishing, *Philosophical Transactions B*, https://www.ncbi.nlm.nih.gov/ pmc/articles/PMC2936168/Accessed October 27, 2020.

[6] "How Human Violence Stacks Up Against Other Killer Animals - Humans get some of their knack for killing each other from the primate family tree, a new study says—but that doesn't let us off the hook, by Erika Engelhaupt, September 28, 2016, *National Geographic*, https://www.nationalgeographic.com/news/2016/09/human-violence-evolution-animals-nature-science/

[7] See "Birds That Eat Mosquitoes," by Meg Campbell, November 22, 2019, *Sciencing*, https://sciencing.com/birds-eat-mosquitoes-6620996.html, Citing Linda deKort of the Flathead Audubon Society.

[8]https://blogs.findlaw.com/law_and_life/2019/12/can-you-stop-neighbors-from-feeding-dangerous-wild-animals-.html

[9] See "In New Mexico, the Birds are Dying at Alarming Rates," by Claire Lampen, *The Cut*, September 14, 2020, https://www.thecut.com/2020/09/new-mexicos-migratory-bird-die-offs-are-unprecedented.html, Accessed October 27, 2020.

[10] See, for example, "Animal rights activists, facing 60 years in prison, to visit farm where they removed baby pig during 'open rescue'," by Taylor Hartman, August 6, 2018, *Fox 13, Salt Lake City*,

https://www.fox13now.com/2018/08/06/animal-rights-activists-facing-60-years-in-prison-visit-farm-where-they-removed-baby-pig-during-open-rescue/

[11] "How Many Birds Disappear Between Migration Seasons? We Now Have a Clue," by Pubrita Saha, *Audubon Magazine*, Winter 2018, https://www.audubon.org/magazine/winter-2018/how-many-birds-disappear-between-migration, Accessed October 27, 2020. See also, "Weather-related mass-mortality events in migrants," by I. Newton, IBIS, May 15, 2007, https://doi.org/10.1111/j.1474-919X.2007.00704.x, https://onlinelibrary.wiley.com/doi/10.1111/j.1474-919X.2007.00704.x

[12] "A Summary and Comparison of Bird Mortality from Anthropogenic Causes with an Emphasis on Collisions," by Wallace P. Erickson, Gregory D. Johnson, and David P. Young Jr., *USDA Forest Service Gen. Tech. Rep. PSW-GTR-191.* 2005, https://www.fs.fed.us/psw/ publications/documents/psw_gtr191/psw_gtr191_1029-1042_erickson.pdf

[13] "Studying the evolution of cooperation and prosciality in birds," by Jorg J M Massen, *Ethology* 126(2): 121-124, doi: 10.111/eth.13001, https://onlinelibrary.wiley.com/doi/ 10.1111/eth.13001. https://www.researchgate.net/publication/338555319_ Studying_the_ evolution_of_cooperation_and_prosociality_in_birds, Accepted December 10, 2019, January 2020.

[14] Lisa Horn, Clara Scheer, Thomas Bugnyar, Jorg J. M. Massen. Proactive prosociality in a cooperatively breeding corvid, the azure-winged magpie (Cyanopica cyana). Biology Letters, 2016; 12 (10): 20160649 doi: 10.1098/rsbl.2016.0649, cited in University of Vienna. "Azure-winged magpies show human-like generosity: Prosocial behavior due to cooperative breeding." *Science Daily. ScienceDaily.*18 October 2016. <www.sciencedaily.com/releases/ 2016/10/161018193150.htm>.https://www.sciencedaily.com/releases/2016/10/161018193150.htm

[15] See *Prosocial and tolerant parrots help others to obtain food*, Max-Planck-Gesellschaft, n.a., January 9, 2020, https://www.mpg.de/14328111/prosocial-grey-parrots, Accessed October 28, 2020.

[16] "Mesotocin influences pinyon jay prosociality," by J F Duque, W Leichner, A Ahmann, and J R Stevens, National Library of Medicine, *PubMed.gov, PMID*: 29643220, PMCID: PMC5938566, doi: 10.1098/rsbl.2018.0105, *Biol Lett* 2018 Apr;14(4):20180105. doi: 10.1098/rsbl.2018.0105. https://pubmed.ncbi.nlm.nih.gov/29643220/

Chapter 4

THE PROPORTION OF HANDSHAKES

It is often noted that Thomas Hobbes (1588-1679) would describe in his autobiography how his mother on April 5th, 1588 gave birth to twins: To the premature Thomas, and to fear.[1] Hobbes' *Leviathan or The Matter, Forme and Power of a Commonwealth Ecclesiasticall and Civil* (London, 1651) would ultimately come down to half-of-a-line characterizing humanity in just such a dualistic state of nature: "continual fear and danger of violent death, and the life of man, solitary, poor, nasty, brutish, and short."[2] Written seventeen years before his *Behemoth: the history of the causes of the civil wars of England, and of the counsels and artifices by which they were carried on from the year 1640 to the year 1660 (The Long Parliament)*, the *Leviathan* has the weight of being mentored by calamities. The seemingly endless affronts to decency and never even a hint of something better looming. The veritable science arising out of terror, pain, so many rational eyewitness accounts of anarchy. And of monarchy itself, sparring throughout history with no surer guidance than all the egotism and cruelty that can possibly be meted out between rivals, from the bloody fields of Edgehill in October 1642, to subsequent sieges, beheadings, betrayals and hideous engagements. Campaigns, uprisings, treasonous resolve and endless conflict. Just two years prior to publication of the *Leviathan*, seared into any consideration of British human nature was Oliver Cromwell's putting down of Irish Royalists – some 3,500 people massacred at the Siege of Drogheda.[3] As if this were the only alternative to the *noble savagery* others who opposed Hobbes' stark geopolitical vision preferred to wistfully imagine and to perceive.

That civic alternative is crucial to representing the ideal that remains steadfast in what today intimates our truly last hope for conceptualizing an opposable thumb of philosophy and ethics. Not savagery, but the nobility of some overarching truth: what it means to be a member of the human species, along with possibly 100 million other species, no more, no less. Hobbes was precise. The antidote to his accusations remains untested for reasons that encompass the *unchallenged* biases, imposturous self-regard, and devastating throw-weight of human psychiatry, demography, superego, and its inflictive ecological carrying capacities.

If we choose to metaphysically circumvent England's and America's Civil Wars, the French Revolution, a thousand other internecine morasses mangled in blood – from the Aztecs to Afghanistan – we might casually alight upon the same tip of an Aspen tree the swallowtail butterfly – first mentioned by our species in Homer's *Iliad* (thus, *Papilio homerus*) – recognizes as a suitable place to lay her eggs. That would be the world we ought to join, not combat. With 550 charismatic species of the family Papilionidae on every continent but Antarctica (as far as is known), this remarkable butterfly (its Latin nominate the substance of all Lepidoptera) relies predominantly upon an equally prodigious plant family for pollination and sustenance, namely, the Rutaceae, with approximately 160 genera across two vast subfamilies, Rutoideae and Cneroideae.[4] This cosmopolitan set of co-symbiotic relationships – swallowtails and plants – dates back biogeographically to Beringia, the Northern Hemisphere, at least, by one comprehensive estimate, to the Eocene some 52 million years ago.[5]

It is worth our while to extrapolate from butterflies of today to the most conspicuous of bipedal vertebrates. If we ask the same questions of both groups, we should expect answers to vary only according to our pre-existing premises, the scenarios we have enshrined as predicated upon concretized metaphors *in our minds*, like evolution, or a Hobbes. To the extent we are fixated on those patterns in nature we have described, defined, outlined and emphasized, we remain prisoners of that imagery-turned-principle, Laws of Nature of our devising. In asserting human identity as the center of this earthly black hole, the multiplication of our presumptions gives only the slightest clues to how we might escape our intrinsic self-referentials in a mirror, much like that "most famous paradox in physics."[6] If the cooperative impulses we witness being phylogenetically demonstrated throughout every ecosystem are to actually impress us – as well as keeping us alive from within (endogenosymbiosis) – why, as branded by Hobbes, do we continue to behaviorally manifest the equivalence of antibodies to the life-sustaining basics of our biology? We note that *our* biology is everyone's biology -the "octopus teacher," Bambi, a barn spider named Charlotte. Whilst aggression is there, "avoidance of competition" and a more generous and complex topology than is casually ascribed to the Serengeti's of the world, appears more likely to be a central key to understanding a "stable" cooperative theory of evolution, whether amongst plants and animals or microbes.[7,8,9]

In contemplating "existential catastrophe"[10] investigators have tended to focus upon neurological architecture oriented to minimizing threats, and the related stress.[11] But the actual cognitive categories are compressed. We do not concern ourselves overtly with timeframes of a century, let alone a billion years hence. We know of no lifeforms that do, even as we celebrate long-lived species like the Great Basin Bristlecone pines or Red Sea urchins.

But the mechanics of getting along; of concerting impulse beyond fright or flight, greed or *mere* personal need, into broader affective arenas at least bordering upon collaboration, afford spellbinding lures of perception that are both premonitory, looking ahead, as well as supremely in the present. This perceptual stage of our evolution reaches beyond the multitude of reciprocal altruisms on display, from symbioses in marine environments,[12] to sympatric and allopatric species warnings.[13] In our species, this conceptual transition marks the changing ratio from handprints to a much larger proportion of handshakes.

ENDNOTES

[1] See "Thomas Hobbes: A grim portrait of human nature," by Lou Marinoff, *The Times Literary Supplement*, April 5, 2020, https://www.the-tls.co.uk/articles/thomas-hobbes-footnotes-to-plato/Accessed October 29, 2020.

[2] Oxford University Press, *Clarendon Edition of the Works of Thomas Hobbes*, Oxford University Press, 2012, XIII.9.

[3] Leniham, Pádraig (2008), *Consolidating Conquest: Ireland 1603–1727*, Pearson Education Ltd., Harlow, UK, p. 128.

[4] *Rutaceae* (Citrus family) – *245 images at PlantSystematics.org images, phylogeny, nomenclature for (Rutaceae)*. plantsystematics.org.

[5] *Global biogeographical pattern of swallowtail diversification demonstrates alternative colonization routes in the Northern and Southern hemispheres*, by Fabien L. Condamine, Felix A. H. Sperling and Gael J. Kergoat, *Journal or Biogeography*, Wiley Online Library, October 19, 2012, https://doi.org/10.1111/j.1365-2699.2012.02787.x, https://onlinelibrary.wiley.com/doi/10.1111/j.1365-2699.2012.02787.x. Of course, this is not to ignore the countering self-preservative mechanisms of defense. One noteworthy botanical example also comes from the Eocene, "Urticaceae leaves with stinging trichomes were already present in latest early Eocene Okanogan Highlands, British Columbia, Canada," by Melanie L. DeVore, Alphonse Nyandwi, Winnie Eckardt, Elias Bizuru, Myriam Mujawamariya, and Kathleen B. Pigg, *American Journal of Botany*, October 22, 2020, https://doi.org/10.1002/ajb2.1548, https://bsapubs.onlinelibrary.wiley.com/doi/abs/10.1002/ajb2.1548

[6] See "The Most Famous Paradox in Physics Nears Its End," by George Musser, *Quanta Magazine*, October 29, 2020, https://www.quantamagazine.org/the-black-hole-information-paradox-comes-to-an-end-20201029/?utm_source=pocket-newtab, Accessed October 31, 2020.

[7] "The evolution of cooperation," by R. Axelrod and W D Hamilton, *Science* 27 Mar 1981: Vol. 211, Issue 4489, pp. 1390-1396, doi: 10.1126/science.7466396, https://science.sciencemag.org/content/211/4489/1390

[8] See Roberto Cazzolla Gatti. A conceptual model of new hypothesis on the evolution of biodiversity**.** *Biologia*, 2016; 71 (3) doi: 10.1515/biolog-2016-0032.

[9] See "Symbiosis as a General Principle in Eukaryotic Evolution," by Angela E. Douglas, *Biology, CSH Perspectives*, 2014, http://m.cshperspectives.cshlp.org/content/6/2/a016113.short. For a short bibliographic overview of numerous microbial symbiotic research papers, see: https://scholar.google.com/scholar?q=symbiosis+among+eukaryotes&hl=en&as_sdt=0&as_vis=1&oi=scholart#d=gs_qabs&u=%23p%3DTobRGg1fj2YJ

[10] "The Survival of Humanity," by Lawrence Rifkin, *Scientific American*, September 13, 2013, https://blogs.scientificamerican.com/guest-blog/the-survival-of-humanity/

[11] See "The ecology of human fear: survival optimization and the nervous system, by Dean Mobbs, Cindy C. Hagan, Tim Dalgleish, Brian Silston and Charlotte Prévost, PMC, *Frontiers in Neuroscience*, 2015; 9: 55. <arcj 19. 2015, doi: 10.3389/finis.2015.00055, PMCID: PMC4364301, PMID: 25852451, https://www.ncbi.nlm.nih.gov/pmc/articles/PMC4364301/

[12] Feder, H.M. (1966). "Cleaning symbioses in the marine environment". In Henry, S.M. (ed.). *Symbiosis*. 1. Academic Press. pp. 327–380. OCLC 841865823.

[13] Olendorf, Robert; Getty, Thomas; Scribner, Kim (22 January 2004). "Cooperative Nest Defence in Red–winged Blackbirds: Reciprocal Altruism, Kinship or By–product Mutualism?". *Proc. R. Soc. Lond. B*. 271 (1535): 177–182.

Chapter 5

VIOLENCE AND UTOPIA

The aforementioned 560+ multilateral treaties (500 of which are environmental treaties, in addition to over 3,716 other environmental agreements from nation to nation) currently in force across the world's political boundaries[1] trace their lineage to the hieroglyphically engraved earliest known peace pact between humans, the "Eternal Treaty" drawn up in the mid-13th century BCE by the Egyptians and Hittites under the rules of Ramesses II and Ḫattušili III. Not surprisingly, a long-delayed ceasefire in Syria was part of that accord.[2] Grumbling consents, protracted negotiations, détentes and all those protocols ordained by the (at present) 116-state parties to the Vienna Convention on the Law of Treaties (VCLT)[3] might suggest that in absence of this countering, biological penchant for resolving conflicts between our communities, there would be far more than the estimated 250-500 million people who have been murdered since the time of the European Renaissance. And, indeed, rates of homicide have diminished in Europe, over the last millennium; but in other parts of the world, including the U.S., have fluctuated, with 5 deaths per 100,000 people remaining an average in America, and 6.2 per 100,000 worldwide.[4,5]

The 240+ known wars in documented human history do not predate the Persian conflicts of the 6th century BCE. For example, it is known that at least 27 people were killed near the end of the Stone Age in Kenya some 10,000 years ago, as unearthed at a site known as Nataruk. Nor do the numbers of battles, purges, genocides, anthropocentrically exacerbated mortalities and co-moralities encompass "highest estimates". For example, some researchers have speculated that the total death toll stemming from the European invasions/colonizations of the Western hemisphere over the course of some five centuries was close to 140 million.[6] It is estimated that for every person alive at present, approximately 15 have died before now, over 107 billion people in the past 50,000 years.[7] Take the number back to our origins, some 330,000 years ago, and a few more billion humans will have come and gone. Parsing through the planet's current biomass reveals that humanity's living presence comprises a small fraction of the numbers of Others. For example, there may be more than ten trillion living ants at present, and ten million previous ant generations.[8]

While as many as 200 million species are yet to be discovered, and as many as 750 million other species have gone extinct since life began (both figures at the upper limits of sound estimates)[9] the ratios of living human biomass to all other wild and captured vertebrates (captured = "domesticated" animals for human exploitation) yield astonishingly disproportionate appropriation coefficients. Since 1970, humans have extirpated at least 60% of all wildlife on the planet, excluding invertebrates and our Net Primary Production rate that, in turn has converted at least "44 percent" of the arable planet to human agricultural plots.[10] Ten-thousand years ago, the human population worldwide probably did not exceed 5 million persons, accounting for, at most, 1% of the weight of all vertebrates. In 2020, our species represents some 32% of all vertebrate weight, whilst those captured animals total another 67% or more. That leaves remaining wildlife at a 1% total, or less.[11] For every remaining Blowout Penstemon, Sequoia or Blue Whale there are some 400,000 people. For each Florida Panther or Woodland Caribou, 156 million people.[12] "For he ignored the Nightmares and annexed their ranges," wrote W. H. Auden in his poem, "The Dirge."[13]

There is no precedent for such problematic and corrosive ecological auditing. Doom scrolling is central to the numeric inflections of human proprietary challenges to the rest of vertebrate, invertebrate and habitat life. It says nothing about the habitats of which we ourselves are host, to hundreds of billions of symbiotic creatures, most notably, the 35-to-100 trillion bacterial cells.[14,15] But it speaks with mathematical gloom to our rank-and-file dominion of biochemical chaos. The many industrial revolutions cumulatively spawned this inordinately destructive ratio. It began in places like Chaldean Ur throughout the third millennium BCE. There, thousands of cattle were slaughtered routinely, the horse was *broken*, countless workers enslaved, meticulous accounting of stocks and surplus carved into stone. Warring paraphernalia adorned tens-of-thousands of troops corralled into one phalanx or other, wearing steel-ribbed skirts and breastplates, marching to do battle. Most celebrated of such superego bloodlusts were the personages of Sargon (ca.24th-23rd centuries BCE) and Naram-Sin from northern Akkadia (2254-2218 BCE). A victory stele of Sin in the Louvre reveals the warrior conquering a mountain, with his army cheering him on.

Then René Descartes (1596-1650) would smugly torture animals to the delight of his colleagues who had formulated a widespread belief that these writhing creatures were responding to pain mechanically, with no actual feelings.[17] Such sentiments coincided with the massive draining of marshes across Europe, the invention of blast furnaces, enormous textile mills with polluting dyes, and Machiavelli's insistence on expedient means and autocratic oversight.[18] Even in Thomas More's *Utopia* (1516) slaves were imported to do those tasks to which the 80,000 residents in each of 45 cities refused to stoop. At the same time, More stressed the notion that his Utopian denizens at the altar of natural religion and abstinence from material greed, were up to the task of lending out their peace-keeping services throughout the world. When Sir Thomas More actually ascended to power as Lord Chancellor, he was not up to the same philosophy of clemency his satire had conveyed, ignoring the various capital punishments meted out in Henry VIII's and Katharine of Aragon's tumultuous reign. Until he himself ended up in the Tower of London, beheaded on charges of treason in 1535.

In 1829, a locomotive named the *Rocket*, and designed in Newcastle upon Tyne by one Robert Stephenson, would travel twelve miles in 53 minutes. It thereby set a silly record, in England and everywhere else. A founder of thermodynamic principles, Sadi Carnot (1796-1832) would write in his memoirs that "Already the heat engine works our mines, propels our ships, deepens our harbors and rivers, forges iron, fashions wood, grinds our grain, spins and weaves our fabrics, pulls the heaviest loads…"[19] By the mid-19th century, as historian Paul Johnson has summarized, "there seemed no escape …[and] as Western ideas spread, increasing portions of the earth's surface required international policing…Would the new world policeman lose control in his own hearth and home? Could the center hold?"[20] Of Carnot's euphoria, John Stuart Mill contemplated more soberly, of a world that had been routed by man, every forest and flowery kingdom brought under his subjugation, "all quadrupeds or birds which are not domesticated for man's use exterminated as his rivals for food…and scarcely a place left where a wild shrub or flower could grow without being eradicated as a weed in the name of improved agriculture."[21]

With some 265,000 protected areas in the world, each the result of some societal value, collective promise, ethical premise, and community codification into law, there is an ingrained tradition of long-standing incremental compromises, first intoned in Sir Thomas More's *utopian paradox*, namely, how to actually protect, on the ground?[22] And what protection actually means?

In our Introduction to our work, *The Metaphysics of Protection*[23], the Authors described the meager task of liberating hundreds of flies that one day were born and took to flight throughout a summery abode in southern California. *Musca domestica*, earlier described, may seem like less a cause célèbre than some, weighing on average 0.00035th of one ounce, a lifespan of 2-4 human weeks. Yet, if we who profess to revere the elephant and the Amazon cannot find the requisite will and ethic by which to protect even the seemingly most diminutive of our kindred here on earth, then what is any of it worth? Etymologies aside, the word protection within legal rubrics remains only partially intelligible, from culture to culture. There are anomalous case briefs, like the court-imposed execution of a pig – declared by morons to have been naughty - in France in 1266, and, nearly 300 years later, a jenny (female donkey) spared by another court, upon the testimony of forgiving witnesses to her virtue.[24]

Our histories of violence, developmental technology and engineering, and the mitigating efforts to diminish humanity's seemingly ineluctable footprint lead us to conclusions that are not new or novel: The many contradictory knots of the human psyche as it grapples with concepts. We both protect and hunt animals (and plants) within national parks. We profess to love rare and charismatic animals, while slaughtering en masse all those others whom we dismiss as unfeeling units of production. It is an ancient story that complicates the through line of any psychoanalysis that hopes to see clear to predictable cause and effect; or extrapolate prospects for a just landscape of esteemed equals and reciprocates under any law of heaven.

Dr. Biruté Mary Galdikas with Rescued Orangutan in Borneo. Photo © J.G. Morrison.

ENDNOTES

[1] See *United Nations Treaty Collection*, https://treaties.un.org/; see also, See Donald K. Anton, "'Treaty Congestion' in International Environmental Law", *International Law Reporter*, January 24, 2012, http://ilreports.blogspot.com/2012/01/anton-treaty-congestion-in.html; See also, https://blogs.un.org/blog/2012/09/24/most-ratified-international-treaties/Accessed November 2, 2020.

[2] Langdon, Stephen H.; Gardiner, Alan H. (1920). "The Treaty of Alliance between Hattusili, King of the Hittites and the Pharaoh Ramesses II of Egypt". *Journal of Egyptian Archaeology.* 6 (3): 179–205. JSTOR 3853914.

[3] "Vienna Convention on the Law of Treaties (1969)". *opil.ouplaw.com.* doi:10.1093/law:epil/9780199231690/law-9780199231690-e1498.

[4] See "Rap Sheet: Why is American history so murderous?" by Jill Lepore, *The New Yorker*, November 2, 2009, https://www.newyorker.com/magazine/2009/11/09/rap-sheet#ixzz0ePRpHvGz, Accessed October 31, 2020.

[5] Intentional homicide count and rate per 100,000 population, by country/territory (2000–2012). Data (in spreadsheet format) for UNODC report titled "Global Study on Homicide 2013". United Nations Office on Drugs and Crime (UNODC). Linked from *Global Study on Homicide – Data: UNODC Homicide Statistics 2013.*

[6] *American Philosophy: From Wounded Knee to the Present*, by Erin McKenna, Scott L. Pratt; Bloomsbury Publishing, New York, 2015, pg. 375.

[7] See "Do the dead outnumber the living?" by Wesley Stephenson, *BBC News*, February 4, 2012, https://www.bbc.com/news/magazine-16870579, Accessed November 1, 2020.

[8] See *Life Counts -Cataloguing Life on Earth*, by Michael Gleich, Dirk Maxeiner, Michael Miersch, and Fabian Nicolay, In Collaboration with UNEP, IUCN, WCMC and Aventis, Translated by Steven Rendall, Atlanticf Monthly Press, New York, 2002, p. 23.

[9] ibid., pp. 24, 261.

[10] See "Humanity has wiped out 60% of animal populations since 1970, report finds," By Damian Carrington, *The Guardian*, October 29, 2018, https://www.theguardian.com/ environment/2018/oct/30/humanity-wiped-out-animals-since-1970-major-report-finds, Accessed February 15, 2019. See also, *All life on Earth, in one staggering chart*, by Brian Resnick and Javier Zarracina, August 15, 2019, VOX, https://www.vox.com/science-and-health/2018/5/29/17386112/all-life-on-earth-chart-weight-plants-animals-pnas, Accessed April 27, 2020, and sourced from the landmark biomass (biosequestration and biogeochemical cycle) quantitative assessment paper, including reflections on abundance, distribution, "broad patterns over taxonomic categories" and "trophic modes" including "global marine biomass pyramid" data that accounts for consumers versus producers: *The biomass distribution on Earth*, by Yinon M. Bar-On, Rob Phillips, and Ron Milo, *PNAS, Proceedings of the National Academy of Sciences of the United States of America*, PNAS June 19, 2018 115 (25) 6506-6511; first published May 21, 2018 https://doi.org/10.1073/pnas.1711842115, https://www.pnas.org/content/115/25/6506, Accessed April 27, 2020. For net primary productivity arrogation rates, see *Global human appropriation of net primary production doubled in the 20th century*, Fridolin Krausmann, Karl-Heinz Erb, Simone Gingrich, Helmut Haberl, Alberte Bondeau, Veronika Gaube, Christian Lauk, Christoph Plutzar, and Timothy D. Searchinger *PNAS* June 18, 2013 110 (25) 10324-10329; https://doi.org/10.1073/pnas.1211349110 Edited by Wolfgang Lucht, Potsdam Institute of Climate Impact Research, Potsdam, Germany, and accepted by the Editorial Board May 5, 2013, received for review July 4, 2012, *PNAS, Proceedings of the National Academy of Sciences of the United States of America*, https://www.pnas.org/content/110/25/10324#: ~:text=The%20human%20 appropriation%20of%20net,managed%20lands%20with%20lower%20productivity.

[11] See *On The Nature of Ecological Paradox*, by M. C. Tobias and J. G. Morrison, Springer Nature, New York, 2021, Ch. 72.

[12] See "Rare – Portraits of America's Endangered Species", by Joel Sartore, Focal Point, *National Geographic*, Washington D.C., 2010.

[13] From *The Age of Anxiety*, by. W. H. Auden, A Random House Book, New York, 1947, p. 105.

[14] See *Microbe Magazine*, February 2014, https://www.asmscience.org/content/journal/ microbe/ 10.1128/microbe.9.47.2citing: J. Fox, *Microbe*, December 2013, p. 494; M. Pollan, *New York Times Magazine*, May 15, 2013; Science 342:1440, 2013; Cell Metab. 12:111, 2010.

[15] op cit., *Life Counts*, pp. 28-29.

[16] See *After Eden: History, Ecology & Conscience*, by M. C. Tobias, Avant Books, San Diego, 1985, pp.36-37.

[17] See "'A Brute to the Brutes?': Descartes' Treatment of Animals," by John Cottingham, *Philosophy*, 53, 1978, http://people.whitman.edu/~herbrawt/classes/339/Descartes.pdf

[18] See *De Principatibus/Il Principe*, 1532.

[19] Quoted in Wilson Clark, *Energy for Survival – The Alternative to Extinction*, Anchor Books, Garden City New Jersey, 1974, p. 10. See op cit., *After Eden*, Ch.6, pp.91-114, "A Tale of Two Centuries".

[20] *The Birth of the Modern – World Society 1815-1830*, by Paul Johnson, HarperCollins Publishers, New York, 1991, p. 355.

[21] J. S. Mill, *Principles of Political Economy with some of their Applications to Social Philosophy, Part IV, Chapter VI*, John W. Parker, London, 1848.

[22] See *2018 United Nations List of Protected Areas – Supplement* on protected area management effectiveness, United Nations Environment Programme (Eric Solheim, Executive Director) and UNEP-WCMC (Director Neville Ash), UNEP-World Conservation Monitoring Centre, Cambridge, UK 2018, in Collaboration with the CBD

(Executive Secretary, Cristiana Pasca Palmer), the IUCN (Director General, Inger Andersen), the WCPA (World Commission On Protected Areas, Chaired by Kathy MacKinnon) and UN environment, Editors: Marine Deguignet, Heather C. Bingham, Neil D. Burgess and Naomi Kingston, https://wdpa.s3.amazonaws.com/UN_List_2018/2018%20List%20of%20Protected%20Areas_EN.pdf.

[23] *The Metaphysics of Protection*, by M. C. Tobias, and J. G. Morrison, A Dancing Star Foundation Book, Waterside Press, Cardiff by the Sea, California, 2014.

[24] See *The Criminal Prosecution and Capital Punishment of Animals*, by E. P. Evans, William Heinemann, London, 1906. See also, *The Trial: Four Thousand Years of Courtroom Drama*, New Random House, New York, 2006.

Chapter 6

THE PUFENDORF HYPOTHESIS

Amid a historic rash of contradictory assertions by individuals and governments, the philosopher/lawyer Hugo Grotius of Delft Netherlands (1583-1645) laid the groundwork for a rational and consistent understanding of natural law (the Greco/Roman "higher law"). Among his foremost conclusions emphatically calling for an end to wars and all belligerency, Grotius made clear that "to deprive another of what belongs to him, merely for one's own advantage, is repugnant to the law of nature."[1] This fundamental fairness underscores reconciliation in every domain, but has yet to be either universally applied, or brought to bear at the court of interspecies rights to life, liberty and justice. The words themselves, when collated, almost invoke a cliché, given the mockery that has largely strewn their frail persistence as concepts embedded in any established *human nature.* Or, as Rousseau declared in his *Social Contract (Du contrat social; ou Principes du droit politique*, 1762), "It is precisely because the force of circumstances tends continually to destroy equality that the force of legislation should always tend to its maintenance." But also volunteers that "He who dares to undertake the making of a people's institutions ought to feel himself capable, so to speak, of changing human nature…"[2] In his study of *Marx and Wittgenstein*, David Rubinstein argued that Marx was most emphatic on the dialectic between nature and the Self at the bifurcation point wherein "Man is in the most literal sense of the word…not only a social animal, but an animal which can develop into an individual only in society."[3]

Each of these considerations invites an endless agitation in the face of the *real nature* that will eventually betray every philosophy and sap the aesthetic jubilation that arises in the same body whose vigor ultimately must wane. Whatever each of us believes, even *unto this last*, as John Ruskin (1819-1900) remonstrates, will, by the force of its conviction, find only echoes in the world. Yet, for all of those same calls in the wild, human wildness and infinitely greater wild, Ruskin satisfied any taxonomic red lines with a concluding argument which held that "all effectual advancement towards this true felicity of the human race must be by individual, not public effort."[4] It was precisely such *effort* that Aristotle utterly misread, comparing it to a "Little Democracy" within an Oligarchy. Hailing pomp and ceremony no less than a Nero as the best means of assuring confidence by the public in what he miscalculated to mean "the stability of Government."[5] Ruskin's contemporary, aesthetics philosopher Philip Gilbert Hamerton (1834-1894) helped solidify those echoes by speculating that it is easier to know the thoughts of a farmer than of a horse; that "This impossibility of knowing the real sensations of animals" can be partly rectified "by careful observation and reflection, aided by a kindly sympathy and indulgence…"[6]

This then is the dialectic with which we have always tended to engage, argue and be distracted: public sentiment, private leaning. Whether the social forces that make us our coffee and hot showers, our books and farmer's markets, rather than our own inner voices are the deciding factors in what we ultimately come to believe, defend, pray for.

General words and thoughts, attuned to the most specific ordination of a lasting dialectic. No mere ricochet of logic, but the steady meditation at the root of all natural science. Hence, Kuo Hsi's (1020-1090) suggestion that "if he can develop a natural, sincere, gentle, and honest heart, then he will immediately be able to comprehend the aspects of tears and smiles and of objects…"[7] Leonardo expressed a different vantage on this summons to become part of the natural world when he advised that "the brilliancy of a landscape" (painting) will never equal that of the real landscape, as illumined by the Sun, unless the painter places the canvas directly in the sunlight, a paean to plein air.[8] These diverse intimations speak of proximity in the course of transforming ideas into some kind of reality that more firmly attunes to the impressions and locality of the senses, whether in the social, aesthetic or psychological realms. How to transcend the reference point, back and still further back into the logical antecedent; or, forward into some incontestable looming truth. Immanuel Kant envisioned it as a condition of *Perpetual Peace*[9] in which a "universal hospitality" reigned supreme. His arguments followed closely upon those of Samuel Von Baron Pufendorf, whose 1672 treatise, *The Law Of Nature And Nations*[10] declared that the natural state of humanity was peace, not Hobbesian war. That this fact alone made us human, all laws following upon a freely adopted predilection towards virtue and goodness. The *Pufendorf Hypothesis* resolves the post-Holocene debate that would set at odds human social predilections and the inner core of humanity herself.

Human aggression has been diagnosed in copious ways, from prehistoric burial sites and murders (e.g., Ötzi, the early 4th millennium BCE victim of either an arrow in the back or human sacrifice), the psychopathologies of genocide to the horror, policework, and anonymous suffering that is everyday harm meted out by our kind. An infinity of impulses and victims. Violence, as much or more than love and kindness, consumes humanity's restless and reckless souls.[11,12] So many hundreds-of-millions of words have been written in an attempt to analyze our very knowledge base, which is, itself, a precursor to the hegemonies of imagination, so easily translated into trespass upon Others. This original transgression was no more insightfully expatiated than in Elias Canetti's haunting investigation into *Crowds And Power* (1962) in which he wrote, "It is those who devote themselves to killing who have power."[13] E. M. Cioran all but combined the revelations of Aristotle on governance, and a Canetti on power, when he acknowledged the following evolutionary conditional: "A great step forward was made the day men understood that, in order to torment one another more effectively, they would have to gather together, to organize themselves into a society."[14]

These collectivities – community franchise, the nation state – are the very instruments of all those invasions which comprise the endless wars of attrition, of a psycho-social predicament that would appear unique to our species. Such human malice and hostility has been comprehensively assessed throughout Jain tradition, the dialectics of hinsa/ahin(m)sa (violence/non-violence).[15] In the rising tide of animal rights discussions during the 18th and 19th centuries, government and its theodicies, as the underlying agency condoning some degree of human aggression towards others, was first provocatively challenged in 1742 by John Hildrop (1682-1756). It is often assumed that French naturalist Georges Cuvier (1769-1832) first proposed the notion of extinction, based upon his interpolations of fossil evidence. But, in fact, it was Hildrop, addressing human hubris and the collective arrogance of humanity's hedonistic interpretation of scripture, who first stated "an entire Extinction of Being" as the direct result of human folly and interspecies imposition.[16]

Hildrop focused upon God's Creation, imputing assumptions prevailing in mid-18th century Europe that the Creation was of far greater magnitude (obviously) than anything humans themselves could engender; and that our compassion towards Others must therefore be subsumed to that same humbling relativity. In 2021, that hubris/humility conflict persists. Semioethic studies speak of "unlimited responsibility" as well as a "profound and pervasive" "difference" between "mono-logic," and "dia-logic". The latter signifies that all important "relation between self and other",[17] seeking to render provable the proposition that humanness is a semiotic *fact*, its ethical outreach inherent to any communicative impulse.

All specificity would thus impel the embrace of *conspecifics*. The question devolves not to the uncertainty of amalgams, but rather, their levels of congruence in political and biological communities; a moral cohesion that relies upon murmuration (e.g., flocking behavior) and its self-organizing stigmergies[18], or forms of direct and indirect coordination, a concept first articulated in the late 1950s by French zoologist, Pierre-Paul Grassé, the E. O. Wilson of termites. Colonies, swarms, the architectonic of pheromones, trails of markings, the very choreography of figure eight fencing duels of Northern Flickers, or other conspicuous dance displays in species like the Birds of paradise, or the peacock, all have a dialogic genetics that is the life-sustaining communication fundamentally driven by two+, not solely by one.

While physicists think of these minions in motion -from dense zooplankton and krill swarms to bats emerging from a cave at dusk - as outside thermodynamic equilibrium states, in fact, within the biosphere there could be no more elegant proof of critical cogency between particles, which just happen to be individuals, local and non-local, which brings us into the land of ecological physics, underscoring the biology of affiliations.

ENDNOTES

[1] See "Natural Law and Peace -A Biography of Hugo Grotius," *Libertarianism.org*, by Jim Powell, July 4, 2000, https://www.libertarianism.org/publications/essays/natural-law-peace-biography-hugo-grotius, See also, Grotius' De iure belli ac pacis libri tres, Buon, Paris, 1625.

[2] As analyzed in T. V. Smith's *Beyond Conscience*, Whittlesey House, New York, 1934, pp. 328-329. See *An Inquiry Into The Nature Of The Social Contract: Or Principles Of Political Right*, Translated From the French of John James Rousseau, G. G. J. and J. Robinson, London, 1791.

[3] Cited from Marx's *The Grundrisse: 17, Marx and Wittgenstein – Social Praxis and Social Explanation*, Routledge & Kegan Paul Ltd., London, 1981, p. 153.

[4] *"Unto This Last" – Four Essays On The First Principles of Political Economy*, by John Ruskin, London, 1862, p. 168.

[5] See Aristotle's *A Treatise On Government*, Translated from the Greek by William Ellis, T. Payne, London, 1776, p. 332.

[6] *Chapters On Animals*, Philip Gilbert Hamerton, Boston, 1882, pp. 12-14.

[7] *An Essay On Landscape Painting*, by Kuo Hsi, Translated from the Chinese by Shio Sakanashi, John Murray, London, 1959, p. 52.

[8] *Treatise on Painting*, by Leonardo Da Vinci, George Bell & Sons, London, 1897, p. 145.

[9] *Zum ewigen Frieden. Ein philosophischer Entwurf* [*To eternal peace. A philosophical draft*], F. Nicolovius, Konigsburg, Germany, 1795.

[10] Or, *A General System Of The Most Important Principles Of Morality, Jurisprudence, And Politics*, 5th edition in English, London, 1749.

[11] See Flannery, D.J., Vazsonyi, A.T. & Waldman, I.D. Editors, *The Cambridge handbook of violent behavior and aggression*. Cambridge University Press, Cambridge UK, 2007.

[12] See the Annotated Bibliography, *Psychology Of Peace And Mass Violence – War, Ethnopolitical Conflict, And Terrorism: Informational Resources*, by Linda M. Woolf and Michael R. Hulsizer, Webster University, 2004, OTRP Online, Office of Teaching Resources in Psychology, http://teachpsych.org/resources/ Documents/ otrp/resources/woolf04war.pdf.

[13] Translated from the German by Carol Stewart, The Viking Press, New York, p. 219.

[14] *History And Utopia*, by E. M. Cioran, Translated from the French by Richard Howard, Seaver Books, New York, 1987, p. 92.

[15] See *That Which Is – Tattvartha Sutra – A Classic Jain Manual for Understanding the True Nature of Reality*, Translated by Nathmal Tatia, With the combined commentaries of Umasvati/Umasvami, Pujyapada and Siddhasenagani, With a foreword by L.M.Singhvi and an introduction to the Jaina fath by Padmanabsh S. Jaini, The Institute of Jainology, Harper Collins Publishers, San Francisco, CA, 1994. In addition, see the three commentaries on the *Tattvartha Sutra* "used as the Basis of the Commentary" in the translation by Nathmal Tatia, namely: Svopajna Bhasya, by Umasvati and recognized as authentic by the Svetambara sect, ed. Pandit Khubchandji Siddhantashstri, Shrimad Rajchandra Jain Shastramala, Bombay, 1932; Svopajna Bhasya Tika, by Siddhasenagani, Svetambara sect, 2 vols., Seth Devchand Lalbhai Jain Pustakoddhara Fund, series nos. 67 & 76; and *Sarvarthasiddhi*, by Pujyapada Devanandi, of the Digambara sect, 4th edition, Bharatiya Jnanapitha, Delhi, 1989. See Uttaradhyayana by Bhadrabahu, English translation by the great German Indologist, Hermann Jacobi, in the *Sacred Books of the East*, Volume XLV, Oxford University Press, Oxford, UK, 1895, Reprinted in 1980, Jaina Sutras, Part II, by Motilal Banarsidass Publishers PVT. LTD., Delhi Lecture XXIII, p. 128. See also, "Ahimsa-Non Violence," PBS Film by M.C.Tobias, KRMA/1987; and the subsequent volume, Life Force – The World of Jainism, by M.C.Tobias, Asian Humanities Press, Freemont, CA, 1991.

[16] See *Free Thoughts Upon The Brute-Creation: Or, An Examination Of Father Bougeant's Philosophical Amusement*, &c, R. Minors, Londo0n 1742, p. 44.

[17] *The responsibility of power and the power of responsibility: From the "semiotic" to the "semioethic" animal,"* by Susan Petrilli, Università di Bari, http://www.susanpetrilli.com/files/The-responsibility-of-power-and-the-power-of-responsibility.pdf, n.d. See also, "Global communication, proximity and responsibility: beyond the logic of identity," by Augusto Ponzio and Susan Petrilli, http://www.susanpetrilli.com/files/Global-communication.pdf, n.d.

[18] Theraulaz, Guy (1999). "A Brief History of Stigmergy". *Artificial Life. 5 (2): 97–116.* doi:10.1162/106454699568700. PMID 10633572. See also *Pages Choisies,* by Pierre-Paul Grassé, Masson Et Cie, Editeurs, 1967.

Chapter 7

The Rudimentary Ethics of Parmenides

In light of the subtle and acute individual and species expressions across the planet, the question of exaptation – shifts in evolutionary traits – becomes of paramount importance, principally as a factor crucial to sustainable change in conceptualization and any potential for ethical behavior. Elsewhere we have examined varieties of real-time evolution, re-evolution, whether in the case of fruit flies, coiled shells and uncoiled limpets, sexuality in oribatid mites, or the life histories of salamanders. But also, human modifications to human cells, gene events that signal reversion to some ancestral or cladistic trait, or aspect thereof, whereby the 6 billion bits of data within the human genome are minutely shifted, deliberately or not.[1]

German polymath, explorer/biologist/philosopher Ernst Haeckel (1834-1919) who was criticized by many for his eugenicism, also embraced a Darwinian concept of human plurality at its origins and declared, "the natural commandments of sympathy and altruism not only arose in human society millennia before Christ; they were to be found characterizing those higher animals that live in herds and social groups. These traits have their oldest phylogenetic roots in the formation of the sexes in the lower animals and in the sexual love and parental care upon which the preservation of the species rests."[2] The problem for genetic theories arising from the beliefs of a Haeckel, or even Jean-Baptiste Lamarck (1744-1829) concerns the notion of Nazism, the implanted concept that degeneracy in people could be weeded out the way any undesirable characteristic allegedly informed by heredity might be extirpated: The ultimate racist fantasy as it stumbled towards some Aryan pinnacle in its Übermensch. But stripping away this historical baggage, one finds in Lamarck's thinking a potent, generous and important facet of our concept of evolution as against the Anthropocene. Namely, the idea that we can, theoretically, change those hereditary traits in ourselves during our own lifetimes, the basis of much recombinant DNA and messenger RNA research. Whether what we do is heritable might matter, but is certainly not key to fathoming critical, life-saving responses. Under countless other names, a human visitation to Mars, the cloning of a sheep and – ineluctably, of a human – are other versions of eugenics. Indeed, everything about the Anthropocene indicts that aggregate of destructive follies undertaken on behalf of the alleged betterment of humankind.

Alternatively, what seems ethically plausible, given the plurality-mechanisms inherent to natural selection (to the extent we can even gauge or imagine them) is some version of biosemiospheric scaffolding[3] as the systems-dominant architecture for buttressing a vision of life that encompasses every layer of real and/or potential feedback. Timo Maran has referred to the "ontogenetic, evolutionary, physiological and cognitive" facets of this communications realm of biological relations.[4] Millions of people on the streets celebrating the triumph of "compassion" over a dictator secures a place in biological response times for patience, conflict resolution, and elation.[5] The elasticity of caudal nerves that enable hummingbirds to migrate in solitaire (as far as 1400+ miles)[6] also accords with other creative communication capacities among avians, and virtually every other studied vertebrate, with unlimited musical sonance, and deeply structured vocalizations that represent a complex infrastructure of intention, reciprocal messaging, and signage that is both conceptual and pragmatic.[7,8]

How do we know that it is conceptual? That warning of potential predators, or remarking upon unanticipated information is as much an ancestral ripple effect as a "piracy" of the horsehair worm whose "interior crowd of mitochondria," as David George Haskell describes it, "is powered by collaboration."[9] Those start-up communities of cell organelles and their molecular energy compounds (adenosine triphosphate, or ATP) comprised what was probably the earliest "community of well evolved clay-based organisms living in shallow waters exposed to the sunlight," "Darwin's pond,"[10] the site of the earliest metabolic symbiosis and, we might adduce, community ethics.

This story of life and death, of suffering and joy is almost entirely about persons who *work with their hands*, whether the worms and snails and cells within, or every other living thing. That we have enshrined the *thing* in the singular, as opposed to the category of *stuff*, underscores the politics and ecology of the one and the many. Parmenides (ca. late 5th century BCE) argued that "*species* have not their dependence on the Mind of Man; on the contrary, they are unknown to human Nature, or Mankind. The Genus and Essence of everything is of itself, not existing in the Singulars, but the support and foundation of the Singulars." And he adds, "The others (i.e., Singulars) proceed from the One, but the One, which is seen in the Others, hath its Essence from a third; that is Individuals exist by those secondary Ideas, yet so as that the secondary Ideas have from the primary Idea their Essence, and the power by which they give to Singulars a Substance."[11] Notwithstanding a difficult translation from a notoriously ambiguous text, in addressing the "Archetypal Idea" Parmenides[12] dissects the Platonic Forms, in which the ability of the one and the many, of humans and of gods, to understand each other is both posited and questioned.[13]

But Parmenides himself lays the groundwork for an ultimate – and critical - rectification by invoking "signposts" proffering multiple "routes" that shall have spanned "appearance" and "acceptance."[14] The philosophical rudiments leave little doubt that ancient Greek epiphanies, oracles, an assessment of the tragic state of mind, were equally infused with odes of joy, and a pantheistic reverence for nature. It was this quality of apprehending other species that mentored the Hellenic genius, whose most demonstrable evidence was her investment in all those Others. The forces of nature and the cosmic propensity for sublime metamorphoses of all Beings into all Beings remains the most telling story of that continuing miracle.[15]

ENDNOTES

[1] See *The Hypothetical Species: Variables of Human Evolution*, by M. C. Tobias and J. G. Morrison, Springer Nature, New York, 2019, pp. 103-106.

[2] ibid., p. 296.die

[3] See "Semiotic scaffolding: a unitary principle gluing life and culture together," by Jesper Hoffmeyer, Studies in Ecocriticism, Volume 19, 2015, Issue 3, "Biosemiotics and Culture," June 18, 2015, https://doi.org/10.1080/14688417.2015.1058175, https://www.tandfonline.com/doi/abs/10.1080/14688417.2015.1058175;

[4] See "Scaffolding and Mimicry: A Semiotic View of the Evolutionary Dynamics of Mimicry Systems," by Timo Maran, *Biosemiotics* 8 (2): 211-222 (2015), PhilPapers, https://philpapers.org/rec/MARSAM-2

[5] 11.25 am, EST, November 7, 2020 in America; also the day the first woman – Jeannette Rankin – was elected to the U. S. Congress.

[6] See Hummingbirds Of North America: Attracting, Feeding, and Photographing, by Dan True, University of New Mexico Press, Albuquerque, NM, 1993, p. 18.

[7] See *Theory and Natural Order from the Renaissance to the Early Twentieth Century*, by Suzannah Clark, Cambridge University Press, Cambridge, UK, 2001.

[8] "Humans recognize emotional arousal in vocalizations across all classes of terrestrial vertebrates: evidence for acoustic universals," Piera Filippi, Jenna V. Congdon, John Hoang, Daniel L. Bowling, Stephan A. Reber, Andrius Pašukonis, Marisa Hoeschele, Sebastian Ocklenburg, Bart de Boer, Christopher B. Sturdy, Albert Newen and Onur Güntürkün, Published:26 July 2017. https://doi.org/10.1098/rspb.2017.0990, *Proceedings of the Royal Society B, Biological Sciences*, https://royalsocietypublishing.org/doi/full/10.1098/rspb.2017.0990?etoc=&, Accessed November 7, 2020.

[9] *The Forest – A Year's Watch in Nature*, Viking Press, New York, 2012, p. 6.

[10] *Genetic takeover and the mineral origins of life*, by A. G. Cairns-Smith, Cambridge University Press, Cambridge, UK, 1982, p. 351.

[11] "Parmenides," in *The History of Philosophy*, by Thomas Stanley, Printed for Thomas Baffett, London 1687, pp. 748-749.

[12] As delivered by Plato in his Dialogues a fictionalized conversation, three times removed, in which Zeno and Socrates discuss Parmenides, as recalled by a yet another individual within the drama, Pythodorus.

[13] S. Peterson, "The Greatest Difficulty for Plato's Theory of Forms: The Unknowability Argument of Parmenides 133c-134c," *Archiv für Geschichte der Philosophie*, 63: 1–16, 1981. See also *The Hypothetical Species*, op cit., p. 27.

[14] *The Route of Parmenides – A Study of Word, Image, and Argument in the Fragments*, by Alexander P. D. Mourelatos, Yale University Press, New Haven, Connecticut, 1970, p. 197.

[15] See *Old Greek Nature Stories*, by F. A. Farrar, George G. Harrap & Company, London, 1914.

Dr. Russell A. Mittermeier atop the Voltzberg, Central Suriname Nature Reserve. Photo © M. C. Tobias.

Chapter 8

BEING-IN-THE-WORLDNESS: *DAS-IN-DER-WELT-SEIN*

Of the 176 anatomical transformations triggered between organisms in Publius Ovidius Naso's (Ovid's) (43 BCE- 17/18 AD) 15 remarkable books of hexameter, *Metamorphoses* (8 AD) – from Actaeon, who turns into a stag by Diana's intercession, Ceyx, who becomes a bird, mushrooms becoming human, the Pigmy queen a crane, Semiramis given wings in order to commune fully with doves – the question that haunts and frolics with human history is: What was his goal in writing this epic? The year of its release, the Emperor Augustus exiled the celebrated love poet to Tomis (the ancient Romanian town of Constanța on the Black Sea). Exile did not sit well with Ovid who lamented his situation during his final decade away from Rome and its fawning, war-weary crowds. He had seen too much, delved into the ambivalent sexual and emotional community standards of the rigidly hierarchical Roman political stage, and clearly felt a mixed defiance in the ancient face of Plato's dialogue the "Euthyphro" in which Socrates labors to understand the independence, or not, of people and their values from those of the Gods.

Ovid was clearly more of the mindset that swept across the brilliant pages of Lucretius' *On the Nature of Things* (50 BCE); of a cosmos ordered by remarkable flux, atomistic reincarnations.[1] A far more naturalistic, wilderness complex than that world addressed by the upscale farmer, Hesiod (750-650 BCE) in his *Works and Days or Theogony*. Citing such historical commentaries as those of Seneca, Varro and Dicaearchus, historian Clarence J. Glacken asserts that Ovid's Golden Age was precisely that earth not yet altered by humankind. A world in which there was everywhere abundance, fertility, the "sweet nectar" and "verdant oak."[2] The metamorphic transitions were all fundamental to similitudes that allowed for each and every personage to engage, or not engage (the professed solitaire); to be themselves irrespective of human presence and intervention. Humanity was but one more actor on that stage, no better, no less for her involvement.

In spite of the advent of scientific evidence mounting on every laboratory table from Peking to London, Ovid's emphatic thread of biological change steeped in alchemical possibilities and moral vicissitude would forever fuel the panpsychic description of the world, known and unknown: of a mind/soul operating throughout the universe. That "everything is full of gods,"[3] according to Aristotle; and that the discussion of "substance" encompasses such diverse propulsions as "Friendship," "Air," and "the Infinite."[4]

Ovid's masterpiece enjoined a fully liberated setting for the interactions of all Beings, ourselves and Others. Everything was a thinking, feeling organism vulnerable to interactivity. This could only be viewed as a fabulously complicated field of moral consideration and precarious identity amid multitudes. Of the fragility of self-interest and the nonsensical insistence upon even the least permanent fealty to biographical specifics.

It would then come as no surprise when, in the late 15th century an artist like Martin Schongauer should engineer a simple woodblock print entitled "The Inconvenient Petitioners," in which diverse quadrupeds with trunks ablaze blasted out their grievances before a human court, a clear precursor to the rearing horse Picasso's "Guernica" (1937). In 1647 Dutch painter Paulus Potter would return absolutism to the animals in his 14 exquisite scenes depicting the "Punishment of a Hunter" (in the Hermitage). Fifteen years later, in the Brussels edition of *Don Quixote*, readers would see the deep emotional apotheosis of Dapple, Sancho Panza's burro, and the great Don's own hack, the sublime equine, Rocinante, in the first illustrations of the original Spanish edition.[5] By that time, Porphyry's famed essays commending a diet of vegetarianism (*Eiusdem De Abstinentia Ab Esu Animalium*), along with that of Plutarch's (De Esu Carnium, Περὶ σαρκοφαγίας) had enjoyed multiple editions in Europe.[6]

In 1945 George Orwell's *Animal Farm* was finally accepted for publication by Secker & Warburg (London), despite the novel's clear condemnation of Britain's ally in the war, the Soviets. It was the same year that Aleksandr Georgīevīch Romm wrote and published his *Protest of Animals* (протест зверей), with Ivan Krylov (Aesop)-like illustrations by Risunki Vladimir Gal'ba, during the siege of Leningrad.[7] A rare show of civil disobedience during on behalf of animals, done like a cartoon scroll, that would clearly inform Orwell's own adroitly animal rights-driven Preface to the March, 1947 Ukrainian edition of his *Animal Farm*.

The politics of an Aesop's rich range of familiar characters has reverberated for at least 2500 years. Their strangely ambivalent relations instructs a philosophical space whose contents describe a biosphere perpetually expanded by the very self-referential consciousness of aesthetic, philosophical and scientific divagations with which humans can so easily relate. There is conversation at every conceivable and inconceivable level of rudimentary exchange. Exchange itself gives unto intent; intention harbors collaborative measures aimed at both being and accepting; or hunting down and dooming. In acceptance is humility; with humility, some clear pathway towards understanding. That which surpasseth all these inroads and corollaries is the same which granteth day and night, the seasons and their senses. From these crossroads it is clear that in aggression, destruction, refusal to collaborate, there is no value, no reason for living. Only in peaceful circumstance can any credo assure a viable interactivity that may be predicted to prosper.

That is the gist of biocalculus, nibbana, the cessation of dukkha, or suffering, and – in Buddhism – the Noble Eightfold Path, or *ariya aṭṭhaṅgika magga*, in ancient Pali; the means by which to attain (reach) that place, situation, frame of devout perception. "Right view, right resolve, right speech, right conduct, right livelihood, right effort, right mindfulness, and right *samadhi* ('meditative absorption or union').[8] Cicero and Quintillian spoke of correctness, much in the line of discussions by Aristotle, and his student Theophrastus on rhetorical doctrine.[9] But a far more substantialized perception of conduct can be fixed ethically in situ (in the world). It is that purpose, ontology, all those meaningful goals whose broad courses of action necessarily encompass at some point one or more of the four causes (*aitiai*) as outlined by Aristotle, namely, material, formal, efficient and final. These provide the basis for consideration of any *effective* goal orientation, as Aristotle viewed the workings of Nature. With an ultimate state of welfare and happiness, or eudemonia.[10] Both Aristotle and Aesop approached the biology of everyday according to similar basics: the familiar encounters and objects of engagement that comprise a human life in concert with all those Others around us.

George Wilhelm Friedrich Hegel (1770-1831) characterized this syllogism (by inference, of the wild and the domestic) as "Reflective Symbolism": "a form of art wherein the idea is not only comprised within itself, but also expressly posited as distinct from the sensuous form by which it is represented."[11] One could take this in many directions. Immanuel Kant (1724-1804) would write, "We are *a priori* conscious of our own complete identity in regard to all representations which can ever belong to our cognition; and this we regard as the necessary condition of the possibility of all representations."[12] While Arthur Schopenhauer's *The World as Will and Representation* (1819) suggested that the world was not independent of a cognizing host of perception, there is no objective, or subjective synonymity that does not inculcate the tautological whole. It is not merely a matter of mathematics, physics or grammar. Neither chemistry nor probability function upholds illusion. Distribution of individuals, populations, aggregates of biomes and ecosystems may vary to the extreme, but the ecological principles, of which every philosophy, theology and aesthetic exuberance is born, are symptomatic of a reality that is the universal faith nestled deep within *in-Itself*. How do we know that? We can't know, but this is not oblivion, either. Every individual harbors a *hunch* about the nature of life, of one's own life, that trembles between empirical supposition and epiphany. Or, languishes in the daunting ache of feeling too much about so vulnerable and exposed a situation that is the known world. Our point is this: As much as Western philosophy has grappled with the representation-in-mind of the world at large, the looming realization that we *are* that world at large, gives loft to the passive *partaking of*; transforming it into true Being, not some hedonistic Dasein, rather, *das-in-der-Welt-sein*, a Being-in-the-Worldness.

The lush swathes of an evergreen dream – the undeviating apotheosis of natural history - linger regardless of fears or calculations; of the power of "Gods" o'er ourSelves; fanciful questions of destiny in the thickets and mud of primeval evolution. Of the hundreds of editions of *Metamorphoses*, all of them most tellingly come to an ultimate headwall in Book XIV, in which the famed tale of the Etruscan Vortumnus and his chicanery with Pomona excites the revelatory scope of changing seasons and the sheer drama of biodiversity.

"She [Pomona] shows, how Stocks invite to their Embrace/ A Graft, and naturalize a foreign Race/ To mend the Salvage Teint; and in its Stead/ Adopt new Nature…"[13] *New Nature*: The final testament to the transforming powers of the Gods and Mortals collaborating, as was conveyed in the overwhelming depicture of Holy Roman Emperor Rudolf II (1552-1612) as Vertumnus, by Giuseppe Arcimboldo (ca. 1590). From the seat of his court in Prague, Rudolf assembled the most formidable private natural history collection in the world, the Walter Rothschild of the Renaissance. His patronage of Roelant Savery, Aegidius Sadeler, Jan Brueghel the Elder, as well as Johannes Kepler and Tycho Brahe, among dozens of others firmly established the European obsession with change in Nature, taxonomic systems, and humanity's role in consciously merging great beauty with empirical studies.

Rudolf's collections – the Kunstkammer – first inventoried in 1607, would eventually be dispersed across Europe, starting in 1612, in a succession of monarchial raids, private and public auctions.[14] Among the thousands of paintings, sculptures, minerals, animals dead and alive, artifacts culled from monasteries, other Royals and from the ages, were the poignant remains of an obsession that had included the last and only known images of the Dodo, as painted from life, and what was rumored to be a dragon. But it was the sheer excess of natural history spilled out across one pirate's imperial domain that clearly suggested to Arcimboldo the utterly ironic and freakish portraiture of human sovereignty overtaken by that which it had respectively conquered, the Amazon devouring the man's face. This was a penetrating insight into Ovid's much purer landscape of interdependent creatures and natural variations. Rudolf, after both Aesop and Ovid, set the groundwork for zoos and other indecencies. For a posterity that had lost track of natural barriers, usurping the traditional compass with a newly envisioned human dominion in the name of Aristotelian science.

ENDNOTES

[1] See *Transformations in Ovid's Metamorphoses*, by Ian Johnston, Translated by Ian Johnston, Vancouver Island University, British Columbia, 2012 (Revised 2017), http://johnstoniatexts.x10host.com/ovid/transformationshtml.html. *Ovid's Metamorphoses in Latin and English*, Translated by the Most Eminent Hands, With Historical Explications of the Fables by the Abbot Banier, Translated by John Dryden, Alexander Pope, John Addison, John Gay, William Congreve, Illustrated by Bernard Picart, Notes by Antoine Banier, Wetsteins

and Smith, Amsterdam, 1732. See G. B. Riddehough, "Man-into-Best Changes in Ovid," *Phoenix*, Vol. 13, No. 4 (Winter, 1959) https://doi.org/10.2307/1086030, https://www.jstor.org/stable/1086030, Published By: Classical Association of Canada, https://www.jstor.org/stable/1086030?seq=1, pp. 201-209. See *Ionian Enchantment: A Brief History of Scientific Naturalism*, by Ignacio Prado, Tufts University, June 2006, in Naturalism, https://www.naturalism.org/worldview-naturalism/history-of-naturalism. Accessed November 16, 2020. See also, *Animal Acts: Configuring the Human in Western History*, Edited by Jennifer Ham and Matthew Senior, Routledge, New York/London, 1997.

[2] In *Traces on the Rhodian Shore – Nature and Culture in Western Thought from Ancient Times to the End of the Eighteenth Century*, University of California Press- Berkeley and Los Angeles, 1967, p.133.

[3] Aristotle. De Anima 411a7–8. See "Aristotle's Psychology," Christopher Shields, *Stanford Encyclopedia of Philosophy*, October 12, 2020, https://plato.stanford.edu/entries/aristotle-psychology/#TextTranComm DeAnim.

[4] *Aristotle's Metaphysics*, Translated with Commentaries and Glossary by Hippocrates G. Apostle, The Peripatetic Press, Grinnell, Iowa, 1979, p. 163.

[5] *Don Quixote*, House of Juan Mommarte, Brussels, 1662.

[6] *Porphyrii De absteinentia ab esu animalium, libri quatuor* [*The porphyrii absteinentia from eating animals in four books*], Ioanne Bernardo Feliciano interprete, Ion. Gryphium, Venice 1547; and again as published by sumptibus Claudii Morillon in Lugduni (Lyon) in 1620.

[7] *Protest Zverei* [*Protest of Animals*], by G. Romm, Illustrated by Vladimir Gal'ba (Galberstad) of The Militant Pencil, Poligrafmasterskaya LSSKh, Leningrad 1945.

[8] Grzegorz Polak, *Reexamining Jhana: Towards a Critical Reconstruction of Early Buddhist Soteriology*, UMCS, Lublin, Poland, 2011, pp.153-156, 196–197. See also, Noble Eightfold Path and Its Factors Explained (Magganga-Dipani), by Ledi Sayadaw, Buddhist Publication Society, Sri Lanka, 1998.

[9] See "Cicero as a Reporter of Aristotelian and Theophrastean Rhetorical Doctrine," by William W. Fortenbaugh, *Rhetorica: A Journal of the History of Rhetoric*, Vol. 23, No. 1 (Winter 2005), pp. 37-64, University of California Press. https://doi.org/10.1525/rh.2005.23.1.37, https://www.jstor.org/stable/10.1525/rh.2005.23.1.37, https://www.jstor.org/stable/10.1525/rh.2005.23.1.37?seq=1, Accessed November 16, 2020.

[10] See Aristotle's *The Nicomachean Ethics*, 1095a15-22. David Ross, Translated with an Introduction. Oxford: Oxford University Press, Oxford, UK, 1925.

[11] *The Philosophy of Art: Being the Second Part of Hegel's Aesthetik in which are Unfolded Historically the Three Great Fundamental Phases of the Art-Activity of the World*, by Wm. M. Bryant, D. Appleton & Co., New York, 1879, p. 29.

[12] *Kant's Prolegomena To Any Future Metaphysic*, by John P. Mahaffy and John H. Bernard, Vol. II, The Prolegomena Translated, With Notes And Appendices, Macmillan And Co., London, 1889, p. 171.

[13] "The Story of Vertumnus, and Pomona," in *Ovid's Metamorphoses In Fifteen Books* – Translated by the most Eminent Hands, Printed for Jacob Tonson, London 1717, p. 500.

[14] See "Rudolph II's art cabinet," *Codex Gigas*, by Michael Gullick, Ivan Hlaváček, Jan Svanberg, Anna Wolodarski, Translators, Göran Bäärnhielm, Bengt Ellenberger, Jana Holá, Janis Kreslins, Christina Manetti, Roger Tanner, http://193.10.12.180/codex-gigas/eng/Long/handskriftens/rudolf-ii/#Rudolph%20II, Accessed November 17, 2020.

Chapter 9

THE METAPHYSICS OF THE 'PEACEABLE KINGDOM'

That the Quaker preacher Edward Hicks (1780-1849) could take a few lines from the Prophet Isaiah, "The wolf shall dwell with the lamb, and the leopard shall lie down with the young goat, and the calf and the lion and the fattened calf together; and a little child shall lead them…" (Chapter XI: 6-9) and, commencing in 1820 proceed to paint at least 62 versions of his "Peaceable Kingdom" must go to the core of that belief in some utterly wonderful, revolutionary possibility for the human condition.[1] Hicks' biographical story-line reads of the perfect ethical disposition for the right moment. Born one century after the Quaker William Penn signed a peace treaty with the indigenous Lenape Indians, his life was situated between the American Revolution, the War of 1812, and the burgeoning social, racial, economic and ecological agitation that was signaled by the rapid population growth of six newly added trans-Allegheny commonwealths to the Union by 1830, and with them, over "3,600,000" people.[2] Initiated into Quaker non-violence philosophies and "the light within," referencing the power of Christ to affect a personal relationship with a world inhabited by the Messiah[3], Hicks transformed the sublime evocations of the earliest landscape traditions in America into pure Quaker interspecies wish-fulfillment.

Whereas the vast majority of animal paintings by the English and French during Hicks' time were largely sportsmen or farming-related showcases for prized heifers, bursting Belted Galloways, and other typically obese "livestock" and obedient equines – depictures by such artists as Thomas Weaver, John Sheriff, Henry Chalon, James Ward, and George Stubbs[4] – Hicks was determined to reconcile controversies within various Quaker (English Quaker Church versus the Society in a colonized America) and other Christian-based metaphysics with an altogether different, day-and-night interpretation of the Creation. To "paint a sermon," as Alice Ford has characterized it in her superb biography of America's most prophetic, ethological painter.[5] It's true that he earned his living as a painter, and did engage in numerous scenes involving "Stock" as in his 1849 "A May morning view of the

Farm and Stock of David Leedom" (Abby Aldrich Rockefeller Folk Art Center).[6] But Hicks' likely final painting was his most subtle, and comprehensive vision of what North America, and the world, could be: "The Peaceable Kingdom," 1849 (Nelson-Atkins Museum, Kansas City, Missouri, lent by the great, great grandson of the artist).[7]

What was fantastically notable in Hicks' response to this vast biophilia predilection, was the representation of three children in the foreground, and a circle of Pennsylvania colonists and local Indians holding hands in a circle at the water's edge – the representation of the signing of the famed peace treaty. This same motif was used in dozens of Hicks' Kingdom paintings. One would be hard-pressed to find anything quite like this in any documented tradition of art previously. The closest precursors, in terms of a human proximity/loving association with other species, come from the tradition of mille-fleurs, notably "The Lady and the Unicorn," ca. 1500 (Musée de Cluny in Paris); and the countless scenes of "The Nativity" -a genre beginning in 1223 with Saint Francis of Assisi, and "Rest on the Flight Into Egypt" which was also magnificently enshrined as early as 1304 in Giotto di Bondone's "The Flight Into Egypt" at the Scrovegni Chapel in Padua. Before that time, one searches throughout the medieval bestiaries[8], some of the thousands of illustrations enshrined in so many Book of Hours (horae)[9], numerous psalters and early renditions of *Aesop's Fables* (which, in spite of the motives behind seemingly benevolent affiliations, like "The Parliament of Birds" graphics, are in fact profuse with images of violence); or, most famed, the illustration of God seated with an abundance of animals in the Holkham Bible Picture book, ca. 500.[10] Prior to that genre, one finds utterly antithetical Achaemenid cylinder seals, for example, figuring heroes battling lions and bulls.[11]

In Hicks' dense and theologically abstruse "Goose Creek Sermon" delivered at a meetinghouse in Loudon County, Virginia February 22, 1837, he echoed the sentiments regarding animals as his late and influential cousin, the Quaker minister Elias Hicks, asserting that there was, indeed, an Inward Light throughout all of nature, manifested in humans whose duty to life must translate into peace, opposition to all oppression, hunting, animal abuse in any form.[12] The spiritual dogma attendant upon Quaker theories of the chain of being tended towards a belief in humanity as the most animated (highest) form of animal life. Carnivores – so critical to the Prophet Isaiah's famed lines prompting Hicks' Peaceable Kingdom paintings – were actually thought of as carnivores, pure and simple, summarized by untoward attributes in Nature which enlightened humans were charged with surmounting. That purports to a difficult contradiction, when simply standing back and appreciating Hicks' work, underscoring an undeniable truth about the dualistic personality at work in a man who preached abstrusely, but painted with startling clarity. There is no ambiguity whatsoever in his painted depictions of a world capable of re-achieving, resolving, returning to the absolute paradise that it is.[13] A similar dialectic can be detected in the Quaker naturalist – gardener, explorer - William Bartram[14] – who has been called "an eighteenth-century Thoreau…"[15] In Bartram's very opening lines of his masterpiece, *Travels…* (1791) he reconciles that Biblically-founded differential:

"…whatever may contribute to our excellence is also of equal importance, whether it be found in the animal or vegetable kingdom…it appears evident, that the great Author has impartially distributed his favours to his creatures, so that the attributes of each one seem to be of sufficient importance to manifest the divine and inimitable workmanship."[16]

Hicks prefigures such iconic Hudson River and Luminist School paintings later in the 19th century as Albert Bierstadt's numerous "Indian Encampments" and his seminal "The Rocky Mountains, Lander's Peak," (1863, The Metropolitan Museum of Art, New York). What Hicks could never have endorsed was a painting like Bierstadt's "The Last of the Buffalo," (1889, The Corcoran Gallery of Art, Washington, D.C.) in which a band of Indians are killing the iconic national animal. Bierstadt himself did not sanction such behavior, writing of his painting that he had "endeavored to show the buffalo in all aspects and depict the cruel slaughter of a noble animal now almost extinct."[17] The reality of portraits of America following Edward Hicks can be generically delineated between romantic nostalgia for a Golden Age, on a par with similar evocations throughout the history of European art, and depictions in one form or other of destruction, as meted out in countless versions of manifest destiny.[18] This developmental paradigm, a graphic plague on our house, was teased open from the grandeurs of Yosemite to the despoiled scenes of forest clearance, mineral extraction, road and city building that figure in the ten-volume masterwork edited by John Muir in 1888 and "Illustrated With Etchings, Photogravures, Wood Engravings, Etc. By Eminent American Artsits."[19]

Artists of every persuasion could not fail to see and reflect upon the continuing abuse of people, other animals, and America's landscapes in spite of the passage of the 14th U.S. Amendment abolishing slavery (ratified July 9, 1868), and the 1866 Civil Rights Act. By the time of post-Civil War Reconstruction, the Indian's "collective world was verging on extinction."[20] Edward S. Curtis' twenty-volume *The North American Indian*, published between 1907 and 1930, was originally going to be titled "The Vanishing Race."[21] Meanwhile, across Europe tens-of-millions of people would be slaughtered, as wars, genocides, rapidly expanding human populations and gas guzzling automobiles concentrated in cities, uncurtailed chemical adulterants and the rapid rise of industrialized agriculture and resulting slaughterhouses, led by turns to the development of the atomic bomb, in concert with the first dramatic data profiling the onset of the Anthropocene. All this rapidly rendered the worldview of an Edward Hicks tragically remote, in a very short period of time.

But what remains of Hicks' vision is his reflex; an amalgam of empirical reactions in one person to *something* that, if in one, must reside somewhere, somehow, in everyone.

ENDNOTES

[1] See *Notes, Critical, Explanatory, and Practical on The Book of the Prophet Isaiah*, by Albert Barnes, Leavitt & Allen, New York, 1860, Volume 1, pp. 243-246.

[2] "The Colonization of the West," by Frederick J. Turner, *The American Historical Review*, Jan. 1906, Vol. 11, No. Oxford Journals, Oxford University Press, p. 304, https://www.jstor.org/stable/1834646, https://www.jstor.org/stable/pdf/ 1834646.pdf

[3] See "The Peaceable Kingdoms of Edward Hicks," by Victoria Emily Jones, *Art & Theology*, December 6, 2016, https://artandtheology.org/2016/12/06/the-peaceable-kingdoms-of-edward-hicks/Accessed

[4] See *Farm Animal Portraits*, by Elspeth Moncrieff with Stephen and Iona Joseph, Antique Collectors' Club, Suffolk, UK, 1996.

[5] *Edward Hicks – His Life and Art*, by Alice Ford, Abbeville Press, New York, 1985, p. 46.

[6] *The Kingdoms of Edward Hicks*, by Carolyn J. Weekley, with the Assistance of Laura Pass Barry, Harry N. Abrams, Inc., Publishers, New York, 1999, 179.

[7] Op.Cit., Alice Ford, p. 247.

[8] See, for example, The Sherborne Missal, Alnwick Castle, in *Birds in Medieval Manuscripts* by Brunsdon Yapp, Schocken Books, New York 1981, p. 153.

[9] See for example, The Master of Mary of Burgundy – A Book of Hours for Engelbert of Nassau, The Bodleian Library, Oxford, *Introduction and Legends* by J. J. G. Alexander, George Braziller, New York, 1970.

[10] See http://www.bl.uk/manuscripts/FullDisplay.aspx?ref=Add_MS_47682

[11] See *Animals and Men – Their relationship as reflected in Western art from prehistory to the present day*, William Morrow and Company, Inc., New York, by Kenneth Clark, 1977, p. 23.

[12] Op. Cit., *Weekley*, pp.223-232, and pp.51-64.

[13] See "Friends and Nature in America: Towards An Eighteenth-Century Quaker Ecology," by Donald Brooks Kelley, *Pennsylvania History: A Journal of Mid-Atlantic Studies*, Vol. 53, No. 4 (October, 1986), pp. 257-272, Penn State University Press, https://www.jstor.org/stable/27773139, https://www.jstor.org/stable/27773139?seq=1.

[14] See "The Quaker Background of William Bartram's View of Nature," by Larry R. Clarke, *Journal of the History of Ideas*, Vol. 46, No. 3 (Jul.-Sep. 1985), pp. 435-448, University of Pennsylvania Press, JSTOR, https://doi.org/10.2307/2709478, https://www.jstor.org/stable/2709478, https://www.jstor.org/stable/2709478?seq=1, Accessed November 19, 2020.

[15] See "The Nature of William Bartram," by Thomas P. Slaughter, *Pennsylvania History: A Journal of Mid-Atlantic Studies*, Vol. 62, No. 4 (Fall 1995), pp. 429-451, Penn State University Press, https://www.jstor.org/stable/27773844, https://www.jstor.org/stable/27773844?seq=1, Accessed November 19, 2020.

[16] *Travels Through North and South Carolina, Georgia, East and West Florida* by William Bartram, 3rd (Dublin) edition, 1793, pp. ix-xii.

[17] See *Landscape Painting – A History*, by Nils Büttner, Abbeville Press Publishers, New York, p. 290.

[18] See *Extended Travels in Romantic America, Chosen and Displayed* by Joseph Jobé, Published by Edita at Lausanne, Switzerland 1966.

[19] *Picturesque California: The Rocky Mountains and the Pacific Slope – California, Oregon, Nevada, Washington, Alaska, Montana, Idaho, Arizona, Colorado, Utah, Wyoming, Etc.*, J. Dewing Publishing Company, New York and San Francisco.

[20] See "Curtis and America's Environmental Social Justice Movement," by M. C. Tobias, in Edward S. Curtis – *One Hundred Masterworks*, by Christopher Cardozo, with Contributions by A. D. Coleman, Louise Erdrich, Eric J. Jolly And Michael Charles Tobias, Delmonico Books/Prestel Verlag Munich, Random House New York 2015, p. 141.

[21] See http://curtis.library.northwestern.edu/

Chapter 10

EQUATIONS OF POTENTIALITY

Purveying biological history with a goal of grasping what it is about Hegel's remark that one salient fact we learn from history is the (overriding) truth that we learn nothing from history, accents what many researchers have called second, and third-order cybernetics.[1] Systems of observation which, in turn, commandeer observing self-referentials. These interior conundrums also conjure second and third-order reciprocities, projected gains, losses and a full circumference of virtue ethics where the very nature of cause and effect must endorse more subtle mechanisms for understanding how the individual rises to the occasion of the world. There are both philosophical and practical difficulties involving that which triggers reactions. Moreover, any special knowledge that might alter the degree of a response by some party to the transaction undercuts the idea of an objective observer or involvement. This has been ably demonstrated by Robert E. Goodin in his analysis of different categories of vulnerability.[2]

The system of switches and circuits maintaining a well-oiled machinery – whether biological or technological - breaks down and defaults, either to an immune system and antibodies, to a buffer system of overrides; an uninterrupted power supply; built-in back-up seals and O-rings, or more comprehensive fail-safes. In most aircrafts this might be a triple modular redundancy. In computers, N-version (multiple functionally equivalent) programming. Reliability engineering, accounting for stochastic (random) parameters, as applied to the risk of entire systems, figures in measures of confidence, plus-or-minus margins of errors in all polls and census taking. This encompasses tolerance thresholds, ultimate tensile strength, circuit breakers, thermostats and safety turn-off signals; compound, complex and Bayesian probability interpolations; the ISO 9000 quality management systems and other procedural gauges of safety, and of the very thresholds of stress in any tribosystem where inputs affect overall resulting efficiency verifications. There is no product guarantee without some means of rapidly tracing global supply chain life cycles, product stewardship, and reasonable responsibility by the manufacturer.[3] What separates all such concerns within the overall sum of machinations of *Homo sapiens* is the reality that our species alone, among all others, leaves waste dumps, obliterates habitats, incurs epidemiological fall-out. We believe ourselves to be somehow *outside of Nature*, as we destroy her - having isolated ourselves, convinced of a surreal immunity. Our parts are not complicit with the idea of a whole. That our biological clocks, absolute physiological requirements, and circadian rhythms have no correlation with our mental constructs means that we not only appear to be slaves to Hegel's prognostication, but that in the interests of ego, greed, moment-by-moment gratification, we are blind to immediate cause-and-effect. Characterized differently, our malevolent penchants – dating to at least 45,000 years ago, during which our ancestors manipulated tropical forests[4] – do not counteract our widespread swathes of destruction with restraint or meaningful mitigation. Or in no observable critical mass, to date.

If we give even meager thought to all the varieties of violence and harm that we mete out to others, and to ourselves, the psychoanalytic trail of tears mounts with exponential weight. Between life, death and all the prospects of reincarnation – about which our knowledge grows faint – determining similitudes between say, Manhattan, and a redwood tree, are imaginative steps, the coining of poetic metaphors and similes that have no basis in anything but our minds. If we were to somehow measure the weight, pace of elicitations, burden of ideation by some abstruse percentage of *all* neural activity, of our ideas and ideals, we know there is a huge gulf, described traditionally as mind versus body. But even taking this differentiation to the second tier, mental derivatives of real-time, real-place interaction with the physical world, we'd likely be astonished by the full scope of our disconnect.

From the full spectrum of virtually uncontrollable passing events in our mind, to the nuts and bolts of preparing and consuming a meal, sleeping and then again standing up and moving, our lives can be said to be controlled – by biochemical switches from within, and sensory stimuli and a rash of environmental stressors and constraints from without – just as our volitions, impulses, needs and wants drive us to do things. But what part of this constellation of our activities do we fail to understand? Proofs, as in mathematical or scientific confirmation, quickly lose relevancy depending on the question. For example, does the movement of a mite in Uruguay alter the periodicity of seismic activity in Japan? If we relegate that kind of question (the conventional "Butterfly Effect") to a certain *ecological category*, we can formulate a gradient of other, countless categories to accord with what we take to be relevant. If personal relevancy is deemed to be the standard, then evolution is seen, in this mind game, to be as fantastically broad and irrelevant as anything could be. *Personal evolution* comes closer to a recognizable mark; connoting something far more personable than, say, the evolution of stripes on a zebra (to control swarms of flies); a general sense of our *own* inner development that can never come however close to being fully ascertained. Even cause-and-effect, unless stated with the clarity of that equation in physics for something as (seemingly) basic as billiards, which prescribes "conservation of momentum," "static" and "kinetic friction," "tangential velocity," "relative slipping," and "obliqueness of the impact(s)" etc.,[5] the causality devolves to a large swathe of *potential equations* and theories: Paul Dirac's 1928 understanding of the movement of atomic particles against the backdrop of the speed of light; Context Framing in a Cumulative Prospect risk-weighted moment of decision; Evolution Clocking to determine our most recent ancestral convergence, etc., each largely irrelevant to our daily cognizant lives.

In other words, theories of evolution, of general relevancy and relativity, of category invention, trespassing, dismissal, and juxtaposition, lose their grounding as we move further away from what we deem essential: air, water, food, sleep. Any proposed basis for understanding the reciprocal functions of life are already multiple steps removed from the most exacting, and obvious imperatives. Does this mean that the entire study of reciprocity is esoteric? Hardly. To take one, perhaps strange, example. When I am parched and help myself to a drink of water I am acting reciprocally, providing my sense of thirst with the required solution to dehydration synthesis. This presupposes, then, that there are *several selves* within myself, which, of course, there are. Tens-of-trillions of cells. We are hosts to as many organisms as conjectures and categories of speculation: bilateral, conspicuous, inconspicuous, competitive, non-competitive, multivariable, solitary-in-mind, conscious, subjective, involuntary, and so forth.[6] At the level of neurons "understanding the set of transformations by which incoming sensory input leads to behavioral responses (i.e., the neural code) remains a central problem in neuroscience" in part because of "strong heterogeneities in neural populations" "even within the same cell type."[7]

Cellular metabolic responses, most appreciably the state of "adrenaline signaling in muscle cells,"[8] but also every aspect of gene expression, presupposes an in situ biological set of evolutionary circumstances concentrated into an individual; poised before the challenge of its kind, clusters, populations, and direct connections. Seen in the context of "cortical microcircuits" neural connections in all known organisms (other than sponges, placozoans and mesozoans) suggest that there are "structures between presynaptic axons and postsynaptic dendrites" whose multiplicities of redundant contact show that it is in this fabric of multiples that optimization of information and other processing occurs.[9] It has been posited that the build-up of discrete, minimal-scope life forms into symbol-procuring beings occurs because direct dependencies between a central nervous system and outside circumstances have been transcended. Language, a command of metaphoric expression, symbolism, ritual, self-consciousness, futurism, culture have all moved from the smallest biological quanta within a finite constellation of cells and nerves, to a density of neural capacity that is free of "corresponding environmental relations."[10]

Such liberation, if indeed conceptual, invites the pure abstraction of compounded logic: the sequence of mental events unfolding with or without precedent – no way to know – but all leading, ultimately, to a deliberate, ecological ending of the Self. Just as the population (a couple) prepares for the beginning, the individual prepares for the end. Think of all the ways. Each is a legitimate ecological coefficient, part of an equation core to the biological sciences. Juxtaposed with one's own invocations, that science, of its own accord, thoroughly filters, becomes, admires and joins the personage, the sum total of which decrees its own storyline and ending. In essence, affords the ultimate reciprocity, giving everything to itself, then returning to the Earth that first gave birth to it. The cycle, immune to all judgement or proof, is complete.

Towards this naturalism, this sequence of inevitable events, the triggering of every conceivable and inconceivable mode of thought is seemingly limitless. In 40 AD Seneca the Younger (4 BCE-65 AD) wrote his "Ad Marciam De consolation," one of his most profound meditations on death and living with death, conceived long before his own suicide. Seneca's death by (allegedly) cutting open his veins and taking poisons, possibly being forcibly suffocated, was obviously not the most delicate way to embrace the end. Handguns - like Emperor Charles V's Wheellock Pistol (1540s) and Georg von Reichweins revolver (1597), which would lead rapidly to the Colt Revolver (1836), were obviously more pain-free methods for accomplishing that overwhelming intention.[11] Friedrich Sertürner in 1805 managed to identify the key effective ingredient of opium poppies although the commercialized product, heroin, would require another 92 years of development, again advancing the cause of self-medication, notwithstanding the obstacles of cultural judgment.[12] On December 1, 2020 the Singapore Food Agency approved the commercial sale of laboratory cultured chicken cells to be sold in the form of chicken nugget-like food dishes in Singapore restaurants, a breakthrough in the rapidly converging global marketplace for vegetarian/vegan product.[13] One could go on adding to such innovations (starting not from zero, but simply some vastly negative value that is the current human overreach): Gatesian green steel, green hydrogen, green concrete, liquid sodium reactor coolants, zero violence, zero emissions, zero trespass, in every action, production and – by turns, intention.

In the human timeline of the amelioration of suffering, such activist, technology endeavors are of prime significance. They underscore the adaptability of human concepts, and suggest that if our imagination can indeed cultivate virtuous behavior, then evolution has every reason to take note of what we're thinking. Thoughts that have no physical barriers. That's a true beginning, as sculptor Henry Moore once said, at every juncture of idealism and praxis.

ENDNOTES

[1] See Ranulph Glanville, (2003), "Second Order Cybenetics," in *Systems Science and Cybernetics*, Ed. Francisco Parra-Luna, in *Encyclopedia of Life Support Systems (EOLSS)*, Developed under the Auspices of the UNESCO, Eolss Publishers, Oxford, UK, http://www.eolss.net.

[2] *Protecting the Vulnerable – A Reanalysis of Our Social Responsibilities*, by Robert E. Goodin, University of Chicago Press, Chicago, Ill., 1985.

[3] See *Waste as Opportunity – A MAHB Dialogue with Materials Scientist Veena Sahajwalla*, by Geoffrey Holland, MAHB, Millennial Alliance for Humanity and the Biosphere, November 12, 2020, https://mahb.stanford.edu/blog/mahbdialogues/waste-as-opportunity-a-mahb-dialogue-with-materials-scientist-veena-sahajwalla/Accessed November 20, 2020.

[4] See "The deep human prehistory of global tropical forests and its relevance for modern conservation," by Patrick Roberts, Chris Hunt, Manuel Arroyo-Kalin, Damian Evans and Nicole Bolvin, *Nature Plants 3*, Article number: 17093 (2017), August 3, 2017, https://www.nature.com/articles/nplants201793,

[5] See *Real World Physics Problems, – The Physics of Billiards*, n.a., https://www.real-world-physics-problems.com/physics-of-billiards.html

[6] See "A Theory of Reciprocity," by Armin Falk and Urs Fischbacher, *Games and Economic Behavior*, Volume 54, Issue 2, February 2006, pp. 293-315, ScienceDirect, Elsevier, https://doi.org/10.1016/j.geb.2005.03.001 https://www.sciencedirect.com/science/article/abs/pii/S0899825605000254

[7] From the Introduction to "Neural Heterogeneities Determine Response Characteristics to Second-, but Not First-Order Stimulus Features," by Michael G. Metzen and Maurice J. Chacron, *The Journal of Neuroscience, J Neurosci.* 2015 Feb 18; 35(7): 3124–3138. doi: 10.1523/JNEUROSCI.3946-14.2015, PMCID: PMC4529327, CAMSID: CAMS4969, PMID: 25698748, https://www.ncbi.nlm.nih.gov/ pmc/articles/PMC4529327/, Accessed November 21, 2020.

[8] *Response to a signal*, Khan Academy, https://www.khanacademy.org/science/ap-biology/cell-communication-and-cell-cycle/signal-transduction/a/cellular-response

[9] See *Redundancy in synaptic connections enables neurons to learn optimally*, by Naoki Hiratani and T Gomoki Fukai, Edited by Terrence J. Sejnowski, Salk Institute for Biological Studies, La Jolla, CA, PNAS July 17, 2018 115 (29) E6871-E6879; first published July 2, 2018; https://doi.org/10.1073/pnas.1803274115, PNAS, https://www.pnas.org/content/115/29/ E6871, Accessed November 21, 2020.

[10] See, for example, *Does Altruism Exist? Culture, Genes, and the Welfare of Others*, by David Sloan Wilson, Yale University Press, Templeton Press, New Haven Connecticut, 2015, p. 53.

[11] *9 Oldest Guns in the World*, n.a., n.d., https://www.oldest.org/technology/guns/Accessed December 2, 2020.

[12] See "Defining a Good Death (Successful Dying): Literature Review and a Call for Research and Public Dialogue," by Emily A Meier, Jarred V. Gallego, Lori P. Montross Thomas, Colin A. Depp, Scott A. Irwin and Delip V. Jeste, *The American Journal of Geriatric Psychiatry*, Volume 24, Issue 4, April 2016, pp.261-271, Elsevier, https://www.sciencedirect.com/science/article/abs/pii/S106474811600138X, Accessed December 4, 2020.

[13] "Singapore becomes the first country to approve lab-grown meat," by Amy Woodyatt and Danielle Wiener-Bronner, *CNN Business*, December 2, 2020, https://www.cnn.com/2020/12/02/business/lab-grown-chicken-intl-scli-scn/index.html, Accessed December 2, 2020.

Three Bhutanese Monks, Rinpung Dzong, Paro, Bhutan. Photo © J.G. Morrison.

Chapter 11

INDIGENOUS CRITICAL MASS: NEURAL AND BIOCULTURAL

From the classic billiard ball model in physics to describe "elastic collision" points in the system of kinetic energy[1] to the use of experimental games and gaming theory to better grasp the fundaments of human exchange, kindness, and fairness – each components of general reciprocity theories,[2] the inherent through story concerns a triggering effect of some kind and extent. "Regime shifts and critical transitions,"[3] as investigated in 2018 (Rocha et al.) revealed that from their study examples of 30 known triggering events amongst ecological targets, "45%" of those changes showed at least "one plausible structural interdependence".[4] The principal dilemma we find in surveying the literature of triggering is that it is so vast a body of commentary and example that one can, at best, cherry pick causes and effects to formulate an argument. But no such proposition, let alone hypothesis, can be independently substantiated because all choices are subjective, even in extremis. Even momentary ulner nerve dysfunction (hitting one's "crazy bone") is by no means a universal elbow-related reaction; nor are untold millions of other physiological voluntary and involuntary cause-and-effect phenomena. For example, the standard suppositions regarding "movement disorder" of one form or other (i.e., "hyperkinetic" "hypokinetic" or "dyskinesia") generally devolve to explanations from the realm of "genetic conditions, traumatic injury, nervous system disease, infections, medication side effects, and other factors…"[5]

Other movements, reactions, quid pro quos as linear or temporal function, quantity and quality of response biochemistry – say, of an advancing or receding glacier, or of excitation when the neurons in the brain are exposed to "aversive cues"[6] - are palpable but conceptually intangible. Such ambiguities were described in Marxist sociologist Henri Lefebvre's (1901-1991) theories pertaining to "trialectics" or, the "dialectics of triplicity" – a new geography of spatial realities for lived lives that are perpetually in flux.[7] Dead, alive, or in transition, there is continuous change. Even Mount Everest continues to grow in height by approximately half-an-inch per year;[8] and ever improving atomic clocks are likely to ascertain a metric more precise, even, than the present caesium fountain clock which can "keep time to within one second over 100 million years."[9] The problem with inches and seconds is the nature of human measurements to begin with. They are arguably arbitrary demonstrations of subjective will presuming to objectivity. The second, for example, obtained a new definition in 1967, the first of seven base units established by the Le Système International d'Unités: "the duration of 9,192,631,1770 oscillations of the microwave radiation absorbed or emitted when a caesium atom jumps between two particular energy states."[10]

The French post-Revolutionary determination (after 1795) to set standards for time, length, weights, electricity currents, temperatures, mass and luminosity arose because of a certain chaos that had persisted in commerce and across different parts of the country, prior.[11] By 1837, the French were insistent on the metric system. But the implications for human adoption of any rules, measures and, by far-reaching implication, laws, principles, theories, constants, judicial and community standards, must contend with a very basic statement by Edmund Burke who remarked, "Reflect how you are to govern a people who think they ought to be free, and think they are not…"[12] Porphyry had insightfully expounded upon the dilemma confronting any attempt to stamp conformity upon uncertainty when he expatiated upon the near impossibility of achieving insight into "incorporeal natures," "peculiarities" in nature, and the very "properties [that] pertain to bodies…" He concluded that the only possible grounds for universalizing a principle around our thoughts and perceptions of the natural world were necessarily bound by a belief – born of multitudinous survey of the world around us – that every organism, thing, power, substance simply existed in a state and place "conformably to its own nature…"[13] In other words, for every principle and law of nature, there is a unique individual who is free to live outside the rule. Lack of oxygen and water will kill us, but in the meantime, nothing can stop our forgiveness, our gifts, our love, the exchange of feelings.

For every ideal, there is a crooked line; ideal circles and spheres also have equal evidence of something other. Many years ago, researchers at the Max Planck Institute for Biological Cybernetics in Germany showed that many people, when sensorily lost, tend to walk more or less in circles.[14] The outgrowth of comparing straight lines to circles is not black and white, but a matter of definitions: measurement and conceptual equivalencies, between cultures and individuals. There is a "goodness of fit" definition that breaks down at the level, for example of IQ or other scoring methodologies.[15] But it is far more interesting than that. At the roots of cybernetic theory, in every action/reaction, systems overview, and prediction; from the mechanisms that drive a living cell, to artificial intelligence, as well as every facet of the sociological character of a species, of a population, and the interaction of individuals within their biome(s), there is great and meaningful variance which, ultimately, clings to the subjective era and vague topologies of its enunciation.[16]

That endlessly fluctuating emphasis upon which our species lives, both through constraint and choice, will entangle complementary insights, revelations and measurements. Paradigms evolve, regress, constantly self-revising. In biological terms, Linnaeus, Wallace and Darwin – if you read and re-read them – are saying very little that could be described as either new or revolutionary. Natural selection and evolution, while couched in a different variety of circumstances than those machinations of late 19th century England, were no less compelling in Titus Lucretius' (ca. 99-50 BCE) book, *On The Nature of Things*,[17] Thomas Aquinas' (1225-1274) *Summa Theologiae*[18] or the proto-evolutionary ideas spawned by Comte Georges Louis Leclerc Buffon's (1707-1788) 44 volume *Histoire Naturelle*. During the span of approximately two-thousand years of most recent human thought, our grappling with sudden change brought about by catastrophe, versus slow development, as perceived over years, decades and centuries, has served to obviously enrich dialogue with noteworthy *facts* deserving of our invented laws. We can all ascribe to the realities of hunger, thirst, gravity and other such intolerant certitudes. But fossils, the notion of extinction, longevity in some species and cell lines, have added a certain tension into this dialogue, teasing certainty with various forms of flexibility; just as the battles we wage against every form of ill-health are punctuated with observable patterns of success and failure.[19]

When professor of forest ecology Suzanne Simard conveys her prodigious insights into the biochemistry of sylvan subterranean ambience, her explanations are rhapsodic. The superorganism relations within intensely cooperative yet opportunistic "mycorrhizal networks"[20] speak to a cornerstone of relatability in every human who has marveled at an old growth forest, or simply dosed in some bliss beneath an expansive shade-lending crown.[21] The primal experience informs without need of science or any detail whatsoever. Passing along empiricized allegory enriches but is no requirement for solace. Yet, if Dr. Simard's focus on "mother trees" can duly inspire those - who have for millennia chopped down trees for one excuse or another -to temper their greed, then we can all agree to agree that the reciprocity networks she and her colleagues around the world have drawn attention to are of crucial importance. Critical to any degree of environmentalist activism; not to mention a re-enchantment: The very clear knowledge that those mythical forest kingdoms – trillions of individual trees – are dominions within dominions, half of which have already been destroyed by our kind. Growing them back, without the ancient seed sources Simard and colleagues describe, doesn't just happen. In fact, in British Columbia where Simard teaches, so called "sequestration dead zones" – 8,064,000 acres of forests clearcut in the past twelve years[22] – are painful testimony to a reality firmly contested by logging interests which claim that "Less than 3% of original [BC] forest is converted to human use."[23]

Our goal here is not to plunge into the morass of numeric contests but to remember that for every ongoing human form of exploitation, there will be lives at stake, with the human species inordinately dominating the discussion. The 86 billion or so neurons in our brains, all working overtime, continue to fail at systematic eco-restoration when multiplied in any largescale community of like-minded humans, no matter how informed, well-intended or indigenously sustainable. At smaller scales, and much leaner numbers, hamlets can make decisions that apply. Or that was the original logic guiding most small-scale populations prior to humanity's global escalation.[24]

Now, the destruction of a forest is described not only in terms of unleashing sequestered carbon into the atmosphere; or the loss of tens of thousands of resident species, potentially hundreds of millions of individuals in any given hectare. In terms of actual soil disruption, it has been posited that there are many billions of organisms in diverse grams of soil. In a square meter, depending upon where, as many as 300 earthworms.[25]

As a matter of *mere* philosophy or orientation, the study of "out-group" "immutable" forces of global "pluralism" reveals the co-evolution of cooperation and competition striking no predictable balance but, rather, the easy likelihood of prejudice, in spite of educational input.[26] Such differentiations suggest that the scale for optimism is indeed a fragile one, subject as much to chance in the human behavioral sphere as it is those fundamental verities science has unearthed. As with any ecosystem, humanity can fathom the critical mass of combinatorial attitudinal breakthroughs in time to back away from an overall crash of ecological tipping points, or not. We cannot rationally manipulate chance, but we *can* set the template within a realm of probabilities that will be primed as a receptor for the collective neurotransmission of that which satisfies the fundamental requirement throughout the natural world for true reciprocity.

ENDNOTES

[1] See "Physics of Billiards – Ball Collision," in *Real World Physics Problems*, n.a., n.d., https://www.real-world-physics-problems.com/physics-of-billiards.html#:~:text=The%20physics%20behind%20billiards%20(or,conserved%20before%20and%20after%20impact, Accessed December 4, 2020.

[2] See "A Theory of Reciprocity," by Armin Falk and Urs Fischbacher, *Games and Economic Behavior*, Volume 54, Issue 2, February 2006, pp. 293-315, https://doi.org/10.1016/j.geb.2005.03.001, Elsevier, https://www.sciencedirect.com/science/ article/abs/pii/S0899825605000254, Accessed December 4, 2020.

[3] See "Cascading regime shifts within and across scales," by Juan C. Rocha, Garry Peterson, Örjan Bodin and Simon Levin, *Science* 21 Dec 2018:Vol. 362, Issue 6421, pp. 1379-1383, doi: 10.1126/science.aat7850, https://science.sciencemag.org/content/362/6421/1379, Accessed December 4, 2020.

[4] ibid.

[5] See "Movement Disorders," *RadiologyInfo.org*, November 25, 2019, n.a.

[6] See "Both Predator and Prey – Emotional Arousal in Threat and Reward," by Andreas Löw, Peter J. Lang, J. Carson Smith, and Margaret M. Bradley, *Psychol Sci.* 2008 Sep: 19(9): 865-873. doi: 10.1111/j.1467-9280.2008.02170.x, PMCID: PMC3612950, NIHMSID: NIHMS451987, PMID: 18947351, HHS Public Access, https://www.ncbi.nlm.nih.gov/pmc/ articles/PMC3612950/Accessed December 4, 2020.

[7] "La production de l'espace" ["The production of space"], in *Editions Anthropops, L'Homme et la Société, 1974, and the production of space*, by Henri Lefebvre, Translated by Donald Nicholson-Smith, Basil Blackwell, Oxford UK, 1991

[8] "Giant Mount Everest Is Still Growing And Shaking," by Hill Williams, *The Seattle Times*, September 17, 1990, https://archive.seattletimes.com/archive/?date=19900917& slug=1093659, Accessed December 4, 2020.

[9] "Precise atomic clock may redefine time," by Phillip Ball, July 9, 2013, *Nature*, doi:10.1038/nature.2013.13363, https://www.nature.com/news/precise-atomic-clock-may-redefine-time-1.13363

[10] ibid.

[11] *La loi du 18 Germinal an 3; la mesure [républicaine] de superficie pour les terrains, égale à un carré de dix mètres de côté* [*The law of 18 Germinal year 3; the [Republican] measure of surface area for land, equal to a square with a side of ten meters*.]. Le CIV (Centre d'Instruction de Vilgénis) – Forum des Anciens.

[12] *First Speech on Conciliation with America, "American Taxation,"* April 19, 1774.

[13] "Auxillaries to the Perception of Intelligible Natures", by Porphyry, in *Select Works Of Porphyry*, Translated from the Greek by Thomas Taylor, Printed for Thomas Rodd, London 1823, p. 223.

[14] "Scientists show that people really walk in circles when lost," *PhysOrg*, August 20, 2009,n.a., https://phys.org/news/2009-08-scientists-people-circles-lost-video.html

[15] See "Examining Equivalence of Concepts and Measures in Diverse Samples," by Tracy W. Harachi, Yoonsun Choi, Robert D. Abbott, Richard F. Catalano and Siri L. Bliesner, *Prevention Science*. 2002 Dec: 7(4): 359-368. doi: 10.1007/s11121-006-0039-0, PMCID: PMC3293252, NIHMSID: NIHMS293849, PMID: 16845592, HHS Public Access, https://www.ncbi.nlm.nih.gov/pmc/articles/PMC3293252/#R37, Accessed Dec.4 2020.

[16] Charles François, "Systemics and cybernetics in a historical perspective". In: *Systems Research and Behavioral Science*. Vol 16, pp. 203–219, 1999.

[17] See *De rerum natura [Nature], Venice edition of 1515*. See also, *Traces on the Rhodian Shore -Nature and Culture in Western Thought from Ancient Times to the End of the Eighteenth Century*, by Clarence J. Glacken, university of California Press, Berkeley and Los Angeles, 1997, pp. 29-30.

[18] See *Summa Theologiae. 61 volumes*, Cambridge University Press, Cambridge, UK, 2006.

[19] For a most general overview, see *Successes and Failures in Fighting Disease*, n.a.. Watchtower Online Library, https://wol.jw.org/en/wol/d/r1/lp-e/102004362, Accessed December 5, 2020.

[20] See "The Social Life of Forests," by Ferris Jabr, Photographs by Brendan George Ko, *The New York Times Magazine*, December 2, 2020, https://www.nytimes.com/interactive/2020/12/02/ magazine/tree-communication-mycorrhiza.html, Accessed December 5, 2020.

[21] *Well over 4 million viewers have watched Simard's June 2016 TED Talk on the subject: "How trees talk to each other,"* https://www.ted.com/talks/suzanne_simard_how_trees_talk_ to_each_other?language=en.

[22] See *Clearcut Carbon – A Sierra Club BC report on the future of forests in British Columbia,* by Jens Wietimg and mapping analysis by David Leversee, December 2019, https://sierraclub.bc.ca/wp-content/uploads/2019-Clearcut-Carbon-report.pdf

[23] "Logging Statistics from the Council of Forest Industry (CoFI)," *Protect Glade Watershed*, https://www.protectgladewatershed.com/forestry-stats-cofi.html

[24] See *"Hope, Risk and Reality: Perspectives from Indigenous Voices on our Relationship with Mother Earth," Interview with Shanny Spang Gion, Art By Brenda Mallory*, February 9, 2021, MAHB (Millennial Alliance for Humanity and the Biosphere), Stanford University, https://mahb.stanford.edu/blog/introduction-to-a-four-interview-cycle-an-interview-with-shanny-spang-gion/

[25] See *Soil Biota*, n.a., https://nematode.unl.edu/gramsoil.htm, Accessed December 5, 2020.

[26] See "Indirect Reciprocity and the Evolution of Prejudicial Groups," by Roger M. Whitaker, Gualtiero B. Colombo and David G. Rand, *Scientific Reports 8*, Article number: 13247 (2018), September 5, 2018, https://www.nature.com/articles/s41598-018-31363-z, Accessed December 5, 2020.

Chapter 12

DEPENDENT TRANSITIONAL STATES

In traditional population demography, ecologists have referred to the most basic "BIDE Factors," namely, "birth, immigration, death, and emigration."[1] This is as non-consciousness related as possible. In other words, a graph, a metric, a change that reveals in overview some kind of change, as simple a concept as the traditional definition of a "species' range," "the smallest area within which the species occurs in appropriate environments and outside which it is absent entirely."[2] The first exception to such a rule could be grabbed out of the sky. In the case of the avian *Calidris canutus* the North American Red Knot, among the earliest documented visitors to the western portions of Lake Erie were "accidental nonbreeding summer visitors" noted the last week of June (1954) and first week of July (1985).[3] The formidable student of hummingbirds, Alexander Skutch, observed as many as one hundred squirrel monkeys of southern Costa Rica feeding beside capuchin monkeys on shared "insects, small ripe fruits such as figs, and nectar" at greater and greater distances through the surrounding forest. Not only were their specific territories flexible, but - noted Skutch – despite "widely separated lineages" demonstrated true mutual "ecological convergence" with "marmosets, tamarins, titis, and birds" only suggesting "exclusiveness" when confronting "contiguous conspecific families at territorial boundaries" delineated by "vocal" signs.[4]

Combining the antiquity of the Talamancan montane and tropical Isthmian-Atlantic moist broadleaf forest ecoregions -north and south America joined between three and five million years ago[5] (and tens-of-thousands of similar rainforest species are found just to the south, in the 55-million-year-old Amazon) with at least 12,000 flowering plants of which nearly 7% are endemic, habitat for herbivores and insectivores is particularly expansive. Hence, a norm that inculcates habitat overlap. From a physics perspective, the sum total of these overlaps represents a quantum superposition. The most basic semiotic signals from within such topologies; or other signage, lays the foundations for a template that is open to whatever interpretation we choose: acoustical, linguistic, pictographic, or any sensory, even metempsychological carrier of information.

The question of biological baselines is of great interest in that we ceased beginning from zero centuries ago. Flattening extinction curves has been overtaken by concepts of resilience, mitigation, and IUCN-quantifications of at-risk status. The underlying premise indicates a no-return situation.

Whereas giving meaning and context to species *not* threatened by humans requires not only tools, but assumptions that are new to the 21st century. When G. V. Hudson (1867-1946) described New Zealand ant migrations of *Atta antarctica* in 1892 his singular opinion was that "the annual migration of the winged males and females of this species usually takes place on a hot day in the last week of March…"[6] While Hudson was something of a farm-bred prodigy, publishing his first observations of insects in an entomology journal when he was 14 and later coming up with the idea of daylight saving time, of the 48 known ant species in New Zealand, there is no member of the *Atta* (leaf-cutting) genus, as was made clear a century after his work by Bert Hölldobler and Edward O. Wilson.[7] It is possible, of course, that for a time leaf-cutters had possibly gotten to New Zealand aboard a ship from South America and then disappeared from Zealandia. Insect distribution in a globally fluctuating system of systems is impossible to read; any and all rules are given to transition.

It is, for example, a fact that across the northern hemisphere "plant photosynthesis largely shuts down" in Winter,[8] which in turn helps explain why over 70% of all known North American birds migrate. Hence, food supply, consumers, the geography of their livelihoods is constantly being altered. Just like every life and landscape. Artist, biologist and paleontologist Laura Cunningham has written, "I began my exploration of Old California by trying to imagine grizzly bears munching fallen acorns, digging brodiaea bulbs among the bunchgrasses, and taking siestas in the willow copses of the oak-dotted East Bay flat that is now car-filled El Cerrito Plaza." Later on, near her home in the vicinity of Death Valley, on August 15th 2004, she experienced a remarkable flash flood in the desert and wrote, "These floods have always happened and always will, and some animals have found new opportunities in them. We can too."[9] Our urban-fostered perceptions of how animals live largely obfuscates the reality of where they *actually* live and where they have come from. In the case of swifts, great grey owls, and such other avifauna as "blackbird, robin and dunnock" many European purported avian city dwellers are actually building their nests in the closest natural forests they can find, amid tree snags or in tree cavities. In their own way, they are relics, no less than the European bison, or a primeval time that we are watching disappear.[10] In North America another Bovidae family member almost completely vanished directly in front of us. Writes Tim Flannery and Peter Schouten, "60 million buffalo roamed the plains in herds a million or more strong. In the nineteenth century, as such a herd passed by, more massed kilograms of a single species were streaming before you than could be seen anywhere else on the planet."[11] By year's end, 1883, a group of defeated Indians, in company with white buffalo hunters, managed to slaughter the last of the northern bison in the Cannonball River area of North Dakota. As Peter Matthiessen wrote in his poignant classic, *Wildlife In America*, "by cutting off access to the water, the hunters accomplished the destruction of the entire herd. By the end of the year stray animals excepted, the buffalo [now the national animal, along with the Bald Eagle] was gone from North America."[12]

All such alterations, whether an ephemeral weather event or actual biologic extinction, have been challenged by human self-reference, the beliefs we wield regarding our impact upon and co-existence with Others. Roger Caras (1928-2001), eighteenth president of the ASPCA, was most outspoken about the role of domesticated animals, declaring emphatically that had humans not fed, welcomed the companionship of, but ultimately oppressed and/or slaughtered as wanted such "beasts of burden" as the goat, sheep, reindeer, moose, dog, bovine, pig, elephant, camelid, equine, cat, rabbit, chicken, turkey and countless others, we would never have populated nearly half the planet, remained entirely rural, with no cities, nor engaged in any major wars or conquests. Our cultures, settlements, indeed most of human progress would have been checked.[13] This relationship was certainly underscored by early colonizations of the Eastern U.S. seaboard. For example, evidence suggests that Virginia might not have survived had colonists not managed, following the Virginia Company bankruptcy of 1624, to have upheld prohibitions against harming animals that could breed, which led by 1629 to a census that included "20,000 head of cattle, 200 horses, 50 asses, 3,000 sheep, 5,000 goats, 'innumerable' swine and poultry." Such numbers represented the dependent equations of early slaughter in North America, the pertinacity of new settlers to a formidable continent.[14]

Human history has ensnared its connections to such a degree that we are *almost* incapable of perceiving, feeling or even remembering how it might have been. Yet, the transitional phase – in chemistry, gas, liquid or solid – is utterly dependent on "intermolecular forces"[15] that are, in turn, reliant upon the constancy of thermodynamics. Ecological transitions, however, are not merely temperature, work, energy and entropy dependent, although ultimately, one could trace everything to that great unknown prior to the heating/expansion and subsequent cooling of subatomic particles prior to and after theorized cosmological events like the Big Bang. Closer to our time, here, on a little rock in space, we note a near infinite array of trigger points and thresholds. Their vast extent encompasses all ecological time, tendencies and punctuation marks. These include: "trophic cascades" that might be triggered by "infectious agents" or "invasive species";[16] statistical variations in species richness within any given area and the multiples of both proposed quantitative and qualitative reasons for abundance or depauperate numbers;[17] "the phenomenon of slowing down" – thought to be in many cases an "early warning" of coming transition or "bifurcation" (the space/time between stability and disturbance).[18]

Other triggering events might encompass the enormous biodiversity losses when measured in forest aggregates that have not accounted for "new forest" transitions;[19] ever changing complexions of "ecological integrity, diversity and resilience"[20] as well as the measures of "rewilding progress"[21] both in light of present day changes, and measures from the past (ecometrics), as in the case of climatic traits favoring "grasslands" as the dominant flora on the Great Plains during the Miocene of North America's.[22] Other diverse transitional measurements include "metabolic rates coupled with niche transitions" (as in the case of "fossoriality and nocturnality in vertebrate ectotherms" (e.g., various lizards);[23] the "toxic effects of microplastics" in any environment;[24] vastly diverse initiations of "adaptive radiation" in species like New World Ratsnakes which evolutionary biologists, utilizing the tool of "time calibrated molecular phylogenies" have been able to examine in terms of the animals' ecological opportunities allowing for "rapid speciation" not just from fossil records but with living organisms.[25]

Such dependent transitions work on either size of zero, or, if one prefers, of the equal (=) sign. They are opportunities, but sharing a single constant, namely, vulnerability to reversals, as when "domino effects" expand into co-dependent specifics, the loss of a plant inflicting a cascade effect on those animal species (or other plant species) which are fundamentally in need of it, an obvious cause of trophic catastrophes.[26] One of the most dramatic of such transformations is now occurring in the Amazon, as anthropogenic forces on all sides and in the air day by day cumulatively undermine its moisture-and-chemical sequestration capacities, thus accelerating its shift to a savannah-like biome.[27]

ENDNOTES

[1] See "Demographic Processes," Ch. 7 of *Principles of Conservation Biology*, by Gary K. Meffe, C. Ronald Carroll, and Contributors, Sinauer Associates, Inc., Sunderland, Mass., 1994, p. 180.

[2] See *The Butterflies Of North America*, by William H. Howe, Coordinating Editor And Illustrator & Twenty Contributors, Doubleday & Company, Inc., Garden City, New York, 1975, p. 33.

[3] *The Birds of Ohio*, by Bruce G. Peterjohn, Original Paintings by William Zimmerman, Indiana University Press, Bloomington, Indiana, 1989, p. 76.

[4] *Trogons, Laughing Falcons, and Other Neotropical Birds*, by Alexander F. Skutch, Illustrations by Dana Gardner, Texas A&M University Press, College Station, TX, 1999, p. 169.

[5] See *Costa Rica – Wildlife of the National Parks and Reserves*, by Michael and Patricia Fogden, Fundacion Neotropica Editorial Heliconia, San José, Costa Rica, 2005, p. 16.

[6] *An Elementary Manual of New Zealand Entomology*, by G. V. Hudson, West, Newman, & Co., London, 1892, p. 37.

[7] *The Ants*, by Bert Hölldolbler and Edward O. Wilson. The Belknap Press of Harvard University Press, Cambridge, Massachusetts, 1990, p. 600.

[8] "State Of The States: Geographic Patterns of Diversity, Rarity, And Endemism," by Bruce A. Stein, Lynn S. Kutner, Geoffrey A. Hammerson, Lawrence L. Master and Larry E. Morse, in *Precious Heritage-The Status of Biodiversity in the United States*, Edited by Bruce A. Stein, Lynn S. Kutner, and Jonathan S. Adams, Graphics by Nicole S. Rousmaniere, Oxford University Press, New York, New York, 2000, p. 143.

[9] *A State of Change – Forgotten Landscapes of California*, Text and art by Laura Cunningham, Heyday, Berkeley CA, 2010, pp. 299-300.

[10] *The Afterlife of a Tree*, by Andrzej Bobiec, editor; and authors Jerzy M. Gutowski, William F. Laudenslayer, Pawel Pawlaczyk, Karol Zub, World Wildlife Fund Poland, Warsaw, Poland, 2005, p. 175.

[11] *A Gap in Nature – Discovering The World's Extinct Animals*, Text Publishing, Melbourne, Australia 2001, p. xx.

[12] *Wildlife In America*, Peter Matthiessen, Revised, Updated Edition, Drawings by Bob Hines, 1987, Viking Press, New Yori, pp. 149-150.

[13] *A Perfect Harmony – The Intertwining Lives of Animals and Humans Throughout History*, by Roger A. Caras, Simon & Schuster, New York, 1996.

[14] See *Creatures of Empire – How Domestic Animals Transformed Early America*, by Virginia DeJohn Anderson, Oxford University Press, New York, 2004, p. 103.

[15] *Chemistry, LibreTexts*, August 15, 2020, n.a. https://chem.libretexts.org/Bookshelves/ Physical_and_Theoretical_Chemistry_Textbook_Maps/Supplemental_Modules_(Physical_and_Theoretical_Chemistry) /Physical_Properties_of_Matter/States_of_Matter/Phase_Transitions/Fundamentals_of_Phase_Transitions, Accessed December 8, 2020.

[16] "Infectious Agents Trigger Trophic Cascades," by Julia C. Buck and William J. Ripple, *Trends in Ecology & Evolution*, Volume XX, 2266, No. of pages 14, CellPress, July 20, 2017, doi: https://doi.org/10.1016/j.tree.2017.06.009, http://trophiccascades.forestry.oregonstate.edu/ sites/trophic/files/ Buck2017_TREE.pdf; See also, "Invasive species triggers a massive loss of ecosystem services through a trophic cascade," by Jake R. Walsh, Stephen R. Carpenter, and M. Jake Vander Zanden, PNAS April 12, 2016 113 (15) 4081-4085; first published March 21, 2016; https://doi.org/10.1073/ pnas.1600366113, *Proceedings of the National Academy of Sciences of the United States of America, PNAS*, https://www.pnas.org/content/113/15/4081

[17] "Measuring Ecological Biodiversity," by R. Hengeveld, *Biodiversity Letters*, Vol. 3, No. 2 (Mar., 1996), pp. 58-65, https://doi.org/10.2307/2999770, https://www.jstor.org/stable/2999770, Wiley, https://www.jstor.org/ stable/2999770?seq=1

[18] "Early Warning Signals of Ecological Transitions: Methods for Spatial Patterns," by Sonia Kéfi, Vishwesha Guttal, William A. Brock, Stephen R. Carpenter, Aaron M. Ellison, Valerie N. Livina, David A. Seekell, Marten Scheffer, Egbert H. van Nes, and Vasilis Dakos, *PLOS ONE*, March 21, 2014, https://doi.org/10.1371/journal.pone.0092097, https://journals.plos.org/plosone/article?id=10.1371/ journal. pone.0092097

[19] See "Forest ecosystem-service transitions: the ecological dimensions of the forest transition," by Sarah Jane Wilson, John Schelhas, Ricardo Grau, A. Sofia Nanni and Sean Sloan, Ecology and Society, 22(4): 38. https://doi.org/10.5751/ES-09615-220438, https://www.ecologyandsociety.org/vol22/iss4/art38/

[20] "Measuring ecological complexity," by Lael Parrott, Ecological Indicators 10(6): 1069-1076, doi: 10.1016/j.ecolind.2010.03.014, *ResearchGate*, https://www.researchgate.net/ publication/248412781_ Measuring_ecological_complexity

[21] "Measuring rewilding progress," by Aurora Torres, Néstor Fernández, Sophus zu Ermgassen, Wouter Helmer, Eloy Revilla, Deli Saavedra, Andrea Perino, Anne Mimet, José M. Rey-Benayas, Nuria Selva, Frans Schepers, Jens-Christian Svenning, and Henrique M. Pereira, *Philosophical Transactions Of The Royal Society B, Biological Sciences*, The Royal Society Publishing, October 22, 2018, https://doi.org/10.1098/rstb.2017.0433, https://royalsocietypublishing.org/doi/10.1098/rstb.2017.0433

[22] See "Measuring Earth-Life Transitions: Functional Traits from Late Cenozoic Vertebrates," by P. David Polly and Jason J. Head, *The Paleontological Society Papers*, Volume 21: Earth-Life Transitions, July 21, 2017, https://www.cambridge.org/core/journals/the-paleontological-society-papers/article/abs/measuring-earthlife-transitions-ecometric-analysis-of-functional-traits-from-late-cenozoic-vertebrates/C0068D2B30ACCF76F415B2EB75F8BEB2

[23] "Shifts in space and time: ecological transitions affect the evolution of resting metabolic rates in microtelid lizards," by Melissa Bars-Closel, Agusti Camacho and Tiana Kohlsdorf, *Journal of Experimental Biology* 2018 221:jeb175661 doi:10.1242/jeb.175661, July 30, 2018, https://jeb.biologists.org/content/221/14/jeb175661

[24] "Microplastics in aquatic environments: Toxicity to trigger ecological consequences," by Hui Ma, Shengyan Pu, Shibin Liu, Yingchen Bai, Sandip Mandal and Baoshan Xing, Envion Pollut. 2020 Jun; 261:114089. Doi: 10.106/j.envpol.2020.114089. https://pubmed.ncbi.nlm.nih.gov/32062100/

[25] See "Ecological Opportunity: Trigger of Adaptive Radiation," by E. A. Myers and F. T. Burbrink, *Nature Education Knowledge* 3(10):23, 2012. https://www.nature.com/scitable/ knowledge/library/ecological-opportunity-trigger-of-adaptive-radiation-84160951/

[26] See "Domino effect – the loss of plant species triggers the extinction of animals," by the Senckenberg Research Institute and Natural History Museum, January 5, 2017, *PHYS ORG*, https://phys.org/news/2017-01-domino-effectthe-loss-species-triggers.html: See Matthias Schleuning et al. "Ecological networks are more sensitive to plant than to animal extinction under climate change, *Nature Communications* (2016). doi: 10.1038/ncomms13965

[27] See "Amazon rainforests could transition to savannah-like states in response to climate change, new study predicts," by Shreya Dasgupta, Mongaby, December 29, 2015, *PHYS ORG*, https://phys.org/news/2015-12-amazon-rainforests-transition-savannah-like-states.html, Accessed December 9, 2020.

"Vision of Saint Eustace," Engraving by Albrecht Dürer, ca. 1501, Private Collection. Photo © M. C. Tobias.

Chapter 13

ALBRECHT DÜRER AND THE NATURE OF ECOLOGICAL CONVERSION

Being and becoming, the sources of all introspective ontology, are obsessed upon in western literature and philosophy. Ruminations said to have been enshrined respectively in the sparse works of the shaman (Iatromantic) Parmenides (515 BC- c.450 BC) and the slightly older Heraclitus (c. 535-c. 475 BC) in their respective works "On Nature," of which only "fragments" exist in both cases. Parmenides provides among the most insightful and revolutionary of speculative baselines when he declared, "*Species* have not their dependence on the Mind of Man; on the contrary, they are unknown to human Nature, or Mankind."[1] Heraclitus presumes that the human mind grasps both being and change, uttering, "What was scattered gathers. What was gathered blows apart."[2] In his Introductory remarks to Plato's (c.424 BC-c.347 BC) "Phaedrus," a dialogue composed at least a century after the death of both Parmenides and Heraclitus, Benjamin Jowett would suggest that "Dialectic may be variously defined, either as the power of dividing a whole into parts, and of uniting the parts in a whole, or as the process of the mind talking with herself."[3] These are metaphysical pillars of cognition that cannot be said, without ultimately flawed hubris, to *mirror* Nature, so much as to render commentary that is, however much its depth, a mathematically randomized selection of things one might say or feel. That is a most tenuous prediction, without consequence. In sooth, we know nothing. The pieces of the tale we put together remain the same pieces. By imprinting them with our contextualization, we backtrack from the baselines we would otherwise need to grasp our situation. Once we have added one plus one, our species is unlikely to ever with clear conscience volunteer subtraction. But that is no reason not to try.

From that basis – knowing nothing - all categories flow, cease and flow again, in endlessly varied renditions that are the substance of all art and its discussion over the primeval campfire. To assert the undying independence of other species, of all Others, offers every polite opposition to the idea of unity within diversity. If biodiversity, specifically, is also aloof from *dialectic* then logically our arbitrary mentations – whether emotional, spiritual or somehow physiologically compulsory – have no bearing whatsoever on anything other than human choice. Neither free will nor creative control are within our power. So then, what kind of choices are we capable of making? This is the only question that can elicit the power of instantaneous post-deliberation; the realm, in other words, of epiphany.

Choice does succeed, in its first principle, at behaving as if the entire world is right there, in front of us. A dramatic stage upon which all options for a completely restorative predilection that may derive from the person and/or her/his community are possible. Analogies with other species may seem appropriate, but – if we are to take Parmenides at his word – their total separation from our kind should inform transitions whose effectualization we are capable of, not those about which we are powerless. This conditional places choice at the center of all.

Enter Voragine, the former Archibishop of Genoa, known as Giacomo da Varazze (c. 1230-1298) whose compilation of saintly epiphanies, *Legenda Aurea* would become one of the most sought-after books of the late 15th century. First translated from Latin into Catalan[4], it would have been required reading for Miguel de Cervantes, and features among its amazing tales the *spiritualem conversionem* of the 2nd century Roman soldier, Placidus. The event is memorialized across the suite of Christian denominations and represents an archetypal moment of revelation. His warring allegiance to the emperor Trajan, then Hadrian, as well as to the Gods of Rome in an instant is transformed by a remarkable stag standing above him in a forest. The deer declares, "I am Jesus, whom you honor without knowing." Placidus' horse and five hunting dogs are astonished, frozen in the bliss of realization. Later, Placidus – who takes the name of Eustace – will be tested, as will his wife and two children; ultimately martyred following the miraculous fellowship displayed by the wild "beasts" in the coliseum sent to devour him.[5] Instead, a bountiful biophilia reigns supreme and the cold brutality of the arena is, before one and all, made into that peaceable kingdom. For Placidus had chosen to put down his bow and arrow, kneel on the ground and convey his adoration for the deer in a unison of prayers shared by all throughout the heavy landscape of romance, love, and outright interspecies veneration. The soldier has become a vegetarian and its variants for the coming millennia emblemized the capacity of humans and all animals to seek harmony rather than discord.

At the height of the St. Eustace rallying cry, the Italian master Pisanello (c.1395-c.1455) painted his "Vision of St. Eustace" (National Gallery of London) sometime between 1438 and 1442. There is the telltale Christ on a crucifix centered in the brace of the stag's broadly spread antlers. Placidus wears a hunter's garb, and oddly, remains in his canonization the patron saint of all hunters, but with the explicit caveat that he is also the patron of protecting fields, of all life. His Vatican-verified modus operandi is that of a Saint Francis exponentially motivated to change from a life of violence and blind subservience to power, to a man whose sole purpose in being has turned in a deer's heartbeat to that of unconditional love and belief in God. In Pisanello's exquisite version, God is Nature, populated with a rich biological credo, both in sweep, detail and everlasting invocation.

To mark the 1300 years of tumultuous history following upon this epic conversion in the forest, the 29-year-old Nuremburg genius, Albrecht Dürer (1471-1528), already one of the most celebrated artists across Europe, created his largest, most detailed and famed of copper engravings dedicated to St. Eustace and his satori in the wilderness. It would instantly be prized as the most important engraving in the Renaissance, roughly 14" by 10 ½".[6] Other German artists would portray a romantic landscape, Carl Wilhelm Kolbe (1759-1835) and Johann Heinrich Wilhelm Tischbein (1751-1829) among them.[7] By the late 18th century the 24-year old Wilhelm Heinrich Wackenroder (1773-1798), remembered by many as Germany's Wordsworth, wrote a small treatise detailing stories of the Renaissance artists as viewed through the German romantic school of the Nazarenes with its attempts to "demonstrate the spiritual nature of true art."[8] Dürer was its most formidable emblem of this paradise found. As has been suggested, to view this engraving in the confines of one's home is to share in "a divine event" privately. "Viewers are privileged to witness this transient moment and are invited to imitate St. Eustace's act of worship."[9] For the Saint, as French philosopher Emmanuel Levinas has persuasively meditated, "...in the full force of their inevitable responsibility. To recognize and name those insoluble substances and keep them from exploding in violence, guile or politics, to keep watch where conflicts tend to break out, a new religiosity and solidarity -in loving one's neighbour... It is on the basis of the existence of the other that my own existence is posited as human."[10] Or, as George Steiner writes, it is "That which comes to call on us...spontaneous visitation and summons..."[11]

Aside from its technical virtuosity, Dürer's "Vision of Saint Eustace" engraving emerges no less cathartically at the heart of the Northern Renaissance than Sandro Botticelli's "Birth of Venus" 120 years earlier (in the Uffizi Gallery in Florence). Like his immediate predecessor, Martin Schongauer (1448-1491) of Colmar, Dürer – in the Saint Eustace engraving as well as hundreds of his other masterworks – makes a decisive and comprehensive choice regarding the power and beneficence of Nature, and the individual's role in re-wilding him/herself within its rapturous ethic that is akin to being reunited with God, what ecologists might characterize as a "critical transition to an alternate state".[12] Despite five centuries of scholarship surrounding Dürer's prolific body of work, it was French historian Elie Faure who most precisely gifts language commensurate with the enormity of its task, perhaps the truest insight, when he writes, "He is a kind of Christ turned scientist who seeks the salvation of the world in an intense study of its aspects...he cut into the copper with the hand of a workman, the heart of a poet, and the brain of a philosopher, [but] never did he arrange nature. He considered it 'the only master'...The massive horses of Germany, its muscular hunting dogs, its deer, its hares, its cows, its pigs groveling in the mud of the villages, all its insects and all its birds participated, almost always, in the adventures of love..."[13]

How similar, in fact, such aesthetic jubilations are to the seemingly far more mundane assemblage of current U.S. Federal land management agencies working under the banner of “Stewarding Ecological Transformation”.[14] That is but one way of thinking about the application of inspired spiritual revelation to pragmatic behavior. Another concerns fundamental reciprocity theory, which holds – as spelled out by Gerald Carter – “that fungi, plants, fish, birds, rats, and primates enforce mutual benefit by contingently altering their cooperative investments based on the cooperative returns…”[15] Carter concludes by pointing to three, by now classical if insufficient “frameworks that have guided empirical studies of cooperation” and they are “inclusive fitness,” “reciprocity theory” and “biological market theory”.[16] It is clear that “A Vision of Saint Eustace” does not fit in to any of these categories of experience and thus calls for some altogether novel approach to recognizing the crucial underpinnings of what theologians know to be *conversion*. An experience outside the bounds of self-interest, or any expectation of gain.

ENDNOTES

[1] “Parmenides,” Chapter III, in *The History of Philosophy*, by Thomas Stanley, Thomas Bassett, London, 1683, p. 748.

[2] *#4 of The Fragments, The Collected Wisdom of Heraclitus*, Translated by Brooks Haxton, With a Foreword by James Hillman, Viking Penguin Group, New York, 2001, p. 27.

[3] *Plato, Lysis, Or Friendship, The Symposium, Phaedrus, Translated from the Greek with Introductory Analyses*, by Benjamin Jowett, with a Preface by Whitney J. Oates and Illustrations by Eugene Karlin, The Easton Press Norwalk, Connecticut, 1979, p. 137.

[4] *Legenda aurea santorum sive Lombardica historia* [*Santorum gold legend or story Lombardica*], by Jacobus de Voragine, et al., Jordanus de Quedlinburg, Strassburg 1486. Some have suggested the year 118AD for the martyrdom, under Pope St. Sixtus I. See “St. Eustace the Hunter and the Magnificent ‘Stag,’” by Larry Peterson, *Catholic365.com*, June 29, 2018, https://www.catholic365.com/article/9024/st-eustace-the-hunter-and-the-magnificent-stag.html

[5] *Martyrologium Romanum* [*The Roman Martyrology*], Libreria Editrice Vaticana, Rome, 2001.

[6] 369 x 270 mm; Bartsch 57; Dogson 32; Hollstein 60 (Meder b); Rowlands 1971, 78; Bartrum 16; Schoch 32, as described by Giulia Bartrum in *Albrecht Dürer and his Legacy – The Graphic Work of a Renaissance Artist*, with contributions by Günter Grass, Joseph L. Koerner and Ute Kuhlemann, Princeton University Press, Princeton New Jersey, 2002, pp. 140-142. See also, *Albrecht Dürer – Paintings, Prints, Drawings*, by Peter Strieder, Translated by Nancy M. Gordon and Walter L. Strauss, Abaris Books, New York, 1982, pp. 174-176.

[7] See *German romantic prints and drawings from an English private collection*, Foreword by Antony Griffiths, Preface by Giulia Bartrum, The British Museum Press, London, 2011.

[8] *Herzensergießungen eines kunstliebenden Klosterbruders* [*Heartfelt Effusions of an Art-Loving Monk*], and L. Tieck, Berlin J F Unger, 1797; See Keith Moxey, “Motivating History,” *Art Bulletin* 77, no. 3 (September 1995), pp. 393–401. See German Romantic Painting, by William Vaughan, Yale University Press, New Haven and London, 1980, p. 163.

[9] From the *Washington University Digital Gateway Image Collections & Exhibitions*, http://omeka.wustl.edu/omeka/exhibits/show/durernatureworks/creatingmasterpieces/steustace

[10] Emmanuel Levinas, *Alterity & Transcendence*, Translated by Michael B. Smith, Columbia University Press, New York, 1999, pp. 87-88, and 167.

[11] George Steiner, *Real Presences*, The University of Chicago Press, Chicago, Ill.,1989, p. 179.

[12] See “Can we detect ecosystem transitions and signals of changing resilience from paleo-ecological records?” by Zofia E. Taranu, Stephen R. Carpenter, Victor Frossard, Jean-Philippe Jenny, Zoee Thomas, Jesse C. Vermaire, and Marie-Elodie Perga, *Ecosphere, Ecological Society of America*, 2018, 9 (10), ff10.1002/ecs2.2438ff. ffhal-01959637f, https://hal.archives-ouvertes.fr/hal-01959637/document

[13] Elie Faure, *History of Art – Renaissance Art*, Garden City Publishing, New York 1937, pp. 356, 358.

[14] *National Adaptation Forum – Action today for a better tomorrow, – Linking the science of ecological transformation to RAD [resist, accept, direct] decisions*, http://www.nationaladaptationforum.org/archive/2019/wednesday/concurrent-sessions-4/linking-science-ecological-transformation-rad-decisions, May 10-22, 2022.

[15] See “The Reciprocity Controversy,” by Gerald Carter, *Animal Behavior and Cognition*, 1(3). 368-386. doi: 10.12966/abc.08.11.2014.http://animalbehaviorandcognition.org/uploads/ journals/3/11.Carter_FINAL.pdf

[16] ibid., p. 379.

Chapter 14

THE DIALECTICS OF EPIPHANY

In his various writings on the topic of "conversion" Mahatma Gandhi famously declared, "I regard no man as polluted because he has forsaken the branch on which he was sitting and gone over to another of the same tree."[1] Implicit in this ethical compass is the literal understanding that despite there being approximately 60,000 (known) species of trees in the world,[2] ultimately, *they* and *we* all exist in the same forest, if by forest we embrace Nature in her entirety. It is a simple, if logical class of conclusions, by no means the most calibrated or sophisticated. Yet, it remains true. Choices from within a context can be made by an individual that change the total context. A fragment may alter the whole. The whole may question itself by any one of its components, members, individuals. If context is viewed through the lens of an equation, then something will always happen, even if equilibrium is an end result. For the individual, the context *undergoes* a state of change.

The power of revelatory experience as stemming from an overwhelming sensation, or idea that has been triggered in some natural setting underscores a continuous spiritual metabolism that can be traced, for one celebrated example, to Buddha sitting for 49 days and nights beneath the shade of a *Ficus religiosa.* But nearly every saint and famous monk can be viewed in the context of their natural epiphanies, moments of revelation involving the wild context and its shifting dimensions.[3]

There is St. Giles, permanently injured whilst saving a doe in his cave hermitage. Christopher, patron of all of mountaineers and difficult river crossings; St. Benedict connected to all farmers but also speleologists; Anthony Abbot, patron saint of domestic animals. The first desert hermit, Paul. John the Baptist and his companion lamb. St. Jerome and his guardian ad lion. St. Francis, patron of all ecology. The Archangel Raphael, patron of all journeyers.[4]

All such examples beg the analyses of human genetic predispositions within the expansive quadrants of biophilia; our emotional and psychological needs, levels of tolerance, thresholds for neurological healing, terror, tumult, passion and calm. The amount of information humans can process, as defined by any ecosystem into which our lives have been thrust; and the concomitant daily tasks before us that are either thwarted or affirmed in co-affiliation by our various attempts to survive. But the saints and sages we admire perennially reveal one critical ecological constant, a trigger within the psyche and its instincts that claims for validity the most sizeable tract of both conscious and pre-conscious faith, namely, that ability and very need to "embrace creatures who cannot reciprocate," as David W. Orr has written, citing Lewis Mumford.[5] Of course, the more finely honed our humility, the greater advantages we begin to recognize in the company of Others who need not speak our language, share our anatomical features, or co-habit anything like the neighborhoods we typically assume to be our grounding on a planet that knows infinity. While we remain fundamentally confined to ourselves and the multitudinous affinities -eye to eye- with other species we are primordially accorded various proximities. Embracing those other sparks is our profoundly rewarding opportunity and, in light of so violent a history, our solemn duty.

Economist, historian, sociologist Francis Klingender (1907-1955) took breathtaking stock of the options enlisted in culture upon culture in his magnum opus, *Animals in art and thought*[6] commencing with a Preface that takes heed of "Lewis Carroll's 'dreamchild in friendly chat with birds and beasts'" and a Chippewayam Arctic "creation myth" that declares, "`At the beginning there were no people, only animals; still they resembled human beings, and they could speak; when the animals could speak it was summer, and when they lost the power of speaking winter followed.'"[7] Klingender's vast illustrated epic is that narrative of binaries presenting to humanity the inclination to revel in the co-habiliment of a zoomorphic plenum. The question underpinning what any one scribe, priest, artist, layperson of any persuasion might choose to think, to do, to represent in company with so many Others is that pivot upon which realism either advances to the ideal, or ignores it. Haunted by the cacophony of hunting myths and the cruelty that marks one civilization after another in the road map prescribing symbolism, domestication, slaughter and/or reverence, it is clear from a 6th century BCE Assyrian perspective that a King maintains his illusion of power by holding two lions upside down.[8] That horses were only a means to an end when Alexander set out to conquer the world. Or, as portrayed in the thinking of the formidable Jewish philosopher Philo (c. 20 BCE -c.50 CE), "the phenomena of nature, as perceived by the senses, are not, therefore, valued for their own sakes, but only as clues to the deeper truths to which human thought must penetrate by the twin roads of rational abstraction and allegorical interpretation."[9] From cave walls to illustrated psalters; etched into marble columns, heraldic kitchenware, and the capitals of crypts; in folktales and bestiaries; Apocalypse manuscripts and literally millions of printed pages in so many Book of Hours and Flemish tapestries, every season and conceivable animal and plant adorns the human story for the very reason, fundamental to our lives, that we are wanting to change; to see the context not marred by humanity, but free of it.

As Klingender is apt to interpret, in the case the illustrations for a Norman hunting party, "By killing the animal-substitute of his lady's husband, the young lover can release the unbearable tension which must arise from restraint which real life, and his loyalty to the lord, impose on him."[10] Described differently but with no less emphasis on the problematic transition involved in relocating probity, concern and preference, the Finnish historian Nils Erik Enkvist suggested that "Christian Anglo-Saxon poets struggling to break away from the terrible winters of the heroics to the ideal climes of the paradise long celebrated on the Continent in religious writings" must implicate our current obsessions with nature. Enkvist could confidently state that "A genius like Chaucer's may infuse even the most worn of common places with a new spirit of impressionistic freshness"; and that this possibility of the human spirit must ennoble even the darkest of days.[11]

We can refine and sum up the tallies of human folly and indignation, as respectively measured against the life of other animals. It was notably done so one year after Enkvist published his study, by Albert Camus. As recounted by J. M. Coetzee, a young boy in Algeria, Camus, was ordered by his grandmother to bring her a hen into the kitchen where she proceeded, to his horror, to cut off its head. "The death-cry of that hen," writes Coetzee, imprinted itself on the boy's memory so hauntingly that in 1958 he [Camus] wrote an impassioned attack on the guillotine. As a result, in part, of that polemic, capital punishment was abolished in France."[12] That deep, post-revolutionary cultural abolition perhaps added to the horror felt in the immediate aftermath of a 47-year-old teacher in Paris, Samuel Paty, who was beheaded by a terrorist in October, 2020.[13]

Moreover, the French Agriculture Minister Didier Guillaume announced that by the end of 2021, France will outlaw the killing of "unwanted male chicks" (grinding up the live birds in shredding machines); a practice that sees at least seven billion baby male chickens slaughtered each year. "Nothing will be like it was," said Guillaume.[14] But considering that nearly 67 million French are raised in a school system that demands that 40% of all meals served in schools contain meat and fish, and that the Coordination Rurale farmers' union hails a slogan, "To save a peasant farmer, eat a veagan,"[15] France is unlikely to see a return anytime soon to that world of Frenchman J. Corbechon's translation of the great Middle Ages natural history encyclopedia, *De Proprietabius rerum* (*On the Properties of Things*). In that 13th century treasure of the real and the imagined, its author, Bartholomeus Anglicus (1203-1272), and the version written at Bruges by Jehan de Ries, contains a page finely illustrated to portray a day in the life of the region in which five chatty men stand beside a canal completely dominated by the birds in their lives: "A white stork, a magpie, a swan, a crane…a cock and hen, a hoody crow and a carrion crow or rook facing it, are immediately identifiable. Some of the others may be guessed. But legend has crept in. The owl on then cliff is a bestiary bird not a natural one…"[16]

The key to this everyday showcase is the conviviality of man and bird. There is no hint of that tension, acutely real in France of an inability to juxtapose opposites outside a restive fiction, or rooted in a faith far from the cities. The Abbey of Clairvaux, for example, the

Cistercian monastery (today in ruins) that was most assuredly located on a site in Ville-sous-la-Ferté that "soothes weary minds, relieves anxieties and cares and helps souls who seek the Lord…"[17] Versus "a medieval dinner" that might "contain more than a dozen different kinds of food heaped high on large platters" including "shields of boiled and pickled boar, hulled wheat boiled in milk and venison, oily stews, salted hart, pheasant, swan, capons, lampreys, perch, rabbit, mutton" Et cetera.[18] A divide in that literary penchant that is all things human and devoted to paradox; what, at the conclusion of Alan Jay Lerner's musical play, "Brigadoon" has the character of Tommy reacting to his friend Jeff and to his blessed memory of Fiona in her Scottish immortal utopia: "Well, get plane reservations right away!.../I know it isn't there, but I want to see where it was…/Who cares if it doesn't make sense?...I want to go…"[19]

ENDNOTES

[1] See *Gandhiji on Religious Conversion*, Selected and Compiled with an Introduction by Sandhya Mehta, Mani Bhavan Gandhi Sangrahalaya, Meghshaym T. Ajgaonkar, Executive Secretary, Mumbai, India 2002, p. 28.

[2] See *The Tree – A Natural History Of What Trees Are, How They Live, And Why They Matter*, by Colin Tudge, Crown Publishers, New York, New York, 2006, p. 20.

[3] See *The Paradise Or Garden of the Holy Fathers Being Histories of the Anchorites Recluses Monks Coenobites and Ascetic Fathers of the Deserts of Egypt…Two Volumes*, Translated Out of the Syriac With Notes & Introduction By Ernest A Wallis Budge, Chatto & Windus, London, 1907.

[4] See *Saints in Art*, by Rosa Giorgi, Edited by Stefano Zuffi, Translated by Thomas Michael Hartmann, The J. Paul Getty Museum, Los Angeles, CA 2003.

[5] "Love It or Lose It: The Coming Biophilia Revolution," by David W. Orr, in *The Biophilia Hypothesis*, Edited by Stephen R. Kellert and Edward O. Wilson, A Shearwater Book, Island Press, Washington D.C., p. 434. Orr cites Lewis Mumford's book, *The Conduct of Life*, Harcourt Brace Jovanovich, New York, N.Y., 1970, p. 286

[6] *Animals in art and thought – to the end of the Middle Ages*, by Francis Klingender, Edited by Evelyn Antal and John Harthan, The M.I.T. Press, Cambridge, Mass., 1971.

[7] ibid., p. xxv.

[8] ibid., p. 43, *Cylinder seal* from the Pierpont Morgan Library, New York.

[9] ibid., p. 150.

[10] ibid., p. 472.

[11] *The Seasons of the Year – Chapters On A Motif From Beowulf To The Shepherd's Calendar*, by Nils Erik Enkvist, Societas Scientlarum Fennica, Commentationes Humanarum Litterarum. XXII. 4. Helsingfors, Ejnar Munksgaards Forlag, Kobenhavn, 1957, p.180.

[12] From *The Lives of Animals*, by J. M. Coetzee, (with commentary by) Marjorie Garber, Peter Singer, Wendy Doniger and Barbara Smuts, Edited And Introduced By Amy Gutmann, Princeton University Press, Princeton, New Jersey, 1999, p. 63. Coetzee cites Albert Camus, *The First Man,Translated* by David Hapgood, Hamish Mailton, London, 1995, pp. 181-183; and Réflexions sur la guillotine," in Essais, edited by R. Quilliot and L. Faucon, Gallimard, Paris, 1965, pp. 1019-1064.

[13] See "France teacher attack: Seven charged over Samuel Paty's killing," n.a., *BBC News*, October 22, 2020, https://www.bbc.com/news/world-europe-54632353, Accessed December 13, 2020.

[14] "France says its poultry industry will stop shredding male chicks alive by 2022," by Michael Brice-Saddler, January 29, 2020, *The Washington Post*, https://www.washingtonpost.com/ science/2020/01/29/shredding-chicks-ban/Accessed December 13, 2020.

[15] "Violent clashes amid the growing beef between vegans and farmers," by Adam Sage, *The Times*, U.K., October 9, 2017, p. 33.

[16] Le Livre des Proprietez des choses [The Book of Ownership of Things]. B.L. MS. Roy. 15 E.iii, f. llr. Flemish, 1482, in *Birds in medieval manuscripts*, by Brunsdon Yapp, Schocken Books, New York, 1982, p. 116.

[17] See *Passing Strange and Wonderful – Aesthetics, Nature, and Culture*, by Yi-Fu Tuan, A Shearwater Book, Island Press, Washington D.C., 1993, p. 218.

[18] ibid., pp. 47-48.

[19] *Brigadoon – A Musical Play*, By Alan Jay Lerner, Coward-McCann, Inc., New York, 1947, Act II, p. 135.

Chapter 15

RE-NEGOTIATING THE LABYRINTH

Neurophysiologists have examined basic decision making in species of mice, flies and worms, looking at the "neurons that control decisions about food," and the "post-ingestive rewarding effect," among numerous other perceived mechanisms at work.[1] There are fundamental difficulties in extrapolating the food choices dependent upon "nutrient sensing" in the brain (or by some more modern "peripheral detection" to "balance metabolic homeostatis") in *Drosophila melanogaster*; or those chemicals *chosen* somehow by bacteria to maintain growth[2], as opposed to other agents and agencies in a primate like ourselves, not that we don't all share many of the same bacteriological building-blocks of life.

Development psychologists discuss free will, the involvement of the brain/mind dichotomy in rendering decision patterns and specific choices, and the resulting moral accountabilities that accrue in social hierarchies.[3] Even the multiple options of any mental action are subject, some believe, to what is termed "choice architecture" and, from the evolutionary perspective, to "presence bias."[4] The choices paradox goes to the core of a Walt Whitman opening his "Leaves Of Grass" with the famed "Song of the Open Road" – "Afoot and light-hearted I take to the open road, Healthy, free, the world before me, The long brown path before me leading wherever I choose."[5]

Aristotle, Thomas Aquinas, Spinoza each mirror thousands of human years of speculation on the nature of free-will and its conditions. If rational organisms cannot choose otherwise and therefore are free of the consequences, conditional analysis suggests that Robert Frost, like Whitman, by taking "the one [of 'two roads'] less traveled" did so under the aegis of some biological compulsion that is both a subjugation of his freedom as well as the ultimate ethical exemption when the poet goes on, "and that has made all the difference." What difference, we might wonder, in a world of myriad causes and outcomes where, regardless of their proliferation, there were no true options to be exercised? Or, conversely, where does our pantheism of unfettered whims, volitions and asseverations rest if there is no actual determinism? This latter liberty is the existentialist pillar of authenticity and true freedom that transcends the Absolutism of a God, or a scenario like that posed by the "best of all possible worlds"[6] It leads everywhere. Aristotle first described the scenario (later to be known as Buridan's Ass), by presenting us with "a man, being just as hungry as thirsty, and placed in between food and drink, [who] must necessarily remain where he is and starve to death."[7] Spinoza saw this predicament as an impossible scenario of reason, a harbinger of a universal irrationality[8] and what one psychologist has referred to as "the paradox of choice,"[9] a calculation run awry which, in any ecological setting is synonymous with consumption and the "tragedy of the commons".[10] There are choice-related, consumerist exceptions to the shared calamities of most *commons*. "Doughnut economics" in the Netherlands; Gross National Happiness in Bhutan; Costa Rica's relatively stellar financial inputs for education and the environment; Suriname's extraordinarily high percentage of protected old growth forest.[11]

The environmental histories of muddled causality shot through by human choice and its implicative anxieties, riddled by self-destruction, are most discernible in those metrics at the intersection of such currents as climate change and exodus – tens, hundreds of millions of ecological immigrant/refugees and the choices made in the face of an increasingly uninhabitable planet. In a set of projections utilizing multidimensional climate/migration models, ProPublica, The New York Times with support from the Pulitzer Center, working on numerous previous reports, including the 2018 World Bank "Groundswell" publication, "fed more than 10 billion data points into [their] model." They subjected assumptions and projections to "past matches" (the way components of international migration patterns played out in empirical hindsight) and then the researchers "looked at how people moved as global carbon concentrations increased in five different scenarios." In Scenario 2, 2050, "there are tens of millions of people more desperate and with fewer options. Misery reigns, and large populations become trapped." This more or less summarized all of the prospects, in one form or other, of humanity.[12]

One could compute the 409,000 deaths from malaria in 2019;[13] the "seven million premature deaths every year" from indoor and outdoor air pollution;[14] or the 4,400-200,000 annual deaths from Schistosomiasis (snail fever).[15] Add to that the "roughly 121 million unintended pregnancies" each year between 2015 and 2019, of which "61% ended in abortion."[16] And finally, the "43 kilograms of meat" consumed on average per capita worldwide in 2014, with the numbers growing rapidly since then.[17] As many choices as inadvertent, painful circumstances out of which no personal compliance with the sum total of realities was ever affirmed, ascribed to, or avoided. Willful destruction, unwitting harm, freely determined options, cultural blinders, oppression, terrible luck, all meshed in an evolutionary moment, person by person, co-dependent species caught out in the tumult of human machinations.

Probability ethics, ecosystem moralities, biosynthetic states of mind each compound the individual's vulnerability to the power of change, the hope and hopelessness of choice, and the catastrophic combinatory algorithms of planetary flux. This is our context, the biological baseline we must acknowledge. Knowing it, in bits and pieces, will context abet redemptive choice? Or are we hardwired genetically to ignore what's best for us? Classical mythology tells us that the engineer of the great Cretan labyrinth, Daedalus, was almost incapable of finding his way back out, which would have left him alone with the Minotaur.[18] Who knows if the two *beasts* might have befriended one another; or if, in fact, our entire undoing is the very fact of a labyrinth, when we could have as a species made it so simple, universally obliging, a family of future contexts more akin to a straight line? One we can all comprehend, ascribe to, honor by our goodness?

Endnotes

[1] See "Cellular and molecular basis of decision-making," by Nilay Yapici, Manuel Zimmer and Ana I Domingos, *EMBO reports*, 2014 Oct: 15(10): 1023-1035, published online 2014 Sep 19. doi: 10.15252emr.201438993, PMCID: PMC24353843, PMID: 25239948, https://www.ncbi.nlm.nih.gov/pmc/articles/PMC4253843/

[2] "Predicting the decision-making chemicals used for bacterial growth," by Kazuha Ashino, Kenta Sugano, Toshiyuki Amagasa and Bei-Wen Ying, *Scientific Reports 9*, Article number: 7251 (2019) May 10, 2019, https://www.nature.com/articles/s41598-019-43587-8

[3] See "Decision Making: Biology, Free-Will and Accountability," by "Spencer," UCLA doctoral student blog, *Psychology In Action*, November 27, 2011, https://www.psychologyinaction.org/psychology-in-action-1/2011/11/27/decision-making-biology-free-will-and-accountability

[4] See "Better decision-making for the planet," detailing research by Princeton professor Elke Weber, by Yasemin Saplakoglu, *Discovery: Research at Princeton*, January 4, 2018, https://www.princeton.edu/news/2018/01/04/better-decision-making-planet, Accessed December 14, 2020.

[5] "Leaves of Grass", by Walt Whitman, Selected and with An Introduction By Christopher Morley, Illustrations By Lewis C. Daniel, Doubleday, Doran & Co., Inc., New York, 1940, p. 3.

[6] See *Free Will as an Open Scientific Problem*, by Mark Balaguer, Cambridge, Mass., M.I.T. Press, 2010. See also, *Deep Control: Essays on Free Will and Value*, by John Martin Fischer, Oxford University Press, New York, 2012. See also, "What Happens When Someone Acts?" by David J. Velleman, *Mind*, 101: 461-81, 1992.

[7] "On the Heavens," by Aristotle, 295b. https://archive.org/details/decaeloleofric00arisuoft/ page/n19/mode/2up

[8] Ethics, Part II, Proposition 49 Scholium. See *"Siris"* -http://branemrys.blogspot.com/2005/10/ spinoza-on-buridans-ass.html

[9] *The Paradox of Choice -Why More Is Less*, by Barry Schwartz, New York, Harper Perennial 2004. See also, Benjamin Scheibehenne, R. Greifendeder and P.M.Todd, "Can There Ever be Too Many Options? A Meta-Analytic Review of Choice Overload" (PDF). *Journal of Consumer Research.* 37 (3): 409–425, 2010, doi:10.1086/651235. Derek Thompson, "More Is More: Why the Paradox of Choice Might Be a Myth". *The Atlantic,* August 19, 2013; See also, Alexander Chernev, U. Böckenhold and J. K. Goodman, (2015). "Choice overload: A conceptual review and meta-analysis". *Journal of Consumer Psychology*. 25 (2): 333–358. doi:10.1016/j.jcps.2014.08.002.

[10] "The Tragedy of the Commons," by Garrett Hardin, *Science* 13 Dec 1968: Vol. 162, Issue 3859, pp. 1243-1248doi: 10.1126/science.162.3859.1243, https://science.sciencemag.org/ content/162/3859/1243

[11] See "Why These Countries Value Happiness Over Endless Economic Growth', by Megha Bahree, Illustrations by Rebekka Dunlap, *FutureProof (HuffPost and Unearthed)*, https://www.huffpost.com/entry/economy-growth-gdp-happiness_n_ 5fad6f57c 5b6d647a39 cf60c, Accessed December 14, 2020.

12 "Where Will Everyone Go?" by Abrahm Lustgarten, *ProPublica*, Photography by Meridith Kohut for the New York Times Magazine, July 23, 2020, https://features.propublica.org/climate-migration/model-how-climate-refugees-move-across-continents/Accessed December 14, 2020.

13 "Malaria," *World Health Organization*, 30 November 2020, https://www.who.int/news-room/fact-sheets/detail/malaria#:~:text=Disease%20burden,411%20000%20deaths% 20in%202018.

14 "Air Pollution," *World Health Organization*, https://www.who.int/health-topics/air-pollution#tab=tab_1, Accessed December 14, 2020.

15 Vos T, Allen C, Arora M, Barber RM, Bhutta ZA, Brown A, et al. (GBD 2015 Mortality and Causes of Death Collaborators) (October 2016). "Global, regional, and national life expectancy, all-cause mortality, and cause-specific mortality for 249 causes of death, 1980-2015: a systematic analysis for the Global Burden of Disease Study 2015". *Lancet*. 388 (10053): 1459–1544. doi:10.1016/s0140-6736(16)31012-1. PMC 5388903. PMID 27733281.

16 "Unintended Pregnancy and Abortion Worldwide," July 2020, *Guttmacher Institute Fact Sheet*, https://www.guttmacher.org/fact-sheet/induced-abortion-worldwide#:~:text=Roughly%20121%20million%20unintended%20pregnancies%20occurred%20each%20year%20between%202015%20and%202019.&text=Of%20these%20unintended%20pregnancies%2C%2061,73%20million%20abortions%20per%20year.

17 See "Meat and Dairy Production," by Hannah Ritchie and Max Roser, *Our World in Data*, August 2017, revised November 2019, https://ourworldindata.org/meat-production

18 Penelope Reed Doob, *The Idea of the Labyrinth: from Classical Antiquity through the Middle Ages*, Cornell University Press, Ithaca, New York, 1992, p. 36.

Todas of the Kerrir Clan, the Nilgiris, Tamil Nadu, India. Photo © M. C. Tobias.

Chapter 16

TEN-CROSSING KNOTS

The catalysts for individual change that might – if picked up by natural selection – affect a critical transformative mass within that person's population as defined by the geopolitical parameters of a cultural community, are the inflexion points most relevant to this converging crisis of the Anthropocene. Our changes have the advantage/disadvantage of being tested for survival relevancy, a primary condition for access to the future, obviously. This test has been the governing constraint for all species. Such a challenge is but the preliminary promontory; what, in faith-based traditions, would be soteriological, oriented to salvation, but in ecological terms signals both habituations to requisite conditions, but also the personal infrastructure to facilitate appropriate choices, multitudes of voluntary actions every day and night. The full dimensions of that imagined odyssey within the human species remains unknown, particularly when our behavior is visible in the mirror of our consciousness, our self-inflicted menace as a species out of kilter with what we actually want and hope for.

Paul Ehrlich writes, "Our natures normally are poorly equipped to register such threats, but we can be trained to be aware of the biological evolutionary defaults that inhibit perception of certain key problems."[1] Ehrlich goes on to suggest that there exists the capacity for "slow reflexes" that, when coupled with rapid response times, are able to "detect" and integrate into more effective strategies of rebuttal the long-term threats that most others never see, or choose to ignore.[2] The act of impeding one's own behavior represents one kind of knot; but entire social knots are of significantly higher orders of complexity, as Scottish mathematical physicist Peter Tait (1831-1901) began to recognize as early as 1876. Within a decade, Tait had described "10-crossing knots" about which he would write, "…they are too complex to be always completely kept in view. Thus, we cannot be absolutely certain that by means of such processes we have obtained all the essentially different forms which the definition we employ comprehends…"[3]

There are flat surfaces, far-off features and polynomial portions with infinite vortices (the astrogeophysical versions of the labyrinth). Their combinations in the ten cosmic dimensions posited by string theory, to date, would mean, if translated in earthly terms, that ecological closure, such as in the case of a climax forest as yet undergoing the changes of a "dead" tree's aftermath, cannot be ascertained, even in so seemingly finite a realm as a planar surface. As Tait anticipated, our biases are the very knots, their unraveling as unclear as Caroline Cocciardi's fascinating analysis of potentially differing knot design, logic and provenance in Leonardo da Vinci's blue robed versus red robed "Salvator Mundi".[4]

Of equal significance is the innocence of any test in ecological terms. We simply cannot hold to any force of habit if, beyond the involuntary nervous system and muscle anomalies, our imaginations take hold of arbitrary or tenacious destinies; of futures that we have invented.

Hence, the full force of reciprocity exists as an innate possibility, at every encounter a virgin of incipient prospects. We (humans) have no idea what that means; of what it bodes; the power inherent to even a de minimis biochemical sequence of events in the name of sui generis spiritual, aesthetic or other non-violent exchange. Granting a mathematical status to free will also connotes the various probability theories that accompany choice or abnegation on their respective journeys toward infinity. What does this mean for the individual? For the biosphere?

The test, as of a Saint Eustace, or Jain Samavasarana, is everything, in the beginning. It may lead, in cultural terms, to an Amirim – the first all vegetarian community in Israel (along with much of Tel Aviv dietary culture, as well as Dimona, "Village of Peace"); to Ghent, vegetarian several days per week; Bhutan, vegetarian two months out of a year; Pushkar, an entire city of such ethical dimensions in Rajasthan; the Todas, Jains, Bishnoi, 7th Day Adventists, millions of followers of PETA, the Inner Badui of Java; and the like. Vegetarianism is but one emblem of this barter system of co-symbiotic revelations that can result from the above-referenced *journeys toward infinity*. They are, at once, ethical advantages in ecological terms.

Other moments of their ilk recall a Dame Jane Goodall with her fellow chimpanzees in Gombé Stream National Park, Tanzania; Sir David Attenborough smiling euphorically amongst a family of gorillas in the Virunga Mountains of the Volcanoes National Park in Rwanda; Elsa, the lioness (1956-1961) bringing her three cubs out of the wilds of Meru National Park in Kenya to show her human friends, the Adamsons ("Born Free"); ten thousand YouTube videos every day revealing those magical encounters of humans joyfully in company with other species. There are no mathematical probability theories to account for the infinite diversity that ever looms, at every instant, within the biosphere. This is no mere double-fisherman's knot, long held to be one of the more insistently difficult knots to untie, like the "Sum of Three Cubes problem" or the as yet unsolved Riemann Hypothesis.[5]

But even the "7 Biggest Unanswered Questions in Physics"[6] – like, "What is matter made of?" or "Why does time seem to flow only in one direction?" – are solvable in the imagination, just as every ecological event, location, context or hope can also be thought of in an ethically pragmatic, if idealistic manner with mathematical precision. In n-fold rotation symmetry, "nested parentheses"[7] and (a)(n), or the sample size from a given population[8] every cause-and-effect, symmetry and asymmetry are liable to the addition of one, a natural number/integer. In other words, something altogether untried, untested, that road untaken. Intuitively, the great mathematician David Hilbert (1862-1943) saw this in his unpublished "Twenty-fourth problem"[9] seeking the numeric equivalent, and an equation to prove it, of Occam's Razor. But in synecological terms – whole biotic processes and the communities involved – every individual represents the possibility of a new equation, an added "one", and, with it, a personage containing a biography, feelings, more imagination. Even a fraction of that person has the equality that must arise from the very recognition of life herself.

Stated with maximum simplicity, every organism has meaning that is independent of any other organism, but gains immeasurably in the context of biological partnerships, which, in any event, are inevitable; the silent, irrepressible, incontestable maxim at the heart of all configurations and philosophy.

And, hence, the prelude to any persuasive formulations leading to the world of reciprocity.

ENDNOTES

1 *Human Natures – Genes, Cultures, and the Human Prospect*, by Paul R. Ehrlich, Shearwater Books, Island. Press, Washington D.C., 2000, p. 327.

2 ibid., p. 327.

3 "On The Unknotting Problem," by Jorge Ramírez Alfonsín, in *Combinatorics, Complexity, and Chance – A Tribute to Dominic Welsh*, Edited by Geoffrey Grimmett and Colin McDiarmid, Oxford Lecture Series In Mathematics And its Applications – 34, Oxford University Press, New York, N.Y., 2007, p. 261.

4 See *Leonardo's Knots*, by Caroline Cocciardi, Published by Mona Lisa Knot, 2019. Reviewed by Scott Taylor in the Mathematical Association of America, December 8, 2019, https://www.maa.org/press/maa-reviews/leonardos-knots, Accessed December 15, 2020.

5 See "The 10 Hardest Math Problems That Remain Unsolved," by Dave Linkletter, September 26, 2019, *Popular Mechanics*, https://www.popularmechanics.com/science/math/g29251596/ impossible-math-problems/ Accessed December 15, 2020.

6 By Dan Falk, MACH, August 8, 2017, *NBC News*, https://www.nbcnews.com/mach/science/7-biggest-unanswered-questions-physics-ncna789666, Accessed December 15, 2020.

7 "Mathematical "N" Words," *themathlab.com*, https://themathlab.com/dictionary/nwords/ nwords.htm

8 See "Sample sizes required," 7.2.2.2., *Engineering Statistics Handbook*, https://www.itl.nist.gov/div898/handbook/ prc/section2/prc222.htm

[9] "Mathematical Problems," by David Hilbert, *Bulletin (New Series) of the American Mathematical Society*, Volume 37, Number 4, Pages 407{436S 0273-0979(00)00881-8Article electronically publishedn on June 26, 2000, https://web.archive.org/web/20090530182730/ http://www.seas.harvard.edu/courses/cs121/handouts/Hilbert.pdf; See also, "Hilbert's twenty-fourth problem," by Rüdiger Thiele, *American Mathematical Monthly,* January 2003.

Chapter 17

SEMIOTIC HOLONS

A few examples suffice to lay ample groundwork for the tangible narrative suggesting what, by human standards, is radical change. More and more people are hugging cows and other "farm" animals for therapeutic benefit.[1] In Oxfordshire, the Kirtlington Wildlife & Conservation Society[2] has brought out the best in nearly 90,000 participants in a program known as the national Hedgehog Street[3] in which small, hedgehog-sized holes are placed in old brick walls and anywhere else that hedgehogs want to travel at night during their normal foraging hours. Like some 25% of all native mammalian wildlife in the UK – from red squirrels and the hazel dormouse to water voles – hedgehog numbers are fast declining and are on the country's endangered species list. The public loves them – as evidenced by the tens-of-thousands of viewers of night camera footage of their nocturnal movements – and this has proved one way to fast escalate a Pax Brittanica with, and unabashed love for this cosmopolitan *Erinaceus europaeus*. Reflecting on Britain's People's Trust For Endangered Species, and locals in general, Chris Powles – creator of the highway, with its more than 12,000 human-assisted hedgehog holes to date – reflects, "Very few people say no…the power of hedgehogs is right there… You feel you can positively do something, and there is satisfaction in that."[4]

Powles' straightforward humility is core biophilia, or whatever you choose to name it. The key here is the decision to make the choice. Whatever neurological theater underpins the series of sensory and cognitive assimilations to get there, the reaction is everything when discussing the future of *Homo sapiens*. The earlier referenced mathematical vortices are of pronounced interest. In math, wake vortex turbulence impinges in a way that creates fluidity, flow, velocity, and an axis line, straight, curved or more complicated than either, around, or in tandem with, existing various flows, the exerted forces of which can be described as ecological. The vorticity of fluid rotation involves particles. But at the scale of environmental flux there is no irrotational (absent) path, because there simply always is. Nothing is empty, and no philosophy can prove otherwise. There are non-steady flows in vortex geometry; pressures, densities, cooling and heating, propagation of vibrations and waves, but at the crux of all such measurable phenomena is the idea of turbulence in nature, at any level, discernible or not.[5]

What this means for us, and all Others, is the migrative power of data, impulse and personal identification; an ecology of metaphors that have traction at the point of individual recognition and response. The topology of these transition-building inputs are endless in their variety. Just consider: marine heatwaves[6]; the inevitability of transitions between stable states[7]; geoecosystem flux on the Tibetan plateau[8]; "signals" "from paleo-ecological records"[9]; forest "ecosystem service transition curve[s]"[10]; the challenges to "conceptuality" in the wake of forest die-offs[11]; vegetation mapping of natural and semi-natural systems[12]; impacts of critical bushfire disturbances on a lake within an ecosystem[13]; species richness data from (controversial) management regime studies of "complex nearshore" California marine ecosystems[14]; assessing "decarbonization" transitions at local, regional and global scales, with ever shifting baselines[15]. Other notable metaphors include: the "hydrological context" for "first-order, primordial determinant[s] of the propensity of ecosystems to undergo transition"[16]; measuring the intensity of multivariant disturbances to Lake Michigan[17]; the "dramatic" replacement of "submersed macrophytes" (aquatic plants) by "filamentous algal mats" in northern Florida springs[18]; the diffuse pathways of recovering forests, an emblem of such vastly diversified biomes from temperate, tropical and boreal forests containing over 3 trillion individual trees[19]; changes that come about as the result of "ecosystem-based management" of Australia's Great Barrier Reef[20]; krill numbers and the Pacific Decadal Oscillation[21]; varying definitions of "components, states, transitions, and thresholds" in modeling disturbances to "plant communities"[22]; transformations along these latter lines of grassland-dominated systems to savanna predominantly comprised of "grasses and woody plants"[23].

Each of these representational, transitional functions of biovortices signifies a process contained by ecological interfaces and turbulence which express the lives of populations and individuals. Their iterative taxonomies belie the sheer biological rapture inherent to the knowledge, as of late 2016, for example, that there are, at a minimum, 60,065 global tree species, the largest number – 8,715 species – found in Brazil; then Columbia, then Indonesia.[24] These living trees are a template for hospices of innumerable other contact, shade and insolation zones that are the sum total of coarse woody debris, or deadwood interacting with the cycles of transition involving millions of organisms, trillions of quadrillions of involved molecules.[25] There is comfort in biological zonation, even as we grapple to imagine and accept unfathomable realities. These great clouds of unknowing that is the biosphere may be more welcome on some days than others. But in the main, every design proffers elegance, beauty and lasting interest, the more churned up their border accounts and cross-fertilizations seem to imply. These inner worlds with their own compounds, margins, cores, redundancies, together inform the succor of responses wherein we meet Others and harbor every reason to believe in the wonderment of being part of something beyond our comprehension.

So fully ubiquitous is the simultaneity of what has been described as "metabolic cooperation" amidst these unbounded feedbacks[26] that the traditional parameters ascribed to stable communities, particularly amongst primates[27] (with an emphasis on the evolutionary preconditions for reciprocal altruism) reveals an incontestable kind of proof of the future nature of relations, particularly in light of the balance of power hanging upon human choices. Hence, John Ruskin's memorable injunction, "If you can paint one leaf, you can paint the world."[28] Diana Beresford-Kroeger describes the walnut family (Juglandaceae) whose trees typically drop "juglone sedatives into the water in the fall... [which] affect the dormancy of fish and water creatures, helping to stabilize and maintain their lowered basal metabolic rate."[29] Persian cultures are famed for their use of walnut to control some types of diabetes.[30]

French philosopher Jacques Maritain (1882-1973) struggled with the ontology, as it were, of Ruskin's leaf, his tree, his entire landscape, by endeavoring to fathom "intelligibility" as it pertained to the "origin of science" and the murky elements of human imagination, confined to "the first order of abstraction".[31] Beyond our senses, what are we capable of knowing, defining, imagining? And what is it worth? Botanists are still confounded by two great mysteries of plant science: the biosynthesis of Gibberellic acid, one of the most important and defining hormones for regulatory functions of a plant, particularly in females – survival, germination, the production of every flowering biome, nearly 392,000 vascular (flowering) plant species on earth.[32] And the other great unknown, namely, the changing of colors of leaves, particularly in Autumn, the "unmasking of the carotenoid or anthocyanin pigments".[33]

We asked Dr. Melanie DeVore, one of the great paleobotanists in the world, what she found amazing about leaves, and her response was nothing less than inspirational: "...You can go collect 30 different dicot leaves, determine the proportion of those leaves with entire leaves, and accurately calculate the mean annual temperature for any flora from the Cretaceous to the present. I have had classes go out, bring back 30 leaves from a campus (both in Milledgeville, Georgia and Jersey City, New Jersey) and they have calculated the correct Mean Annual Temperature (MAT) to three significant digits."[34] DeVore, and colleague K. B. Pigg have extensively reviewed the carbon cycling regime of the Paleocene-Eocene Thermal Maximum (PETM) 56 million years ago, particularly, during a 200,000 year period in which the detection of carbon's presence during the past 500 million years on earth appears to have reached an all-time high, at least until the Industrial Revolution and its aftermath. Relying upon records predicated from the analysis of "fossil leaf, pollen, and wood", DeVore and Pigg summarized the methodological framework paleobotanists have built to see how plant structures like leaves, "serve as proxies for temperatures, precipitation, and atmospheric concentrations of carbon dioxide" over vast timeframes utilizing an "approach, known as the nearest living relative (NLR)"; as well as the critical "coexistence approach".[35]

In addition, DeVore and Pigg rely upon an equation for calculating "mean annual precipitation... from a set of fossil leaves by sorting morphotypes into categories based on [a] Raukier-Webb size classification scheme" which suggests "that the drier the climate, the higher proportion of woody dicots with small leaves in the flora…"[36] Tracking the various plant zones and stratigraphic transitional points over millions of years in the past, is an imprecise but nonetheless instructive guide to that botanical regime likely to dominate the Anthropocene in coming years and decades, and the leaf, in all of its phenomenal complexity and beauty is as good a roadmap as humanity is likely to inherit, from scientists like DeVore and Pigg. Their human culture of commitment to the leaf, and all that it represents to life on earth will prove crucial to the underlying question as to whether humanity can continue to coexist with monocots and dicots into the future. Says DeVore, Leo Hickey (1940-2013) and Jack Albert Wolfe (1936-2005) were the brilliant minds who really "set the stage for making leaves the tools we use today."[37]

But then, there is the ultimate human issue itself, our intrusiveness – however benign - within the fixated scope of our Anthropic definitions across every suite of the ecological sciences. In their study of self-asserting holons – biological entities aware of (experiencing) both their inward and outward roles in any ecosystem - as distinct from the abstract category of taxon, Timothy F.H. Allen and Thomas B. Starr incisively called into question "the imposition of the scale of the individual onto our conceptual models of the present crisis".[38] The holon remains a human definition, but one providing for the idea that it is a word representing a transparency that is real in the natural world, given the transmission of material and sensory information between itself and the environment. In other words, a Self, a member of the Others. A community that suffers from no top-down authoritarian biological protocols.

ENDNOTES

[1] "Cow-hugging, an alleged wellness fad, has people cuddling farm animals to relieve stress," by Alexandra Deabler, *Fox News*, October 17, 2020, https://apple.news/AAABgH-QARZiP_bTCmRrtFA.

[2] http://kirtlingtonvillage.co.uk/kwacs

[3] https://www.hedgehogstreet.org/

[4] *The age of extinction - Prickly business: the hedgehog highway that knits a village together*, by Phoebe Weston, October 17, 2020, https://www.theguardian.com/environment/2020/ oct/17/prickly-business-hedgehog-highway-knits-a-village-together-kirtlington-oxfordshire-aoe

[5] See Rupert Klein and Lu Ting, *Viscous Vortical Flows, Lecture Notes in Physics*, SpringerLink, New York, 1991. See also, *Quantum Ecodynamics*, unpublished manuscript by Michael Charles Tobias, George Fitzgerald Smoot III, and Jane Gray Morrison, 2021.

[6] See T. Wernberg, S. Bennett, R. C. Babcock, T. de Bettignies, K. Cure, M. Depczynski, F. Dufois, J. Fromont, C. J. Fulton, R. K. Hovey, E. S. Harvey, T. H. Holmes, G. A. Kendrick, B. Radford, J. Santana-Garcon, B. J. Saunders, D. A. Smale, M. S. Thomsen, C. A. Tuckett, F. Tuya, M. A. Vanderklift, S. Wilson. Climate-driven

regime shift of a temperate marine ecosystem. *Science*, 2016; 353 (6295): 169 doi: 10.1126/science.aad8745, American Association for the Advancement of Science. "Marine heatwave triggers dramatic ecosystem transition." *ScienceDaily. ScienceDaily.* 7 July 2016. www.sciencedaily.com/releases/2016/07/160707151113.htm, http://dx.doi.org/10.1126/science.aad8745

[7] "Inferring critical thresholds of ecosystem transitions from spatial data," by Sabiha Majumder, Krishnapriya Tamma, Sriram Ramaswamy and Vishwesha Guttal, First published: 03 May 2019, https://doi.org/10.1002/ecy.2722, Corresponding Editor: Tom E. X. Miller., *ESA, Ecological Society of America*, Volume 100, Issue 7, July 2019, e02722, https://esajournals.onlinelibrary.wiley.com/doi/10.1002/ecy.2722

[8] *Geo-ecosystems in Transition on the Tibetan Plateau (TransTiP)*, Technische Universität Braunschsweig, https://www.tu-braunschweig.de/en/irtg-transtip, Jan.10, 2020.

[9] Zofia E. Taranu, Stephen R. Carpenter, Victor Frossard, Jean-Philippe Jenny, Zoe Thomas, et al.. "Can we detect ecosystem critical transitions and signals of changing resilience from paleo-ecological records?" *Ecosphere*, Ecological Society of America, 2018, 9 (10), ff10.1002/ecs2.2438ff. ffhal-01959637, https://hal.archives-ouvertes.fr/hal-01959637/document

[10] "Viewing Forest Transition Outcomes through the Lens of Ecosystem Services," by Wilson, S.J., J. Schelhas, R. Grau, A. Sofia Nanni and S. Sloan 2017. Forest ecosystem-service transitions: the ecological dimensions of the forest transition. *Ecology and Society*, 22: 38. https://partners-rcn.org/viewing-forest-transition-outcomes-through-the-lens-of-ecosystem-services/Accessed December 16*, 2020*

[11] *Ecosystem dynamics and management after forest die-off: a global synthesis with conceptual state-and-transition models*, Cobb R.C., Ruthrof K.X., Breshears D.D., Lloret F., Aakala T., Adams H.D., Anderegg W.R.L., Ewers B.E., Galiano L., Grünzweig J.M., Hartmann H., Huang C., Klein T., Kunert N., Kitzberger T., Landhäusser S.M., Levick S., Preisler Y., Suarez M.L., Trotsiuk V., and Zeppel M.J.B (2017), Critical Zone Observatory (Collaborative Network), Santa Catalina Mountains and Jemez River Basin, https://czo-archive.criticalzone.org/catalina-jemez/publications/pub/cobb-et-al-2017-ecosystem-dynamics-and-management-after-forest-die-off-a-gl/Accessed December 16, 2020

[12] *Monitoring Vegetation under Global Change*, Humboldt-Universität zu Berlin, 03.07.2020, https://www.enmap.org/data/projects/HU%20Berlin.pdf

[13] *Bushfires may have driven a critical ecosystem transition*, ANSTO Staff, http://onlinelibrary.wiley.com/doi/10.1002/2017JG004135/full, https://www.ansto.gov.au/news/bushfires-may-have-driven-a-critical-ecosystem-transition

[14] *Bulletin of Marine Science*, 74(3): 693–708, 2004 Bulletin of Marine Science 693 © 2004 Rosenstiel School of Marine and Atmospheric Science of the University of Miami, *Transition From Low To High Data Richness: An Experiment In Ecosystem-Based Fishery Management* From California, by Les Kaufman, Burr Heneman, J. Thomas Barnes, and Rod Fujita, http://fisherysolutionscenter.edf.org/sites/catchshares.edf.org/files/Transition%20from%20Low%20to%20High%20Data%20Richness%20%20An%20Experiment%20in%20Ecosystem-based%20Fishery%20Management%20from%20California.pdf

[15] "Keeping up with an evolving energy transition ecosystem," *ARDIAN Media Coverage*, April 1, 2020, https://www.ardian.com/news-insights/keeping-evolving-energy-transition-ecosystem

[16] *Ecohydrology of ecosystem transitions: a meta-analysis*, by Viglizzo, Ernesto F.; Nosetto, Marcelo D.; Jobbágy, Esteban G.; Ricard, M. Florencia; Frank, Federico C., URI: http://hdl.handle.net/10625/54836*, 2014, IDRC Digital Library, Canada,* https://idl-bnc-idrc.dspacedirect.org/handle/10625/54836

[17] "Seasonal zooplankton dynamics in Lake Michigan: Disentangling impacts of resource limitation, ecosystem engineering, and predation during a critical ecosystem transition," by Henry A. Vanderploeg, Steven A. Pothoven, Gary L. Fahenstiel, Joann F. Cavaletto, James R. Liebig, Craig A. Stow, Thomas F. Nalepa, Charles P. Madenjian, and David B. Bunnell, *Journal of Great Lakes Research,* Volume 38, Issue 2, June 2012, Pages 336-352, Elsevier, ScienceDirect, https://www.sciencedirect.com/science/article/abs/pii/S0380133012000408

[18] "Flow reversals as a driver of ecosystem transition in Florida's springs," Robert T. Hensley, and Matthew J. Cohen, *Freshwater Science*, Volume 36, Number 1, 2017, The University of Chicago Press Journals, https://www.journals.uchicago.edu/doi/full/10.1086/ 690558?mobileUi=0

[19] "Forest ecosystem-service transitions: the ecological dimensions of the forest transition," by Sarah Jane Wilson; John Schelhas; Ricardo Grau; A. Sofía Nanni; Sean Sloan, 2017, *Ecology and Society Publication Series:*

Scientific Journal (JRNL), Southern Research Station, doi: https://doi.org/10.5751/ES-09615-220438, https://www.fs.usda.gov/treesearch/pubs/56948, USDA, U.S. Forest Service, Treesearch.

[20] "Navigating the transition to ecosystem-based management of the Great Barrier Reef, Australia," Per Olsson, Carl Folke, and Terry P. Hughes, *PNAS* July 15, 2008 105 (28) 9489-9494; https://doi.org/10.1073/pnas.0706905105, https://www.pnas.org/content/105/28/9489

[21] See "Analysis of Abrupt Transitions in Ecological Systems," Bestelmeyer, B. T., A. M. Ellison, W. R. Fraser, K. B. Gorman, S. J. Holbrook, C. M. Laney, M. D. Ohman, D. P. C. Peters, F. C. Pillsbury, A. Rassweiler, R. J. Schmitt, and S. Sharma. 2011. Analysis of abrupt transitions in ecological systems. *Ecosphere* 2(12):129. https://jornada.nmsu.edu/bibliography/11-036.pdf, The Jornada-Rangeland Research Programs, https://jornada.nmsu.edu/ltar/abrupt-transitions

[22] "State and transition modeling: An ecological process approach," by Tamzen K. Stringham, William C. Krueger, and Patrick L. Shaver, *J. Range Manage.* 56: 106-113 March 2003, https://www.nrcs.usda.gov/Internet/FSE_DOCUMENTS/16/nrcs143_023865.pdf

[23] "An Ecosystem in Transition: Causes and Consequences of the Conversion of Mesic Grassland to Shrubland," by John M. Briggs, Alan Knapp, John M. Blair and Jana L. Heisler, *ResearchGate*, March 2005; BioScience 55:243-254, doi: 10.1641/0006-3568(2005)005[0243: AEITCA]2.0 CO;2, https://www.researchgate.net/publication/ 269038536_An_Ecosystem_in_Transition_Causes_and_Consequences_of_the_Conversion_of_Mesic_Grassland_to_Shrubland

[24] *There Are More Than 60,000 Tree Species Worldwide, Scientists Say*, by Merrit Kennedy, April 12, 2017, NPR, https://www.npr.org/sections/thetwo-way/2017/04/12/523616033/there-are-60-000-species-of-tree-worldwide-scientists-say, Accessed December 16, 2020.

[25] See *The Afterlife Of A Tree*, by Jerzy M. Gutowski, Karol Zub, Pawel Pawlaczyk and William F. Laudenslayer, edited by Andrzej Bobiec, WWF Poland, Warszawa-Hajnowka 2005, pp. 18-27.

[26] See "Reciprocal Fitness Feedbacks Promote the Evolution of Mutualistic Cooperation," by Daniel Preussger, Samir Giri, Linéa K. Muhsal, Leonardo Oña, and Christian Kost, *Current Biology*, July 23, 2020, doi: https://doi.org/10.1016/j.cub.2020.06.100, Article 1, Volume 30, Issue 18, P3580-3590.E7, September 21, 2020. https://www.cell.com/current-biology/pdfExtended/S0960-9822(20)30986-6#back-bib3

[27] See *The Symbolic Species – The Co-Evolution of Language and the Brain*, by Terrence W. Deacon, W. W. Norton & Co., New York and London, 1997, pp. 397-401.

[28] Modern Painters, by John Ruskin, Volume 5, Ch. 5, p. 61, *The Works of John Ruskin in 25 Volumes*, Merrill and Baker, New York, 1890.

[29] *The Global Forest*, by Diana Beresford-Kroeger, Viking Publishers, New York, 2010, p. 31.

[30] "Chemical Composition and the Effect of Walnut Hydrosol on Glycemic Control of Patients With Type 1 Diabetes," by Hossein Moravej, Alireza Salehi, Zahra Razavi, Mahmood Reza Moein, Hamed Etemadfard, Forough Karami and Forough Ghahremani, *International Journal of Endocrinology and Metabolism*, 14 (1): e34726, January 30, 2016, doi: 10.5812/ijem.34726, https://sites.kowsarpub.com/ijem/articles/17732.html, Accessed December 18, 2020.

[31] *Philosophy of Nature, To which is added, Maritain's Philosophy of the Sciences*, by Yves R. Simon, Philosophical Library, New York, 1951, p. 29.

[32] "Gibberellic acid in plant: still a mystery unresolved," by Ramwant Gupta and S K Chakrabarty, *Plant Signal Behav.* 2013 Sep:8 (9):e25504 doi: 10.4161/psb.25504. Epub 2013 Jun 28, PMID: 23857350 PMCID: PMC4002599, https://pubmed.ncbi.nlm.nih.gov/23857350/ Accessed December 18, *2020.*

[33] *The Biological Significance of Leaf Color Change*, by Harvard Forest, n.a., n.d., https://harvardforest.fas.harvard.edu/leaves/biological, Accessed December 18, 2020.

[34] *Personal communication via email with Dr. DeVore*, December 18, 2020.

[35] See DeVore, Melanie L., and Kathleen B. Pigg, "The Paleocene-Eocene Thermal Maximum: Plants as Paleothermometers, Rain Gauges, and Monitors," by Melanie L. DeVore and Kathleen B. Pigg, in *Nature Through Time – Virtual field trips through the Nature of the past*, Edited by Edoardo Martinetto, Emanuel Tschopp and Robert A. Gastaldo pp. 109-128, Spring, Cham, Switzerland, 2020, https://link.springer.com/book/10.1007/978-3-030-35058-[1]

[36] ibid., 4.3.

[37] See "East meets West: The contrasting contributions of David L. Dilcher and Jack A. Wolfe to Eocene systematic paleobotany in North America," November 2007, CFS *Courier Forschungsinstitut Senckenberg* 258(258): 85-94, https://www.researchgate.net/publication/ 242342988_East_meets_West_The_contrasting_contributions _of_David_L_Dilcher_and_Jack_A_Wolfe_to_Eocene_systematic_paleobotany_in_North_America

[38] *Hierarchy – Perspectives for Ecological Complexity*, by Timothy F.H. Allen and Thomas B. Starr, The University of Chicago Press, Chicago, Ill. 1982, p. 217.

Sacred Lake in the Vegetarian City of Pushkar, Rajashtan, India. Photo © M. C. Tobias.

Chapter 18

FUTURE SYMBOLS

Structural networks supporting one form or other of mutualism across time and geography purport to obvious symbioses – some species of amphipods (*Podocerus jinbe*) in the mouths of whale sharks[1], tarantulas and frogs, sloths and algae, some insects (lacewing larvae) and lichens[2], the mongoose and the hornbill, cattle egrets and their numerous hosts, from bovines to hippos, leafcutter ants and leaves, every plant-pollinator from flies to hummingbirds, as earlier referenced examples[3]. Conversely, interspecific competition – e.g., predator/prey - following along the traditional Lotka-Volterra equations[4] is also a fact of enormous planetary consequence and trauma. We have suggested that human imagination is a force planted upon transitional foundations that is perfectly capable of inventing an ideal to transcend ecological sadness. But this concept is set forth in absence of any algorithms that might project the impacts upon the current physiological average of biodiversity in general. Rather than suggesting an assault on a statistically contested constellation of genetically modified organisms, and by implication, whole communities – a form of aggression akin to a striking asteroid – we confine our notion to human communities, more than a challenge.

Given that the *Podocerus* cleaning the Whale Shark's teeth weighs 0.014084507 pounds, and the whale shark up to 20,000 pounds, one must assume that scale is irrelevant to evaluating some meaningful metric of feeling, consciousness, vulnerability, innocence, even sensorial authority. There is no doubting that both organisms -shark and amphipod- feel pain deeply concentrated in their brains and/or neural clusters in other parts of their bodies. In fact, shrimp and numerous other groups of crustacea, have been the subjects of extensive nociceptor trials.[5] The number of neurons means nothing, when grasping at comparisons amongst organisms at the previously referenced holon level.[6] The quantification of neurons per species has proved a meaningless set of mismatches. The basal group of multicellular trichoplax exhibit not one neuron; no synapses and "only six cell types"' yet, it can target its secretions, coordinate feeding behavior upon algae, and suggests "the presence of mechanisms for cellular communication and integration."

Moreover, this remarkable creature possesses "similarities to the walls of the digestive tracts of mammals…"[7]

Mammalian radiation, extinct and extant species, the examination of molecular clocks providing insights into "divergence-time estimates"; and the application of the latest advances in "molecular methods" have revolutionized even this level of taxonomy, to which we all belong. Only since 2018 have mammals had their own "Mammal Diversity Database," long after previously compiled digital repositories for other groups of organisms. Not surprisingly, between 2005 and 2018, "1251 new [mammalian] species" were recognized. Currently, the number is "6,485" mammalian species.[8] The total number of species in the world is a complete unknown. But using a multivariate lognormal distribution analysis[9] that encompasses microbial worlds, Kenneth J. Locey and Jay T. Lennon describe, "a unified scaling law that predicts the abundance of dominant species across 30 orders of magnitude to the scale of all microorganisms on Earth. Using this scaling law combined with the lognormal model of biodiversity, we predict that Earth is home to as many as 1 trillion (10^{12}) microbial species."[10] Locey and Lennon point out that in discussing laws of scale in ecology (least understood amongst microbes), "Among the most widely known are the scaling of metabolic rate… with body size…and the rate at which species richness…. scale with area." And that such formulae "reveal how physiological, ecological and evolutionary constraints hold across genomes, cells, organisms, and communities of greatly varying size."[11]

Like the number 1 trillion, which still does not account for all of the Others, the concept of infinity is tempting to pair with that of biodiversity. Neither afford the luxury of identifiable scale. Add *one* to the idea, and it has a tempting, if scarcely discernible impact. Spell out that one, and at once a human being will devise symbols, adjectives and other characteristics by which to represent it. A single cell that can divide, for example. Take this evidentiary singularity to the level of a human community, and already there will be debate regarding policies that this *one beyond infinity* should merit. Stem cells? Should it be cloned? What is its taxonomic, cladistic place on some Tree of Life? But left undescribed, voiceless, and fully outside the realm of deliberate human investiture, to quote Robert Kaplan, it is nothing more or less than "the nothing that is."[12]

German mathematician/philosopher Georg Cantor (1845-1918), best known for set theory, also devised the notion of transfinite numbers, as well as an infinity of infinities, in the tradition of the Jain mathematician, Surya Prajnapti (c. 4th-3rd century BCE) who envisioned three qualities of infinity: nearly, truly and infinitely infinite.[13] In the Book of Proverbs, we are told that "the way of an eagle" is far beyond human understanding. We don't require the conception of infinity to prove that; nor can we accept the simple adage that a bird's wing, as dissected morphologically, is nothing more than the evolution of flight engulfed in the "struggle for existence," an idea propounded in British zoological quarters four years after the conclusion of World War I[14], which saw over 100,000 planes shot down between Germany and the Entente nations. It was that War which induced

Brigadier General Billy Mitchell – who had led the first air offensives in the history of warfare– to declare that "The main power of defense and the power of initiative against an enemy has passed to the air."[15] We recognize that such cultural memes as human-induced violence readily turns to symbols. They may materialize in the form of flight, or fire power, or the arbitrary lessons of a Tower of Babel or some other engineering chimera. But it remains to be discovered how our species is best placed – in *absence* of symbol-making - to interact with any *other* species.

This may seem like an infertile line of discussion, yet, it appears to us to be the crucial social, psychological and material pivot upon which our future as individuals and a species clearly hinges. How might interdependency, or independence of thought and feeling be sustained, and to what ends? - in the face of our facile metaphors of conquest; infantile paradigms of infinite growth; self-satisfaction; and insufferable exploitation of Others? Reaching the lunar surface had become an ultimate source of exploratory desire for our species for many years. Then, its power was eclipsed by other sensational cravings shared across the countless mores and boundaries of singular human experience. If it takes a bright object in the dark night, a pandemic, or the fantasy of personally amassing a few hundred billion dollars to somehow galvanize the unified interest of our selfish species, what else can be offered in support of an entirely different category of moral magnets? Biological coefficients that speak to the future perpetuation of a biosphere largely unmolested? How much comfort do you take in the discovery, far off, unseen, at the ocean floor, of a thriving community of Antarctic glass sponges? Would you give your life for them?

ENDNOTES

1 "A New Species of *Podocerus* (Crustacea: Amphipoda: Podoceridae) Associated with the Whale Shark *Rhincodon typus*," by Ko Tomikawa, Makio Yanagisawa, Takuo Higashiji, Nagisa Yano, Wim Vader, *Species Diversity*, Online ISSN: 2189-7301, Print ISSN: 1342-1670, 2019 Volume 24 Issue 2 Pages 209-216, doi: https://doi.org/10.12782/specdiv.24.209, See: "[Research] New species of shrimp-like creatures discovered in a whale shark's mouth," Hiroshima University, Associate Professor Ko Tomikawa, https://www.hiroshima-u.ac.jp/en/news/55023, December 10, 2019.

2 See *Lichens of North America*, by Irwin M. Brodo, Sylvia Duran Sharnoff, and Stephen Sharnoff, with Selected Drawings by Susan Laurie-Bourque, published in Collaboration with the Canadian Museum of Science, Yale University Press, New Haven, Connecticut, 2001, p. 52.

3 Bascompte, J.; Jordano, P.; Melián, C. J.; Olesen, J. M. (2003). "The nested assembly of plant–animal mutualistic networks". *Proceedings of the National Academy of Sciences.* 100 (16): 9383–9387. Bibcode:2003PNAS..100.9383B. doi:10.1073/pnas.1633576100. PMC 170927. PMID 12881488. See also, Ollerton, J. 2006. "'Biological Barter': Interactions of Specialization Compared across Different Mutualisms". pp. 411–435 in: Waser, N.M. & Ollerton, J. (Eds) *Plant-Pollinator Interactions: From Specialization to Generalization.* University of Chicago Press. See also, Wright, David Hamilton (1989). "A Simple, Stable Model of Mutualism Incorporating Handling Time". *The American Naturalist.* 134 (4): 664–667. doi:10.1086/285003. See also, Bastolla, U.; Fortuna, M. A.; Pascual-García, A.; Ferrera, A.; Luque, B.; Bascompte, J. (2009). "The architecture of mutualistic networks minimizes competition and increases biodiversity". *Nature.* 458 (7241): 1018–1020. Bibcode:2009Natur.458.1018B. doi:10.1038/nature07950.

PMID 19396144. Breton, Lorraine M.; Addicott, John F. (1992). "Density-Dependent Mutualism in an Aphid-Ant Interaction". *Ecology*. 73 (6): 2175–2180. doi:10.2307/1941465. JSTOR 1941465. See p. 15 of Introduction.

[4] "Evolution of predator-prey systems described by Lotka-Volterra equation under random environment," by Y. Takeuchi, N. H. Du, N. T. Hieu, and K. Sato, *Journal of Mathematical Analysis and Applications*, Volume 323, Issue 2, 15 November 2006, pp. 938-957, https://doi.org/10.1016/j.jmaa.2005.11.009, Elsevier, *ScienceDirect*, https://www.sciencedirect.com/science/article/pii/S0022247X0501142X, https://www.jstage.jst.go.jp/article/specdiv/24/2/24_240213/_article/-char/enAccessed December 18, 2020.

[5] Maldonado, H. & Miralto, A. (1982). "Effect of morphine and naloxone on a defensive response of the mantis shrimp (Squilla mantis)". *Journal of Comparative Physiology*. 147 (4): 455–459. doi:10.1007/bf00612010. S2CID 3013237. See also, Hanke, J., Willig, A., Yinon, U. and Jaros, P.P. (1997). "Delta and kappa opioid receptors in eyestalk ganglia of a crustacean". *Brain Research*. 744 (2): 279–284. doi:10.1016/s0006-8993(96)01114-6. PMID 9027387. S2CID 26232087. See also, Casares, F.M., McElroy, A., Mantione, K.J., Baggermann, G., Zhu, W. and Stefano, G.B. (2005). "The American lobster, Homarus americanus, contains morphine that is coupled to nitric oxide release in its nervous and immune tissues: Evidence for neurotransmitter and hormonal signaling". *Neuro Endocrinol. Lett*. 26 (2): 89–97. PMID 15855878. See also Lüschen, W., Buck, F., Willig, A. and Jaros, P.P. (1991). "Isolation, sequence analysis, and physiological properties of enkephalins in the nervous tissue of the shore crab Carcinus maenas L." *Proceedings of the National Academy of Sciences. 88 (19): 8671–8675.* Bibcode:1991PNAS...88.8671L. doi:10.1073/pnas.88.19.8671. PMC 52571. PMID 11607217. Lozada, M., Romano, A. and Maldonado, H. (1988). "Effect of morphine and naloxone on a defensive response of the crab Chasmagnathus granulatus". *Pharmacology Biochemistry and Behavior*. 30 (3): 635–640. doi:10.1016/0091-3057(88)90076-7. PMID 3211972. S2CID 45083722. See also, Dyakonova, V.E. (2001). "Role of opioid peptides in behavior of invertebrates". *Journal of Evolutionary Biochemistry and Physiology*. 37 (4): 335–347. doi:10.1023/A:1012910525424. S2CID 23134631.

[6] See Calabrese, M. (2011). *Hierarchical-Granularity Holonic Modelling*, Doctoral Thesis. University of Milan, Italy, https://air.unimi.it/retrieve/handle/2434/155499/138859/ phd_unimi_R07647.pdf

[7] "Coordinated Feeding Behavior in *Trichoplax*, an Animal without Synapses," by Carolyn L. Smith, Natalia Pivovarova, and Thomas S. Reese, *PLOS ONE*, September 2, 2015, https://doi.org/10.1371/journal.pone.0136098, https://journals.plos.org/plosone/ article?id=10.1371/journal.pone.0136098, Accessed December 19, 2020.

[8] See https://www.mammaldiversity.org/; *Illustrated Checklist of the Mammals of the World, 2 Volumes*, by Connor J. Burgin, Don E. Wilson, Russell A. Mittermeier, Anthony B. Rylands, Thomas E. Lacher, Wes Sechrest, with Albert Martinez-Vilalta, Marc Olivé, David M. Leslie, Jr., Foreword by Razan Al Mubarak, Texts based on the initial contributions from the authors of the Handbook of the Mammals of the World Volumes 1-9, Color plates by Toni Llobet (main artist), Ilian Velikov, Lluis Sogorb Faansie Peacock, Alex Mascarell, Francesc Jutglar, Blanca Martí, Stephen D. Nash and Jesus Rodriguez-Osorio, Lynx Edicions, Barcelona, 2020, Volume 1, p. 24.

[9] Johnson, Norman L.; Kotz, Samuel; Balakrishnan, N. (1994), 14: "Lognormal Distributions", Continuous univariate distributions. Vol. 1, *Wiley Series in Probability and Mathematical Statistics: Applied Probability and Statistics* (2nd ed.), New York: John Wiley & Sons, ISBN 978-0-471-58495-7, MR 1299979

[10] "Scaling laws predict global microbial diversity," by Kenneth J. Locey and Jay T. Lennon, *PNAS* May 24, 2016, 113 (21) 5970-5975, Edited by David Michael Karl, University of Hawaii, https://www.pnas.org/content/113/21/5970, Accessed December 19, 2020.

[11] ibid.

[12] *The Nothing That Is – A Natural History of Zero*, by Robert Kaplan, Illustrations by Ellen Kaplan, Oxford University Press, New York, NY, 2000.

[13] Ian Stewart, *Infinity: a Very Short Introduction*, Oxford University Press, New York, NY, 2017, p. 117.

[14] See *Birds In Flight*, by W. P. Pycraft, Illustrated by Roland Green, Gay & Hancock Limited, London, 1922, p. 18.

[15] https://www.britannica.com/biography/William-Mitchell; See also, *Time Magazine*, June 22, 1942.

Chapter 19

A NEW NATURAL POSITIVISM

Darwin spoke of some "general plan" throughout the habits and organization of life and was insistent that "nothing can be more hopeless than to attempt to explain this similarity of pattern in members of the same class, by utility or by the doctrine of final causes." The scientific basis upon which he speculated concerned "modifications" and modifier correlations within the *organization*. While his focus in these meditations concerned "morphology" "embryology" and "rudimentary organs" he was never adamant on the subject of "same class" and left open entirely the manner by which any finality was to be defined or discerned. In fact, Chapter Fourteen of *On The Origin Of Species*[1] was titled "Mutual Affinities Of Organic Beings".[2] Biophilial and physiolatrical to the core, this is also very much in keeping with what Anne Buttimer has described as a "root metaphor," the philosophical agility which empowers any conceivable diversity.[3] The *organization*, as Darwin referred to it, is that *root metaphor*. As such, a community theoretically mirrors the individual, at the combined phenotypic level, but is still not the rank of an individual's life. As such, given the absolute multiplication of varying perspectives and lived experiences, John R. Stilgoe writes in his remarkable book *What Is Landscape?* "Groups of people share certain general angles of vision and agree on certain rough-hewn simplicities." But all else is an individual matter: "Each inquirer realizes something unique…"[4] Therein exists the fantastic dialectic, fabulist hermeneutics; commentaries that reside within, articulate genetic predilection, suggest wholeness and fracture. Both are essential.

Monophyletic and apomorphic distinctions of heritable traits from shared ancestors going back one, or more generations have distinct taxonomic boundaries, but genus and species, even entire lineage divisions may merge.[5] But they are more likely to separate, particularly in light of the sheer volume of their phylogenies.

Darwin's grandfather, Erasmus Darwin (1731-1802) famously rhapsodized these individualities to the extent that in his commentary on Bishop George Berkeley's (1685-1753) *An Essay Towards a New Theory of Vision* ("Treatise on Vision," 1709) he suggested that "the colours which we see, are only a language…"[6] So as to adduce that *everything* offers up a language, unique to itself. This realization was core to Maurice Maeterlinck's (1862-1949) revelatory study "The Intelligence Of The Flowers" in writing that "the genius of the Earth…gropes, it hesitates, it corrects itself time after time;…it fights against the huge, ponderous, obscure mass of its being. It knows no more than we do whither it is going; it seeks and finds itself gradually. It has an ideal…"[7]

No less a metaphorical ponderation attains to the level of population ecology than that continuing uncertainty with respect to just what an individual means or is; of what it (she/he/otherwise) bodes. Morphology, patterns, numeric separations, reproductive boundaries, assemblage dynamics, biochemical construct and contexts, each present their foibles and blurs in terms of strict definition. In 1917 the botanist H. A. Gleason asserted "that species had individualistic ecological characteristics and were assembled into local communities with a large stochastic component according to the vagaries of dispersal and the available environments."[8] In the case of conventional stochastic processes in which a function comprises at least two variables -say, an individual, and its community – the distribution of any estimated relationship between the two entities, when translated to the species level, is enough to recognize a mathematical uncertainty principle teeming with variation. Whether it is random or some other operative is a matter of grammar. White noise, distribution functions, oscillations, mutual probabilities, "path properties," convergence and periodicity all compound the dilemma of ascertainment[9]. These variables represent a curve on any type of graph. A bend is a transition. It may be fixed geographically for a certain timeframe, but change is fundamental. No second in a river is ever the same; no cloud forms like its predecessors. Euclid (mid-4th century BC) thought of a curve as the trace remaining after some imaginary motion, either determinate (finite) or indeterminate. In either case the curve defines a surface; the surface, a topology or landscape of moveable points across which beings transit, or exist inchoate. One might imagine complex zeros and polynomial (multiple) identities within the suggested space(s), but these are merely the manipulated fractions, the emblems, delegates, representative identities, that human deliberation fixes upon, irrespective of fleeting probity.

We can assign them individual or group affiliations and names, but ultimately, it is an abstract: call the space mathematical, physical, existent or non-existent; within a dimension or extra-dimensional. The point, if we are to *relate* as living beings to the experience of these identities, then we are positing an ecological recovery. That is, a biological set of coordinates, relationships and – most important – conceptual *possibilities*.

It may be a fundamental exchange of labor and goods with a given system.[10] But ultimately, there is, as Freya Mathews writes, "the greatest moral urgency" to the question of just what, exactly, qualifies *self* and "selves"?[11] Her answer, or one of them, achieves, in our minds, tremendous vitality, in a manner like that of Henri Bergson or Albert Schweitzer. She writes of "meaningfulness" that "contemplation of sublime Nature" or all of our varied and assiduous efforts to "save" Nature, may not be as strong a force as that basic effort involved "by teaching our hearts to practice affirmation, and by awakening our faculty of active, outreaching, world-directed love."[12]

Self, world; subjective, objective: the dyads are vulnerable to all characterization, but it does not matter. We come closer and closer to neutrality, even as conspirators in Zeno's Paradox know that there is no there. What it means for us is that our fixations are not, in fact, *fixed*. All is number, density, distribution, probability function, the space in which our ideas court these singular or multiple destinies; the timeframes that implicate what has been called Zeno Quantum Behavior.[13] We take away from these obscurities the reality that human behavior is fickle, pernicious, but ecologically open, no matter how dire the horizon events closing in upon our collective follies. Thus, the remaining prospect, as outlined by Dr. Russell A. Mittermeier and friends, for "conserving at least 75% of the planet's terrestrial biodiversity in only about 8% of its land area," the method of choice predicated on the determination of the 36 known "hotspots".[14]

No matter how intensive our delineations of species and habitat at risk, the Anthropic Paradox – what we know is through our own lens, not that of a tree's – confines our queries and commitments to humanity's preoccupations, borne of all of the same unanswered questions that are the product of the individual/community interactions; hybridizing alloys of thought, gene, instinct, and mosaics of process. The virtue component, while human-allied and self-referential, is not a closed ecosystem of contemplation, however. We know that because we know that we don't know. All of earthly and cosmic attention-grabbing phenomena, however received, are not ignored. To accomplish good we need not subject the motion and motives to mathematical paradox. We can and must bypass the laborious theoretical positioning of what it is or means to be a human being, and seek, instead, a wild, collaborative course of heart, a change of compass however aliquot the sliver of sentience directed towards this unknown quantity of mechanism, quality of a soul.

Ecologists speak of RLKs – rare or little-known species – within systems of knowledge known to host gaps[15]. Social structures, in mammals or any other class of organisms are as variable as geographical size or intellection. Emotional vagaries can be discerned across the physical Himalayas, every jungle, sea, desert… everywhere. As has been declared vis à vis Jainism, "The Jain Dharma ontology states that the soul can be found not only in humans but in gusts of wind, sparks of fire, in flowing waters, in plants, in microbes, and in rocks and stones and soil, as well as in insects, reptiles, birds, marine life, and mammals."[16] This pantheistic embrace is the most persuasive orientation, in absence of any firm understanding, worthy of our volitional spectrum. It is at the heart of the reciprocal altruism at a planetary scale, infusing every discipline with a human predictability that can be turned towards fellowship.

How that *turning* might actually affect the ethical suppositions at the heart of natural positivism, is what concerns us.

Unknown People with Shaman, in a Canoe, ca. 4500 BCE, Tamaulipas, Mexico. Photo © M. C. Tobias.

ENDNOTES

[1] *On the Origin of Species by Means of Natural Selection, or the Preservation of Favoured Races in the Struggle for Life,* With a New Preface by Charles G. Darwin, Illustrated with Wood Engravings by Paul Landacre, A Limited Edition. The Easton Press, Westport, Connecticut, p. 397.

[2] ibid., pp. 376-418.

[3] *Geography and the Human Spirit*, by Anne Buttimer, With a Foreword by Yi-Fu Tuan, The Johns Hopkins University Press, Baltimore, MD, 1993, p. 85.

[4] *What Is Landscape?* By John R. Stilgoe, The MIT Press, Cambridge, Massachusetts, 2015, p. 17.

[5] See Podani, János (2010). "Taxonomy in evolutionary perspective. An essay on the relationships between taxonomy and evolutionary theory". *Synbiologia Hungarica. 6: 1–42.*

[6] "'The Botanic Garden Part II Containing The Loves Of The Plants', A Poem, With Philosophical Notes", 3rd Edition, J. Johnson, London, 1791, p. 131.

[7] *In Hours Of Gladness*, by Maurice Maeterlinck, Translated by A. Teixeira De Mattos, Illustrated by E. J. Detmold, George Allen & Co., London, 1912, p. 94.

[8] See H. A. Gleason, "The structure and development of the plant association," *Bulletin of the Torrey Botanical Club*, 43, 463-81, discussed by Robert P. McIntosh in The Background Of Ecology – Concept And Theory, Cambridge Studies In Ecology, Cambridge University Press, 1985, p. 265.

[9] *Stochastic Processes*, in http://physics.gu.se/~frtbm/joomla/media/mydocs/LennartSjogren/ kap4.pdf, by Lennart SJÖGREN, Associate professor (retired) of University of Gothenburg,Sweden; See also, *Non-equilibrium Stochastic Processes in Physics, Chemistry, and Biology*, by T. Löfwander, May 27, 2020.B. Mehlig

[10] See, for example, "Labour relationships and the role of reciprocity," from *The Andean economic system and capitalism*, by Rodrigo Sánchez, in Ecology and Exchange in the Andes, Edited by David Lehmann, Cambridge University Press, Cambridge, UK, 1982, pp. 167-183.

[11] *The Ecological Self*, by Freya Mathews, Barnes & Noble Books, Savage, Maryland, 1991, p. 129.

[12] ibid., p. 160.

[13] W. M. Itano; D. J. Heinsen; J. J. Bokkinger; D. J. Wineland (1990). "Quantum Zeno effect," *Physical Review A. 41* (5): 2295–2300. Bibcode:1990PhRvA..41.2295I. doi:10.1103/ PhysRevA.41.2295. PMID 9903355.

[14] See "Forests And Primates Worldwide," Chapter 9, in *Saving Endangered Species – Lessons in Wildlife Conservation from Indianpolis Prize Winners*, Edited by Robert W. Shumaker, Foreword by Harrison Ford, Johns Hopkins University Press, Baltimore, MD, 2020, p. 292. See also, www.hotspots-thefilm.com

[15] See "Social Considerations," by John D. Peine, Chapter 9 in *Conservation of Rare or Little-Known Species – Biological, Social, and Economic Considerations*, Edited by Martin G. Raphael and Randy Molina, Foreword by Nancy Molina, Island Press, Washington-Covelo-London, 2007, p. 236.

[16] Comment by Sulekh Jain, See *Ecological Tragedy, Memory and the Birth of Hope – A Special MAHB Feature - An Ultimate Collision*, by M. C. Tobias, https://mahb.stanford.edu/library-item/ecological-tragedy-memory-and-the-birth-of-hope-a-special-mahb-feature-from-michael-charles-tobias/December 19, 2020.

Chapter 20

NON-ASYMPTOTIC BOUNDARY SOLUTIONS

That turning could be likened to the paramount class of biocybernetics, recognition of feedback relations ("reflexive," "symmetric," "transitive" etc.)[1] – integrating systems rather than permanently isolating them. This would mentor the most telling divide in evolutionary analytics: the nurturing of boundaries, divisions, opposition and dyads whose behavioral responses are called upon to perpetuate difference, relinquish conversation, abrogate coalitions. Versus that transitive calling to herald all unions, mergers, collectivities and collaboration. The latter concord calls for an appreciation of elliptical return[2] (as opposed to a strict, but implausible pure circularity) that has insinuated protocols, priorities, the gravitas of imperatives into our daily lives.[3] It comprises non-asymptotic boundary solutions. As such, this is an evolutionary model whose predictive collations presuppose a vision that has at least rudimentarily identified particulars in this global coalescence of reciprocal causality.

It extends beyond planetary boundaries. Arctic Terns, for example, expend much of their energy migrating to take advantage of Summer sunlight, from the Arctic to the Antarctic.[4] This solar system's star fuels nearly all food production on earth, reminding us that our human affairs are of no more consequence than the most humble photoautotroph, busy propagating oxygen, converting chemical energy into carbohydrates, emitting and maintaining oxygen levels, engendering the most crucial process on earth: the transformation of light energy by proteins within chloroplasts, whose dominant hosts are all those leaves, dating back at least 1.6 billion years.[5] They write no epitaphs.

We recognize that every number, alleged piece of *evidence*, perspective and portion of analysis, rests upon our human bias and arbitrary metrics. The edifice of criteria amassed to assert every confidence in purported facts can be brushed aside without apology; or, granted the apotheosis of scientific legitimacy. These determinants of reality are the consensus coefficients of a hit-or-miss parliament, loyalists to themselves and the simplistic *idea* of their aggregate civilizations. The mathematical paradoxes are worlds within worlds trapped in yet more of them. For example, one could pose the following quandary: If a passenger sitting in the rear of an aircraft that is in flight walks to the forward area of the carrier, is it the same plane? We may from an abundance of practicality dismiss such Zenotic hairsplitting, though they involve a number of intersecting disciplines, each with potential formulations, principles of obscurity. We may even layer those principles with full ranges of obscurity and applaud the dissections. For purposes of simplicity, let us designate that particular kind of scenario as a *first order simile* (FOS). Such comparisons populate all of Aesop and his successors. "The Crow Who Wished To Imitate The Eagle," in La Fontaine. "The Countryman and Several Rats," in Pilpay. "The Jackall, Leopard, and other Beasts," in John Gay. The many paradises of "Saint Augustine," "Isidore," "John Mandeville," "Saint Brendan," "Walter Raleigh," etc.[6]

But it is the *second order simile* (SOS) that closes the non-asymptotic hypothesis, is counter-dissective, or, one could add, anti-oblivious. This means that rather than be stumped, foot, inch, half-an-inch, quarter-of-an-inch, on the way to Zeno's finish line, never (theoretically) reaching there, that the resolution to this tortoise-hare scenario, one of ten paradoxes proposed by Parmenides' great student, is solvable by a rather non-scientific irritation with the very idea of a "transfinite infinity of indivisible parts." To which Aristotle proclaimed a simple solution: "there are only potential infinities."[7] This assertion in face of the so called Achilles, and Dichotomy Paradoxes, has been refreshed after two millennia by the *Standard Solution*, as it is casually described, meaning that one can, actually step over sequences that go on forever (an infinity of parts of inches, for example) within any finite time frame.[8] In other words, practical considerations are indeed relevant to accomplishing something, let alone moving from one area of the rug to another; or touching a tree. And with that modest conclusion to theoretical pointillism, one can get on with it, unfettered by meteorite showers or sunspots of logic.

This, without fanfare, represents a category breakthrough. Think of the esteemed Cesare Bonesana di Beccaria (1738-1794) whose treatise entitled *On Crimes and Punishments* (1764) actually laid out the groundwork for distributive justice, greatly relied upon by America's founders; a template for ordering cause and effect that was juridical, and gave absolute preference to reciprocity that commenced with a premise: the presumption of innocence. In ecological terms, it grants that any innocence is possible, and that laws should serve as a preamble to this generous approach to nature.

We've mentioned first, and second order similes, the recognition of division and unity as manifested in the way we interpret natural science, events, singularities. But there is a critical *third order simile* which reverses alleged incomprehensibility between objects, sensate, insensate, abstract, manifested, idea, ideal. It is no Order, Law, Power or other pretense of regularity. Rather, this third order smile portends of actual ecological resolution and is critical to accepting certain principles, albeit ones proposed by a specific species. If we can imagine a *post-Anthropic, trans-species first order metaphor* (PATSFOM, for simplicity sake), then we are in a position to embrace the blind but altogether pragmatic equation that looks like this: (patsm) x 1… = 1+?, where "?" is equivalent to any object of affinity, affection, interest, interactivity (the words become inconsequential within a metaphoric space that is unbounded). This would include, necessarily, observation and deliberation, accenting the realizable concept that by interacting with the world, the "1+?" principle allows for any connection, transcending the debilitating infinite series with actual manifested relations, no matter what we term them: social, solitary, virtuous or other.

At least we have established that *patsm* allows for 1+ any other Being or community. Admittedly awkward in its attempt to delineate that which empowers *H. sapiens* to step outside its isolation, it is a useful tool in fathoming how we can raise an idea to an ideal and act upon it. First as a linguistic model, a deep structure of wish fulfilment, a lexicon of ideals, and then the compressed alignment of basic empirical fundaments to fill in the gaps between ideals and observations. The resulting possibilities offer no spectacle; no unblushing eurekas. Only the quiet realizations of our potential to connect with anything, at any level, at any time, in a naturally positivistic, virtuous manner.

No one favors blunt equations as acceptable (or elegant) explanations for behavior. This is even more cumbersome when presuming to embrace multiples of life forms, about whom we know next to nothing. Nothingness that promotes itself is, literally, a senseless exercise in futility. But by acknowledging that exponential nonsense, we can step into what every pit of numbers one cares to nominate. We don't have to open that telegram. We can choose to bypass ugliness, hostility, stupidity and whole civilizations of the past, should we care to. And we usually do. There are endless theories to explain that, as well.

But this *patsm* is interesting, as is the simple addition of "one" to infinity. Our goal in wallowing through the mud of such pretentious circumlocutions hinges upon the frustrated desire to present a truth looming inside the metaphors and similes of human experience; not a single truth but a category of truths that are available to those who would avail of such possibility. Namely, the ethological connectivity that can be filtered by all it is that makes up a person. Rather than simply hailing the ethical, or spiritual, or cognitive… let us grant some combinatory power which, by virtue of its existence at whatever rank in the neural, subconscious and conscious edifice which happens to be our individual lives, grants an equally combinatory alliance, alliances, between Selves and Communities of all cladistic, taxonomic, phylogenetic inheritance and futurism. This mouthful is meant to adduce the sum total of far simpler adages that have bolstered the prospects for humanity since its inception, in spite of so much negative forensics.

Most recently, we see that primal truism arrayed in the Columbian jungles of Cerro Azul south of Bogota painted throughout eight miles of cliff at La Serranía La Lindosa (Department of Guaviare) between 12,600 and 11,800 years ago, in which several of the last species and various hybrids of the Ice Ages are portrayed, as well as a scene that suggests a group of people worshipping large animals.[9] We have personally surveyed other expansive sites of Late Pleistocene and Holocene petroglyphs and rock art across the American southwest, California, the Burgos region in Tamaulipas (Mexico), in southeastern France and in southern Suriname. But the spectacular finds at Cerro Azul,

Limoncillos and Cerro Montoya appear without exception to be the most revealing ethological works of art ever discovered, certainly since those of Lascaux (found in 1940) and La Chauvet (discovered in 1994).[10] Almost immediately UNESCO added the sites – part of Columbia's Chiribiquete National Park – to its list of World Heritage Sites.[11] We will never decipher the codes displayed so richly and emphatically across these colorful smooth rock walls, for the very reason, as Melody Jue describes in her book *Wild Blue Media* that we are, today, "embodied observers".[12] That containment in our *present* renders a startlingly clear judgment on our bias. André Leroi-Gourhan calls it "the Paleolithic paradox" [which] "consists to an appreciable extent in the fact that images of masterly technique and of appealing form can seem to represent nothing of any coherence." Leroi-Gourhan refers to it as "the nonfigurative". And elucidates: "To the modern eye, the swarms of animals and signs signify no action except of a most fragmentary kind, and no narration can be detected. We may wonder whether this is so simply because we cannot read…"[13]

The La Serranía La Lindosa storyline is no less emphatic than any paradise scene by Roelandt Savery or Jan Brueghel the Elder; no less narrative driven than Picasso's "Guernica." We are unlikely to discover what happened 12,000+ years ago in those jungles of Columbia. It may well be that *nothing* out of the ordinary happened. Perhaps we should simply be asking: what was happening, century after century? To that conjecture we see the world of those artists who were inspired to create these outdoor museums. Given their height and prominence, *museum* may be an appropriate designation. If that is the case, then we feel at once a relational resonance between the 21st century humans and ancestors 480 generations ago. The relation is not one of unknowing. One could argue unsettled, unclear divination of meaning standing before, say, Jackson Pollock's "#1 Number 5, 1948".[14] There are modern stories we read what we want to decipher. This is a deliberate modus operandi of the imagination that transcends objective commandeering and is fundamental to all art. We still debate the truncated pillar in Giorgione's "La Tempesta" (1506/1508) or Piet Mondrian's paintings in red, yellow and blue.[15] To most non-physicists, $E=mc^2$ cannot be *read.*[16] We doubt very seriously that .00001 percent of all human beings alive can read (explain) in detail how a Central Processing Unit, the television, the International Space Station, the Covid vaccine or a lightbulb actually works.

But in attempting to *read* painted tales from 10,000 or 14,000 years ago, we can be reasonably assured that the instincts elicited in our purviews; the sensorial power of persuasion, irrespective of the actual rungs on a ladder or length of a camelid's neck portrayed on the outdoor museum walls of jungle rock, can be assimilated into our current psyches. We hesitate to interpret only because we know that we are looking back in time at a civilization that revered other species, and ours does not. That aboriginal conflict, as it reverberates today throughout the human world, is shocking, painful and so unsettling as to dismantle our confidence at the most fundamental levels. The challenge across millennia *has* to be absorbed. Anyone who has spent the night cuddled up with a yak, sleeping under a starry firmament, or giant sequoia, knows that these otherwise uncountable margins of error can be narrowed.

Marieta van der Merwe and Friend at Their Harnas Sanctuary in Namibia. Photos © M. C. Tobias.

ENDNOTES

[1] *Cause and Effect*, by John Denker, https://www.av8n.com/physics/causation.htm, 2003. See also, "Decision-Theoretic Foundations for Causal Reasoning," by David Heckerman and Ross Shachter, *Journal of Artificial Intelligence Research* 3 (1995) 405-430, December 1995, https://arxiv.org/pdf/cs/9512104.pdf

[2] Jason Merchant, *The syntax of silence: Sluicing, islands, and the theory of ellipsis*. Oxford: Oxford University Press, Oxford UK 2001.

[3] "Systems biology ... is about putting together rather than taking apart, integration rather than reduction. It requires that we develop ways of thinking about integration that are as rigorous as our reductionist programmes, but different. ... It means changing our philosophy, in the full sense of the term." Denis Noble, (2006). *The music of life: Biology beyond the genome*, Oxford University Press, Oxford, UK, p. 176.

[4] "To the Ends of the Earth," by Tara Ramroop and Kara West, January 16, 2011, *National Geographic*, https://www.nationalgeographic.org/article/ends-earth/

[5] "Oldest plant-like fossils discovered are 1.6 billion years old," by Mattias Karisson, *NewScientist*, March 14, 2017, https://www.newscientist.com/article/mg23331172-900-oldest-plantlike-fossils-discovered-are-16-billion-years-old/Accessed December 22, 2020.

[6] See *The Book of Legendary Lands*, by Umberto Eco, Translated by Alastair McEwen, Rizzoli ex libris, New York, 2013, pp. 167-177.

[7] See "Zeno's Paradoxes," by Bradley Dowden, *Internet Encyclopedia of Philosophy*, https://iep.utm.edu/zeno-par/Accessed December 23, 2020. See also, "Why Mathematical Solutions Of Zeno's Paradoxes Miss The Point: Zeno's One And Many Relation And Parmenides' Prohibition," by Alba Papa-Grimaldi, http://philsci-archive.pitt.edu/ 2304/1/zeno_maths_review_metaphysics_alba_papa_grimaldi.pdf, *The Review of Metaphysics* 50 (December 1996): 299-314,Accessed December 23, 2020

[8] ibid, Dowden.

[9] "Tens of Thousands of 12,000-Year-Old Rock Paintings Found in Colombia," Livia Gershon, *Smithsonianmag.com*, Dec.1, 2020. https://www.smithsonianmag.com/smart-news/tens-thousands-12000-year-old-rock-paintings-found-colombia-180976427/Accessed December 23, 2020.

[10] "Colonisation and early peopling of the Colombian Amazon during the Late Pleistocene and the Early Holocene: New evidence from La Serranía La Lindosa," by Gaspar Morcote-Ríos· Francisco Javier Aceituno, José Iriarte, Mark Robinson, Jeison L.Chaparro-Cárdenas, *Quaternary International*, April 29, 2020, https://doi.org/10.1016/j.quaint.2020.04.026

[11] See *UN honors site of prehistoric Colombia cave paintings hidden for millennia*, by Jaime Ortega Carrascal, July 3, 2018, efe-epa, https://www.efe.com/efe/english/technology/un-honors-site-of-prehistoric-colombia-cave-paintings-hidden-for-millennia/50000267-3671693, Accessed December 23, 2020.

[12] *Wild Blue Sea – Thinking Through Seawater*, by Melody Jue, Duke University Press, Durham, NC, 2020.

[13] *Gesture and Speech*, by André Leroi-Gourhan, Translated from the French by Anna Bostock Berger, and Introduced by Randall White, An October Book, The MIT Press, Cambridge, Massachusetts, 1993, p. 396.

[14] See https://learnodo-newtonic.com/jackson-pollock-famous-paintings, Accessed December 23, 2020.

[15] See "How Mondrian Won the Art World with Just Three Colors," Jessica Stewart, *My Modern Met*, May 22, 2019, https://mymodernmet.com/piet-mondrian/Accessed December 23, 2020.

[16] See David Bodanis, *$E=mc^2$: A Biography of the World's Most Famous Equation*, Bloomsbury Publishing, New York, 2001.

Chapter 21

PARASPAROPAGRAHO JĪVĀNĀM

No proverbs, prayers, sutras, religious verses, spiritual texts, ethical common principles, theological affirmation, rule of Dharma, or cornerstones of compassion have ever deviated, as a first principle, from some composited version of the Golden Rule. Yet, the science and psychology of those battlefields across the Nature we think we understand, in fact only partly substantiates the record of virtuous affiliations. Hyper-, hypo- and mesocarnivores, mucophagous parasites, avivores that eat other birds, durophagous and paedophagous fish that eat others of their kind, myrmecophagous insects like termites and ants, and numerous terrestrial vertebrates, ovophagous egg-consuming sharks, even those lepidophagous marine species that specialize in consuming scales of fish, are all widespread examples of anything but the dietary behavior of a Golden Age. Science implicates intraguild predation as the primary motive – eating one's competitors, often seen in ophiophages, e.g., mongooses, snake eagles and great blue herons, that go after snakes; just as the otherwise pastoral-like Hawksbill sea turtle turns violent when it confronts a sea sponge; or the otherwise gentle salamander or warbler would kill for a worm. At least 12% of all mammals, and over 90% of all birds are voracious consumers of insects. 270 known mammal species are pure carnivores. And since 1500 BCE, 92% of human history has witnessed at least one war, some of which, like the Second Sino-Japanese conflict resulted in as many as 25 million casualties. World War I was double those mortalities; World War II more than triple, or some 85 million possible deaths.

So what should we glean from all this carnage? There are many ways to absorb the stark contradictions of the ideal of the Golden Rule. Friedrich Nietzsche (1844-1900), never one to mince his words, declared, "Far too many live…" Rüdiger Safranski ponders this Nietzschean penchant: "To prevent the 'madness' of past generations from erupting in current and future generations and landing all of history in an awful state of 'degeneration', specific measures must be undertaken. Which ones?" His answer, "Pay careful attention, my brothers, to every hour in which your spirit wants to speak in parables: that is the origin of your virtue."[1]

Mahavira (599-527 BCE) was far more generous in his exposition on the nature of reality when he described the "twenty-two troubles" a monk must tolerate and, by direct implication, must renounce "suffering" no matter its cause. "Suffering from insects a great sage remains undisturbed."[2] The sage Umaswati (2nd-century to 5th-century CE), in his famed *Tattvartha Sutra*, declared, "Abstinence from violence, falsehood, stealing, carnality and possessiveness – these are the vows,"[3] an utterance stated with the full underpinnings and cognitive philosophy of mind that could also represent in fullest confidence that, "All life is bound together by mutual support and interdependence," "Parasparopagraho Jīvānām."[4]

Though this sampling of verities represents a miniscule fraction of opinions, some altered and obscured over time (Nietzsche's sister, Elisabeth, thoroughly misunderstood and misrepresented his work in her curatorial editing of Nietzsche's oeuvre after his death, as happened to Percy Shelley), they remain pillars of an antinomy forever unresolved in the annals of life. Near the conclusion of his own definitive book, *Biophilia*, Edward O. Wilson is frank in divulging the unabashed truth of ourselves. In his reference to Christopher D. Stone's 1972 classic *Should Trees Have Standing?*[5] Wilson refers to the concept as a realm of "enlarged generosity" and goes on, "Human beings, for all their professed righteousness and brotherhood, easily discriminate against strangers and are content to kill them during wars declared for relatively frivolous causes. So it is much easier to find an excuse to exterminate another species."[6] And Wilson concludes that this is "that astonishing blend of achievement and farce that has always defined our species."[7]

In the curves, lines and parabolic equations intimated in the previous chapter, we suggested multi-dimensional space, vortices, and mathematical conundrums on either side of the zero; of infinities, in other words, that are always capable of the addition of "1=?". That vortex of Koans leads us nowhere or everywhere. To embrace the latter is to believe in 1; to recognize the power of 1+1; to know that in spite of so much agony ecologically splayed across time and space, biology favors originality, élan as much as quiet introspection; beauty, restraint, love. That these seasons of unknowing belief rise to the level of unshakeable ideals, by which the weight of infernos and despair can be resurrected out of Faustian fashion. That we harken to the tender call, not in random quagmires of tragedy, but with deliberate faith-turned-activism, as the credible narrative guiding us from this moment on: the "patsfom" principle discussed earlier, or post-Anthropic, trans-species first order metaphor that elicits real faith plus action, not unrealistically amongst Congresses and municipal bodies, where change has always cottoned to the sluggish and safe. But in the individual where triumph confronts no obstacle, none.

If this "individual" is to reach beyond the clutches of Darwinian evolution, past the Dark Ages of mechanistic thinking; to step out the door without dependence on technological delusions of perfect self-empowerment, and monetary cravings for endless gain; but to seek new friendships and alliances that do not hinge upon or require a response, a language, the approbation of our endeavors, then we can be certain of what, elsewhere, we have described as "reciprocity potential".[8] RP = the square root of compassion, times the square root of the population mean divided by the sum of the entire group or any population (n). Mahatma Gandhi termed one form of it as "The Gospel of Selfless Service."[9] Does it fly in the face of biochemistry and natural selection? Or imagine that it can shake up whole ecosystems, tilting at windmills?

We think not. Rather, it is the only unflappable truth that can save our species and most others, from utter ruin. However complex the systems we decipher across Nature, when cozied up to by the roughshod, withered humility that must be the individual, we can assay any principle of offer and receive; love and be loved. Some may claim for it a miracle. But in truth it is merely that key to open complexity.

ENDNOTES

1 *Nietzsche – A Philosophical Biography*, by Rüdiger Safranski, Translated by Shelley Frisch, W.W. Norton & Company, New York, London, 2002, p. 267.

2 Uttaradhyayana, Lecture II, p. 11, *The Sacred Books of the East* Translated By Various Oriental Scholars And Edited By F. Max Müller, Vol. XLV, At The Clarendon Press, Oxford, UK, 1895, p. 11.

3 *Tattvartha Sutra – That Which Is*, Translated by Nathmal Tatia, with a Foreword by L. M. Singhvi and an Inroduction to the Jaina faith by Padmanabh S. Jaini, The Institute Of Jainology, HarperCollins Publishers, Manchester UK, 1994, p. 169.

4 Sangave, Dr. Vilas A. (2001). *Facets of Jainology: Selected Research Papers on Jain Society, Religion, and Culture. Mumbai: Popular Prakashan.* p.123. See *Tattvartha Sutra*, 5.21 –"Souls render service to one another."

5 *Should Trees Have Standing? Law, Morality, and the Environment*, Oxford University Press, Oxford and New York, 3rd edition, 2010.

6 *Biophilia – The human bond with other species*, by Edward O. Wilson, Harvard University Press, Cambridge, Massachusetts, and London, UK, 1984, pp. 130-131.

7 ibid., p. 145.

8 See *The Theoretical Individual – Imagination, Ethics and the Future of Humanity*, by M. C. Tobias and J. G. Morrison, Springer, New York, p. 52. We first applied this notion in our work, *Anthrozoology: Embracing Co-Existence in the Anthropocene*, by M. C. Tobias and J. G. Morrison, Springer, New York, 2017, https://www.springer.com/gp/book/9783319459639

9 See *"Gandhi Memorial Center" – "Selfless Service,"* http://www.gandhimemorialcenter.org/ selfless-service, Accessed December 24, 2020.

Chapter 22

THE FOURTH ORDER SIMILE

Both time and inherent structure, temporal vulnerabilities, and the multitude of other variables across all biological regimes, shrewdly contest ecological homeostasis. Adaptation inculcates friction. Tension ramifies throughout stability such that all thermodynamic traits in the universe appear to be conditioned to oppositional forces. They render the extensive (e.g., entropy, mass, volume), intensive (e.g., pressure, temperature, thermal conductivity), and specific (e.g., heat capacity) properties that exert continuing change, before, during and after human perception.[1] Within a mathematical context, one could multiply compounded complexities of nature in much the manner of the century's old grazing goat problem. This quite arbitrary, if fickle quest was recently solved by recognizing "multidimensional space rather than a field of grass" as the means to determining the length of a tether on a hungry goat within an acre of grass where the intention is to allow the goat access to half the acre.[2] The complex analysis relies upon both real and imagined components and this, not surprisingly, is the only resolution to human nature aspiring to meld with nature, whilst supporting ideal behavior as preferable to a lesser, ethical threshold. What does it all mean? Are all half-acres, or goats, the same? Knowing that they are most certainly not lends yet a fourth order simile to the biological bounty of non-boundaried dimensions.

Nietzsche's book *Beyond Good and Evil: Prelude to a Philosophy of the Future* (1886) employed several of his principal concepts intended to transcend the stumbling blocks of both human and worldly nature. The "will to power," "imaginary counterworld," "truth from the grave," and "absolute value, in contrast to man's smallness and coincidental status in the flux of becoming and passing on."[3] This very dichotomous circumstance, he says, prompts humanity "alone" to have "fabricated causes, succession, reciprocity, relativity, compulsion, number, law, freedom, motive [and] purpose."[4] One queasily reacts to this contradictory effect in such cultures as the Maya whose repeated collapses have been attributed to, among other things, "a chronic emphasis on war and erecting monuments rather than on solving underlying problems…"[5] "The Maya warn us," Jared Diamond points out, "that crashes can also befall the most advanced and creative societies"[6] echoing the oft'iterated lines 'from poet Percy Shelley's poem "Ozymandias," (1817): "'My name is Ozymandias, king of kings: Look on my works, ye Mighty, and despair!/Nothing beside remains…" Nowhere is this cautionary tale more poignantly conveyed across centuries than in the "Palenque, Sculpture, Interior Courtyard, Principal Palace, Chiapas, Mexico" stonework profile of a proudly enigmatic man, lips solemnly sealed, giving away no artistic information as to what was happening, and what was likely to go wrong – either to himself or his culture - as captured in the brilliant photographic achievements of William Frej.[7] Frej writes of his photograph, "Like the ancient walls of the palace surrounding it, the carved stone face, probably of a prisoner, reminds us of the fleeting nature of once mighty civilizations… Pakal, the greatest king of the ancient Maya city of Palenque in Chiapas, Mexico, constructed a magnificent city-state that flourished in the 7th century. Yet, this bas-relief of a prisoner represents to me the beginning of the end of a Maya kingdom fraught with war, lost to the tropical jungle before its rediscovery a thousand years later."[8]

Did the sculptor know what we know? Does knowing change anything? Is change, at the bequest of *H. sapiens*, a worthy topic of consideration? What does *worthy* mean? Does meaning translate from individual to individual, across species, between holons, through time and space? The questions are annoying, the answers, if any, unreliable. So then what? And who said "if-then" should even be recommended as the conditional of choice? There are others, like "why-not?" "Either-or"; "true-or-false?" "Greater than or equal?"

There are "assignment operator" and "equality operator" predicates that form a dizzying rash of juxtapositions each matrix conditioned by the same logical thrust – the code of "if" statements[9], all suggesting human consensus. But ethological commons demand an entirely different cognitive or emotional modus operandi.

At this point, we best step back. Intellectual weariness can become a very comfortable couch, particularly in the act of trying to resolve the unresolvable. Roland Barthes, in referring to "hyperconsciousness" states that "The right minimalist ethic would help bring harmony between the maximum internal intensity."[10] But Barthes is actually probing the logic of Spinoza's "*conatus* (will, appetite, indefinite exigency of existence, effort to persevere in one's being)" and does so by raising the stakes to the ethological point of departure. He writes, "as we are dependent on all the other beings of the world" there is a fluctuation, a differing lens, a changing landscape, what he refers to as the "plasticity of the *conatus*."[11]

Neural plasticity has been cited in song birds as that capacity, among other things, to deliberately add neurons into a circuit that specifically acquires new language-song, or promote nurturance in both male and female.[12] "Synaptic plasticity" has been viewed as instrumental to the brain's "new learning and memory formation."[13] First described in rhesus monkeys as early as 1923 by behaviorist Karl Spencer Lashley (1890-1958), evidence now supports the notion that various forms of neural plasticity assist neurogenesis throughout the brain, well into old age; the data sets supporting "neural Darwinism" first proposed in 1978 by Gerald Edelman.[14] Model species, like the fruit fly, have been studied to determine how, and what level numerically natural selection works upon neural regeneration. Nerves are said to mature with time[15] – why wouldn't they? And given the brain's ecosystem(s) of neurons, as in any other biome given the option to maintain a living organism as the reciprocal agent of its own recipes for perpetuation, we should expect a real-time selection mechanism oriented towards cooperative success, as measured by the pertinacity of its own efforts.[16] Neural clusters over time will select for themselves, if they *choose* to. Or if the underlying conatus has the necessary pertinacity to preserve original impulses and commandeer its own evolution. At the biovortices of epiphany, we have no reason to doubt such real-time evolution, or re-evolution. At the horizon event of a fourth order simile, consciousness is capable of anything.

Hence, the changing biological landscape within as the mind endeavors to determine new conditions that would facilitate the exchange of fundamental freedoms, pleasures, tendernesses and primal kindness. The classical summary of ecological exchange has treated the content of information and associations as the historic, "emergent property of the species' evolution" under the doctrine that "given an autotroph, herbivore and carnivore, we can predict the ecological association that will emerge."[17] But this type of discourse presumes to know on the basis, strictly of how and what one eats. "Eats" is a narrative class, like "gravity" but cannot substitute for actual identity. A Jain monk went for many months in Mumbai without eating. Yet, we do not categorize his identity as nothing more than one who fasts. Nor would we suggest that the Greater Roadrunner (*Geococcyx californianus*) does only that; and Leonardo da Vinci, only this. Michel Foucault dealt effectively with this problem by seeking "the system that links all these fields of co-existence together" by deconstructing statements that have "a bearing on groups of signs" which, in turn, subsequent to their expression or hidden intent, see the evolution of a "position that may be filled in certain conditions by various individuals…"[18]

Returning to the dyad that differentiates according to expense, but also gain, there is the more fundamental and endearing tangency which moves from the asymptote to integrative being. As Lorraine Daston and Peter Galison have written, "For Goethe, objective and subjective phenomena were complementary and equally essential to the science of colors."[19] The merging of obvious factors, among others, a "mutual preoccupation," as Kierkegaard described something like it,[20] reflected on that tendency in humans (by association, all life forms) to *fall in love*; to engage in "*immanent thinking*" so that "the whole reality can be complete in the idea…"[21] But, thinking as much, having plunged into this concrete reality of love, all of that immanence also recognizes from deep within the intangible, something beyond, something more. This *more* gives rise to a remarkable occasion of occasions, namely, the transcription of every possible essence into a moral imperative that guides life beyond the hindrances we know too clearly throughout the natural world. We arrive upon that plateau given to grace and unconditional love.

ENDNOTES

[1] *Thermal Engineering*, by Nick Connor, May 22, 2019, https://www.thermal-engineering.org/what-is-thermodynamic-property-definition/Accessed December 26, 2020.

[2] "After Centuries, a Seemingly Simple Math Problem Gets an Exact Solution," by Steve Nadis, *Quanta Magazine*, December 9, 2020, https://www.quantamagazine.org/mathematician-solves-centuries-old-grazing-goat-problem-exactly-20201209/Accessed December 25, 2020.

[3] *Nietzsche – A Philosophical Biography*, by Rüdiger Safranski, Translated by Shelley Frisch, W. W. Norton & Company, New York, London 2002, pp. 294-297.

[4] *Beyond Good and Evil*, Translated by Walter Kaufmann, New York: Random House, 1966; reprinted in Vintage Books, and as part of Basic Writings of Nietzsche, New York: Modern Library, 2000, §21.

[5] See *Collapse – How Societies Choose To Fail Or Succeed*, by Jared Diamond, Viking Publishers, New York, 2005, pp. 159-160.

[6] ibid., *Collapse*, p. 159.

[7] See *Maya Ruins Revisited – In The Footsteps Of Teobert Maler*, by William Frej, With Essays By Stephan Merk, Alma Durán-Merk, Tomás Gallareta Negrón, Jeremy A. Sabloff, and Khristaan D. Villela, Peyton Wright Gallery Press, Santa Fe, New Mexico, University of Oklahoma Press Distributer, 2020, pp. 4-5. Frej provided a tumblr link to us, "Pakal the Great: Architect of Palenque," "A digitized version of an exhibition on K'inich Janab Pakal, ruler of Palenque. This exhibition was a final project in ARTH 321 Maya Art and Architecture at St. Mary's College of Maryland, Fall 2016," https://pakalthegreatexhibition.tumblr.com/post/ 154741734763/patio-of-the-captives-late-classic-period-palace

[8] William Frej was kind enough to email us his comment on his photograph, December 27, 2020.

[9] Khan Academy, *Review: Logic and if Statements,* n.a., n.d.., https://www.khanacademy.org/ computing/computer-programming/programming/logic-if-statements/a/review-logic-and-if-statements, Accessed December 26, 2020.

[10] *The Neutral, Lecture Course at the College de France (1977-1978)*, by Roland Barthes, Translated by Rosaline E. Krauss and Denis Hollier, Text established, annotated, and presented by Thomas Clerc under the direction of Eric Marty, Columbia University Press, New York, NY, 2005, pp. 200, and 257.

[11] ibid.

[12] See "Neural Plasticity," by J. H. Kaas, in the *International Encyclopedia of the Social & Behavioral Sciences*, Edited by N. J. Smelser and P. B. Baltes, Pergamon, Oxford, UK, 2001.

[13] See "Synaptic Plasticity," by Ruth Heidelberger, John H. Byrne, et al., in *From Molecules to Networks*, 3rd Edition, Academic Press, Elsevier, Cambridge, Massachusetts, 2014, https://doi.org/10.1016/C2011-0-07251-4

[14] Gerald Edelman, Neural Darwinism. The Theory of Neuronal Group Selection, Basic Books, New York, 1987. See also, *Yan Sun, Wang Chao, Dai Jiapei (2010).* "Biophotons as neural communication signals demonstrated by in situ biophoton autography". *Photochem. Photobiol. Sci.* 9 (3): 315–322. doi:10.1039/b9pp00125e. PMID 20221457.

[15] van Praag, Henriette; Schinder, Alejandro F.; Christie, Brian R.; Toni, Nicolas; Palmer, Theo D.; Gage, Fred H. (February 2002). "Functional neurogenesis in the adult hippocampus". *Nature.* 415 (6875): 1030–1034. doi:10.1038/4151030a. ISSN 1476-4687. PMID 11875571. S2CID 4403779.

[16] Alvarez-Buylla, A; Lim, DA (4 March 2004). "For the long run: maintaining germinal niches in the adult brain". *Neuron.* 41 (5): 683–6. doi:10.1016/S0896-6273(04)00111-4. PMID 15003168. S2CID 17319636.

[17] See *Evolution as Entropy – Toward a Unified Theory of Biology*, by Daniel R. Brooks and E. O. Wiley, The University of Chicago Press, Chicago and London, 1986, p. 283.

[18] *The Archaeology of Knowledge*, by Michel Foucault, Translated from the French by A. M. Sheridan Smith, Pantheon Books, Random House, New York, NY, 1972, p. 115. First published under the title, L'Archéologie du Savoir, by Éditions Gallimard, Paris, France, 1969.

[19] *Objectivity*, by Lorraine Daston and Peter Galison, Zone Books, New York, NY, 2007, p. 277.

[20] *The First Love – A Comedy in One Act by Scribe*, Translated by J. L. Heiberg, in Søren Aabye Kierkegaard's Either/Or – A One-Volume Abridgment In A New Translation, Edited By Steven L. Ross, Translated by George L. Stengren, Perennial Library, Harper & Row, New York, N.Y., 1986, p. 65.

[21] ibid., p. 63.

Tashi Payden Tshering, Executive Director of Bhutan's Royal Society for Protection and Care of Animals, with Several Resident Dogs. Photo © M. C. Tobias.

Chapter 23

BIOLOGICAL SCALING FOR A NEW NATURE

"The plane of immanence has two facets as Thought and as Nature, as Nous and as Phusis," write Deleuze and Guattari.[1] Phusis, bound etymologically to "growing" tells something of transitive thinking in an ontological phase. We seek not to inform but stimulate; to echo whatever dreams, paradigms, ideals and tenable concepts might have traction outside the disastrous envelope of an historic solipsism that has dominated humanity's first harmonic of engine speed, what an auto engineer would think of as reciprocating and rotating imbalances.

This is, in fact, the anharmonicity of human history which, in turn, authenticates current fundamental questions. What are well-founded conjectures in an epoch of the Anthropocene? Is our personal and community self-absorption sufficiently at biological scale to have the highest spread rate, to paraphrase the obvious model of contagion? Nature is in everything. But are the big questions simply human questions, or a kind of query posed by some natural force, *asking* the questions independent of our own meager voices as a species, because it matters to that dimension which is its Dasein, or Cogito, but these are old, very old adages.

Is there a relevant mathematical point of entry into this field or plateau of relevancy? Not by rote, by numbers or by any preordained set of assumptions, how can we think without human thought? Transcend the cliché of silence, of being still? Is our outcry enough to obviate rhetorical flaws, self-obsessed and miscreant weathers dogging the last few thousands of years of our history?

Let us take the proposition that there are mathematical *quantities* in this twenty-kilometer thin biosphere that are critically differentiated from the biological *qualities* therein. The numbers meet only at the biovortices – inside the xylem, phloem, parenchyma, sclerenchyma and enveloping cellular bunders within leaves, for example; vertex propagators one would find in Richard Feynman diagrams.[2] We could ask: do the mathematical properties pertain at levels of the atom, cell and molecule in ways that contribute not only to the biology, but to the *biography* of each of these life forms? It's a simple matter to simply suggest, "yes"; but are there relevant implications to such a yes?

Can we think of the leaf as a complex system prone to stochastic probabilities that behave within any thermodynamic model but without Euclid, Newton, Euler, Darwin, etc. to prove their volitional status, sponges beneath a star, transforming all that light into life? That is to say, is the natural world the same "nature" we observe if we are not around to observe it? The same kapok tree crashes in the Amazon and billions of organisms, trillions of cells, molecules, neurons are aware of the collision, obviously. That awareness is no less the equation maker, the lens. Whether humans care more about trees than rocks, babies more than phytoplankton debris, does not impede a novel harmonic/thermodynamic: The Observer(s) and the Observed(s) are not subject to judgement. Encephalization quotients, even the neocortex itself, do not obscure the all powerful equal (=) sign. Across its infinite breadth, a Santa Clause, a Trojan Horse, a soothsayer, an Alice may pass. But, it appears, only our species does so with an ambiguous agenda that, from a sub-atomic perspective, is clearly biologically entangled in a mass of contradictions. That mass contests the aforecited "yes" and demands some kind of rectification.

Theoretical physicist Frank Wilczek (b. 1951) devised the notion of "asymptotic freedom"[3] (for which he, David Gross and H. David Politzer) received the 2004 Nobel Prize in Physics), an intriguing example that could account for the energy inherent to biosemiotics signaling between species within a general distribution of so called Sobolev Space (named after the Soviet mathematician, Sergei Lvovich Sobolev, 1908-1989). The distribution he theorized (numbers, not signals but equally potent application to biovortices) comprises modern calculus with its infinite range of differentiations (read: biodiversity)[4] in an equally and unfathomably diverse constellation of matrices (e.g., the Toeplitz matrix and organ polarities in all life forms)[5].

By according the widest behavioral anomalies, as well as norms to the physics of biology and vice versa, we detect novel measures and new functions – as observers with a will – that, whilst acknowledging false positive negentropies (e.g., the nearly still eye of a hurricane) may freely embrace original paradigms. Politicians call them "possibilities"; the public, "hope"; economists, "stimulus"; philosophers, "ontology". But derived biology enables us to pose some very basic causal systems we believe to be proven throughout all evolutionary phylogeny. No life form wants to suffer; no life form knows it's a life form except in the most fleeting, time-dependent, sense; feelings matter but only to other feelings; the aggregate of feelings in any biological domain can be assessed as to the probability of a similarly diffuse parity in other aggregates because the feelings represent a uniform propagator. Even though the codes of life within remain obtuse as to origins and causes. Hence, biosemiotics research must contend with coefficients that are neither nominal nor ordinal; that eschew intervals and can re-invent ratios.

The scale of these operants appears to act within assemblages both organic and inorganic, in-tact or fractured. Their constructive adaptation and resilience combines the multivariable functions of both open and closed sets; of topologies that function in any direction, thus obviating the taxonomic compulsion to find hierarchies.

Moreover, as in the case of re-evolution, these hierarchical templates provide holonic reversibility. We see, as does the leaf and its tree, in all directions. Imagine the literal holonic sight inherent to the peacock mantis shrimp (*Odontodactylus scyllarus*), a 340-million-year old crustacean, Order Stomatopoda, sub-Order Unipeltata, with the most powerful eyesight of any animal -12 to 16 photoreceptor cells. Humans have but 3.

Importantly, we can determine some idea of scale for biological patterns and periodicity on a finite planet where emblematic junctures stand out at the largest biological interfaces. For example, the Himalayas and the Indian sub-continent; any of the 36 terrestrial hotspots; or, in miniature, the Mulberry Grove Plantation cemetery and remaining Tillie Fowler forest in the Jacksonville Florida area, a human "ghost town" within a thriving ecosystem. We can apply the unilateral Laplace–Stieltjes transform[6] to the issue of "ghost species"[7] as well as those organisms described under the rubric of the Lazarus Effect[8]. Species once thought missing forever do, on occasion, come back. Take the Takahe (*Porphyrio hochstetteri*) on New Zealand's South Island, assumed extinct in Lord Walter Rothschild's seemingly definitive book, *Extinct Birds*[9] of 1907, but rediscovered in the Murchison Mountains by Geoffrey Orbell in 1948. Or the return of wildlife at Chernobyl; of real-time re-evolution occurring at the probability vertex of any of the long-exploited model organisms, like *E. coli*, *D. melanogaster*, or *P. ubique*, thought of as a kind of life force with magnitude, direction, and acceleration capacities.

What we are therefore looking at is a critical sensory and conscience-driven *tactic* to effectively circumvent the anthropic discrepancies that must result from an unquestioning observer/observed paradox, with its compounding biases by which nothing can be said to be real, including the mathematical scaling that presently estimates that there have been approximately "4 billion species" that have come and gone in the history of biology[10]. Does that human lens distort impact projections, such as "Climate and Land-Use Change on the Global Diversity of Birds"?[11] Or perceptions in the past about biochemical disruptions?[12] Determinations regarding habitat perturbation "thresholds";[13] other borders delineating extinction of entire populations?[14] Or even fellow members of our own species who commit suicide?[15]

Whatever mosaic of transdisciplinary inquiries into the psychology of *Homo sapiens* is employed to better grasp who we are, in whatever historic and ecological context, there will persist a clouded mirror, what in much Asiatic ethical traditions are thought of as *karmic impediments*, like dust and particulate matter obscuring an otherwise clear surface, thereby impeding our ability to see with pellucid, unprejudiced panclarity. There is no alchemical suite of philosophical or scientific chicaneries to address our biological limits in the framework of reaching out to understand. By acknowledging those parameters, we begin the recognition of our humble situation, even if we have no way to place it within a range of sensory or cognitive *powers*. No scale, no harmonic, no energy pulse or measurement that is relevant to ethical rectification.

Only an openly moral causality, embraced without scaling, can see us to the beginning of this *new nature*.

ENDNOTES

[1] In Gilles Deleuze and Félix Guattari, *What is Philosophy?*, Columbia University Press, New York, NY, 1994, p. 38, cited on p. 143, of "Gilles Deleuze: A Philosophy of Immance," by Fredrika Spindler, in: Phenomenology and Religion: New Frontiers, [ed] Jonna Bornemark, Hans Ruin, Huddinge: Södertörns högskola, 2010, p. 149-163 https://www.diva-portal.org/smash/get/diva2:406664/FULLTEXT01.pdf;Gilles; See also, Daniel. W. Smith, "The doctrine of univocity. Deleuze's ontology of immanence" in Deleuze and religion, ed. Mary Bryden, London/New York: Routledge, 2000, p. 160; See also François Zourabichivili, Deleuze. Une philosophie de l'événement, Paris: PUF, 1994.

[2] See *"Diagramology' Types of Feynman Diagram,"* by Tim Evans, January 2, 2018, https://www.imperial.ac.uk/media/imperial-college/research-centres-and-groups/theoretical-physics/msc/current/qft/handouts/qftdiagramtypes.pdf

[3] See https://www.nobelprize.org/prizes/physics/2004/summary/

[4] Sergei Lvovich Sobolev (1908-1989). Bio-Bibliography, Edited by S. S. Kutateladze, Novosibisrk, Sobolev Institute, 2008.

[5] Ye, Ke; Lim, Lek-Heng (2016), "Every matrix is a product of Toeplitz matrices", *Foundations of Computational Mathematics, 16 (3): 577–598,* arXiv:1307.5132, doi:10.1007/s10208-015-9254-z

[6] Feller, William. *An introduction to probability theory and its applications*. Vol. II. Second edition, John Wiley & Sons, Inc., New York-London-Sydney 1971 xxiv+669 pp.

[7] See "Mysterious 'ghost' populations had multiple trysts with human ancestors," by Ann Gibbons, February 20, 2020, *Science*, https://www.sciencemag.org/news/2020/02/mysterious-ghost-populations-had-multiple-trysts-human-ancestors, Accessed January 3, 2021.

[8] See "The Lazarus effect: can the dynamics of extinct species lists tell us anything about the status of biodiversity?" By David A. Keith and Mark A. Burgman, *Biological Conservation*, Volume 117, Issue 1, May 2004, pp. 41-48, https://doi.org/10.1016/S0006-3207(03)00261-1, Elsevier, https://www.sciencedirect.com/science/article/abs/pii/S000632070300261l#: ~:text=In%20general%2C%20the%20lists%20are,their%20distribution%20in%20remote%20locations.

[9] Hutchinson, London, p. 142, 1907.

[10] See "The Mathematics of Extinction Across Scales: From Populations to the Biosphere," by Colin J. Carlson, Kevin R. Burgio, Tad A. Dallas, & Wayne M. Getz, *PeerJPreprints,* p. 2, | https://doi.org/10.7287/peerj.preprints.3367v2 | CC BY 4.0 Open Access | rec: 11 Sep 2018, publ: 11 Sep 2018, https://peerj.com/preprints/3367.pdf

[11] See "Projected Impacts of Climate and Land-Use Change on the Global Diversity of Birds," by Walter Jetz, David S. Wilcove, Andrew P. Dobson, *PLOS Biology*, Published: June 5, 2007, https://doi.org/10.1371/journal.pbio.0050157

[12] See "Methanogenic burst in the end-Permian carbon cycle," by Daniel H. Rothmana, Gregory P. Fournierc, Katherine L. Frenchb, Eric J. Almc, Edward A. Boyleb, Changqun Caod, and Roger E. Summons, *PNAS*, September 27, 2013, https://www.pnas.org/content/pnas/ 111/15/5462.full.pdf

[13] See "Methods and models for identifying thresholds of habitat loss" by Deyi Yin, Shawn J. Leroux, and Fangliang He, First published: 10 November 2016, *Ecography*, Wiley Online Library, https://doi.org/10.1111/ecog.02557, https://onlinelibrary.wiley.com/doi/full/ 10.1111/ecog.02557

[14] See "Recent responses to climate change reveal the drivers of species extinction and survival," by Cristian Román-Palacios and John J. Wiens, *PNAS* February 25, 2020 117 (8) 4211-4217; first published February 10, 2020; https://doi.org/10.1073/pnas.1913007117, PNAS, https://www.pnas.org/content/117/8/4211

[15] See "Firearms and Suicide," Ch. 7 of *Firearms And Violence – A Critical Review, Committee to Improve Research Information and Data on Firearms*, Edited by Charles F. Wellford, John V. Pepper, and Carol V. Petrie, National Research Council of the National Academies Press, Washington D.C., 2004, p. 152, https://www.nap.edu/read/10881/chapter/9. See also, "More and More Americans are dying by suicide. What are we missing?" by Alla E. Dastagir, USA Today, January 30, 2020, https://www.usatoday.com/story/news/nation/2020/01/30/u-s-suicide-rate-rose-again-2018-how-can-suicide-prevention-save-lives/4616479002/

Chapter 24

RECIPROCITY PARADOXES

We speak of a humble situation in the context of a new nature. If this is merely a function of grammar, as bespoke in consciousness, then we are only saying what we have been saying, whether in a so called "global workspace" or as part of an "Integrated Information Theory".[1] Combine "satellite cameras with a convolutional neural network" and we can locate individual elephants from space.[2] As but one example of the tools we have amassed, which -if concentrated on delaying and impeding the sixth mass extinction underway - represent the humility and peril of our times.[3] Grappling with the idea that all insects could be extinct by 2119[4] sets an effective timeframe for declaring the ultimate barrier for further human existence, because without insects pollination of plant life, and the biological infrastructure for microbiota would collapse. Our species would die out, having paid witness to the loss from the world of all large mammals, the vast majority of avifauna, and disappearance of most marine shoals of vertebrates. Consciousness in *humans* within a biological span of relatively very few millennia, would have proved to be no more than the testimony to glaring causes-and-consequences by a band of doomed, ecocidal narcissists.

Neither physicists nor economists can solve this most important of all scaling problems. The duration stressors. They only imagine them as causal curiosities that manifest with no more immediate interest than the volatility of stock exchanges. Cosmological reciprocity is Newtonian, asteroids colliding and repulsing. Economic trading, while theoretically a benevolence born of benign bargaining – envisioned by Immanuel Kant in his treatise *Perpetual Peace – A Philosophical Sketch* (1795), does not solve the bigger issues of international human thresholds for unstinting altruism at great variance.[5] Peaceful patterns of bartering, or townhall theatre may suggest that we can maintain a power grid, the tenacious threads of democracy. But we know that all is wrought of fragility. There have always been two driving points of view on solution finding paradigms, "evolutionary and functional ecology" wherein we strive to "envision it ["abundance and distribution of an organism" or for that matter any symptom of life on earth] as the product of a history of adaptation to the community in which it is found."[6] That does little if anything for our cause, however. Even if we take "adaptation" to heart, which few humans have ever, to our knowledge, done. Perhaps the few remaining Jain Digambara munis in India are best case scenarios; or cenobites at Mt. Athos, or those at Mount Saint Catherine's Monastery in the Sinai. The last remaining members of those few Scheduled Tribes of the Andamans, or the Todas of the Nilgiris.[7,8]

A crucial impediment to grasping the biological properties and prospects for unselfish (a crucial corollary), altruism is the widely differentiated approach to reciprocity, in whose fertile crescent lies the roots of both compassion and tolerance. From these two instinctive and/or intuitively good norms arises the inevitable realization that every species is indeed a separate, though not necessarily removed, iteration of the principle. For example, a study comparing "the Norway rat, common vampire bat and brown capuchin monkey" underscores what two researchers have described as "probably one of the most debated theories in evolutionary research" ("Reciprocity").[9] Part of the challenge is semantics, suggests one group of researchers; a "confusion" over the use of terms, as well as "the need to distinguish between proximate (mechanism) and ultimate (survival value) explanations of behaviours."[10]

Coordinated, cooperative behavior amongst unrelated cichlids (with Cretaceous origins, the more than 3,000 fish species in the order Perciformes, including the tilapia) was recently "experimentally demonstrated."[11] Such ad hoc random scatter shot research defines several thousand years of human inquiry in the name of natural science. In 1987, Robert Trivers published an essay entitled "The Evolution of Delayed Return Altruism"[12] in which he formulated the accrual of net benefits to the altruist over the non-altruist, a relationship that is contrived and accelerated by natural selection. He considered the levels of "mutual dependence," time-frames, even theoretical alleles at differing loci in the genotypes of individuals. Domination, game theory, the symbioses that arise when "the wrasse, Labroides dimidatus, cleans another organism (e.g., the grouper, *Epinephelus striatus*) of ectoparasites (e.g., caligoid copepods)" that get in the mouth or gill chambers of the "host".[13]

Trivers considered warning calls in birds, the functionality of distinct patterns of assistance, and, in humans, suggested that there were at least 5 forms of "altruistic behavior" that would surely be deemed outright principles. These included "helping in times of danger," "sharing food," "helping the sick," "sharing implements" and "sharing knowledge."[14] These "preconditions for the evolution of reciprocal altruism," he argued, may have been in place even before the "Pleistocene,"[15] meaning, potentially, prior to 2.6 million years ago. The actual failure -beyond common sense - to account for widespread distributions of friendship, symbiosis, mutually beneficial partnerships that will result in gain to both parties to the relation, over a much broader timeframe, say billions of years, and involving single-celled organisms, is more difficult to delineate.

Isolated clusters of individuals (demes) may have little reason to change primordial patterns, unless a sudden reward from strangers enters the equation. What was the original prompting that induced two organisms to "hold hands"? Or, for that matter, to kiss, feed together, supply nutrients the one to the other? The selfish versus unselfish nature of relationships has long been subjected to the microscopes of natural selection with little resolution.[16] One might simply ask, What is love? And, more crucially, how does unrequited love operate within any evolutionary rubrics?

Martin Nowak was right to reject British/Indian scientist JBS Haldane's (1892-1964) assertion ("Rule") that – exclusively - "natural selection can favor cooperation if the donor and the recipient of an altruistic act are genetic relatives." Obviously, "unrelated" individuals are generous to one another in countless instances and categories of instance ("The Prisoner's Dilemma," for example). "Direct," "Indirect," "Network," "Group" and "Evolutionary" altruism are as real as family members sharing scarce supplies.[17] The problem, then, is not the stewarding power, or ethical realization that is reciprocity, but the underlying consciousness that moves towards the conscience, as a higher functionality of awareness in general. In fact, a most obvious case of this was in the life of Haldane himself. What reciprocal reward was he anticipating when he willed that his body, once dead by cancer, be given over to science? One could argue that such afterlife altruism was a fundamental form in the nature of isomorphism, of a profound pairing effect. There are nearly two dozen well known cases across the studies of mathematics and the physical sciences of such pairings, and all are described within the orbit of reciprocity; whether in classical electromagnetism (circuits, points and relatable voltage), an antenna transmitting and receiving signals, properties of restricting and inducting (Frobenius reciprocity), the reciprocal light intensities and their causes and exposure effects in photography (aspects of Helmholtz reciprocity), mutual links in neural networks, right-handed B-DNA and left-handed Z-DNA, etc.

But of all pairings, there is nothing like "Rembrandt's Eyes," as Simon Schama's great illumination of the old master commends, both eyes always different in each of the painter's portraits.[18] Similarly, there are numerous optical studies of non-conventional mirrors in which "two people [who would traditionally be able to] look at each other through their reflection" would suddenly see such "symmetry" broken.[19] The break represents a universal vulnerability, whether in what F. Scott Fitzgerald labeled "The Crack-Up,"[20] or within the "perpetual oscillations between cooperation and defection" – the disintegration of "equilibrium strategies" when the "tension between private and common interest" exceeds the tensile strength of the conventional physical or mental bonds shaping communities.[21]

Other manifestations of that tension exist in the ever-uncertain gray zones of direct and indirect reciprocity. Some years ago a very odd study of the lovely little passerine of the tit family, *Parus major*, was conducted. The idea was to determine where these gorgeous and cosmopolitan birds combined direct and indirect experience in forming impressions of potential rivals.[22] It turns out that they, like all living beings at some level, do so. The precise conjecture was levied some years later: "If birds can combine the two…then it is probable that humans will be able to as well."[23] The supposition is particularly relevant in the context of reputation, such that it has been suggested that "IR [indirect reciprocity] cannot explain a concern for reputation in typical societies unless reputations provide as reliable a guide to cooperative behaviour as does experience."[24] A public space, public goods, are involved. The likelihood of future encounters between the specific individuals who are party to the observation and/or transaction. At the most interior level, this is the iteration of Self condemned to its own dialogue. Or, *how many times must I tell you…* It has its own repetitive politic, and parsimony; a brash sociology that dispenses with poetry in search of something reliable in a world prone to fits and the unexpected.

But these summations assume that rivalry is everything in the world; when there is no Special Law of Rivalry. Indeed, in the late 18th century, Cesare Beccaria's Academy of Fists was inspired by a passion for "waging relentless war against economic disorder, bureaucratic tyranny, religious narrow-mindedness, and intellectual pedantry."[25] While the tensions referenced above may be applicable in physics and engineering, from billiard ball collisions to the wind shear on aircraft and all thermodynamics, with true reciprocity within the biosemiosphere, we are *feeling* beings at toying with general impulses that include love and gratitude; curiosity and friendship; passion and forgiveness; the joy of giving. These qualia will muck up any mathematic, flatten statistics, undermine prediction and all but destroy extrapolation. With true conscience we must start anew.

ENDNOTES

1 See "What is Consciousness? Scientists Compete to Find Out," by Andreas Kluth, January 23, 2021, *Bloomberg Opinion*, https://www.bloomberg.com/amp/opinion/articles/2021-01-23/what-is-consciousness-scientists-around-the-world-compete-to-find-out, Accessed January 26, 2021.

2 "New AI system counts endangered elephants from space," by Thomas Macaulay, January 20, 2021, *Neural*, https://thenextweb.com/neural/2021/01/20/new-ai-system-counts-endangered-elephants-from-space/amp/

3 *How the Sixth Mass Extinction Affects the U.S. Economy?* By Kimberly Amadeo, Treehugger, August 21, 2020, https://www.treehugger.com/mass-extinction-definition-events-economic-effect-of-6th-extinction-4589742

4 "Last year, 40% of honey-bee colonies in the US died. But bees aren't the only insects disappearing in unprecedented numbers," by Aylin Woodward, June 21, 2019, *Business Insider*, https://www.businessinsider.com/insects-dying-off-sign-of-6th-mass-extinction-2019-2, Accessed January 26, 2021.

5 *Behavioral Economics.com*, "Reciprocity," https://www.behavioraleconomics.com/resources/ mini-encyclopedia-of-be/reciprocity/Accessed January 26, 2021.

6 "A Critical Look at Reciprocity in Ecology and Evolution: Introduction to the Symposium," by David N. Reznick, Editor, Judith L. Bronstein, The University of Chicago Press Journals, *The American Naturalist*, Vol. 181, Supplement, May 2013, https://www.journals.uchicago.edu/ doi/abs/10.1086/670030?journalCode=an

7 See *The Toda Landscape – Explorations In Cultural Ecology*, by Tarun Chhabra, Harvard University Press, Orient BlackSwan, 2015.

8 See "Ahimsa – NonViolence," PBS one-hour documentary, Written and Directed by M.C.Tobias, Direct Cinema Ltd., 1987.

[9] "Reciprocity: Different behavioural strategies, cognitive mechanisms and psychological processes," by Martin K. Schweinfurth and Josep Call, *Learning & Behavior* 47, 284-301(2019), https://doi.org/10.3758/s13420-019-00394-5, December 2019, https://link.springer.com/article/10.3758/s13420-019-00394-5#citeas

[10] See "Social semantics: altruism, cooperation, mutualism, strong reciprocity and group selection," by S. A. West, A. S. Griffin, and A. Gardner, *Journal of Evolutionary Biology*, 16 October 2006, https://doi.org/10.1111/j.1420-9101.2006.01258.x, https://onlinelibrary.wiley.com/doi/full/10.1111/j.1420-9101.2006.01258.x

[11] "Coaction vs. Reciprocal Cooperation Among Unrelated Individuals in Social Cichlids," by Michael Taborsky and Thomas Riebli, *Frontiers in Ecology and Evolution*, 21 January 2020 https://doi.org/10.3389/fevo.2019.00515, https://www.frontiersin.org/articles/10.3389/fevo.2019.00515/full

[12] In *The Quarterly Review of Biology*, Vol. 46, No.1, Mar. 1971, pp. 35-37, The University of Chicago Press, http://www.jstor.org/stable/2822435, https://greatergood.berkeley.edu/ images/uploads/Trivers-EvolutionReciprocalAltruism.pdf; See also, "Reciprocal altruism: 30 years later," in Cooperation in Primates and Humans, Edited by P. M; Kappeler, and C. P. van Schaik, Springer, Berlin Heidelberg, https://doi.org/10.1007/3-540-28277-7_4, doi: https://doi.org/10.1007/3-540-28277-7_4, https://link.springer.com/ chapter/10.1007%2F3-540-28277-7_4

[13] ibid. p.40.

[14] ibid. p.45.

[15] ibid. p.45. See also, A. Gouldner, "The norm of reciprocity: a preliminary statement," *Am. Sociol. Rev.*, 47: pp.73-80, 1960.

[16] See The Evolution of Reciprocal Sharing," *Ethology and Sociobiology* 5: 5-14 (1984), https://pages.ucsd.edu/~jmoore/publications/Recip.html; See also *Effects of multiple gene control on the spread of altruism by group selection*, by Tom´aˇs Kulich, Jaroslav Flegr, https://arxiv.org/pdf/1002.4204.pdf

[17] "Five rules for the evolution of cooperation," by Martin A. Nowak, *Science*. 2006 Dec 8; 314(5805): 1560–1563.doi: 10.1126/science.1133755. https://www.ncbi.nlm.nih.gov/pmc/ articles/PMC3279745/; See also, "Upstream reciprocity and the evolution of gratitude," by Martin A. Nowak and Sébastien Roch, *Proc Biol Sci.* 2007 Mar 7; 274(1610): 605–610. Published online 2006 Dec 19. doi: 10.1098/rspb.2006.0125, The Royal Society Publishing, *Proceedings B, PMCID*: PMC2197219, PMID: 17254983, https://www.ncbi.nlm.nih.gov/pmc/articles/PMC2197219/

[18] *Rembrandt's Eyes*, Simon Schama, Alfred A. Knopf, New York, NY, 1999.

[19] See "Non-reciprocal Optical Mirrors Based on Spatio-Temporal Modulation," by R. Fleury, D. L. Sounas, and A. Alù, 10.1088/2040-8986/aaaa3e, *Physics, Optics*, Cornell University, https://arxiv.org/ftp/arxiv/papers/1709/1709.00702.pdf

[20] *The collection of posthumously published essays by F. Scott Fitzgerald*, Edited by Edmund Wilson, New Directions Publishers, New York, NY 1945.

[21] See Direct Reciprocity in structured populations," by Matthijs van Veelen, Julián Garcia, David G. Rand and Martin A Nowak, *PNAS, Proceedings of the National Academy of Sciences of the United States of America*, PNAS June 19, 2012 109 (25) 9929-9934; https://doi.org/10.1073/pnas.1206694109, https://www.pnas.org/content/109/25/9929

[22] Peake T.M, Terry A.M.R, McGregor P.K, Dabelsteen T. Do great tits assess rivals by combining direct experience with information gathered by eavesdropping? *Proc. R. Soc. B.* 2002; 269:1925–1929. doi:10.1098/rspb.2002.2112 [PMC free article] [PubMed] [Google Scholar]

[23] "Evolution of direct and indirect reciprocity," by Gilbert Roberts, *The Royal Society Publishing Proceedings B, PMC, Proc Biol Sci.* 2008 Jan 22; 275(1631): 173–179. Published online 2007 Oct 31. doi: 10.1098/rspb.2007.1134, PMCID: PMC2596181, PMID: 17971326, https://www.ncbi.nlm.nih.gov/pmc/articles/PMC2596181/#bib17

[24] ibid.; See also, "Evolution of indirect reciprocity," by Martin A. Nowak and Karl Sigmund, *Nature*, 2005/11/01, 1291, 8, Vol. 437, DO – 10.1038/nature04131, https://www.researchgate.net/figure/Direct-and-indirect-reciprocitya-Direct-reciprocity-means-that-A-helps-B-and-B-helps-A_fig2_7514734

[25] See *Cesare Beccaria (1738-1794), Biography*, July 16, 2020, https://www.biography.com/ scholar/cesare-beccaria, Accessed February 17, 2021.

The Quintessential Bond, Equine Couple, Socotra Highlands, Yemen. Photo © M. C. Tobias.

Chapter 25

THEORIES OF PAIRING IN A WORLD OF DIFFERENCES

Einfühlungk, empatheria, empathy[1]. We believe this quality is the physiological trigger for lasting reciprocity, as opposed to a false efficacy that relies on entirely other motivations which are not inter-ecologically beneficial. The Latin and Old French origins of the word, empathy, suggest suffering *with*, underscoring the Buddhist first noble truth, Dukkha, equated with life itself as inchoate pain. We could argue that the Old French *joie*, from the Latin *gaudium*, or *gaudere* to 'rejoice' might equally qualify in any truthful biological setting, the exact opposite as that preached by the Gautama Buddha. But there is a third, potentially crippling psychological reality to the dyad. We feel guilty, or shame if the slightest cause for joy erupts amid so much sadness; but we do not suffer amid those joys of a laughing child or adored companion. Both experiences are contextualized by ecological awareness which knows that a community has birds in sweet flight, but also - without pausing - absolute hells meted out to every physiology.

A tempered psychology acknowledges the odds of pain against the ephemeral ruse that is happiness. By substantiating these experiences in every living being, all who want joy, not cruelty, we have this idée fixe that looks out upon a world of fresh wounds, or choses to ignore them in its own time, its own way. What then, by venturing into a geography of reciprocity can we expect to rationalize, with a goal of interspecies conviviality in mind? An approach to kindness that transcends all the theories?

Recent literature on the nature of reciprocity is a mixed affair. What are the rudiments for establishing a system of *moral resources and cordial goods*, asks Patrici Calvo?[2] Does every species have its *own* economy, tit-for-tat, short-sighted self-interest that is daily given more weight than the idea realized of an external authority calling upon greater acts of altruism that will help the community far beyond the Thomas Hobbes in each individual? Let us ask that question again, differently: Is kindness manifest in 100 million different versions per species of the same volition? Is that sameness distributed similarly across the map of every individual, trillions of them? Why should that question matter to us? Does it matter to all of them? We would simplify this with arbitrary attributions. Mammals are all said to nurture in some version of the same nurturance. We know this is never 100%, certainly not among humans where violence within households represents the largest realm of aggression, particularly towards women and girls. Animal and plant behaviorists will readily cite vast literatures of contradictions to any portrait of a peace-loving biosphere, as we have earlier discussed in numerous cases. So then, if we focus strictly upon the kindness of humans, what can we expect in terms of any predictable generalizations? Only that the penchant exists, is clearly widespread, universally applicable, but never guaranteed. There are no definitive studies within our own species that calibrate the degree to which various stressors (environmentally inclement triggers like hunger, unrelenting heat, air and water pollution, etc.) diminish the statistical rate, or very capacity for generosity, compassion, let alone tolerance. Combine the variables with even a single other species, and we are at sea.

Ultimately, argue Ostrom and Walker, with activation of the human frontal lobe when trust and cooperation are involved, it is clear that "the importance of reciprocal relationships in explaining the origins of trust and trustworthy behavior"[3] go far beyond selfish genes, game theory and mathematical abstractions.[4]

Much of human reciprocation waffles between RSVP (the French *Répondez s'il vous plait* now rendered obsolete by the far less friendly *Réponse attendue avant le....*, or "regrets only." Its economic underpinnings spring from gift giving and barter systems whose imprecision was bolstered by monetary subjectivism: What is the value of a glass of water versus a coconut palm? Clearly, that depends on the time element of one's thirst. Situational reciprocity, engendering the narratives that bend context to fit a time/space singularity, are crucial to recognizing how self-interest must fluctuate, thereby obviating the need to generalize about principles of giving and receiving. Economic reciprocity is not subject to the boom-and-bust characteristics of ecosystems, but the sense that there is a fairness at the heart of exchange.[5,6] But kindness is not one of the long-standing components of economic theory or markets.

Justice as a practical point of departure will find it challenging to explain how, for example, an allegedly impoverished community can still be characterized as astonishingly charitable?[7] Or what mechanisms have tended genetically and culturally to prompt strangers to help one another, even at great inconvenience, expense, even risk to themselves? And, moreover, to uphold such behavior as necessary "ethical norms"?[8] This has been noted for centuries amongst the Japanese, whose lives have been so buffeted by earthquakes, volcanoes, and – most recently – the horror of the nuclear disaster/tsunami at Fukushima. In spite of a history of tragedies, Japanese communities have always displayed remarkable stoicism in the way people come to one another's assistance, bearing all manner of selfless heroism. Can it be said of Nazi Germany? The differences are stark, illuminating a first responder gene in some people, and the opposite in others. Justice, as a means of acquitting manifestly non-reciprocal behavior, rises at the occasion of legal means and some, but not all recourse.[9]

The differentiation of reciprocal rules from dispositions are organically distributed in the sense that evolution confers rewards and penalties, all of which may be hoped for, or ignored.[10] If we come to view choices as acts of redemption for past wrongs, we will see a very different moment from those who cannot empathize, or even recognize pain in others (the most prominent example being a Descartes [1596-1650] describing animals as mere machines with no feeling of pain). How we lay out the arguments for a predilection to commit good rather than inflict misery remains an unknown realm of relations, personality, attitude, personal history. The sociological overview of communities may capture some aptitude, but fall apart at the level of individuals who are capable of anything.[11] Such capacities are abetted by any number of stimuli. Great leadership, inspiring friendship, the drowning animal the means of whose rescue are self-evident; faith, hope, honor, self-esteem, the primacy of paired feelings.[12]

Pairings are communication at some phase. Which phase we cannot ascertain in a vacuum, though the ideal warrants demonstration at the point of collegial, shared endeavor. The Green New Deal, for example, excites the reciprocity of ecological goals, a mutual politic that fundamentally favors a specific co-belief in one version of reality over another.[13] Every charged *issue* does so. A community exists by its most pressing constraints, anomalies, and manner of resolving them. In absence of such tension, nothing holds a community together. A fully liberated individual, or group of them might be equated with rural designs oriented to strict independence, as calibrated by the number of acres per household, which is to say, the economic status. Conversely, marginal circumstances will dictate the look and topology of dependency. At both extremes, there is still plumbing and electricity, or lack of a reliable infrastructure, as in a State like Texas, whose energy grid is largely dependent upon natural gas, but unprepared for massive cold fronts. Food and medical acquisition strategies. Eventually, most solitaires wander into a warmly lit pub on harsh nights. No social animal or plant will readily forego some aspect of togetherness or signaling. This is also true at the single-celled level. Indeed, since 2012, researchers – recognizing the remarkable amount of storage capacity in DNA – have been endeavoring to manipulate single DNA strands to encode more and more data.[14] But a world of signals and information invites reciprocal rules of translation that are raw, pre-filtered to ignite ecological causes and consequences, as much for the individual as a monastery or megacity of individuals.

Hence, the biology of pairings – those reciprocal translations – is the quintessence of ecological community. Fundamental *base* pairing gives us DNA and inherently quantum mechanical surprises.

In the case of climate change, or consumption, or any other environmental debacle, the enemy of *empirical data* encoded naturally in DNA, is selfishness, which, of course, has its own steadfast codes subject to change. Its counterpart is that reciprocity penchant – also subject to great volatility – that can be shared by a majority, or, when necessary, implanted by some method (or instinctive neural function) equivalent to *executive order*, assuming the instilled legislative or biological hierarchy supports it. The alternative, too easily, is the mob. In the case of multi-generational virtue societies, it would be unthinkable for an individual to undermine long-standing ethical accords, as in the case, for example, of the Salish and Pend d'Oreille Indians who have lived sustainably along the Jocko river, with its indigenous bull trout, for thousands of years within the Flathead Indian Reservation of northwestern Montana.[15] A vast ethnographic literature on this topic (sustainability) is supported by the mechanisms described throughout an expansive history of social psychology and peace research;[16,17] as well as the ecological ethics describing any number of polluter-pays, and "tragedy of the commons" scenarios.

One readily turns here to Leviticus, "The stranger who resides with you shall be to you as one of your citizens; you shall love him as yourself, for you were strangers in the land of Egypt: I the LORD am your God."[18] Perceiving Earth to be God, all feeble human orchestrations and orientations change direction radically, or would be wise to, in keeping with doctrines of reciprocity that bypass the awkward anthropic lens of self-interest. In the case of an Algonquin elder, the man suggests that "sustainability" seems to be nothing more than continued perpetuation of exploiting, with the simplistic calculation that if you exploit enough you can also give some of the surplus back, or ideologies to that effect. But that doesn't really work, does it? He declares, "…among our people our concern is not what we can take from the land, but what we can give."[19]

It is a most powerful focal point for consideration of what and how things must be done; and amongst which constituencies, if we are to be true to one another, in the most generous, unconditional sense.

ENDNOTES

[1] See *Center for Building a Culture of Empathy*, http://cultureofempathy.com/ References/History.htm

[2] See *The Cordial Economy – Ethics, Recognition and Reciprocity*, by Patrici Calvo, Springer International Publishing, New York, 2018.

[3] See *Trust and Reciprocity: Interdisciplinary Lessons for Experimental Research*, Edited by Elinor Ostrom and James Walker, The Russell Sage Foundation Series on Trust, New York, NY, 2003, https://www.jstor.org/ stable/10.7758/9781610444347; See also, *Trust and Reciprocity*, by Gary Charness and Valentin Shmidov, Now Publishers, Norwell, Mass., Delft, Netherlands 2014.

[4] See discussions regarding the reciprocity laws for quadratic and cubic equations, residues, and rational reciprocity laws in *Reciprocity Laws From Euler to Einstein*, by Franz Lemmermeyer, Springer Berlin Heidelberg, 2000.

[5] *The Economics of Reciprocity, Giving and Altruism*, Ed. By J. Ythier, S. Kolm, L. Girard-Varet, Palgrave Macmillan UK, Camden, UK, 2000.

[6] See *Justice and Reciprocity in Aristotle's Political Philosophy*, by Kazutaka Inamura, Cambridge University Press, Cambridge, UK 2015. See also, Reciprocity in Ancient Greece, Ed. By Christopher Gill, Norman Postlethwaite and Richard Seaford, Oxford University Press, Oxford UK 1998. See also, Resilience, Reciprocity and Ecological Economics: Northwest Coast Sustainability, by Ronald Trosper, Taylor & Francis, *Routledge Studies*

in Ecological Economics, Abingdon-on-Thames, UK, 2009; and, Giving Back: Research and Reciprocity in Indigenous Settings, ed. By R. D. K. Herman, Oregon State University Press, Corvallis, OR, 2018.

[7] *The Culture of Giving in Myanmar: Buddhist Offerings, Reciprocity and Interdependence*, by Hiroko Kawanami, Bloomsbury Academic, London, UK, 2020.

[8] *A Cooperative Species – Human Reciprocity And Its Evolution*, by Samuel Bowles and Herbert Gintis, Princeton University Press, Princeton, NJ, 2011.

[9] *The Persistence of Reciprocity in International Humanitarian Law*, by Bryan Peeler, Cambridge University Press, Cambridge UK, 2019. See also, *Reciprocity and the Art of Behavioural Public Policy*, by Adam Oliver, Cambridge University Press, Cambridge, UK, 2019; See, in addition, *Reciprocity in International Law: Its impact and function*, by Shahrad Nasrolahi Fard, Taylor & Francis, Abingdon-on-Thames, UK, 2015.

[10] See *Altruistically Inclined? The Behavioral Sciences, Evolutionary Theory and the Origins of Reciprocity*, by Alexander J. Field, University of Michigan Press, Ann Arbor, MI, 2004. See also, *The Philosophers' Gift: Reexamining Reciprocity*, by Marcel Hénaff, Translated by Jean-Louis Morhange, Fordham University Press, New York, NY, 2019.

[11] *See Reciprocity, Altruism and the Civil Society: In praise of heterogeneity*, by Luigino Bruni, Taylor & Francis, Abingdon-on-Thames, UK, 2008.

[12] See *Reciprocity in Human Societies: From Ancient Times to the Modern Welfare State*, by Antti Kujala and Mirkka Danielsbacka, Springer International Publishing, New York, 2019, in which Marcel Mauss's "gift-theory" and Barrington Moore's studies of mutual obligations between people caught out in very different societal situations are examined. See also, *Emotion, Identity, and Religion – Hope, Reciprocity, and Otherness*, by Douglas J. Davies, Oxford University Press, Oxford, UK, 2011, with an emphasis on a multi-variety of persuasive reasonings for exchange that can include God and its myriad interpretations; or that belief in an entity of alterity – of Otherness – that commands with a voice dating back to our own personal ancestors; See also, *Reciprocity and Dependency in Old Age – Indian and UK Perspectives*, by Sue Thompson, Springer New York, 2015; and Reciprocity, by Lawrence C. Becker, University of Chicago Press, Chicago, Ill., 1990.

[13] See *Contemporary Collaborative Consumption: Trust and Reciprocity Revisited*, Ed. By Isabel Cruz, Rafaela Ganga, and Stefan Wahlen, Springer Fachmedien Wiesbaden, Germany, 2018.

[14] See *DNA could store all of the world's data in one room*, by Robert F. Service, Science, March 2, 2017, https://www.sciencemag.org/news/2017/03/dna-could-store-all-worlds-data-one-room, Accessed January 30, 2020.

[15] *Bull Trout's Gift: A Salish Story about the Value of Reciprocity*, by Confederated Salish and Kootenai Tribes, Flathead Indian Reservation, Montana, 2011. See also, *The Natural Model of Reciprocity*, by Madronna Holden, Our Earth/Ourselves, https://holdenma.wordpress.com/ culture-and-environment/reciprocity/2008. Of particular interest, see Holden's inclusion of Winona LaDuke's discussion of reciprocity from the point of view of her Anishinabe tradition.

[16] See "Social Psychology and The Study of Peace: Personal Reflections," by Herbert C. Kelman, in *The Oxford Handbook of Intergroup Conflict*, Edited by Linda R. Tropp, Oxford Handbooks Online, Nov 2012doi: 10.1093/oxfordhb/9780199747672.013.0021, https://www.oxfordhandbooks.com/view/10.1093/oxfordhb/9780199747672.001.0001/oxfordhb-9780199747672-e-21

[17] See *Learning Non-Aggression: The Experience of Non-Literate Societies*, Edited by Ashley Montagu, Oxford University Press, Oxford UK, 1978. See also, *Mutual Service And Cooperation*, by Charles T. Sprading, Libertarian Publishing Company, Los Angeles, CA, 1930.

[18] *Leviticus* 19:34

[19] Kimmerer R. (2011) "Restoration and Reciprocity: The Contributions of Traditional Ecological Knowledge." In: Egan D., Hjerpe E.E., Abrams J. (eds) *Human Dimensions of Ecological Restoration*. Society for Ecological Restoration. Island Press, Washington, DC. https://doi.org/10.5822/978-1-61091-039-2_18, https://link.springer.com/chapter/10.5822/978-1-61091-039-2_18

Chapter 26

THE PROCACCINI EFFECT

"Familiar Quotations" from John Bartlett are the stock and trade of ecology, particularly in the sense of *giving*. "He doubly benefits the needy who gives quickly," declares Publilius Syrus (1st century BC).[1] "We make a life by what we give," stated Winston S. Churchill, a "Brainy Quote"[2] that rhymes with the history of philanthropy, as articulated by so many of history's powerful abettors and galvanizers. Whether in the form of love, time, patience, laughter, the most mundane thoughtfulness and ethereal artform. To the tune of money or some other charity we think, as only Shakespeare said it, that "The quality of mercy is not strained, It droppeth as the gentle rain…"[3] Alfred Russel Wallace, writing in 1863 of the remarkable endemic biodiversity on Sulawesi, echoed this sentiment in the form of a challenge: "Future ages will certainly look back on us as a people so immersed in the pursuit of wealth as to be blind to higher considerations," foremost among which he cited all that remarkable life force which, at the present course of things, could "perish irrecoverably from the face of the earth, uncared for and unknown."[4] In his masterpiece of biological philosophy and discovery, *The Malay Archipelago*, researched between 1854 and 1862, first published in 1869, Wallace challenged English civilization, Christianity, and jurisprudence entirely, inciting the strongest discontent by pointing towards British ownership of property as the ultimate "state of barbarism" in its utter dispossession and destruction by the few of his "fellow-creatures" who actually have an evolutionary reason to be on each of those usurped acres.[5] Wallace wrote with the moral force of Tolstoy, Gandhi, and knew throughout the entire ecological warp and weft that natural selection was not connecting with human ways and means. Our own brute force intelligence was commandeering its better angels and leaving riot where previous coordination and grace had prevailed since well before any fossils told us so.

But this is the province of leisure time. Only those for whom selection has been benevolent have such long-winded minutes on their hands to idly speculate. For so many, trapped in generations of poverty, hunger, illness and malaise, natural selection is only the latest vogue of cruelty out of so many other bad ideas. How easily the romantic landscape quickly turns on its advocates. How lavish memory crumbles as easily as neglected floors, roofs, walls. Until the standard for romance relocates from the present to past tense; from warm hearths to gale-bitten ruins. Between swarms of midges and blinding blizzards, and multiple histories of bloodshed between clans, and battlefields vying for national pride, a place so otherwise cherished like the Scottish highlands, to those caught out, or hunted, is no haven, naught else than nightmare. We cannot pretend that it is otherwise. As in the case of the most gruesome death of the Pharoah Seqenenre Taa II, ruler of southern Egypt between 1558-1553 BC, whose mummified facial wounds, as analyzed with computed tomography, revealed the numerous slicing off of various parts of his face and skull (apparently because the occupying pharaoh was unhappy with Taa's hippos making so much noise at night in their sacred pool).[6]

This stops us in our tracks. Conventional progress, crowds angling for protection, unfairness in any guise or season, what then? Which force to petition, whose solace be thy guide out in the open, where thickness of coat or plumage is rarely distributed in equal measure, and no ecology or mathematic can make it so?

As gods and ontologies are fueled against such checkered ill-ratios, errant geographies and congeries that collide, only giving back against all such odds and sure violence can supply some other ledge upon which to climb, and lodge our inner hopes. To defend such myriad underdogs. Herein reigns an altogether separate sense of the world, by which our greatest instincts are attuned to one another, species by species, day by night. With neither greed nor want of self to prompt any other response. It thrives by instinct and the great regard that holds no grudge; advances no calamity. It gains by pure desire to feel that which one softly intuits; to see as we are seen; to hold to itself as if it were never anything but that Other, alive within her body of origins. Which is *us*; us in all our same needs and hopes and despairs.

Such mysteries are easily diffused in rapture, or the wanton excitement of natural history. But to bring back some form of calibrated truths, rejecting none, embracing all, requires more than mere theory to stem the tears or many treasons of self-interest.

This aboriginal reciprocity is a goal worthy of every halo between life and death. A perfect sense of that sublime bokeh-effect-like sfumato can be gleaned in a study of Camillo Procaccini's (1561-1629) uniquely rendered in his etching of "The Transfiguration." (1587-1590).[7] Jesus absorbs all conflict in this depicture of what must be one of the most elusive sections in all of the synoptic gospels from Matthew 17. There, upon a mountain where Jesus, Peter, James and John and come to pray, they encounter Moses and Elias. A "bright cloud overshadowed them" and the "voice out of the cloud" as well comes with a message translated to Christ: "Tell the vision to no man, until the Son of Man be risen again from the dead."[8] Mount Tabor, in Israel, is traditionally accorded the topographic location. Others contend that the precise locus is strictly symbolic. Images, predominantly fashioned between the 12th and 16th centuries, from Giovanni Bellini to the Russian Novgorod School, are core to perceiving a graphic, linguistic and philosophical debate. The Koine Greek "ὁ υἱὸς τοὺ ἀνθρώπου" (*ho huios tou anthropou*); and the Hebrew בן–אדם- (ben-'adam) are repeated 132 times, enshrining every conceivable link between humanity and God; man and nature; nature and nature. Every bridge between dyads is envisioned but no one crosses that bridge. Instead, there is iteration upon iteration, superpositioning of force fields, of waves and particles. A figure of Christ transcends before us. "And as he prayed, the fashion of his countenance was altered, and his raiment was white and glistering."[9]

Biblical scholars and people of some faith see the Transfiguration as evidence of immanence, of a world soul emanating from *nous* out into each individual mortal. These doctrines linking the earth and our perception of her to the *heavens above* are core to most theological odysseys. Buddha twice experienced a transfiguration, during enlightenment and during death, the Shussan Shaka narrative.[10] Similarly, Mahavira leaves his mortal body at death but describes, as if in equally loving flesh, the transcendent experience, pure and classical cosmology.[11] Reference points in that anthropic journey from Self to Other. The great diffusion. Everything we think, do and dream rests upon both memories of the past and future; transcendent equations remade in the everyday soils and weathers of our experience. Whoever the representative of this Son, this Daughter, this God, rebirth upon rebirth, it remains the most fitting of all tasks, the amelioration of all suffering that is the transfiguration. Realizing it is no glory, as such; but our sole duty, within the great drama of this ecological timebomb we have created.

Endnotes

1 "Maxim 6, Familiar Quotations -A collection of passages, phrases and proverbs traced to their sources in ancient and modern literature", by John Bartlett, Fourteenth Edition, Emily Morison Beck, Editor, Little, Brown and Company, Boston, Toronto, 1968, p. 125.

2 See https://www.brainyquote.com/quotes/winston_churchill_131192

3 From "The Merchant of Venice," Act IV, Scene I. See https://poets.org/poem/merchant-venice-act-iv-scene-i-quality-mercy-not-strained

4 *He Helped Discover Evolution, And Then Became Extinct*, by Anthony Kuhn, NPR, April 30, 2013, https://www.npr.org/2013/04/30/177781424/he-helped-discover-evolution-and-then-became-extinct

5 Alfred Russel Wallace, *The Malay Archipelago – The Land Of The Orang-Utan And The Bird Of Paradise; A Narrative Of Travel With Studies Of Man And Nature*, Macmillan And Co. Limited, London, 1906, p.458.

6 See *Egyptian pharaoh was executed on the battlefield, mummy reveals*, by Stephanie Pappas, February 16, 2021, https://www.livescience.com/egyptian-pharaoh-battlefield-death.html

7 "The Transfiguration. Medium: etching. Museum: National Gallery of Art, Washington DC. Author: Camillo Procaccini, Image ID: R2R41F, https://www.alamy.com/the-transfiguration-medium-etching-museum-national-gallery-of-art-washington-dc-author-camillo-procaccini-image225472171.html; See also, See RKD/Artists, https://rkd.nl/nl/explore/artists/record? filters%5Bsterfplaats%5D=Milaan&query=&start=40; See also, *Christ-Life, Death, & Resurrection*, University of San Diego, in Collaboration with the British Museum and the Timken Museum of Art in San Diego, CA, aby Hugo Chapman, Foreword by Hartwig Fischer, Director of the British Museum, 2019, p. 64.

8 *Holy Bible – King James Version, Containing The Old And New Testaments*, The Christian Science Publishing Society, Boston MA, n.d., pp. 1213-1214.

[9] ibid., p. 1284, St. Luke 9:29.

[10] E.W. Hopkins, "The Message of Buddhism to Christianity," *The Biblical World*, Vol. 28, No. 2, Aug., 1906, pp. 94-107. See also, Robert S. Cohen, *Beyond Enlightenment: Buddhism, Religion, Modernity*, Routledge, Abington, UK, 2006, pp. 1-3. See Chapter 47 on the double-enlightenment in *Buddhism, in On The Nature of Ecological Paradox*, by M.C.Tobias and J.G.Morrison Springer Nature, New York, NY 2021.

[11] *Mahavir - Prince of Peace, Ranchor Prime*, Illustrations by B. G. Sharma, Foreword by Vinod Kapashi; Preface by Michael Tobias, Mandala Publishing, 2006, p.133, https://jainqq.org/explore/020465/133.

Chapter 27

BEREITSCHAFTSPOTENTIAL

Hannah Arendt writes that "Philosophic freedom, the freedom of the will, is relevant only to people who live outside political communities, as solitary individuals…[and] We were talking here about the partiality of the actor who, because he is involved, never sees the meaning of the whole."[1] But we would contend otherwise, ennobling connectivity with a solitaire's full bounty of interiorization made manifest in a teeming, confusing world. Where both will *and* representation compound the transcendence of individual love for nature with nature herself.

In the context of the Great Dying (this Anthropocene), we can easily retrench at an instant to revivify our built-in connections without losing everything all at once. A "Garden Photographers' Exhibition 2005" entitled *PrivatePlaces* conveyed choices that people in their solitude as well as communities had made to uphold the virtues and enduring promise of biodiversity linked to our souls.[2] The black and white, and mostly color photographs from many countries demonstrated the famed line from Sir Thomas Browne (1605-1682), "…in the more obvious and neglected pieces of Nature, which without further travel I can do in the Cosmography of my self; we carry with us the wonders we seek without us: there is all *Africa* and her prodigies in *us*."[3] Gracing the cover of *PrivatePlaces* is Peter Baistow's remarkable image of a couple in their small, brick ensconced back yard, she seated at a little round table chatting on her cell phone, he pruning the potted plant on that table in East Ham, London. But most fundamentally, there is a man caring for another lifeform. Behind him, hundreds of other humans, gone to some other world, their tombstones increasingly disheveled. But from within this "private garden"[4] these two people, very much alive, can remind us that we are all made of the same impulses and particulars. The same actors, to all sides.

What separates the actions and motives of people may be the scale of their ambitions, or the baggage from which they cannot escape. Perhaps other residual effects of so brief a maturation time, the most harried evolution of any of the more than twenty species of *Homo*. Some choose to garden, in one fashion or other, the most ancient through story of our almanac. Others, like the rare John D. Rockefeller gives away 100 billion dollars in philanthropy. By comparison, total charitable giving in the United States for 2019 reached nearly $450 billion.[5] In his work *The Poetics (*c.335 BC), Aristotle wrote, "Metaphor consists in giving the thing a name that belongs to something else; the transference being either from genus to species, or from species to genus, or from species to species, on the grounds of analogy."[6] To "transfer or carry" was the original Greek (μεταφορική έννοια) definition of metaphor, a concept that has been embraced as a general pattern of descriptive choices for landscape, biodiversity, and biocomplexity, not to mention philanthropy, far in excess of what Aristotle narrowly envisioned.[7]

Yet, given over to metaphor as the most crucial state of consciousness and apperception in *H. sapiens*, all imaginative Being and Giving stems from its primal fundament: the recognition of Others and of Self among them, within their very center. It all seems oddly simplistic. If it were so simply, we would all surely *do unto others* in accordance with the devout strictures we all grew up with. What transpires in the mind *before* the metaphor? Benjamin Libet (1916-2007) studied the gap in time between the commission of an action and the thought of it, what neurobiologists term the Bereitschaftspotential (BP) or readiness potential.[8] The 0.35 second gap is ultimately of little near-term importance when considering the mechanisms that might engender human ethical commons, an unimpeded, widespread generosity that requires no monetizing of nature, or terms and conditions of material cause and consequence to shake us from that stupor which would otherwise deny the inner prognosis of a primordial quality of unconditional virtue.

The Laugerie Basse (3 1/8 inches) of a pondering bear-like figure, a Middle Magdalenian (17,000-12,000 BP) sculpture hewn from reindeer horn[9], represents far more than the *chaîne opératoire*, as André Leroi-Gourhan (1911-1986) described the life cycles that have gone in to the making of an artifact. It is a reciprocity-in-mind that conjoins deep meditation with expression. This bear anticipates Giacometti's diffusion of genetic and cultural memory backwards in time, all the way to those moments where anticipatory anxiety is overcome with something else entirely. An aesthetic reverence that changes everything in the human experience. Where it is located within the person, matters not. Only that the personage *is* this change across all borders and barriers between Aristotle's species and genus, genus and species. As metaphors for life, such taxonomic distinctions, in and of themselves, become irrelevant at once. Obsolete science, in terms of the urgencies of our time

No ecodynamic simile suffices to carry forth the exquisite promise of this revelatory perspective, the deep veneration for Others which – just ten, twenty thousand years ago – still dominated the entire human Weltanschauung.

Although our susceptibilities to the lure of wasteful power have all but replaced it, the reciprocity potential remains ever rooted in us. Continuous commentary for so many tens-of-thousands of years lends engaged credence to involvement by our kind with those for whom we expect nothing in return, thus changing the entire landscape of reciprocity from conventional paradigms of giving/taking, offering/receiving, to some third level of living metempsychosis. Leroi-Gourhan's extensive charts detailing "Frequency" of "Themes," "Compositional Subjects," of "Individuals," "Complementary Animals," various pairings and "Paired Signs," and an abundance of human other animal styles,[10] are the same substrata of a vast history of landscape art and human obsessions with other species culminating presently in the defining ambiguity of our age. Indeed, the species we are today overwhelmingly elucidates the problem of being human in a manner we can only now appreciate. The fullness of our population crisis – its density-laden violence and disparagements - propels no anodyne reconciliations other than in the intensity and purpose of our remaining biophilial commitments[11]. The apotheosis at all levels of Others. We see and feel this orientation embodied in the artwork of a contemporary photographer like Macedonian Goran Anastasovski;[12] or contemporary films by the likes of Carroll Ballard[13] or, more recently, Glendyn Ivin's "Penguin Bloom," chronicling two entirely reciprocating recoveries, one by a woman (Mrs. Sam Bloom) who has been paralyzed from a bad fall, and an orphaned magpie (Mrs. Sam Bloom's boys think the chick resembles a penguin) whom she comes to love.[14] We gravitate to the passions and shared premises of those desperately working to save biodiversity around the world.[15] But the global realm in fact bears downs only to the extent of the metaphor that shadows our immediate surroundings, no matter how worldly or clever we might think of our beyond borders altruism. Oddly, individuals call to us with greater urgency than entire populations. No matter what the numbers hide, we relate immediately to the singular; to ones rather than quadrillions.

And so the question naturally devolves: To what singularity might we attribute transformations in the human psyche, given this readiness-to-give potential; (unshakeable) readiness-to-relate-to-nature? Are we more prone to goodness than something other? Psychologists have long argued that the majority of humanity has always been engaged peacefully, most of the time. Crime statistics certainly underscore this fact[16]. As does the reporting from Medical Corps psychiatrists who, investigating combat during World War II, determined that the "fear of killing, rather than fear of being killed [was] the most common cause of battle failure."[17] Cynics will point out that in the first three decades following World War II, another 25 million+ people were slaughtered in 133 small wars. That the split mind entropy of a My Lai massacre or one-in-three chance that the Bay of Pigs might easily have accelerated into nuclear war, should give us no cause for optimism. Indeed, historians and behaviorists alike, such as Konrad Lorenz, Erich von Holst, Karl von Frisch and Niko Tinbergen have been divided on the Hobbesian notion of "the war of all against all" (or "eat and be eaten" -bellum omnium contra omnes)[18]. Raymond Dart was perhaps the most unambiguous in his famed acknowledgment of "The blood-bespattered, slaughter-gutted archives of human history from the earliest Egyptian and Sumerian records to the most recent atrocities of the Second World War [which] accord with early universal cannibalism… in proclaiming this common blood-lust differentiator, this predaceous habit, this mark of Cain that separates man diatetically from his anthropoidal relatives and allies him rather with the deadliest of Carnivora."[19]

We are keen to build up a case from the world of other species – like Hyponomeuta caterpillars – which take such strong and positive social cues. Van Tieghem cells of the slime mold Dictyostelium that maintain their individuality even as they aggregate like coral; or sea urchins, herring gulls and every flocking bird[20]. Yet the real catharsis is far more than the sum of all such case studies. To truly persuade is to become a family member. Hence, the power of true epiphanies;[21] of compassion borne of one's own suffering;[22] or of those who have struggled to negotiate to peace through that very metaphor, once again, compelling us to change over, translate, see Others as we would see ourselves.[23] A belief pattern that peers out, feels intensely or imagines one's own evolution guided by what must be the most supra-rational series of observations and choices in sync with their interdependent callings. In other words, to engage at every instant of interspecies opportunity with the love and patience of a saint, or mere angel, brought down to earth to use this *liveness*, every breath in our bodies, towards the beginning of love, nothing less.

ENDNOTES

[1] See *The Life Of The Mind*, Volume Two, "Willing," by Hannah Arendt, Harcxourt Brace Jovanovich, New York and London, 1978, pp.99, 272.

[2] *PrivatePlaces – International garden and plant photography*, Foreword by James Alexander-Sinclair, Introduction by Henrietta Van den Bergh and Helen Fickling, "A Patch of Earth," Chris Young, Editor; Garden Design Journal, Third Millennium Publishing, Ltd., London 2005.

[3] *Religio Medici*, Part 1, Section 15, p. 31, 1669.

[4] op cit., PrivatePlaces, p. 95.

[5] See, https://www.forbes.com/sites/jenniferwang/2020/01/20/from-buffett-to-bloomberg-the-25-philanthropists-in-america-making-the-biggest-donations/?sh=6d5c1e3f4e8f; See *Giving USA 2020*, June 16, 2020, https://givingusa.org/giving-usa-2020-charitable-giving-showed-solid-growth-climbing-to-449-64-billion-in-2019-one-of-the-highest-years-for-giving-on-record/. More than half of that money went to religion, education and other human services, with more than half given by individuals; whilst less than 3% of the total went to environmental causes.

[6] Translated by Ingram Bywater, Chapter 21, 1457b1-30, in "Aristotle on Metaphor," by Dr. Jan Garrett, March 28, 2007, http://people.wku.edu/jan.garrett/401s07/arismeta.htm

[7] See "Ecology, Complexity, and Metaphor," by James D. Proctor and Brendon M. H. Larson, *BioScience*, Volume 55, Issue 12, December 2005, Pages 1065–1068, https://doi.org/10.1641/0006-3568(2005)055 [1065:ECAM]2.0.CO;2, Published: 01 December 2005, Oxford Academic, https://academic.oup.com/ bioscience/article/55/12/1065/407139

[8] See "Biological Consensus Mechanisms: The Future of Coexistence," Chapter in *The Hypothetical Species: Variables of Human Evolution*, by M.C.Tobias and J.G.Morrison, Springer Nature, New York, NY, 2019, p. 254.

[9] See Chapter 4 "Objects of Religious Significance," in *Treasures of Prehistoric Art*, by André Leroi-Gourhan, Translated from the French by Norbert Guterman, Harry N. Abrams, Inc., Publishers, New York, NY, 1967, p. 89.

[10] ibid., Chapter 12, "Charts Supporting The Hypotheses," pp. 491-523.

[11] See *The Biophilia Hypothesis*, Edited by Stephen R. Kellert and Edward O. Wilson, Island Press, Shearwater Books, Washington D.C., Covelo, California, 1993; See also, biophilia -Johanna Pegler, Sarjeant Gallery Te Whare O Rehua, Whanganui, New Zealand, 2007. See also *Art & Nature – Three Centuries of Natural History Art from Around the World*, by Judith Magee, Greystone Books, Natural History Museum, London, UK 2009; and *The*

Exultant Ark – A Pictorial Tour of Animal Pleasure, by Jonathan Balcombe, University of California Press, Berkeley Los Angeles London, 2011; and Birds & People, by Mark Cocker, Photographs by David Tipling, With specialist research by and the support of Jonathan Elphick and John Fanshawe, Jonathan Cape London, UK 2013.

[12] "Wildlife Photographer Captures What Love Looks Like in the Animal Kingdom," by Margherita Cole, February 3, 2021, *My Modern Met*, https://mymodernmet.com/goran-anastasovski-affectionate-animals-photography/See https://www.gorananastasovski.com/Other/ABOUT-ME/

[13] Carroll Ballard, Director of: "The Black Stallion" (1979), "Never Cry Wolf" (1983), "Fly Away Home"(1996) and "Duma"(2005).

[14] See, https://www.rogerebert.com/reviews/penguin-bloom-movie-review-2021

[15] See *Saving Endangered Species – Lessons in Wildlife Conservation from Indianapolis Prize Winners*, Edited by Robert W. Shumaker, Foreword by Harrison Ford, Johns Hopkins University Press, Baltimore, MD, 2020; See also, *Sanctuary: Global Oases of Innocence*, by M.C.Tobias and J.G.Morrison, A Dancing Star Foundation Book, Council Oak Publishers, Tulsa OK and San Francisco, CA, 2008. See also, "Ten conservation success stories when species came back from the brink," by Adam Vaughan, *The New Scientist*, February 17, 2021, https://www.newscientist.com/article/mg24933223-400-ten-conservation-success-stories-when-species-came-back-from-the-brink/Accessed February 18, 2021.

[16]https://www.unodc.org/unodc/en/data-and-analysis/statistics/reports-on-world-crime-trends.html

[17] See *Men Against Fire*, by S. L. A. Marshall, William Morrow, New York, 1947.

[18] See *After Eden: History, Ecology & Conscience*, by M.C.Tobias, Avant Books, San Diego, CA, 1985, p. 245.

[19] See "The Predatory Transition from Ape to Man," by Raymond Dart, *International Anthropological and Linguistic Review*, 1, 1953.

[20] See "A Descriptive Study of the Development of the Slime Mold Dictyostelium discoideum," by John T. Bonner, *American Journal of Botany*, Vol. 31, No. 3 (Mar., 1944), pp. 175-182, Published By: Wiley, https://doi.org/10.2307/2437641, https://www.jstor.org/stable/2437641, https://www.jstor.org/stable/2437641?seq=1

[21] See McDonald, M. G. (2008). The nature of epiphanic experience. *Journal of Humanistic Psychology*, 48(1), 89–115. https://doi.org/10.1177/0022167807311878, https://psycnet.apa.org/record/2008-00641-006

[22] See "Transformation Through Suffering: A Study of Individuals Who Have Experienced Positive Psychological Transformation Following Periods of Intense Turmoil," by Steve Taylor, *Journal of Humanistic Psychology* 52(1): 30-52, doi: 10.1177/0022167811404944, https://www.researchgate.net/publication/241653920_Transformation_Through_Suffering_A_Study_of_Individuals_Who_Have_Experienced_Positive_Psychological_Transformation_Following_Periods_of_Intense_Turmoil

[23] F. Aquilar, F., and M. Galluccio, (Eds.). (2011). *Psychological and political strategies for peace negotiation: A cognitive approach.* Springer Science + Business Media. https://doi.org/10.1007/978-1-4419-7430-3, https://psycnet.apa.org/record/2010-25371-000

First Contact. Photo © M. C. Tobias.

Chapter 28

ZOOSEMIOTIC TRANSLATIONS

The continuing loss of what economists like to designate as "natural capital" has, most recently plummeted to some 27% per person across the planet.[1] What does that actually mean? The nearly eight billion human interconnections have engendered an aggressive network of topologies far beyond their geometrical shaping, imposing what Randolph T. Hester as called "priority framework" atop "unchanged structure". The outcome, in stark, ethical terms, is what Hester, quoting Japanese landscape scholar Nagahashi Tamesuke has said is "A good man [who,] with bad information is a bad man."[2] In addition, well in advance of the 15th Conference of the Parties to the Convention on Biological Diversity (COP15) and the 26th Conference of the Parties to the UN Framework Convention on Climate Change (COP26), *The Dasgupta Report* suggested that human subsidies valued at between US$4 and 6 trillion – out of an approximately $84 trillion global economy – were supporting the undermining of the natural world.[3] Such numbers and percentages, while chilling, are also misleading, even in the most rudimentary sense. By 2005, it was projected, for example, that globally there were 1 billion species of bacteria.[4] By 2016 estimates for the number of microbial species (bacteria, fungi, archaea, protists, but not viruses and prions)[5] had risen to 1 trillion,[6] with numbers continuing to increase. But because there is absolutely no relationship between 1 dollar and 1 species, the challenge to manifesting a logic equal sign (=) requires more than mere metaphor. As Hester writes in the Epilogue of his treatise pertaining to design and ecological democracy, "While I am overwhelmed by the despair around us and the meanness we confront in our habitation, ecological democracy takes root, sprouts, and blooms here and there. I have reported some of the cases."[7]

We should apply a similar expectation for statistical drift and ambiguity to the full biological map of reciprocities.

Our deliberations on such "cases" give us little or no pause. Approximately 150,000 people die every day. The number swirls far around us in a poetic distance. Like the burial of the famed ornithologist (and "collector") Henry Seebohm (1832-1895). The former steel magnate eventually became obsessed with the skins and eggs of birds, most of which were eventually donated to the British Museum. His adventures took him across Siberia, Japan, South Africa and all over Europe as he devoted his life to cataloguing birds, embracing the trinomial advent of sub-species categories. As his coffin was lowered into the earth in Hitchin just north of London, those present apparently watched an arriving thrush singing in a nearby tree and took this as a confirming sign, presumably of a lifetime commitment to birds that was now being rewarded, acknowledged by an ambassador from the Land of Avifaunia.[8] In his Preface to the 1855 Edition of "Leaves of Grass" Walt Whitman himself lapsed into explanatory prose: "The land and sea, the animals fishes and birds, the sky of heaven and the orbs, the forests mountains and rivers, are not small themes... but folks expect of the poet to indicate more than the beauty and dignity which always attach to dumb real objects...they expect him to indicate the path between reality and their souls."[9]

That very path has been laid out in nearly every *original* human community, some with more long-lasting effect than others. Near Tassajara, the Excelen peoples of the Esselen territory along the Big Sur Coast of California left some 250 elegantly painted handprints in a rock shelter ("CA-MNT-44")[10] utilizing some white mineral, perhaps limestone, and a binder that may have been "heated bear fat" or possibly some kind of tree sap. Fingertips were delicately conveyed. Writes Pewter Friedberg, "The name Esselen most probably derives from a tribal location known as Ex'selen, 'the rock,' which is in turn derived from the phrase Xue elo xonia eune, 'I come from the rock.' Like the picto-graphs on a rock wall that today proclaim their ancient existence..."[11] Having lived in Big Sur for a few years and exploring many of the most remote archaeological sites over the course of many decades, we can attest to the sheer elegance of mind, beauty of community, that permeated every aspect of Esselen bioheritage. It hasn't dissipated from the deep coastal redwood canyons. As one moves ever deeper into their past, through contemporary real-life natural history assemblages, it is overwhelmingly apparent that those peoples remain. We just need to be attuned. And to try and ignore Highway One which bifurcates not just the all-vulnerable ecosystem, but is the most glaring ambassador of conquest.

Two Spanish expeditions beginning in the early 17th century, managed by their invasive influence to overrun a 6,000-to-10,000-year-old collective of numerous indigenous peoples along Big Sur. There were more than 750 square miles of native territory, with Ohlone/Costanoan, Esselen, Salinan and Yokut cultures intermingling, but probably numbering fewer than 1,500 individuals total within the rugged coastal canyons. With Spanish missions built just south and north of Big Sur during the late 18th century, and the Monterey Presidio (one of four military forts in California at that time) becoming the very capitol of Alta California[12] it is little wonder that despite the relative inaccessibility of the Big Sur coastline, those locals were quickly assimilated, many forcibly baptized. By the 1880s, American ethnologist Hubert Howe Bancroft (1832-1918) after whom the University of California-Berkeley main library was named, took great pains to point out the maltreatment of the Indians by the Franciscans, writing, "It never occurred to [Junipero Serra] to doubt his absolute right to flog his neophytes for any slight negligence in matters of faith…" While yet other critics point to a system that condoned "genocide, slavery, and a wide range of atrocities."[13] Yet, the earliest European observation in 1602 by Spanish sea captain Sebastian Vizcaino, who had sailed into Monterey Bay, described the Esselen as a "gentle and peaceful people..."And "according to Fray Antonio de la Ascencion, who accompanied Vizcaino: 'The port is all surrounded with...affable Indians, good natives and well-disposed, who like to give what they have...They go naked at this port.'"[14]

California's biocultural history is now easily visible from space; the increasing dustbowl that is the Central Valley, and that Highway One rudely bifurcating the Big Sur Coast; and by the megacities up and down the State; the fact that at night in Southern California there is but one dark (17-mile) stretch, Marine Base Camp Pendleton, the only turf that is not aglow from all those urban congeries, from Tijuana to Santa Barbara. Where the Green and Olive Ridley Sea Turtles, Pacific Pocket Mouse, the Least Bittern and Golden Eagle can sleep without the glare of lights all around. Their last stand.

Ecologists have borrowed from experimental economics such phrases as "noncooperative equilibrium conditions" and "nuisance variables";[15] and from the literature of ethics such concepts as "aggregative well-being" and "the general welfare."[16] Such stretches of darkness as Camp Pendleton at night, brashly distorted to all sides by bright lights, is, in geological time, a recent phenomenon, of course. In 1855 when the first Western envoys reached Japan, soft quiet lantern lights had not yet been overrun by the overwhelming electricity whose origins would, in fact, emanate from a bamboo glen chosen by Thomas Edison adjoining a Shinto Temple not too many miles south of modern-day Kyoto. In that 1855 foray to a mysterious nation, naval officer Commander John Rodgers was astonished to learn that there were no guns in the country. They'd all been melted down centuries before. Unlike demographic conditions at that time in the U.S., the Japanese maintained a low fertility rate and had existed in a steady-state bubble for hundreds of years. Townsend Harris, the first U.S. Ambassador to that nation would say of her, "It is more like the golden age of simplicity and honesty than I have ever seen in any other country." And Harris' assistant, Henry Heusken would write in his diary, describing his first visit to the Japanese Imperial Palace, "The simplicity of the Court of Edo, the noble and dignified bearing of the courtiers, their polished manners which would do honor to the most illustrious court, cast a more dazzling splendor than all the diamonds of the Indies… Not one bayonet follows them… I fear, Oh, my God, that this scene of happiness is coming to an end and that the Occidental people will bring here their fatal vices."[17]

Ironically, Heusken, in his late 20s, was actually killed by a samurai during a period that saw the new Emperor Mutsuhito ordering his largely obsolete armies to turn in their swords. Japanese art historian turned curator of the oriental art section of the Boston Museum of Fine Arts, Okakura Kakuzō, described the chaos of cultural misunderstandings (one thinks of the death of Captain James Cook in 1779 during his third visit to the

Hawaiian Pacific Island group): "When will the West understand, or try to understand, the East?" wrote Kakuzō. "Strangely enough, humanity has so far met in the tea-cup. It is the only Asiatic ceremonial which commands universal esteem. Let us stop the continents from hurling epigrams at each other, and be sadder if not wiser by the mutual gain of half a hemisphere."[18]

We may draw certain conclusions from these varied examples of first contact and subsequent change and aggression. As Aldous Huxley wrote, "The best of constitutions and preventive laws will be powerless against the steadily increasing pressures of over-population and of the over-organization imposed by growing numbers and advancing technology."[19] Conservation strategies have always considered "focal species" whose "distribution and abundance…provide inference on the status of other species.[20] In the case of so called "umbrella species," we are now confined to a planet of one such creature, notwithstanding the millions of follicle mites teeming in our eyelashes, or bacteria on our hands and in our guts. If our strategies to contain the fall-out from an umbrella organism (ourselves) to all other focal species – those whose relations bear all the semblances of interactivities unchanged for hundreds of millions of years – our violence must be curtailed to the extent that its character has not shaped our observations. That, alone, is proving most difficult. Only unstinting reciprocity between neighbors, strangers, friends and allies, foes and those we might rather have nothing to do with; to each of them, as well as all those multitudes, unknown, throughout the biological kingdoms about which *we* are the strangers. To each of them such reciprocity, kindness, curiosity and love, are the only answer.

Because all anthropic lenses are fogged over by those very human observers inescapably linked to their own kind, we are confronted by the most incipient of all mortalities: everything we think, feel, perceive and do remains locked within its own orbit. But we also know that so many of our kind past, and present have shown the way towards love. Towards Others. Samuel Beckett characterized a group of fated people, the human community in his short work, *The Lost Ones*, about whom he wrote, "So on infinitely until towards the unthinkable end if this notion is maintained a last body of all by feeble fits and starts is searching still."[21] We can embrace that with fatalism or a re-ignited enchantment with ecology, and all her ethnographic, mystical parts; unwavering, in balance, and with a universally non-violent resolve.

ENDNOTES

[1] "Estimates show that between 1992 and 2014, produced capital per person doubled, and human capital per person increased by about 13% globally; but the stock of natural capital per person declined by nearly 40%...At their core, the problems we face today are no different from those our ancestors faced: how to find a balance between what humanity takes from Nature and what we leave behind for our descendants. While our ancestors were incapable of affecting the Earth system as a whole, we are doing just that." Estimates show that between 1992

and 2014, produced capital per person doubled, and human capital per person increased by about 13% globally; but the stock of natural capital per person declined by nearly 40%." From *The Economics of Biodiversity: The Dasgupta Review,* Headline Messages, February 2021, Crown copyright, www.gov.uk/official-documents, https://assets.publishing.service.gov.uk/government/uploads/system/uploads/attachment_data/file/957629/Dasgupta_Review_-_Headline_Messages.pdf, p. 4.

[2] See *Design for Ecological Democracy*, by Randolph T. Hester, The MIT Press, Cambridge, MA, 2006, pp. 264 and 325.

[3] op cit., Dasgupta, p. 2.

[4] "Species Numbers in Bacteria," by Daniel Dykhuizen, *Proc Calif Acad Sci.* Author manuscript; available in PMC 2011 Aug 24. Published in final edited form as: Proc Calif Acad Sci. 2005 Jun 3; 56(6 Suppl 1): 62–71, PMCID: PMC3160642, NIHMSID: NIHMS222630, PMID: 21874075, https://www.ncbi.nlm.nih.gov/pmc/articles/PMC3160642/

[5] *What are microorganisms?* Centre for Geobiology, University of Bergen, https://www.uib.no/en/geobio/56846/what-are-microorganisms#:~:text=Microorganisms%20can%20be%20bacteria%2C%20fungi,generally%20classified%20as%20non%2Dliving.

[6] See "Earth May Be Home to a Trillion Species of Microbes," by Nicholas Bakalar, May 23, 2016, *The New York Times*, https://www.nytimes.com/2016/05/24/science/one-trillion-microbes-on-earth.html

[7] op cit., Hester, p. 419.

[8] See *The Travelling Naturalists*, by Clare Lloyd, Croom Helm, London, 1985, p. 119.

[9] *Leaves Of Grass* – Selected And With An Introduction by Christopher Morley, Illustrations by Lewis C. Daniel, Doubleday, Doran & Co., Inc., New York, 1940, p. 299.

[10] See *The Esselen Indians of the Big Sur Country – The Land and the People*, by Gary S. Breschini and Trudy Haversat, Coyoto Press, Salinas CA, 2004, pp. 140-167.

[11] *The Esselen Hands*, by Peter Friedberg, Wayback Machine, Esalen Institute, https://web.archive.org/web/20111115115734/http://www.esalen.org/air/essays/esselenhands.html

[12] ibid., Breschini and Haversat, p. 173.

[13] ibid., p. 172.

[14] op cit., Friedberg.

[15] *Experimental Economics*, by Douglas D. Davis and Charles A. Holt, Princeton University Press, Princeton New Jersey, 1993, p. 524.

[16] *Moral Differences – Truth, Justice and Conscience in a World of Conflict*, by Richard W. Miller, Princeton University Press, 1992, p. 292.

[17] *Japan Journal 1855-1861*, By Henry Heusken, Translated and edited by Jeannette C. van der Corput and Robert A. Wilson, Rutgers University Press, Brunswick, New Jersey, 1964, cited on page 84 of *After Eden: History, Ecology & Conscience*, by M.C.Tobias, Avant Books, San Diego, CA 1985.

[18] *The Book of Tea*, by Kakuzo Okakura, Dover Publishers, New York, NY, 1964, p. 3.

[19] *Brave New World Revisited*, by Aldous Huxley, Harper & Brothers, 1958, New York, NY, p. 137.

[20] "Species-Level Strategies for Conserving Rare or Little-Known Species," in *Conservation of Rare or Little-Known Species – Biological, Social, and Economic Considerations*, Edited by Martin G; Raphael and Randy Molina, by Bruce G. Marcot and Curtis H. Flather, Foreword by Nancy Molina, Island Press, Washington, Covelo, London, 2007, Chapter 6, p. 135.

[21] *The Lost Ones*, by Samuel Beckett, Evergreen Press, New York, NY, 1972; originally published by Les Editions de Minuit in Paris in 1971 under the title *Le Depeupleur*, p. 60.

Chapter 29

Conscious Speciation

Every week new species are described. In 2018, at least 270 were discovered; at least 71 in 2019; 503 in 2020[1]. Most if not all of these new species are newly *found*, not *new*. Though our genus is over 2.8 million years old, our species is now dated to between 250,000 and 334,000 years ago, from finds at the Moroccan site of Jebel Irhoud[2], and South African Florisbad Skull[3]. We rank, as far as can be determined, among the newest species on earth. The apple maggot (*Rhagoletis pomonella*, or railroad worm) is even newer: approximately 150 years old. It is actually a fruit fly that shifted hosts from the hawthorn tree to the *Malus domestica*, an apple tree throughout the northeastern United States. The process is called a sympatric shift. British biologist Sir Edward Bagnall Poulton (1856-1943) first described this process whereby sister species can evolve from shared ancestral lineages within overlapping ranges. In bacteria, rapid horizontal gene migrations can easily account for such new species, requiring far fewer mutations prior to origination within distinct new niches. Homozygous-dominance in the copies of alleles specific to advantaged traits in a genotype are most likely to accelerate natural selection in their favor.

The occurrence of newly formed species may be determined by various distributions of alleles within a population genetics set of parameters. For example, sympatric speciation can readily occur at the genetic fringes of panmixial conditions that make for random mating within macro-sympatric geographics, overlapping ecological niches. There are other conditioning factors, as well. For example, the myriad non-spatial dimensions that can accelerate a new organism, such as diurnal or nocturnal characteristics altering the rate and scale of interdependencies. Or the degree to which habitat has become "patchy" – a gene diffusion coefficient given to a higher heteropatric speciation probability; or, in turn, species coexistence where gene flow may be expedited by assortative mating or some other altered force of allure and/or disaggregation. The speciated novelties have been noted in snails, toads, orca whales, East African cichlids, various leaf beetles, and a large variety of songbird species, among many others[4].

The cumulative weight of these new life forms, mostly hybrids, but also biosynthetic (often patented) creatures[5] presents no guidance with respect to humankind. The "bio-bricks" or biological standard units of the synthetic commercial world of biotechnology are manifestations of something beyond Frankenstein genes. In our opinion, they represent an unsurprisingly new height of pollution arrogance; the existential hubris of engineers playing god, with any number of self-serving attitudes, motives and goals.[6] The rush to secure biological patents – typically codified for 20 years – has included embryonic stem cells, thousands of genetically modified organisms, genes from rice, pigs, transgenic mice, and entire DNA sequences; perhaps more than 6,000 human gene patents (at least 20% of all 30 known human genes) and 50,000+ other genetic-related patents.[7] Originally, in the 1906 *Parke-Davis v. Mulford* case, the judge determined that human modified adrenaline was better than in its natural state and therefore its chemical additives patentable.[8]

So not only have we usurped the planet's thin biosphere and primary products of photosynthesis, but we have aggressively sought to proclaim ownership of the very genetic materials, biological pathways and mechanisms by which life has, until our presence, perpetuated and diversified herself safely and by her own rules.

The questions we had hoped to pose at the beginning of this chapter remain unasked: What can we, a new species, learn from other new species? How are they doing it? What wisdom have they gleaned from trial and error? Such efforts to profile comparable experiences between species has been ignored; dissembled, alienated from its cause and concept. The only questions we have asked of Others is what can be gotten from them, taken away by butchery or in the mob rule for natural resources, exploited as fully as possible, Bible in hand. Yet, it is there, in Job, Chapter 12, 7-8: "But ask now the beasts, and they shall teach thee; and the fowls of the air, and they shall tell thee:/Or speak to the earth, and it shall teach thee: and the fishes of the sea shall declare unto thee."[9]

We thus return empty handed to a biological first base. We neither understand the persistence of some genes over others – unrelated to reproductive barriers or not,[10] nor have we made sense of what is known as a "genomic island" with respect to chromosomal distributions and the multidisciplinary in situ and ex situ studies necessary to yield greater clarity on patterns of survival.[11]

However, at least one study, over a decade ago, has provided evidence that chromosomal evolution can occur in multiple geographical regions close to one another, forming viable reproductive clusters, an evolutionary process that is typically slow and steady. In this instance, that statistically projected time-frame has been entirely altered due to the behavioral movements of a co-existing species: humans living in isolated valleys from one another across the Portuguese archipelago of Madeira, where one 500-year-old species of mice has evolved into "at least six different species," as "some chromosomes have fused together" placing reproductive constraints upon an isolated founder group.[12] Instead of succumbing to a genetic cul de sac, the exact opposite has occurred, inducing rapid mammalian speciation.[13] The wild house mice are thriving, not as hybrids but distinct, viable species, a classic instance of what is characterized as "Robertsonian (Rb) fusion, a chromosome rearrangement involving centric fusion of two acro-(telo)centric chromosomes to form a single metacentric…"[14]

The wild and more domesticated house mice on Madeira (*Mus musculus domesticus* Rutty, and *M. sprelus* Lataste) have become international scientific celebrities in as much as they have rapidly learned how to embrace their locations, co-habitants, food options, and living quarters. In no other quadrant of the planet is "of mice and men" a more apt expression for the community.[15]

In 2006 a new subspecies of mouse was verified on Cyprus.[16] Globally, there are presently a known 42 species of *Mus*, a Tribe known as Murini.[17] Of the Family Cricetidae (Suborder Myomorpha and Superfamily Muroidea, comprising True Hamsters, Voles, Lemmings and New World Rats and Mice), it has been written that "Cricetidae constitutes one of the largest and most successful experiments of the major mammalian order Rodentia;" with "142 genera, 765 species, at least 1505 extant taxa." Approximately 3 species of Cricetidae per century have gone extinct since 1600, while 32 remain Endangered, 40 species Vulnerable.[18] This remarkably adept group of mammals, though originating throughout the Eocene between 33.9 and 56 million years ago, have been greatly facilitated by the human population explosion. It is worth citing at some length Thoreau's commentary on one individual in particular, whom he wrote of in *Walden* (1854): "It probably had never seen a man before; and it soon became quite familiar, and would run over my shoes and up my clothes…At length, as I leaned with my elbow on the bench one day, it ran up my clothes, and along my sleeve, and round and round the paper which held my dinner, while I kept the latter close, and dodged and played at bo-peep with it; and when at least I held still a piece of cheese between my thumb and finger, it came and nibbled it, sitting in my hand, and afterward cleaned its face and paws, like a fly, and walked away."[19]

The New York Society Library has published on "The Great Mice of Literature".[20] Certainly Beatrix Potter's *The Tale Of Two Bad Mice*[21] advances that pantheon of persuasive sentiment, as did Jan Brueghel the Elder's oil painting of a "Mouse, Rose, and Butterfly" in 1618 commissioned and collected by the Archbishop of Milan, Federico Borromeo.[22] Pliny the Elder, in his *The Natural History* found the mice of the Nile to be capable of presaging the future; marvels of creation, heraldic and all-encompassing.[23] John Constable painted his pet mouse, Jack, in an 1824 portrait, "A Mouse with a Piece of Cheese"; whilst Hieronymus Bosch conveyed a most remarkable image of a mouse and a human staring at one another in a horizontal cylindrical glass tube within the central panel of his "The Garden of Earthly Delights"(1490-1510). In her wonderful tribute to the mouse, Orna Owen points out some of the more incisive behavioral qualities of this prolific gang of sweethearts, including the fact of its ultrasonic singing repertoire,[24] and an abundance of other qualities as perceived by André Kertész, Picasso, Claes Oldenburg, Franz Marc, Gustav Klimt and Hokusai and Walt Disney. For Disney, his character of Mickey Mouse first appeared in 1928 and it is clear that but for Mickey there would never have been a Disneyland.[25]

The mouse, culturally enshrined in our own species, provides one "additional vantage point"[26] from which to gaze upon real-time evolution, and the genetics that make for rapid speciation on an island like Madeira, or in the animated brain that calls Hollywood home. But it is now believed that new species are more likely to emerge in areas that have lost species abundance; where richness has turned depauperate, particularly amongst birds.[27] Hence the terms "hotspots" and "coldspots," with their equivalencies in terms of the forces of speciation.

We know that those forces are activating two types of causes and consequences: drivers/stressors undermining existing genera and populations, fragmenting habitat, homogenizing biodiversity; and co-eval pressures pertaining to boom populations, like that of the Covid virus variants, ants, to a lessening extent, krill, and of *H. sapiens*. This means that very little, if anything is being gleaned at the matrix between these two exigent gene flow mechanisms. There is clear evidence of various adaptations occurring in novel (human-mediated) ecosystems. Mosquitos, for example, are attracted to certain odorants applied by humans with a disease vector consequence.[28/29] Within fifteen years following the construction of a dam in central Brazil, geckos nearby, *Gymnodactylus amarali*, grew larger heads in order to consume more of the wide-ranging insects that came as a result of the dam. But, as J.W. Bull and M. Maron have written, "While it has been shown that humans can drive contemporary evolution to a degree that is significantly higher than that from natural causes, estimates of speciation attributable to human activities do not exist for most organisms."[30] This is particularly the case among microbiota. What has been interpolated is that between human agriculture, hunting/poaching, killing in general, humanity has "outpaced" (perhaps by more than 300%) all other equivalent heritage traits existing on Earth prior to our presence. That is a conservative, but nonetheless startling realization.[31]

Despite a current renaissance in empirical natural histories and futures, most conservation strategies are aimed at proximate, human-determined outcomes. While zoosemiotic translation research is at the near beginning of its chronicling the natural world of interspecies communication data, as far as our active participation in the biosemiosphere, it remains incipient, if not disastrous, despite a myriad of good intentions.

That leaves only one practical ideology in the animal liberation and conservation biology arsenals: to continue doing exactly what humble and virtuous individuals have been doing, in spite of their species.

ENDNOTES

[1] University of Turku. "New species described in 2020." *ScienceDaily*. *ScienceDaily*, 1 July 2020. www.sciencedaily.com/releases/2020/07/200701100030.htm,

https://www.sciencedaily.com/releases/2020/07/200701100030.htm; Sources from: Zamani A. and Marusik Yu.M. A new and easternmost species of Loureedia (Aranei: Eresidae) from Iran. *Arthropoda Selecta*, 2020 doi: 10.15298/arthsel.29.2.09; Diego Galvão de Pádua, Ilari Eerikki Sääksjärvi, Ricardo Ferreira Monteiro, Marcio Luiz de Oliveira. Seven new species of spider-attacking Hymenoepimecis Viereck (Hymenoptera, Ichneumonidae, Pimplinae) from Ecuador, French Guiana, and Peru, with an identification key to the world species. ZooKeys, 2020; 935: 57 doi: 10.3897/zookeys.935.50492 ; Valter Weijola, Varpu Vahtera, André Koch, Andreas Schmitz, Fred Kraus. Taxonomy of Micronesian monitors (Reptilia: Squamata: Varanus): endemic status of new species argues for caution in pursuing eradication plans. *Royal Society Open Science*, 2020; 7 (5): 200092 doi: 10.1098/rsos.200092

[2] Hublin, JJ., Ben-Ncer, A., Bailey, S. et al. New fossils from Jebel Irhoud, Morocco and the pan-African origin of *Homo sapiens*. *Nature* 546, 289–292 (2017). https://doi.org/10.1038/nature22336, June 8 2017, https://www.nature.com/articles/ nature22336#citeas

[3] *J Anthropol Sci*, 2020 Dec 31;98. doi: 10.4436/JASS.98014. "The skull from Florisbad: a paleoneurological report," Emiliano Bruner, Marlize Lombard· *PMID:* 33130632, doi: 10.4436/JASS.98014, National Library of Medicine, *PubMe*d, https://pubmed.ncbi.nlm.nih.gov/33130632/

[4] Smith, J. Maynard (1966). "Sympatric speciation". *The American Naturalist.* 110 (916): 637–650. doi:10.1086/282457. See also, M. Malinsky et al. "Genomic islands of speciation separate cichlid ecomorphs in an East African crater lake," *Science* (2015). doi: 10.1126/science.aac9927

[5] See "Synthetic Life: Made from Scratch," by Niko McCarty, *synbiobeta*, November 22, 2019, https://synbiobeta.com/synthetic-life-made-from-scratch/; on the legality, see, "BRIA 23 4 c Patenting Life," Constitutional Rights Foundation, Winter 2008 (Volume 23, No. 4), https://www.crf-usa.org/bill-of-rights-in-action/bria-23-4-c-patenting-life; See also, "Intellectual Property in Higher Life Forms: The Patent System and Controversial Technologies," by Robert P. Merges, https://cyber.harvard.edu/IPCoop/88merg1.html. See also, "The Supreme Court Holds Genes Are Patent-Ineligible Products of Nature," by Teige P. Sheehan, NYSBA Bright Ideas, Fall 2013, Vol. 22, No. 2, New York State Bar Association, https://www.hrfmlaw.com/wp-content/uploads/2018/04/The-Supreme-Court-Holds-Genes-Are-Patent-Ineligible-Products-of-Nature-TPS-Fall-2013-H1209113xD16FB.pdf

[6] See *Mendel's Ark – Biotechnology and the Future of Extinction*, by Amy Lynn Fletcher, Springer New York, NY 2014, p. 96.

[7] "Gene Patents," by Robert Cook-Deegan, MD, The Hastings Center September 23, 2015, https://www.thehastingscenter.org/briefingbook/gene-patents/#:~:text=There%20are%203% 2C000%E2%80% 935%2C000%20U.S.,and%20genetic%20therapies%20unacceptably%20high. See also, "Gene Patenting," *AMA*, https://www.ama-assn.org/delivering-care/precision-medicine/gene-patenting#:~:text=Gene%20 Patenting%20background&text =The%20total%20number%20of%20human,the%20government%20and %20even%20individuals.

[8] Parke-Davis & Co. v. H.K. Mulford Co., 189 F. 95 (C.C.S.D.N.Y. 1911), *Learned Hand, J.* See also, Dutfield, Graham. "DNA Patenting: Implications for Public Health Research." *Bulletin of the World Health Organization*, May 2006, Volume 85, Issue 5. See also, Eisenberg, Rebecca S (December 2002). "Why the Gene Patenting Controversy Persists: Academic Medicine". *Academic Medicine*. 77 (12): 1381–1387.

[9]"Job," 11: 7-8, *Holy Bible, King James Version*, The Christian Science Publishing Society, Boston, MA, n.d. p. 709.

[10] See "Macroevolutionary speciation rates are decoupled from the evolution of intrinsic reproductive isolation in *Drosophila* and birds," by Daniel L. Rabosky and Daniel R. Matute, *PNAS*, September 17, 2013 110 (38) 15354-15359; https://doi.org/10.1073/pnas.1305529110, https://www.pnas.org/content/110/38/15354

[11] "Speciation: The Origin of New Species," by Rebecca J. Safran, and P. Nosil, *Knowledge Project, Nature Education Knowledge* 3(10): 17, 2012, https://www.nature.com/scitable/knowledge/library/speciation-the-origin-of-new-species-26230527/; See also, "Ecotypes and the controversy over states in the formation of new species," by David B. Lowry, Linnean Society, *Biological Journal*, 21 December 2011, https://davidbryantlowry.files. wordpress.com/2013/01/lowry_david_3.pdf

[12] "Are new species still evolving?" by Abbie Thomas, an interview with Michael Schwarz and Arthur Georges, *ABC Science*, March 10, 2010, https://www.abc.net.au/science/articles/ 2010/03/10/2820949.htm, Accessed February 9, 2021.

[13] See Britton-Davidian, J., Catalan, J., da Graça Ramalhinho, M. et al. Rapid chromosomal evolution in island mice. *Nature* 403, 158 (2000). https://doi.org/10.1038/35003116, https://www.nature.com/articles/35003116#rightslink

[14] See "Telomeres and mechanisms of Robertsonian fusion," by P. Slijepcevic, *Cromosoma*, 1998 May;107(2):136-40. doi: 10.1007/s004120050289. https://pubmed.ncbi.nlm.nih.gov/9601982/

[15] "On the origin and colonization of house mice in the Madeira Islands," Maria Da Luz Mathias and António Mira, *Biological Journal of the Linnean Society,* Volume 46, Issue 1-2, May 1992, Pages 13–14, https://doi.org/10.1111/j.1095-8312.1992.tb00846.x, 14 January 2008, https://academic.oup.com/biolinnean/article-abstract/46/1-2/13/2654404, Accessed February 9, 2021.

[16] T. Cucchi, A. Orth, J.-C. Auffray, S. Renaud, L. Fabre, J. Catalan, E. Hadjisterkotis, F. Bonhomme, J.-D. Vigne (23 June 2006). "A new endemic species of the subgenus Mus (Rodentia, Mammalia) on the Island of Cyprus". *Zootaxa*. Magnolia Press. 1241 (1).

[17] See *Handbook of the Mammals of the World*, 7, II, Rodents, Chief Editors, Don E. Wilson, Thomas E. Lacher, Jr., and Russell A. Mittermeier, Published by Lynx Edicions in association with Conservation International and IUCN, Barcelona, Spain, 2017, pp. 788-798.

[18] ibid., p. 204.

[19] See "Henry and the Mouse," by Corinne H. Smith, Thoreau Farm, October 8, 2015, https://thoreaufarm.org/2015/10/henry-and-the-mouse/Accessed February 9, 2021.

[20] https://www.nysoclib.org/collection/great-mice-literature

[21] Frederick Warne & Co., London, UK, September 1904.

[22] See *Nature and Its Symbols*, by Lucia Impelluso, Translated by Stephen Sartarelli, The J. Paul Getty Museum, Los Angeles, 2004, p. 231. See also, "Jan Brueghel Complete Catalogue," https://www.janbrueghel.net/object/mouse-rose-and-butterfly-paris

[23] See "Pliny the Elder, The Natural History," John Bostock and H.T. Riley, Chapter 84 (58.) –"The Mice of the Nile," http://www.perseus.tufts.edu/hopper/text?doc=Perseus%3Atext%3A1999.02.0137%3Abook%3D9%3Achapter%3D84.

[24] *Mouse Muse: The Mouse in Art*, by Lorna Owen, Monacelli Press, New York, NY, 2014.

[25] Holliss, Richard; Brian Sibley, *Walt Disney's Mickey Mouse: His Life and Times.* New York: Harper & Row, New York, NY, 1986, pp. 40–45. See also, Inge, M. Thomas (September 30, 2014). "Mickey Mouse". In Apgar, Garry (ed.). A Mickey Mouse Reader. University Press of Mississippi. pp. 341–342.

[26] See "Natural selection in action during speciation," by Sara Via, Proc Natl Acad Sci U S A. 2009 Jun 16; 106 (Suppl 1): 9939–9946. Published online 2009 Jun 15. doi: 10.1073/pnas.0901397106, PMCID: PMC2702801, PMID: 19528641, https://www.ncbi.nlm.nih.gov/pmc/articles/PMC2702801/

[27] "Diversity hotspots: Coldspots of speciation?" by Hélène Morlon, *Science* 11 Dec 2020:Vol. 370, Issue 6522, pp. 1268-1269, doi: 10.1126/science.abf0830, https://science.sciencemag.org/content/370/6522/1268

[28] See "Evolution of mosquito preference for humans linked to an odorant receptor," by Carolyn S. McBride, et al., *Nature*. Author manuscript; available in PMC 2015 May 13, Nature. 2014 Nov 13; 515(7526): 222–227. doi: 10.1038/nature13964, PMCID: PMC4286346, NIHMSID: NIHMS635307PMID: 25391959, https://www.ncbi.nlm.nih.gov/pmc/articles/PMC4286346/

[29] See *Geckos evolve rapidly in Brazil after new dam constructed*, by Katie Langin, Science, July 31, 2017, https://www.sciencemag.org/news/2017/07/geckos-evolve-rapidly-brazil-after-new-dam-constructed, Accessed February 9, 2021.

[30] See "How humans drive speciation as well as extinction," by J. W. Bull and M. Maron, The Royal Society, *Proceedings of the Royal Society B, Biological Sciences*, 29 June 2016, https://doi.org/10.1098/rspb.2016.0600, https://royalsocietypublishing.org/doi/full/10.1098/rspb.2016.0600

[31] See "Human predators outpace other agents of trait change in the wild," by Chris T. Darimont, Stephanie M. Carlson, Michael T. Kinnison, Paul C. Paquet, Thomas E. Reimchen, and Christopher C. Wilmers. PNAS January 20, 2009 106 (3) 952-954; https://doi.org/10.1073/pnas.0809235106, PNAS, *Proceedings of the National Academy of Sciences of the United States of America*. https://www.pnas.org/content/106/3/952. For an excellent commentary on these issues, see "Are Humans Creating More Species Than We're Killing?" by Emily Anthes, *The Atlantic*, October 11, 2016, https://amp.theatlantic.com/amp/article/503597/

Young Orangutan, Tanjung Puting National Park, Central Borneo, Indonesia. Photo © M. C. Tobias.

Chapter 30

CODA

Everything about the year 1802 in England now seems as pivotal as it was familiar. Absolutist George III was deep into his monarchial middle years while Western Europe watched as Napoleon systematically undermined the ideals that had propelled the French Revolution. Popular culture was mesmerized by the first wax museum, mention of an asteroid, photographic experiments, the Rosetta Stone and a record-setting balloon flight with two men in the basket. Charles Darwin's grandfather, poet/philosopher/botanist Erasmus Darwin died, and the great Scottish geologist Hugh Miller committed suicide, just after writing a poem, "Strange But True".[1] Whilst Jean-Baptiste Lamarck and Gottfried Reinhold Treviranus were proposing radically original overviews of the tree of life and evolution of species, in Paris Francois Antoine Rauch was warning the world against deforestation. Charles Bell published the first comprehensive "Anatomy of the Brain" with spectacular engravings, and in Exeter, George Bodley patented the first true kitchen stove.

It is the year *popular science* is said to have commenced, with as much public sentiment roused over fossils – the first dinosaur footprints found by a boy in Massachusetts - as over Richard Trevithick's newly invented "high pressure engine." For all that, it was the moment poet William Wordsworth (1770-1850) vacillated between his love of daffodils, his outspoken, cynical disdain for the increasing ethical and ecological deterioration in London and British culture in general – "London, 1802" – in which he summoned forth the genius of yesteryears, of Golden Ages, John Milton; and the time when he wrote his weariful assessment of life, in a 14-line despondent sonnet, "The World Is Too Much With Us": "Getting and spending we lay waste our powers;…We have given our hearts away, a sordid boon!... For this, for everything, we are out of tune;"[2]

Just two years prior, the last remaining Blue Buck Antelope (*Hippotragus leucophaeus*) had been shot to death by European settlers, making it the first known extinction on the African continent.

Ample metaphorical precursors of the Anthropocene were realized by 1802, their eco-psychological fall-out represented across poignant scientific, political and artistic disciplines. It was only a matter of time before we would be asking about the biochemistry of philanthropy, the physiological mechanisms involved in love, and the specific neurotransmitters signaling the performance we call the kindness to strangers; possible antidotes to our age, vague glimmers of hope from within ourselves.

By 1936 a Nobel Prize had been given to Otto Loewi and Henry Hallett Dale for their research into chemical neurotransmission, the so called "coexistence principle" suggesting that nerve pulses and each neuron contained multiple chemical, and some electrical synaptic signals.[3] In 1906 Dale had discovered the protein hormone/neuropeptide known as oxytocin. By the 1950s it was understood that it was involved in a strange and wondrous journey – seen in nearly all vertebrates – from the hypothalamus and posterior pituitary into the bloodstream during, among other things, childbirth in mammals, but also somehow key to socializing, bonding, sex and love.[4] In the spirit of that enquiry, Patricia S. Churchland recently discussed a variety of the brain's chemistry as an underlying template for what we think of as the expression of human conscience. Her research took her into the realm of "monogamous pair bonding" "and receptor density for specific neurochemicals," particularly the hormones "oxytocin and vasopressin," as seen in the differentiated densities of "receptors for oxytocin" in prairie versus montane voles, the former far more monogamous than the latter.[5] "…new knowledge of the properties of oxytocin has proven useful in explaining several enigmatic features of love," write Sue Carter and Stephen Porges.[6] Kindness itself remains the subject of no one discipline, and the circumference of them all:[7] from the anthropological insights gleaned from the very nature of gift-giving,[8] "food sharing" and "hospitality rituals"[9] to the deeply rooted psychology of *all* exchanges.[10]

Giving, in general – the ultimate proxy for kindness, empathy, tolerance, empathy and love – is the final precursor to that which Wordsworth found lacking in his age: ecological liberation. It is estimated that satisfaction of the 17 United Nations Sustainable Development Goals[11] requires between US$5-and-7 trillion of annual investment,[12] though such numeric translations are postpositivist calculations, unknown by any real experiential and microanalytical tools from geographical domain to culture.[13] But if even close, monetarily, then we can assume an easily-achieved rationale for the distribution of such dollars – all dollars being arbitrary within any puzzle of ecosystems harboring evolutionary needs and proactive measures that are all too clear at this late stage in the rapid deterioration of the biosphere. You just print money if human lives and those of all other species and habit are hanging in the balance. Let future generations deal with the debt, once the lives have been saved, in this generation, while there is still a chance to do so.

Just as we cannot rationally correlate any particular value with nature, we must not succumb to algorithms in place of common sense or ultra-compassion. It has been suggested that "high levels of uncertainty produce high rates of reciprocity."[14] But this notion also suggests that *giving* is predicated on some level of insecurity, desire for some axiomatic equity or re-assurance; whereas true giving, like true conscience, is unambiguous: there are no half-truths, no half-virtue, half-goodness philosophies of any logical traction. We do not seek to combine reciprocity with any hint of poetic or retributive justice; to give only by means of the promise of reward. That is not the pure license we are heralding herewith. Indeed, such transactional states of mind -give in order to obtain - have thoroughly impeded the premise, and staunched the flow of any unconditional interest and generosity. Broken by ties that bind without a free consensus, there has been no intergenerational rite of passage granting kindness at large; kindness that asks for and needs nothing other than its own selfless proliferation amongst all those with whom we are so honored and fortunate to share in this blessing of a life force.

It is this interspecies lack of requests that forms the basis for wild companionship and honest accounting of our deeds. A true prognosis for and reckoning with the future.

Endnotes

[1] Robert Farquharson Sharp, *A Dictionary of English Authors: Biographical and Bibliographical*, K. Paul, Trench, Trübner & Company, Limited, London, UK, 1904, p. 198.

[2] Tianyu Ma, "Boons, authority, and imagination: A reading of 'The World Is Too Much with Us'". *ANQ: A Quarterly Journal of Short Articles, Notes and Reviews*, 2017; doi:10.1080/0895769X.2017.1385377.

[3] R.M.E. Sabbatini, "Neurons and synapses. The history of its discovery. IV. Chemical transmission," *Brain & Mind*, 2004, https://cerebromente.org.br/n17/history/neurons5_i.htm

[4] See D.D. Chiras, *Human Biology* (7th ed.). Jones & Bartlett Learning, Sudbury, Massachusetts, 2012, p. 262.

[5] *Conscience – The Origins of Moral Intuition*, by Patricia S. Churchland, W.W. Norton & Company, Inc., New York, NY 2019, p. 46.

[6] "Biochemistry of Love," by Sue Carter and Stephen Porges, NOBA, https://nobaproject.com/modules/biochemistry-of-lovenew%20knowledge%20of%20the%20properties%20of%20oxytocin%20has%20proven%20useful%20in%20explaining%20several%20enigmatic%20features%20of%20love, Accessed February 11, 2021. See also, "The biochemistry of love: an oxytocin hypothesis," by Sue Carter and Stephen Porges, EMBO Rep, v.14(1); 2013 Jan, PMC3537144, Journal List.

[7] See "On the Biology of Kindness: is Kindness the Antidote of our Time?" by Nadia Waber, June 15, 2020, Association of Women in Business, http://awib.org.et/newsite/on-the-biology-of-kindness-is-kindness-the-antidote-of-our-time/Accessed February 11, *2021.* See also, Zeki, S (June 2007). "The neurobiology of love". *FEBS Lett.* 581: 2575–9. doi:10.1016/j.febslet.2007.03.094. PMID 17531984. See also, McGregor IS, Callaghan PD, Hunt GE (May 2008). "From ultrasocial to antisocial: a role for oxytocin in the acute reinforcing effects and long-term adverse consequences of drug use?". Br. J. Pharmacol. 154 (2): 358–368. doi:10.1038/bjp.2008.132. PMC 2442436. PMID 18475254.

[8] "Gift Giving in Anthropological Perspective," by John F. Sherry, Jr., Journal of Consumer Research, Volume 10, Issue 2, September 1, 1983, pp. 157-168, Oxford Academic, https://academic.oup.com/jcr/article-abstract/10/2/157/1801227?redirectedFrom=PDF

[9] Chris Hann, Hann, Chris, "The Gift and Reciprocity: Perspectives from Economic Anthropology," *Handbook on the Economics of Giving, Reciprocity and Altruism*, in: S. Kolm & Jean Mercier Ythier (ed.), 2006, edition 1, volume 1, chapter 4, pages 207-223, Elsevier, https://ideas.repec.org/h/eee/givchp/1-04.html

[10] *The Psychology of Gift Exchange*, by C. Mayet and Karen J. Pine, University of Hertfordshire, 2010, https://karenpine.com

[11] *Take Action for the Sustinable Development Goals*, https://www.un.org/sustainabledevelopment/ sustainable-development-goals/Accessed February 11, 2021.

[12] *Shaping Philanthropy*, by Josef Stadler, UBS Wealth Management, UBS Philanthropy, https://www.ubs.com/global/en/ubs-society/philanthropy/experiences/philanthropy-reports/shaping-philanthropy.html

[13] See, for example, Colin Robson, *Real World Research. A Resource for Social Scientists and Practitioner-Researchers* (Second ed.), p. 624, Malden: Blackwell, Malden MA, 2002.

[14] See Richard L. West and Lynn H. Turner, *Introducing communication theory: analysis and application*, pp.147–165, McGraw-Hill Publishers, New York, NY, 2014.

ABOUT THE AUTHORS/TRANSLATOR

AUTHORS

Michael Charles Tobias, long-time president of the Dancing Star Foundation (dancingstarfoundation.org), earned his PhD in the history of consciousness from the University of California-Santa Cruz, specializing in global ecological ethics and biological systems, as well as the interdisciplinary humanities. His wide-ranging work embraces the global ecological sciences, art, comparative literature, the history of ideas, philosophy, and natural history. Dr. Tobias is the author of over 60 books – fiction and non-fiction – and has written, directed, and/or produced over 50 films (mostly documentaries, docudramas, and TV series). Tobias' works have been read, translated, and broadcast throughout the world, and he has, in addition, conducted ecological field research in nearly 100 countries. Most recently, his book, *The Earth In Fragments: A Memoir* was published by Nova Science Publishers (2021).

Jane Gray Morrison is an ecologist whose work has taken her to over 40 countries. As a filmmaker, Ms. Morrison has produced numerous films for such networks as Discovery, PBS (where she co-directed "A Day in the Life of Ireland" for Irish Television and WNET/New York), "Hotspots" (www.hotspots-thefilm.com), and Turner Broadcasting, for which she served as senior producer for "Voice of the Planet," a 10-hour dramatic series based upon the history of life on Earth. As a Goodwill Ambassador to Ecuador's Yasuni National Park, she produced the short film "Yasuni – A Meditation on Life," which premiered at the United Nations Rio+20 Summit in 2012. Her books include *Sanctuary: Global Oases of Innocence* (www.sanctuary-thebook.com), *Donkey: The Mystique of Equus Asinus*, *God's Country: The New Zealand Factor*, and *No Vacancy* (co-edited). She has co-written seven books published by Springer Nature. Since 1999, Jane Morrison has served as the executive vice president of Dancing Star Foundation, a nonprofit organization that focuses on animal rights, the interdisciplinary humanities, and social and ecological justice movements as they concern humankind's relationship to the natural world.

TRANSLATOR

Niki Stavrou is the CEO of "Kazantzakis Publications" (est. 1964), the publishing house based in Athens, Greece, which publishes the works of the national author of Greece, Nikos Kazantzakis. She is also head of the Kazantzakis Estate, managing its public relations and foreign rights division, collaborating with publishing houses and literary agencies in Europe, Asia and North and South America, such as Simon & Schuster in the US and Faber & Faber in the UK. Her foreword is featured in the current edition of *Zorba the Greek*, published by Simon and Schuster. Niki organizes conferences and events globally delivering lectures on author Nikos Kazantzakis nationally and internationally. She has given numerous interviews on T.V. and radio shows, has participated in documentaries, and adapted and performed plays for the theater, presented in various foundations, schools, and the Athens Concert Hall. She was Assistant Professor of English Literature at Webster's University, Athens, and Senior Lecturer at the University of Indianapolis, Athens Campus.

Index of Names

A

B

C

I

J

K

L

M

N

O

Index of Terms

#

A

B

C

D

E

F

G

H

I

J

K

O

P

Q

R

S

T

U

V

W

Y

Z

Στα Ελληνικα

Εργο Οικολογικη Αμοιβαιοτητα

Μια Πραγματεια Σχετικα Με Την Καλοσυνη

Μαικλ Καρολος Τομπαιας
και
Τζειν Γκρει Μορρισον

(Μεταφραση Στα Ελληνικα
Νικη Σταυρου)

Milford Sound, Εθνικό Πάρκο Fiordland, Νότια Νέα Ζηλανδία. Φωτογραφία © M. C. Tobias.

Ευχαριστιες

Οι συγγραφείς ευχαριστούν την Νάντια Κολούμπους, Προέδρου του Nova Science Publishers, για τη φιλία και την πίστη της σε αυτό το έργο. Είμαστε απίστευτα ευγνώμονες στην Νίκη Σταύρου για τις εξαιρετικές προσπάθειές της στη μετάφραση αυτού του κειμένου στα ελληνικά. Και είμαστε επίσης εξαιρετικά ευγνώμονες στον Δημήτρη Μουτάφη για τη διόρθωση, επιμέλεια και την ενσωμάτωση της επιστημονικής ορολογίας στα ελληνικά.

Michael Charles Tobias
(*Ο Μάικλ Κάρολος Τομπάιας*)
Jane Gray Morrison
(*Η Τζέιν Γκρέι Μόρρισον*)

Εισαγωγη

Μια από τις πρώτες ηθικές προσταγές που μαθαίνουμε και επαναλαμβάνεται σε όλη την Παλαιά και την Καινή Διαθήκη, είναι η ρήση "Ἀγάπα τόν πλησίον σου ὡς σεαυτόν" (Λευιτικόν, 19:18). Πρόκειται για ένα μήνυμα το οποίο μας εφιστά την προσοχή για κάτι το οποίο, στην πραγματικότητα, αποτελεί βιολογικό κανόνα: την επίδειξη εκ μέρους μας συντροφικότητας και καλοσύνης χωρίς καμία προσδοκία ή ελπίδα για οποιαδήποτε ανταμοιβή. Έχει διατυπωθεί και επαναδιατυπωθεί ως ένας από τους πολλούς καθολικούς Χρυσούς Κανόνες που περιέχονται στον Όρκο του Ιπποκράτη, καθώς και σε αυτό που έχουν παρατηρήσει οι βιολόγοι να διαδραματίζεται στο μεγαλύτερο μέρος του φυτικού βασιλείου, στις σχέσεις δηλαδή οι οποίες μπορούν να χαρακτηριστούν αμοιβαίες. Παρόμοιες σχέσεις έχουν μελετηθεί σχεδόν σε όλους τους τομείς της ζωής, μεταξύ άλλων στα είδη των αμφίποδων (*Podocerus jinbe*) που ζουν στο στόμα των φαλαινοκαρχαρίων, στις ταραντούλες και τους βατράχους, στους βραδύποδες και τα φύκη, σε ορισμένα έντομα (προνύμφες χρυζόπης) και τις λειχήνες, στη μαγκούστα και τον βούκερο, στους γελαδάρηδες ερωδιούς και τους πολυάριθμους ξενιστές τους, από βοοειδή έως ιπποπόταμους, στα φυλλοκόπα μυρμήγκια και τα φύλλα, καθώς και σε όλους τους επικονιαστές των φυτών, από τις μύγες έως το κολιμπρί.

Πράγματι, ολόκληρος ο βιολογικός κόσμος λειτουργεί σε αρμονική συμφωνία.

Σπάνια αναλύουμε την αγάπη για τη φύση με τον ίδιο τρόπο που αναλύουμε την αγάπη μας για ένα συγκεκριμένο ποντίκι ή τα βακτήρια με τα οποία είναι γεμάτο το σώμα μας, για τα οποία μάλιστα αποτελούμε τους πλέον γενναιόδωρους ξενιστές. Στην πραγματικότητα, ωστόσο, η φυσική επιλογή είναι εκείνη που επιλέγει τι θα αγαπηθεί, και μάλιστα όχι μόνο από το είδος μας. Η επιλογή είναι μια διαδικασία μυστικιστική και στοργική, και ταυτόχρονα αγνή, πρακτική και ιδεαλιστική. Δεν πρόκειται ωστόσο για μια βίαια, εκδικητική ή προμελετημένη διαδικασία. Και, παρόλο που μπορεί να θεωρούμε ότι εμείς οι ίδιοι μπορεί να είμαστε κάπως αποκομμένοι από τα επίμονα σχέδια της εξέλιξης, η ίδια η τάση μας να επιδεικνύουμε ενσυναίσθηση πρέπει να μας υπενθυμίζει ότι, όπως κάθε άλλος οργανισμός, υπηρετούμε έναν ανώτερο σκοπό ο οποίος βρίσκεται, γενικά, σε αρμονία με όλα τα άλλα είδη. Όσες υπερβάσεις και αν κάνουμε σε επίπεδο ηθικής, όσα εμπόδια και αν ξεπερνάμε, ο κόσμος το παρατηρεί. Παρατηρούμε λοιπόν μια δυναμική η οποία ενεργοποιεί μια αλλαγή όσον αφορά τη στάση μας σε θέματα οικολογίας, είτε μέσω της ίδιας της φιλοκοινωνικής συμπεριφοράς με την οποία διεξάγουμε βασικές φιλικές συναλλαγές, είτε μέσω του πολλαπλασιασμού των χειραψιών, όπως περιγράφεται από τους αναπτυξιακούς ψυχολόγους, είτε μέσω των άνω των 560 πολυμερών συνθηκών (εκ των οποίων οι 500 αφορούν το περιβάλλον), πλέον των άνω των 3.716 άλλων περιβαλλοντικών συμβάσεων που βρίσκονται εν ισχύ στα διάφορα έθνη.

Παρά το γεγονός ότι οι συμβάσεις αυτές για την προστασία της βιοποικιλότητας ενδέχεται να είναι εξαιρετικά ημιτελείς, οι ίδιες αντιπροσωπεύουν την ανάπτυξη προς ένα ιδανικό το οποίο υπήρξε κοινό για όλη την ανθρωπότητα στο μεγαλύτερο μέρος των 330.000 ετών της αρχαιότητας.

Και ενώ τα αριθμητικά στοιχεία που διαθέτουμε για την εποχή που ονομάζουμε Ανθρωπόκαινο – την έκτη περίοδο μαζικής εξάλειψης των ειδών όπως αυτές έχουν καταγραφεί σε ολόκληρη την ιστορία της ζωής στη γη, με έναν καταιγισμό οικολογικών δεινών τα οποία επιφέρει αποκλειστικά το δικό μας είδος – είναι δυσοίωνα, παρατηρείται επίσης ένα νέο σχήμα και μια νέα ψυχολογία που καθοδηγεί την ύστατη προσπάθειά μας να αναστήσουμε την περιβαλλοντική υγεία και μια θεμελιώδους σημασίας γενναιοδωρία η οποία υποκινείται από την αγάπη μας για τους συγκατοίκους μας στον πλανήτη Γη.

Ναι, έχουμε καταστρέψει περίπου το 60% του συνόλου της άγριας ζωής, έχοντας καταδυναστεύσει με τον εγωισμό μας σχεδόν το ήμισυ του χερσαίου πλανήτη και ρυπαίνοντας με οποιονδήποτε τρόπο το άλλο ήμισυ, κυρίως το θαλάσσιο περιβάλλον.

Ωστόσο, με την θέσπιση σχεδόν 240.000 προστατευόμενων περιοχών και την κινητοποίηση συνασπισμών σε όλη τη γη με σκοπό τη διάσωση όσων έχουν απομείνει από τα πολύτιμα βιολογικά "κομβικά σημεία" (τις περιοχές εκείνες που φιλοξενούν τα πλέον άφθονα ενδημικά είδη και τη γενετική ποικιλομορφία), οι πρόσφατα διακηρυχθέντες στόχοι μας ενέχουν ένα πλήρως αναζωογονημένο αίσθημα συμπόνιας. Ένα κοινό όνειρο, το οποίο βασίζεται στον πόθο εκατοντάδων εκατομμυρίων ευφυών και εμπνευσμένων ανθρώπων να δουν την ανθρώπινη πείνα, τη φτώχεια, τον ρατσισμό, την καταπίεση των φύλων και την απάνθρωπη κατάχρηση αγαθών και υπηρεσιών να εξαλείφονται σε αυτήν και την επόμενη γενιά, όσο προλαβαίνουμε πριν την ολοκληρωτική κατάρρευση της βιόσφαιρας.

Το έργο *Οικολογική αμοιβαιότητα: Μια πραγματεία σχετικά με την καλοσύνη* εξετάζει συνοπτικά πληθώρα στοιχείων σχετικά με αυτό που στη θεωρία της εξέλιξης τείνουμε να αποκαλούμε αληθινή συνείδηση, η οποία εκδηλώνεται στους κανόνες συμπεριφοράς που έχουν γαλουχήσει και σταθεροποιήσει τις βιολογικές συνδέσεις εδώ και εκατοντάδες εκατομμύρια χρόνια. Οι Συγγραφείς συνεργάστηκαν με την κυρία Νίκη Σταύρου στην Αθήνα, η οποία μετέφρασε αυτό το βιβλίο στα ελληνικά για την πρώτη δίγλωσση έκδοση των εκδόσεων Nova Science Publishers, για έναν προφανέστατο λόγο. Στην Ελλάδα εκδηλώθηκαν για πρώτη φορά οι πρώτες παρορμήσεις της Δύσης προς τη φυσιολατρία, την αγάπη για τη φύση, με τόμους γραπτού περιεχομένου σχετικά με την οικολογική φιλοσοφία και τα δικαιώματα των ζώων, και με δραματική μορφή. Στα έργα του Θαλή, του Αισώπου, του Ηράκλειτου, του Πυθαγόρα και του Παρμενίδη, του Αριστοτέλη, του Πλωτίνου, του Πλούταρχου και του Πορφυρίου. Από τους μοναχούς του Αγίου Όρους μέχρι τον μεγάλο Κρητικό φιλόσοφο και συγγραφέα του 20ου αιώνα, Νίκο Καζαντζάκη, η ελληνική ποίηση αντηχεί τα χρονικά της κοσμολογίας και της βιολογίας με μια πνευματική αναζήτηση που είναι σχεδόν απαράμιλλη στην παγκόσμια λογοτεχνία. Η κυρία Σταύρου είναι βαφτισιμιά της Ελένης Καζαντζάκη.

Η *οικολογική αμοιβαιότητα* επικεντρώνεται πρωτίστως στην ίδια τη μεταφυσική που ενυπάρχει στην ατομική και συλλογική μας βούληση να προστατεύσουμε τον κόσμο. Ο λόγος που το κάνουμε παραμένει συνήθως άγνωστος και μυστηριώδης, έως ότου διαπιστώσουμε ότι όλοι οι βιολογικοί οργανισμοί παροτρύνονται από το ίδιο κίνητρο. Πόσοι άραγε να είναι εκείνοι οι οποίοι έχουν ασχοληθεί επί μακρόν και έντονα με τη διάσωση μιας μύγας, της *Musca domestica*, η οποία ζυγίζει περίπου 12 χιλιοστογραμμάρια; Με τη εξαιρετική της ικανότητα όρασης, την απαράμιλλη αεροδυναμική της και το πάθος της για ζωή και επιβίωση, έστω και για 2-4 εβδομάδες σε ανθρώπινο χρόνο, κάθε μύγα είναι πολύτιμη. Προσπάθησε να φτιάξεις μία! Μια μύγα, ή ένα ποντίκι, ή μια αράχνη, ή μια φάλαινα, ή ένα τριαντάφυλλο. Τα αγαπάμε όλα και αυτή η ανησυχία και αφοσίωση δεν αποτελούν παρά τον θεμελιώδη κανόνα της ζωής, εκείνες τις δυνάμεις της ζωής των οποίων η αφοσίωση δεν είναι διαιρεμένη, όχι σε αυτόν τον πλανήτη. Πράγματι, είναι η ίδια η λατρεία και η λαχτάρα μας για σύνδεση με την άγρια φύση που εγγυάται ότι *ο Homo sapiens* είτε θα επιβιώσει ως είδος, είτε θα εξαφανιστεί με τον συντομότερο επικήδειο από οποιοδήποτε γνωστό μεγάλο σπονδυλωτό στη γη.

Στο έργο αυτό αναφέρονται πολυάριθμοι ποιητές, επιστήμονες, μηχανικοί, μαθηματικοί, καλλιτέχνες και φιλόσοφοι. Όλοι τους έχουν αναλύσει διάφορα μονοπάτια και οδούς προς μια αρμονική ένωση φυσικού θετικισμού και εναγκαλισμού. Ο Ιμμάνουελ Καντ, ο Φίλιπ Γκίλμπερτ Χάμερτον, ο Σάμουελ φον Πούντερντορφ, ο Τζόρτζ Όργουελ, ο μεγάλος Κουάκερος ζωγράφος Έντουαρντ Χικς με τις περίπου 62 διαφορετικές εκδόσεις του "Ειρηνικού Βασιλείου" του, και η ιδιοφυΐα της Αναγέννησης στη Βόρεια Ευρώπη, ο Άλμπρεχτ Ντύρερ, του οποίου ο πρωταγωνιστής, ο Ρωμαίος στρατηγός Πλάκιδος θα μαρτυρήσει και θα αγιοποιηθεί ως Άγιος Ευστάθιος, ο κυνηγός που έγινε λάτρης του πανίσχυρου ελαφιού, ανάμεσα από τα κέρατα του οποίου ο Πλάκιδος είδε σε όραμα τον Χριστό ο οποίος του μίλησε για την αγάπη. Η αγάπη με την οποία ο Βούδας αναζήτησε την Ευγενή Οκταπλή Ατραπό του, τον βιολογισμό δηλαδή για την εύρεση ενός τόπου στον οποίο θα τερματίζονταν όλα τα δεινά.

Παρόμοια με τις παιδικές αναμνήσεις του Αλμπέρ Καμύ στις οποίες περιέγραψε πώς η γιαγιά του σκότωνε μια κότα, τις ηθικολογικές αποκαλύψεις του διάσημου ορνιθολόγου Αλεξάντερ Σκουτς, ή τις βοτανολογικές παρατηρήσεις του βραβευμένου με Νόμπελ έργου του Μωρίς Μαίτερλινκ "Η Νοημοσύνη των Φυτών", το έργο αυτό εξετάζει τη συμπεριφορά των ζώων και των φυτών μέσα από το φακό της αυτόχθονης κρίσιμης μάζας εκείνης η οποία θα μας βοηθήσει να κατανοήσουμε τα νευρικά και βιοπολιτισμικά όρια που μπορούν να διευκολύνουν τις συνενώσεις μας με τον φυσικό κόσμο. Αυτός είναι ο μοναδικός μας στόχος. Να βρούμε με ποιον τρόπο μπορούμε να μετατρέψουμε την ασυδοσία της βίας μας σε μια εντελώς διαφορετική πολυμορφική νοοτροπία η οποία να είναι ευθυγραμμισμένη με την τρυφερότητα και τη γενναιοδωρία προς όλους. Πρόκειται για μια τάση η οποία ταιριάζει απόλυτα στην ηθική της εποχής μας, και για την οποία θα πρέπει τελικά να επιστρατεύσουμε δυνάμεις από όλες τις επιστήμες, ώστε να περιγράψουμε ουσιαστικά τη Φύση και να τη σώσουμε.

Τα αμέτρητα μεταβατικά στάδια τα οποία θα περάσουμε μέχρι να ξεπεράσουμε εντελώς το στάδιο της οικολογικής διαλεκτικής περί ανθρώπινης συμπεριφοράς θα οδηγήσουν αναγκαστικά -αν παραμείνουμε σταθεροί- σε διακηρύξεις όπως εκείνη ενός πρόσφατου γάλλου υπουργού Γεωργίας ο οποίος ζήτησε τη διακοπή το 2021 όλων των θανάτων αρσενικών νεοσσών σε ολόκληρη τη Γαλλία (το άλεσμα στην κυριολεξία ζωντανών πτηνών σε μηχανές τεμαχισμού επειδή οι αρσενικοί νεοσσοί είναι ανεπιθύμητοι). Ακόμη και η περιγραφή του γεγονότος αυτού προκαλεί σε αυτή τη γενιά μια φρίκη η οποία μόλις πριν από λίγες γενιές δεν αποτελούσε παρά έναν λεκέ στον φακό του είδους μας. Βελτιωνόμαστε ως δυνητικά ενσυναίσθητο είδος, και οι αλλαγές αυτές θα πρέπει να επιταχυνθούν αν θέλουμε να αποτρέψουμε και να βελτιώσουμε αποτελεσματικά την αχαλίνωτη καταστροφή η οποία αποτελεί τη σημερινή οικολογική μας κληρονομιά.

Για να διασφαλίσουμε τη μεγάλη αυτή κλιμάκωση της καλοσύνης, έχουμε μεγαλύτερη υποχρέωση από ποτέ -σε συναισθηματικό, ψυχολογικό, πνευματικό επίπεδο- σε όλα αυτά τα μεγαοικοσυστήματα τα οποία οι προκάτοχοί μας δεν δίστασαν καν να καταδυναστεύσουν με τα μεγάλα γεωπολιτικά τους παιχνίδια για την απόκτηση και εποπτεία εδαφών. Μια άγρια μάχη εγωισμών και εθνικοτήτων που αναζητούσαν πόρους οι οποίοι δεν ήταν καν φυσικοί. Όπως και οι άλλοι αυτόχθονες άνθρωποι καταπιέστηκαν ή/και σφαγιάστηκαν, έτσι και το ένα οικοσύστημα μετά το άλλο μας άφησαν σε αυτή τη γενιά μια ευκαιρία διαφορετική από οποιαδήποτε άλλη στην ιστορία μας: την ευκαιρία να επιδείξουμε την απόλυτη άνευ όρων αγάπη. Ο Δαρβίνος περιέγραψε την εξέλιξη ως απόκτηση ενός "γενικού σχεδίου" και όντως αυτό ήταν. Ο μεγάλος σοφός του Τζαϊνισμού, Ουμασβάτι (2^{ος} αιώνας έως 5^{ος} αιώνας Κ.Ε.) έγραψε στο διάσημο έργο του *Tattvartha Sutra* ότι "οι ψυχές υπηρετούν η μία την άλλη", αντηχώντας τον 24^{ο} σοφό του Τζαϊνισμού, Μαχαβίρα, ο οποίος μίλησε για τη βιολογική αλληλεξάρτηση, ή αλλιώς "Parasparopagraho Jīvānām".

Η μετάβαση προς την αρετή της ηθικής συνδυάζεται με μια απαραίτητη νευρωνική πλαστικότητα η οποία ευνόησε τα ιδανικά έναντι των περιστασιακών ιδεών, υπερασπίστηκε την αλήθεια για όλα τα είδη, όχι μόνο για μερικά, και δημιούργησε τις προϋποθέσεις για μια βιολογική κλιμάκωση που δεν περιλαμβάνει τίποτα λιγότερο από *μια νέα φύση*. Μια μαρτυρία για τα αίτια και τις συνέπειες η οποία συνεκτίμησε όλα τα ηθικά μεγέθη και αναγνώρισε ότι στην εθιμοτυπία, την ευγένεια και τη συμβίωση βρίσκει κανείς την επιμονή, την επιβίωση και τη χάρη. Δεν πρόκειται για κάποια φιλοσοφία αλτρουιστικής αποκλειστικότητας ή καθαρών κερδών. Αντίθετα, μιλάμε για μια αιτία δράσης που εξαρτάται αμοιβαία από μια άλλη αιτία: το να βοηθάμε τους άλλους, να μοιραζόμαστε, να δίνουμε, χωρίς όρους.

Οι αιτίες αυτές, τις οποίες αποκαλούμε "συζεύξεις", αποτελούν την ουσία της ενσυναίσθησης. Ανάμεσα σε κάθε δύο, υπάρχει ένα, και αυτό το ένα αποτελεί το εξελικτικό αποτέλεσμα κοινών προσπαθειών που σημαίνουν κάτι και είναι ζωτικής σημασίας για τα 4 δισεκατομμύρια και πλέον χρόνια ζωής. Το άθροισμα όλων αυτών των κοινών προσπαθειών περιγράφει μια κβαντική οικοδυναμική που κρύβεται πίσω από το λεγόμενο "δυναμικό ετοιμότητας" το οποίο έχουν εδώ και καιρό περιγράψει οι νευροφυσιολόγοι. Ένα σωρευτικό συστατικό στοιχείο του απέραντου εργαστηρίου του ανθρώπινου πνεύματος, δηλαδή, της ικανότητάς μας να συγχωρούμε, να αγαπάμε, να καθοδηγούμε άλλους με την αγάπη μας. Είναι το ίδιο ηθική όσο και επιστημονική, και αυτή είναι η πραγματικότητα του 21ου αιώνα. Μια μη βίαιη απόφαση που δεν μπορεί απλά να ποσοτικοποιηθεί στα βιοχημικά χρονικά της χιλιετούς έρευνας, επειδή υπάρχει μια πεμπτουσία πνευματικότητας μέσα της η οποία ενισχύεται από τις γνώσεις κάθε παρελθόντος και ζωντανού πολιτισμού.

Καθώς το βάρος της ανθρώπινης επίδρασης καθίσταται ολοένα και πιο εμφανές, γνωρίζουμε πλέον ότι η γενιά αυτή έχει το καθήκον να τα κάνει όλα σωστά. Να βρούμε την ισορροπία που θα μας επιτρέψει να συνεχίσουμε ως είδος, αλλά με τις επιφοιτήσεις και την πνευματική ανάταση όλων εκείνων των ατόμων που αντιλαμβάνονται εύκολα αυτό που διακυβεύεται στην παρούσα φάση: μια συντροφική σχέση μεταξύ των ανθρώπων και όλων των άλλων, η οποία αποτελεί την πραγματική βάση μιας ρεαλιστικής πρόγνωσης και εκτίμησης των καταστάσεων του μέλλοντος. Ενός μέλλοντος καλοσύνης.

Κεφάλαιο 1

Η ΒΑΣΗ ΤΗΣ ΦΥΣΙΟΛΑΤΡΙΑΣ

Αναζητήστε εκατοντάδες εκατομμύρια παραπομπές, άρθρα, περιοδικά και βιβλία[1] σε όλες τις βιβλιοθήκες και βάσεις δεδομένων, και αφιερώστε το μεγαλύτερο μέρος της ζωής σας εξερευνώντας όλες τις ηπείρους. Επισκεφθείτε πολλά και διάφορα μουσεία τέχνης, φυσικής ιστορίας και επιστήμης. Προσπαθήστε να μετρήσετε και να φανταστείτε μέσα σας όσο το δυνατόν περισσότερα δέντρα και λουλούδια, καθώς και όλες τις άλλες μορφές ζωής που μπορεί να υπάρχουν. Ή, τουλάχιστον, μελετήστε έναν τεράστιο αριθμό αντιπροσωπευτικών από κάθε είδος οργανισμών από όλο τον πλανήτη. Σας αναφέρουμε ενδεικτικά ότι το έχουν ήδη κάνει ο Μαχαβίρα, ο Πλίνιος ο Πρεσβύτερος, ο Αριστοτέλης, ο Λουκρήτιος, ο Γκέσνερ, ο Λινναίος, ο Μπουφόν, ο Χούμπολτ, ο Οντιμπόν, ο Γουάλας και ο Δαρβίνος, οι οποίοι συνδύασαν όλα τα σχετικά στατιστικά στοιχεία και διατύπωσαν με αυτόν τον τρόπο κάποια μέθοδο, άποψη, σχέδιο ή πεποίθηση. Συγκεντρώστε όλα αυτά τα δεδομένα και τις δικές σας σχετικές αντιλήψεις σε ένα αποθετήριο στο οποίο να μπορείτε να προσθέτετε νέα επίκαιρα στοιχεία και πληροφορίες. Ο αείμνηστος φιλόσοφος και ιστορικός της επιστήμης, Τόμας Κουν, εξέτασε σε βάθος τα δομικά εμπόδια και τα πλεονεκτήματα της συμμετοχής σε κάτι εφάμιλλο και κατέληξε στο συμπέρασμα (μεταξύ άλλων) ότι τελικά οι επιστήμονες “μπορούν να βλέπουν τη φύση με διαφορετικό τρόπο”, ωστόσο τέτοιου είδους καινοτομίες ενδέχεται να έχουν και δυσάρεστες επιπτώσεις, όπως η πιθανότητα να “προκύπτουν νέα παραδείγματα με καταστροφικές αλλαγές στις πεποιθήσεις σχετικά με τη φύση”. [2]

Στην περιγραφή τόσο των ιστορικών γεγονότων, όσο και της τρέχουσας κατάστασης στην επιστήμης της οικοδυναμικής, παρατηρείται διαρκώς η χρήση προβληματικών σημασιολογικών προσδιορισμών όσον αφορά τη ζωή των μεμονωμένων όντων και ειδών και την τεράστια συλλογή των ευάλωτων παγκόσμιων οικοτόπων που αποκαλούμε βιόσφαιρα. Το πρόβλημα αυτό σχετίζεται με την αντίληψή μας σχετικά με τον αριθμό και τις ιδιότητες των όντων με τα οποία συμβιώνουμε στη βιόσφαιρα. Δίνουμε σημασία περισσότερο στο ίδιο “το παιδί” παρά στην πολυποικιλότητά του. Λέγεται μάλιστα ότι ο

Μάο Τσετούνγκ είχε κάποτε δηλώσει σχετικά με αυτό ότι κάθε παιδί είναι πολύτιμο, όμως τα πάρα πολλά παιδιά ενδέχεται να αποτελέσουν πρόβλημα. Η καθημερινή ειδησεογραφία ενισχύει την άβολη αλήθεια της ικανότητάς μας να συντονιζόμαστε και με την ίδια ευκολία να αποσυντονιζόμαστε από τις τόσο έντονα δυσάρεστες αιτίες και τις συνέπειές τους, στα ψυχικά και φυσικά τοπία στα οποία όλοι συγκατοικούμε.

Σε ορισμένες περιπτώσεις, ο μόνος τρόπος να συντονιστεί κάποιος πραγματικά με κάτι είναι να αποσυντονιστεί από αυτό. Τόσο ο Άγιος Αυγουστίνος όσο και ο Τζον Μιούιρ υποστήριξαν, σε σχέση με τις βιβλιοθήκες της φύσης, ότι άμεση προσωπική εμπειρία σημαίνει πλέον τόσο την εξωτερική αλληλεπίδραση όσο και την εσωτερική αξιολόγηση. Από τις πρώτες μέρες που βρέθηκε στην οροσειρά της Σιέρρα Νεβάδα, στα τέλη της δεκαετίας του 1860, και μέχρι τον θάνατό του το 1914, ο Μιούιρ εξελίχθηκε από βοσκό προβάτων και εργάτη μύλου σε μυστικιστή, ποιητή, ορειβάτη/εξερευνητή, βοτανολόγο, δοκιμιογράφο, πολιτικό ακτιβιστή, ιδρυτή του Sierra Club, καθώς και σύζυγο και πατέρα.[3] Βέβαια, όπως δείχνει η ιστορία του, είχε και τις αδυναμίες του. Κατά τη διάρκεια της ζωής του έγινε μάρτυρας γεγονότων καθοριστικής σημασίας για την υλοποίηση στην πράξη μιας διαλεκτικής καίριας σημασίας: από τη μια της καταστροφής των οικοσυστημάτων από την ανθρώπινη απληστία, τον θυμό, το "εγώ" και την αδιαφορία, και από την άλλη τη δημιουργία δεσμών και συμμαχιών από καλούς ανθρώπους, που συμφώνησαν όλοι να ενωθούν μέσα από πράξεις ηθικής εξυγίανσης και αυστηρής νομοθεσίας για την προστασία του κόσμου που μας περιβάλλει. Στα μέσα του 19ου αιώνα, τα προστατευόμενα μέσω της πολιτικής προστασίας οικολογικά πάρκα και δρυμοί σε όλη τη γη ήταν λιγότερα από δώδεκα. Στις μέρες μας είναι σχεδόν 250.000[4].

Η προστασία της φύσης στα προστατευόμενα μέρη με αυτόν τον τρόπο αντικατοπτρίζει ένα αρχέγονο σύνολο κωδίκων που έχουν καταγραφεί κατά διάρκεια της εξέλιξης όλων των ειδών στο γονιδίωμά τους. Από πλευράς γενετικής, πολυάριθμες μελέτες συμπίπτουν στο συμπέρασμα ότι οι πολυεπίπεδες αντιδράσεις όλων των μεμονωμένων ατόμων, πληθυσμών και ειδών διασταυρώνονται συνεχώς, καθώς εναλλάσσονται γύρω από ένα σύνολο φαινομενικά σταθερών αξιωμάτων, το οποίο οι άνθρωποι διαμόρφωσαν σε αυτό που ενδεχομένως αποτελεί το παλαιότερο, παγκοσμίως καθιερωμένο σύμβολο πίστεως, και συγκεκριμένα, έναν θεμελιώδη μονοδιάστατο Χρυσό Κανόνα[5]. Φαίνεται πως όλες ανεξαιρέτως οι πνευματικές παραδόσεις διαθέτουν ενσωματωμένη κάποια εκδοχή της ρήσης "Ἀγάπα τόν πλησίον σου ὡς σεαυτόν" (Λευιτικόν, 19:18). Επαναλαμβάνεται στο πρώτο βιβλίο της Καινής Διαθήκης, στο Ευαγγέλιο του Ματθαίου (22: 35-40), καθώς και στα Ευαγγέλια του Μάρκου και του Λουκά, ενώ και ο Χριστός λέγεται ότι θεωρούσε αυτή τη δήλωση το πλέον θεμελιώδες μήνυμα του Θεού: την αμοιβαιότητα με πρόθεση, χωρίς διακρίσεις, προσδοκίες ή ίδιον συμφέρον. Αντιθέτως, συζητάμε για έναν πανθεϊσμό ο οποίος, χωρίς ουδόλως αυτό να αποτελεί έκπληξη, εξελίχθηκε στο κεντρικό χαρακτηριστικό γνώρισμα τόσο της οικολογίας όσο και της ψυχολογίας. Η επίδειξη καλοσύνης προς τους ξένους, και αντίστοιχα από εκείνους προς εμάς, προκαλεί ταυτόχρονα τον κυνισμό και το δέος· τη δυσπιστία που διέπει όλα τα παραδείγματα της "επιβίωσης του ισχυρότερου", αποσαφηνίζοντας με πικρή αλλά ελπιδοφόρα συνειδητοποίηση την καθόλου αμερόληπτη λογική μας σχετικά με το Εθνικό Πάρκο Σερενγκέτι (για παράδειγμα, ο αριθμός των σαρκοφάγων έναντι αυτού των φυτοφάγων)· και ρητά οδυνηρές αλήθειες άξιες κάθε αισιοδοξίας, παρά την όποια στάση μας. Με μια αρχική αναζήτηση στη βάση δεδομένων του αρχείου του Παγκόσμιου Καταλόγου βρίσκει κανείς πάνω από 3,1 εκατομμύρια βιβλία και άρθρα που χρησιμοποιούν τη λέξη "καλοσύνη". Σε αυτά περιλαμβάνεται και ένα δυνατό πρώιμο έργο σχετικά με τα δικαιώματα των ζώων της πολυγραφότατης συγγραφέως του 19ου αιώνα Σάρλοτ Ελίζαμπεθ Τόνα (Charlotte Elizabeth Tonna, 1790-1846), με τίτλο "Kindness To Animals: Or, The Sin of Cruelty Exposed and Rebuked" (Φιλαδέλφεια, 1845). "Η αίσθηση της ευαισθησίας των συναισθημάτων τους", λέει, αναφερόμενη στις αρκούδες, τα σκουλήκια, τις νυχτερίδες, τα ψάρια και οποιοδήποτε άλλο ον, θα πρέπει να μας προκαλεί δέος και μια βαθύτερη, ανώτερη και πιο στοργική στάση απέναντι στον κόσμο.

Στην Παλαιο-Αγγλική γλώσσα του 14ου αιώνα, η χρήση της αγγλικής λέξης "kindness" (η οποία στα ελληνικά αποδίδεται ως "καλοσύνη" ή "ευγένεια") παρέπεμπε στη σημασία των λέξεων "έθνος" και "κέρδος", ενώ κατά την εποχή της Μέσης Αγγλικής γλώσσας διαμορφώθηκε ώστε να αναφέρεται στις ευγενείς πράξεις. Η ετυμολογία της περιλαμβάνει όλα τα *γένη ("kinds")*, υποδηλώνοντας ότι κανείς και τίποτα δεν πρέπει να μένει απ' έξω. Στην Ιουδαιοχριστιανική Βίβλο και στο Θιροκουράλ (திருக்குறள்), ένα κείμενο του 5ου αιώνα μ.Χ. στη γλώσσα Ταμίλ που έγραψε ο διάσημος ποιητής και φιλόσοφος Βαλουβάρ, η καλοσύνη (και η ηθική χορτοφαγία στο Θιροκουράλ) βρίσκεται στον πυρήνα όλων των διδαχών. Τόσο στην Παλαιά όσο και στην Καινή Διαθήκη, αποτελεί την υψηλότερη αρετή, όπως και η Αχίμσα, η μη βία, η οποία κυριαρχεί σχεδόν σε όλα τα θρησκευτικά κείμενα της Ανατολής. Στη Βόρεια Αφρική, στα παλαιά αιγυπτιακά *Κείμενα των Πυραμίδων*, του 2375 π.Χ. περίπου, αναφέρεται για πρώτη φορά η έννοια του *maat*, που συμβολιζόταν με τη στρουθοκάμηλο, και αποτελούσε την ουσία της αλήθειας και των έντιμων πράξεων και μια από τις παλαιότερες γνωστές θρησκείες σε όλη την Αφρική. Οι γεωγραφικές προεκτάσεις της καλοσύνης στους ανθρώπινους πολιτισμούς δεν έχει διερευνηθεί και μελετηθεί επαρκώς. Παρόλα αυτά, στα τέλη του 2019, το Πανεπιστήμιο UCLA ίδρυσε το διεπιστημονικό Ινστιτούτο Bedari Kindness Institute στη Σχολή Δημοσίων Σχέσεων Luskin, με σκοπό τη διεξαγωγή μιας μεγάλης έρευνας που περιλαμβάνει προσεγγίσεις της γενετικής, της ψυχιατρικής, της ανθρωπολογίας, καθώς και τις πλέον ευρύτερες ιατρικές και κοινωνιολογικές προσεγγίσεις, με σκοπό την κατανόηση της εξελικτικής σημασίας και του μελλοντικού δυναμικού της απλής καθημερινής καλοσύνης.

Δεδομένου ότι τα κείμενα αυτά αποτελούν τα πλέον αντιπροσωπευτικά σωζόμενα κείμενα, όσον αφορά την ουσία και τη βαρύτητα της ανθρώπινης φύσης, ορθώς ένα από τα πλέον προσβάσιμα και λειτουργικά συνώνυμα που εφαρμόζονται για την πράξη της καλοσύνης είναι αυτό που συνθέτει τις τεχνικές, τα κίνητρα και την εφαρμογή σε πράξεις της *αμοιβαιότητας*. Η ίδια η λέξη, όπως και η λέξη "καλοσύνη", πηγάζει από τα χρονικά ενός ιδεώδους, παρόλο που οι βασικοί μηχανισμοί της ωθούνται περισσότερο - αν και όχι όλοι – από τη βιολογική φροντίδα. Τη βλέπουμε στον αμοιβαίο αλτρουισμό που επιδεικνύουν όλα τα είδη, από τα βακτήρια και τα πουλιά έως και τον άνθρωπο[6]. Πρόκειται για πολυάριθμες μορφές "άμεσης", "χωρικής", "αντίστροφης" ευγνωμοσύνης, εφάμιλλες με τις στοιχειώδεις μορφές συνεργασίας, καθεμία από τις οποίες έχει εξερευνηθεί όσον αφορά τη φυσική επιλογή και την εξέλιξη της οργανωτικά αποτελεσματικής συνεργασίας.[7] Αυτό περιλαμβάνει όλες αυτές τις άφθονες δραστηριότητες της αμοιβαίας ανταλλαγής, της κοινοκτημοσύνης και της θυσίας.[8] Ακόμα και μεταξύ των αρουραίων στη Νορβηγία έχει αποδειχθεί ότι υφίσταται ένα καθεστώς ακόμα πιο "γενικευμένης αμοιβαιότητας", γεγονός από το οποίο συνάγεται το συμπέρασμα ότι οι ερευνητές θα πρέπει να μελετήσουν την πιθανότητα να πρόκειται στην πραγματικότητα για "κοινό φαινόμενο". [9]

Μια άλλη εκδοχή αυτής της διαδραστικής σχεδόν σταθεράς είναι η βιολογική αμοιβαιότητα η οποία παρατηρείται σχεδόν στο ήμισυ, έως και το 90% όλων των σχέσεων μεταξύ των φυτών, είτε στο επίπεδο της ανταλλαγής χημικών ουσιών, είτε όσον αφορά στην αμοιβαία ανταλλαγή υπηρεσιών για τη διασπορά σπόρων[10]. Η συνεξέλιξη, οι αμοιβαίως συμβιωτικές σχέσεις και οι συμβιώσεις που ευνοούν τουλάχιστον ένα είδος χωρίς να βλάπτουν το άλλο[11], οδηγούν στον κοινωνικό και ατομικό εναγκαλισμό των Άλλων, με οποιαδήποτε κίνητρα. Ο Έντουαρντ Όσμπορν Γουίλσον έχει γράψει ότι "Το βιοφιλικό ένστικτο μπορεί να θεωρηθεί ως ένας από τους ευτυχείς παραλογισμούς της ανθρωπότητας..."[12]. Ας υποθέσουμε ωστόσο, αντιθέτως, ότι υπάρχει ένας υπερλογικός προσανατολισμός, ο οποίος εκτείνεται σε όλα τα σημεία του βασιλείου των έμβιων όντων, από τη δέσμευση του ριζοβιακού αζώτου, κατά την οποία τα βακτήρια του πλανήτη μετατρέπουν το άζωτο της ατμόσφαιρας σε αμμωνία, παρέχοντας έτσι στα φυτά τη δυνατότητα να συνθέτουν σειρές αμινοξέων - πρωτεϊνών[13], μέχρι την εγγενή ομαδική εργασία, τις θετικές συνεργασίες, την παροχή βοήθειας που υπαγορεύει η κοινή λογική και τις διάφορες εφαρμογές του Όρκου του Ιπποκράτη (460-370 π.Χ.): οι στοιχειώδεις οικολογίες κάθε είδους θεραπείας και βελτίωσης. Ένας από τους παππούδες μας ήταν καρδιολόγος και έκανε στους ασθενείς του ταχυδακτυλουργικά, γιατί αυτό συνέβαλε, όπως έλεγε, στην ταχεία ανάρρωσή τους. Με άλλα λόγια, τους έκανε να χαμογελούν. Η αισιοδοξία αυτή είναι θεραπευτική και πηγάζει από τη φαντασία, την προσωπική επιλογή, την περιέργεια και την αγάπη. Σε συνδυασμό μεταξύ τους, οι εν λόγω τάσεις και στάσεις ζωής αποκαλύπτουν αυτό που αποτελεί την πραγματική σταθερά στην καρδιά της φυσιολατρείας, τη λατρεία δηλαδή της φύσης. Ένα διαρκώς ανοιχτό δηλαδή παράθυρο για τα απόλυτα αντικείμενα της καλοσύνης: τα μεμονωμένα άτομα και ολόκληρο τον κόσμο από τον οποίο εξαρτόμαστε. Εάν μπορούμε να αντιμετωπίζουμε την καθολική αυτή βιόσφαιρα ως ένα συνεχές χωρίς κενά, τότε η στάση μας αυτή δεν αποτελεί τίποτα άλλο παρά μια ειλικρινή πράξη αφοσίωσης και ακλόνητης συντροφικότητας.

Εάν η φυσιολατρική συμπεριφορά βρίσκεται (όπως, προφανώς, πιστεύουμε) στη ρίζα *όλων* των συμπεριφορών, τότε η εξέλιξη και η φυσική επιλογή τείνουν να ευνοούν αποτελέσματα που διαιωνίζουν τη ζωή, με δεδομένο ότι η ζωή η ίδια επαινεί και καθοδηγεί τη συνεργατική συμπεριφορά. Όταν μελετήσαμε για πρώτη φορά τις σχέσεις αρπακτικών και θηραμάτων στην Ανατολική Αφρική, στα τέλη της δεκαετίας του 1980, για να προσδιορίσουμε τους κατά προσέγγιση πληθυσμούς κάθε είδους, η αναλογία των σαρκοφάγων προς τα φυτοφάγα ήταν της τάξης του ενός στα 4.000 άτομα. Στις μέρες μας η μείωση των φυτοφάγων φαίνεται να έχει φτάσει στο δεκαπλάσιο. Με άλλα λόγια, η αναλογία των αρπακτικών προς τα θηράματα βρίσκεται πλέον πιο κοντά στο ένα προς 400. Σε κάθε μεγάλη γάτα αντιστοιχούν περίπου 400 οπληφόρα ζώα βοσκής. Οι στρεσογόνοι παράγοντες της Ανθρωπόκαινου (η κλιματική αλλαγή, οι πυρκαγιές και άλλα αίτια της καταστροφής των βιοτόπων, κυρίως από τον ανθρώπινο υπερπληθυσμό και την επεκτατική ανάπτυξη, τη σταδιακή εξαφάνιση του γλυκού νερού και τη λαθροθηρία) άφησαν τα σημάδια τους με τρόπους που δεν υποστηρίζουν πλέον την υπόθεση της φυσιολατρίας, σε σημείο που θα μπορούσαμε να υποστηρίξουμε ότι ο ανταγωνισμός και η σαρκοβόρα συμπεριφορά δεν πρόκειται να εξαφανιστούν κάποια στιγμή στο μέλλον, όσο και αν το θέλουμε. Ό,τι και αν επιφυλάσσει το μέλλον ως προς τις βασικές βιολογικές αρχές που θα διέπουν όλους μας, το παρελθόν έχει σίγουρα δημιουργήσει αμφιβολίες ως προς την ουσία οποιασδήποτε καθολικής λογικής σαν εκείνη που στις τζαϊνιστικές, βουδιστικές ή ινδουϊστικές παραδόσεις αναφέρεται στα σανσκριτικά ως "αχίμσα" ή μη βία.

Η Χίνσα, η βία, δεν αφαιρεί τίποτα από τον Χρυσό Κανόνα. Δημιουργεί όμως στους ανθρώπους του 21ου αιώνα πλήθος προβληματισμών, ακόμα και για την ίδια μας την επιβίωση, βλέποντας κάθε μέρα τις αδυσώπητες, ολοένα και περισσότερες και χειρότερες καταστροφές που προκαλεί το ανθρώπινο είδος στον φυσικό κόσμο. Τα στοιχεία αποτυπώνονται με μεγάλους αριθμούς που προκαλούν ζάλη, και τους οποίους ο εγκέφαλός μας είναι αδύνατο να αφομοιώσει. Για παράδειγμα, τα 349 δισεκατομμύρια γαλόνια νερού των υπογείων υδάτων που χρησιμοποιούν οι Αμερικανοί ανά 24ωρο. Πρόκειται για έναν τεράστιο όγκο, ακόμη και αν αναλογιστεί κανείς ότι η ποσότητα του νερού το οποίο ρέει αέναα γύρω από τη γη εκτιμάται στα 2,78 επτάκις εκατομμύρια γαλόνια, εκ των οποίων ποσοστό 30,1% αποτελείται από γλυκό νερό, το οποίο, με τη σειρά του, αποτελεί λιγότερο από το 1% του συνολικού *διαθέσιμου* γλυκού νερού[14]. Στην πλειοψηφία του αυτό (μαζί πάνω από το 40% όλων των δημητριακών) χρησιμοποιείται για τη σίτιση των ζώων - που αποτελούν πάνω από το 60% της συνολικής βιομάζας των θηλαστικών - τα οποία με τη σειρά μας εμείς εκμεταλλευόμαστε και σφάζουμε. Ταυτόχρονα, το 2019 διοχετεύτηκαν από τον άνθρωπο στην ατμόσφαιρα 43,1 δισεκατομμύρια τόνοι CO2.[15] Παρά το γεγονός ότι, αν και με όχι πολύ ασαφή τρόπο, οι αριθμοί είναι αποκαλυπτικοί, γνωρίζουμε σίγουρα πως οτιδήποτε κάνουμε - σε διαφορετικά ποσοστά ο καθένας, ανάλογα με τη σύνθεση του μωσαϊκού της εκάστοτε κοινωνίας στην οποία ζούμε και τον τρόπο με τον οποίο αντιλαμβανόμαστε στο πλαίσιο αυτό τις έννοιες της εξόρυξης και εκμετάλλευσης των φυσικών πόρων - πολλαπλασιάζεται με σχεδόν ακόμα οκτώ δισεκατομμύρια ανθρώπους. Κανείς μας δεν έχει δει ποτέ οκτώ δισεκατομμύρια ανθρώπους. Δεν ξέρουμε καν πώς μοιάζει να δει κανείς μαζεμένους τόσους ανθρώπους. Μπορούμε, όμως, έστω αμυδρά, να υπολογίσουμε την τεράστια κλίμακα της παλινδρομούσας μάζας τους. Το τεράστιο αυτό μέγεθος "χτυπάει" στον εγκέφαλο ο οποίος, με τη σειρά του, απορρίπτει αυτή την αίσθηση αντανακλαστικά.

Η νοητική αυτή απόρριψη δεν προσαρμόζεται εύκολα σε καμία εξελικτική αρχή. Για παράδειγμα, στην περίφημη δήλωσή του, ο Ντέιβιντ Σλόαν Γουίλσον αναφέρει ότι "η λειτουργική οργάνωση σε ομαδικό επίπεδο πρωταρχικά εξελίσσεται από τη φυσική επιλογή μεταξύ [διαφορετικών] ομάδων" και όχι "εντός των ομάδων"[16]. Δεν υπάρχει κάποια άμεση επιστημονική απόδειξη για τη θεωρία του Υπεροργανισμού (που υποστηρίζει την έννοια της λειτουργίας της φυσικής επιλογής τόσο σε ομάδες όσο και σε μεμονωμένα άτομα), παρά μόνο η σημασιολογική μετάθεση της έννοιας αυτής στη συμπεριφορά και την επιτυχία παρόμοιων τεράστιων συνόλων και την αποτελεσματικότητα του φαινομένου της συσσωμάτωσης που μαρτυρείται στους κοραλλιογενείς ύφαλους, τα νεοτροπικά δάση, το ηλικίας 80.000 ετών κλωνικό δάσος "Pando" της Γιούτα με τις 47.000 γενετικά πανομοιότυπες λεύκες[17], τις κυψέλες των μελισσών, καθώς τις αποικίες των τερμιτών και των μυρμηγκιών. Ωστόσο, η εννοιολογική αναπαράσταση των βιολογικών οντοτήτων - ως πληθυσμών ή ως ατόμων, καθώς και μεταξύ συστάδων – με αυτόν τον τρόπο αποτελεί μια αυθαίρετη ταξινόμηση, με έμφαση σε διαφοροποιημένες ιδέες, που θολώνουν τα όρια μεταξύ της υποκειμενικότητας και της αντικειμενικότητας της διαδικασίας ταξινόμησης. Πράγματι, οι διαφορετικές κλίσεις αποτελούν τόσο μια εντυπωσιακή αντίδραση στους μεγάλους αριθμούς όσο και στην ίδια την έννοια της λειτουργικότητας σε εκτεταμένες περιοχές. Αυτή η συζήτηση, μεταξύ της άυλης γραμμής που διαχωρίζει κάποιο μεμονωμένο άτομο από ένα είδος, δεν έχει σταματήσει ποτέ, ιδίως σε κυτταρικό επίπεδο. Ούτε όμως και όσον αφορά τη διαχείριση από τον άνθρωπο – με διάφορες μορφές άσκησης ελέγχου – μεγάλης κλίμακας μεγάλου όγκου φυσικών πόρων που τυχαίνει να τους αντιλαμβανόμαστε ως αγαθά που απολαμβάνουμε από κοινού, ή ως κοινά αγαθά, δηλαδή, του αέρα, του νερού, του εδάφους, της βιοχημείας[18]. Πρόκειται για ζήτημα σημασιολογικού προσδιορισμού και δεν υφίσταται απαραίτητα στην πραγματικότητα (όπως η εξίσωση του George Price το 1972, ο οποίος θεώρησε ότι υπάρχει μια μαθηματική "σχέση ισοδυναμίας" μεταξύ "της επιλογής συγγενών και της επιλογής σε πολλαπλά επίπεδα...")[19]

Οι αριθμοί, μεγάλοι και μικροί, ρητοί ή άρρητοι, αντιπροσωπεύουν ιδέες οι οποίες, σε εναλλαγή, αποκαλύπτουν διάφορα κίνητρα που προέρχονται από την ιστορία της βιολογίας. Εκείνα, με τη σειρά τους, υποδηλώνουν τις γεωπολιτικές και οικονομικές βασικές αρχές, οι οποίες διέπουν τόσο τις συνεργατικές σχέσεις όσο και τις καταστροφές. Τόσο την τυραννία όσο και την καλοσύνη. Οι πολυσύνδετες αυτές σχέσεις καταδεικνύουν την πορεία μας προς την απομόνωση, θα είναι όμως επίσης και η ταφόπλακα που θα βάλουμε οι εμείς οι ίδιοι στη βιοποικιλότητα, τουλάχιστον όσον αφορά το μεγαλύτερο μέρος της.

Σημειωσεις

[1]Βλ. για παράδειγμα, *Environmental Resources*, NJ State Library, https://libguides.njstatelib.org/ environment/general; and https://libguides.njstatelib.org/ environment/environmental_libraries

[2] Βλ. Frank Pajares, "Outline and Study Guide," σχετικά με το *The Structure of Scientific Revolutions,* του Thomas S. Kuhn, https://www.uky.edu/~eushe2/Pajares/Kuhn.html. Βλ. *The Structure of Scientific Revolutions,* του Thomas Kuhn, University of Chicago Press, Chicago, Ill., 1962.

[3] Βλ. "Going out, I found I was really going in: John Muir's spiritual and political journey," του Tim Flinders, 18 Απριλίου 2016, *openDemocracy*, https://www.opendemocracy.net/en/ transformation/going-out-i-found-i-was-really-going-in-john-muir-s-spiritual-and-politi/

[4] Βλ. "Protected Areas," *WDPA*, https://www.protectedplanet.net/en/thematic-areas/wdpa; βλ. επίσης: *On the Nature of Ecological Paradox*, των M.C.Tobias και J. G. Morrison, Springer Nature, New York and Switzerland, 2021. Βλ. επίσης, *Sanctuary: Global Oases of Innocence*, by M.C.Tobias and J. G. Morrison, Μία έκδοση του Dancing Star Foundation, Council Oaks Books, San Francisco and Tulsa, 2008.

[5] Βλ. για παράδειγμα, *The Golden Rule – The Ethics of Reciprocity In World Religions*, Συντάκτες: Jacob Neusner και Bruce Chilton, Continuum Books, New York, 2008.

[6] Βλ. Trivers, R.L. (1971). "The evolution of reciprocal altruism". *Quarterly Review of Biology*. 46: 35–57. doi:10.1086/406755. S2CID 19027999.

[7] Βλ. "Upstream reciprocity and the evolution of gratitude," των Martin A. Nowak και Sébastien Roch, PMC, Proc Biol Sci. 2007 Mar 7; 274(1610): 605–610. Εκδοθέν διαδικτυακά στις 19 Δεκ. 2006. doi: 10.1098/rspb.2006.0125, PMCID: PMC2197219, PMID: 17254983, *The Royal Society Publishing, Proceedings B*, https://www.ncbi.nlm.nih.gov/pmc/articles/ PMC2197219, Πρόσβαση στις 19 Οκτωβρίου , 2020. Βλ. επίσης, "Five rules for the evolution of cooperation," του Martin A. Nowak, Science. 2006 Dec 8; 314(5805): 1560–1563. doi: 10.1126/science.1133755, PMCID: PMC3279745 NIHMSID: NIHMS49939 PMID: 17158317, https://www.ncbi.nlm.nih.gov/pmc/articles/PMC3279745, Πρόσβαση στις 19 Οκτωβρίου 2020.

[8] Βλ. "The Natural Model of Reciprocity," by Madronna Holden, *Our Earth/Ourselves*, https://holdenma.wordpress.com/culture-and-environment/reciprocity/ Πρόσβαση στις 19 Οκτωβρίου 2020.

[9] "Assortment and the Evolution of Generalized Reciprocity," by Daniel J. Rankin and Michael Taborsky, doi:10.1111/j.1558-5646.2009.00656.x, January 10, 2009, *Evolution* 63-7: 1913-1922, https://onlinelibrary.wiley.com/doi/pdf/10.1111/j.1558-5646.2009. 00656.x

[10] Bronstein, JL (1994). "Our current understand of mutualism". *Quarterly Review of Biology*. 69 (1): 31–51. doi:10.1086/418432.

[11] Mikula P, Hadrava J, Albrecht T, Tryjanowski P (2018). "Large-scale assessment of commensalistic-mutualistic associations between African birds and herbivorous mammals using internet photos". *Peer J. 6: e4520.* doi:10.7717/peerj.4520. PMC 5863707. PMID 29576981.

[12] Βλ. *The Future of Life*, by Edward O. Wilson, Alfred A. Knopf, New York, 2001, p.140.

[13] Βλ. "Nutritional Adaptations of Plants," in *Nitrogen Fixation: Root and Bacteria Interactions*, https://courses.lumenlearning.com/boundless-biology/chapter/nutritional-adaptations-of-plants/; Βλ. επίσης, *Biological Nitrogen Fixation*, by Stephen C. Wagner, 2011, Nature Education Knowledge, 3(10):15, https://www.nature.com/scitable/knowledge/ library/biological-nitrogen-fixation-23570419/

[14] "Groundwater Facts," *Groundwater Summit*, 8-10 Δεκεμβρίου 2020, NGWA, https://www.ngwa.org/what-is-groundwater/About-groundwater/groundwater-facts

[15] See *The World Counts – Tons of CO^2 emitted into the atmosphere*, https://www.theworldcounts.com/ challenges/climate-change/global-warming/global-co2-emissions/story; Άλλες πηγές αναφέρουν μικρότερη ή μεγαλύτερη συνολική ποσότητα. Βλ. "Feeding the World Vertically," by Geoffrey Holland, *The Planetary Press*, Οκτωβρίου 2020, https://www.theplanetarypress.com/2020/10/feeding-the-world-vertically/

[16] *Does Altruism Exist? Culture, Genes, and the Welfare of Others* του David Sloan Wilson, Yale University Press/Templeton Press, New Haven, Conn., 2015, σελ. 21.

[17] Βλ. "Pando, One of the World's Largest Organisms, Is Dying," της Brigit Katz, *Smithsonianmag.com*, 18 Οκτωβρίου 2018, https://www.smithsonianmag.com/smart-news/pano-one-worlds-largest-organisms-dying-180970579/

[18] Βλ. για παράδειγμα τις έννοιες "optimal scale" (βέλτιστης κλίμακας) και "optimal governance" (βέλτιστης διακυβέρνησης) που διασαφηνίστηκαν από τη βραβευμένη με το Νόμπελ 2009 οικονομολόγο Elinor Ostrom στο βιβλίο της *Governing the Commons: The Evolution of Institutions for Collective Action*, Cambridge University Press, Cambridge, UK, 1990.

[19] Βλ. "The Biology of Being Good to Others," H. Allen Orr, *The New York Review*, 19 Μαρτίου 2015, https://www.nybooks.com/articles/2015/03/19/biology-being-good-others/

Κεφάλαιο 2

ΟΙ ΕΠΙΛΟΓΕΣ ΤΗΣ ΦΥΣΙΚΗΣ ΕΠΙΛΟΓΗΣ

Από πολύ παλιά, την ύπαρξή μας συνοδεύουν ή ακολουθούν διάφορες λέξεις, όπως η αμοιβαιότητα, η καλοσύνη, ο αλτρουισμός, η συνείδηση και η ενσυναίσθηση. Η ετυμολογική ανάλυση καθεμιάς από αυτές τις έννοιες γίνεται εξαιρετικά σχολαστικά, ενώ χιλιάδες είναι τα βιβλία και δοκίμια που αναλύουν λεπτομερώς τις κοινωνικές, οικονομικές, ηθικές και πολιτικές βάσεις και όρια των συγκεκριμένων λέξεων και ιδανικών. Η λέξη "αμοιβαία" αναφέρεται με την έννοια της παλίνδρομης κίνησης το 1632: "δυο δέντρα αψηλά κουνιούνται μπρος πίσω αμοιβαία στον αγέρα", ή το 1622: "Τα πλεμόνια, που μαζεύονται ως ένα ζευγάρι φυσητήρων, αμοιβαία γιομίζουν από αγέρα, και έξω πάλι τον διώχνουν". Στα μαθηματικά, με τις έννοιες "Αντίθετα" ή "Αντίστροφα". Το 1776, ο Τόμας Πέιν έγραψε στο βιβλίο του "*Δικαιώματα του ανθρώπου*" το εξής: "Η Διακήρυξη Δικαιωμάτων αποτελεί, με βάση την αρχή της αμοιβαιότητας, και Διακήρυξη Υποχρεώσεων."[1] Στα γαλλικά και στα ιταλικά, ο αλτρουισμός αναφέρεται σε "κάποιον άλλο"· συνείδηση = λατινικά, conscientia, "το να έχει κάποιος πλήρη γνώση", συμπόνια = compati, "σύν + πονέω". Όταν αυτές οι αυτές οι εκφράσεις διατυπωθούν συνδυαστικά, τότε η ανθρωπότητα βγάζει τον καλύτερό της εαυτό.

Τρεις φιλόσοφοι του Διαφωτισμού και της Αναγέννησης, ο Έντμουντ Μπερκ, ο Φράνσις Μπέικον και ο σερ Τόμας Μπράουν, μας κληροδότησαν με κατατοπιστικές δυϊστικές ερμηνείες σχετικά με τους όρους αυτούς και τη δυνατότητα εφαρμογής τους στα ανθρώπινα ζητήματα. Οι ίδιες οι λέξεις από μόνες τους αφήνουν να διαφανούν αχαλίνωτες και αλλόκοτες ερμηνείες και επινοήματα. Για παράδειγμα, η λέξη "συμπάθεια" (ελληνικά + λατινικά = σύν + πάσχω), η οποία μοιάζει να ενοποιεί και τις τέσσερις παραπάνω συγκεκριμένες κατηγορίες γλώσσας και σκέψης. Στην πραγματικότητα, πολλά έχουν γραφτεί για τις ιστορικά διαφοροποιημένες διαφορές μεταξύ της συμπάθειας, του οίκτου και "του μηχανισμού αλλαγής τόπων με το αντικείμενο της παρατήρησής μας", όπως στα γραπτά του Πλάτωνα και του Αριστοτέλη, του Άνταμ Σμιθ) και του Ντέιβιντ Χιουμ.[2] Ωστόσο, ο Έντμουντ Μπερκ θεώρησε ότι σημαίνει κάτι τελείως απροσδόκητο, όταν έγραφε για "τις στενοχώριες των άλλων" και για "την ευχαρίστηση που νιώθουμε για αυτά τα πράγματα, [η οποία] μας εμποδίζει να αποφεύγουμε να βλέπουμε σκηνές δυστυχίας· και ο πόνος που νιώθουμε, μας ωθεί να ανακουφιζόμαστε ανακουφίζοντας εκείνους που υποφέρουν· και όλα αυτά προϋπήρχαν πριν από κάθε λογική, από ένα ένστικτο που μας κινεί για τους δικούς του σκοπούς χωρίς τη συναίνεσή μας".[3] Ένα χρόνο πριν ο Μπερκ γράψει το "*Περί του υψηλού και του Ωραίου", δημοσίευσε το "A Vindication Of Natural Society*" ("Δικαίωση της φυσικής κοινωνίας"), στο οποίο υπέθεσε ότι η ανθρώπινη βαρβαρότητα στα πεδία μάχης της ιστορίας είχε οδηγήσει στη σφαγή "τριάντα έξι εκατομμυρίων" πολεμιστών, συμπεριλαμβανομένης "μιας χαλαρής εκτίμησης που πραγματοποιήθηκε από τη δολοφονία δεκάδων εκατομμυρίων ειδών" από την κατάκτηση της Λατινικής Αμερικής"[4]. Μάλιστα, συνεχίζει επισημαίνοντας ότι "Αναντίρρητη αλήθεια αποτελεί το γεγονός ότι μεγαλύτερος είναι ο όλεθρος που έχουν προκαλέσει σε έναν χρόνο οι άνθρωποι στους ανθρώπους, παρά οι καταστροφές που έχουν προκαλέσει όλα τα λιοντάρια, οι τίγρεις, οι πάνθηρες, οι λεοπαρδάλεις του χιονιού, οι λεοπαρδάλεις, οι πάρδοι, οι ύαινες, οι ρινόκεροι, οι ελέφαντες, οι αρκούδες και οι λύκοι στα διάφορα είδη τους, από την αρχή του κόσμου…" [5]

Στο έργο του "*Πολιτικά και ηθικά δοκίμια*" (1597), ο Φράνσις Μπέικον αναφέρει δύο βασικά στοιχεία της ανθρώπινης φύσης. Πρώτον, ότι "η κλίση προς την καλοσύνη είναι βαθιά εντυπωμένη στη φύση του ανθρώπου, σε τέτοιο βαθμό, που αν δεν εκδηλωθεί προς τους ανθρώπους, θα διοχετευθεί σε άλλα έμβια όντα…" [6]. Και δεύτερον, ότι "Παρά το γεγονός ότι παρατηρούμε πως η πλειοψηφία των ανθρώπων παραδίδεται σε φαντασιώσεις ελπίδας και τέτοιου είδους ανησυχίες του νου, με τόση αγνωμοσύνη για τα γεγονότα του παρελθόντος, και με τρόπο που αγνοεί τα γεγονότα του παρόντος, σαν να μην υπήρξαν ποτέ παιδιά και πρωτάρηδες, συνεχίζουν να λαχταρούν για αυτό που πρόκειται να γίνει" [7]. Ενώ στο έργο του "*Religio Medici*" ο σερ Τόμας Μπράουν (1643) γράφει: "Η φιλανθρωπία ξεκινά από το σπίτι, είναι η φωνή του Κόσμου…Δεν είναι τα δάκρυα των δικών μας ματιών μόνο, αλλά και των φίλων μας, που εξαντλούν τη ροή της θλίψης μας…"[8].

Σε όλους αυτούς τους στοχασμούς βλέπουμε μια παρόμοια διαμάχη με μια διαλεκτική που περιστρέφεται γύρω από την ιδέα της ύπαρξης των φυσικών νόμων, ότι δηλαδή οι άνθρωποι παγιδευόμαστε στη σπείρα της εξέλιξης, μπλέκοντας τις προδιαθέσεις μας με τις πράξεις μας, χωρίς να γνωρίζουμε ακριβώς πού οδεύει όλο αυτό.

Ας ξεκινήσουμε, λοιπόν, με την ενσυναίσθηση, μια λέξη που επινόησε ο ψυχολόγος Έντουαρντ Τίτσνερ (1867-1927)[9] το 1909, από τη γερμανική λέξη "Einfühlung" ("αισθάνομαι εντός"). Νωρίτερα, ο Γερμανός φιλόσοφος Τέοντορ Λιπς (1851-1914) την είχε στοχαστεί ως την "προβολή του εαυτού μας στη θέση του αντικειμένου της αντίληψης", δηλαδή τους κατοπτρικούς νευρώνες[10] που διερευνώνται σήμερα από επιστήμονες πολλών κλάδων. Ή, όπως γράφει ο Μόνταγκ, "η θεωρία του Λιπς για την "εσωτερική μίμηση" μπορεί να αντικατοπτρίζεται με κάποιον τρόπο στις σύγχρονες έννοιες της συναισθηματικής επιδεκτικότητας και στα μοντέλα αντίληψης-δράσης της ενσυναίσθησης"[11]. Ωστόσο, η φιλοσοφία και η ψυχολογία της ενσυναίσθησης έχουν υποβαθμιστεί από ακτιβιστική στάση σε μια κυρίως ακαδημαϊκή και σε μεγάλο βαθμό απογοητευτική σειρά διδακτικών αναλύσεων που εγείρουν πολλά και διάφορα, κατά τα άλλα θεμελιώδη, ερωτήματα, με βασικότερα των οποίων τα εξής: Γιατί πρέπει να μας νοιάζει; Ποιες είναι οι προκαταλήψεις μας κατά τη λήψη ηθικών αποφάσεων; Ποια κίνητρα καθοδηγούν στην πραγματικότητα τα υποτιθέμενα ένστικτά μας; Αποτελεί όντως "πρόβλημα" η "γνώμη των άλλων" ή μήπως είναι η βάση για την εύρεση λύσεων; Υπάρχει διαφορά μεταξύ του να αισθανόμαστε άσχημα και του να αισθανόμαστε άσχημα για κάτι συγκεκριμένο, κάποιον άλλο άνθρωπο ή κάτι Άλλο; Η φιλανθρωπία γεννιέται από εγωισμό, αλτρουισμό ή πραγματική πίστη στη δύναμη της ενσυναίσθησης στην πράξη; Γεννά η ενσυναίσθηση τον αλτρουισμό, αλλά και άλλους βασισμένους στην ενσυναίσθηση τρόπους αισθάνεσθαι και σκέπτεσθαι; Εάν ναι, το θυμικό αυτό είναι καθολικό, επιλεκτικό ή περιορισμένα κατανεμημένο; Υπόκεινται οι ηθικοί κανόνες σε αναπτυξιακές πτυχές της ψυχολογίας ή πρόκειται για μια απλή απεραντολογία που ενισχύεται από τη βιβλιογραφία της απεραντολογίας; Τελικά, το ερώτημα παραμένει: γιατί θα πρέπει καν να μας ενδιαφέρει οποιαδήποτε από αυτές τις πτυχές του λεγόμενους *ηθικού κώδικα* των ανθρώπων;[12] Έχει καμιά σημασία να νοιαζόμαστε; Και έχει αυτή *η σημασία* καμιά σημασία;

Δίνει καθόλου σημασία η φυσική επιλογή σε αυτές τις σημασίες;

Η αποβολή της άμεσης επαφής, γεγονός που επιβάλλεται για την αποστασιοποίησή μας από άλλους ανθρώπους και οργανισμούς γίνεται όλο και λιγότερο αναγκαία όσο περισσότερο διασφαλίζεται η κυριαρχία μας στον χρόνο και στον χώρο, όντας προστατευμένοι από την έλλειψη, την απώλεια, τον πόνο και την παρατεταμένη ταλαιπωρία για την οποία μπορεί να ενημερωνόμαστε με οποιονδήποτε τρόπο, από τις ειδήσεις, από στόμα σε στόμα, ή ακόμα πιο απρόσωπα, όταν διαβάζουμε κάποια περιγραφή ή εξιστόρηση γεγονότων τα οποία συνέβησαν παλαιότερα και/ή σε κάποια άλλη περιοχή. Εάν τα γεγονότα αυτά δεν έχουν επηρεάσει πρόσωπα του άμεσου περιβάλλοντός μας, όπως κάποιους από τους προγόνους μας, την οικογένειά μας ή τους φίλους μας, είναι πιο δύσκολο να αισθανθούμε "πόνο" για όλους τους ανθρώπους που χάθηκαν κατά τη διάρκεια των πανδημιών της Μαύρης Πανώλης ή για τους δεινόσαυρους που εξαφανίστηκαν. Η συμπόνοια μας για όλους εκείνους οι οποίοι πέθαναν κάποια εποχή που βρίσκεται πιο κοντά στη δική μας σύγχρονη πραγματικότητα ενισχύεται περισσότερο όσο πιο κοντά σε εμάς διαδραματίζονται τα θανατηφόρα περιστατικά. Δεν υπάρχει καμία εξίσωση για τη μέτρηση του χάσματος μεταξύ των συναισθημάτων μας, της μεταδοτικότητας της θλίψης και των γεγονότων που προηγήθηκαν της δικής μας παρουσίας στο χρόνο. Είναι αυτό καν πρόβλημα;

Ο Έντμουντ Μπερκ και ασφαλώς τα ατυχή μαθήματα ιστορίας που παρέδωσε ο σύγχρονός του Έντουαρντ Γκίμπον στο εξάτομο έργο του "*Ιστορία της παρακμής και της πτώσης της Ρωμαϊκής Αυτοκρατορίας*" (1776), επικοινωνούν κάτι διαφορετικό από την στάση ουδετερότητας όσον αφορά το κλείσιμο του χάσματος. Ο Γκίμπον καταλογίζει ευθύνες στον "καταιγισμό των Βαρβάρων" και στη συνέχεια, από αιώνα σε αιώνα, στην τεράστια και εγωιστική υπερβολή, τη σπατάλη και την εσωτερική παρακμή έναντι της οποίας πρέπει να επικρατήσει ο μελλοντικός ορθολογισμός. Ωστόσο, ακόμα και από την εποχή του Πετράρχη που θαύμαζε τα αρχαία ερείπια, γράφει ο Γκίμπον, οι Ρωμαίοι είχαν ξεπεράσει εντελώς αυτή τη νοσταλγία, είχαν ξεχάσει κάθε αίσθηση ομορφιάς, και είχαν εγκαταλείψει κάθε νόημα ή μεγαλείο των παρατημένων απολειφαδιών της ιστορίας των δώδεκα προηγούμενων αιώνων. Είχαν απαρνηθεί την ίδια τους την ιστορία ("νωχελική αδιαφορία", όπως το έθεσε ο Γκίμπον)[13] ή τη σημασία των ερειπίων που έβρισκαν τόσο συναρπαστικά οι επισκέπτες στη Ρώμη και τα οποία εξακολουθούμε να εκτιμούμε ως ζωντανά μουσεία. Εμείς τρέχουμε να χωθούμε για να σταθούμε και να φωτογραφηθούμε μέσα σε αυτά τα μνημεία του μεγαλείου, της απόλυτης ματαιότητας, του πόνου και της ελπίδας. Ο Γκίμπον γνώριζε ότι οι επόμενες γενιές θα καταδικάζουν πάντα την ανθρώπινη ανοησία, ακόμα και αν οι ίδιοι οι άνθρωποι συνεχίζουμε απτόητοι να επαναλαμβάνουμε επίμονα τα ίδια λάθη. Και αυτό είναι που προκαλεί τη μεγαλύτερη αμφιβολία σχετικά με το αν όλα αυτά μας έχουν γίνει μάθημα. Πράγματι, το χάσμα μεταξύ παρατήρησης και μνήμης δεν αποτελεί σε καμία περίπτωση αντικείμενο πρόβλεψης για την επιστήμη της συμπεριφοράς. Οι εργαζόμενοι στα σφαγεία αποτελούν το πλέον κραυγαλέο παράδειγμα αυτής της αυταπόδεικτης αλήθειας. Σε ένα από τα μεγαλύτερα γνωστά σφαγεία των Ηνωμένων Πολιτειών, κάθε μέρα 5.000 εργάτες ακούν, εισπνέουν και παρατηρούν τη σφαγή περίπου 32.000 χοίρων. Πολλοί από αυτούς τους εργάτες περιφέρονται μέσα στο αίμα και τα εντόσθια των θυμάτων που σφαδάζουν από τον πόνο, παίζοντας ενεργό ρόλο στην εξολόθρευση αυτών των ζώων. Αν *όλοι* είμαστε άνθρωποι, με σχεδόν παρόμοια σύσταση DNA, πώς μπορεί κάτι τέτοιο να συμβαίνει;

Τι είναι έμφυτο, υποτυπώδες, ενστικτώδες και βασικό στοιχείο του γονιδιώματός μας. Ας δεχθούμε, προσωρινά, ότι το γονιδίωμά μας ισούται με μηδέν, με όρους στατιστικής. Ότι καμία εξελικτική προτίμηση δεν επηρεάζει την προσωποποίηση των βασικών στοιχείων της φυσιολογίας μας, είτε ως προς την ατομική συμπεριφορά είτε ως προς την κατανομή τέτοιων συμπεριφορών. Ας δεχθούμε, αντίθετα, ότι η καταγωγή, όπως υπέθεσε ο Δαρβίνος, συνεπάγεται κάθε απαραίτητη τροποποίηση, το "μυστήριο των μυστηρίων", όπως το αποκάλεσε. Μια κατάσταση εξέλιξης σε πραγματικό χρόνο βάσει της οποίας κάποιος εκδηλώνει τη δύναμή του να επιταχύνει, να ανακοινώνει, να προασπίζεται και να υπερασπίζεται κάποιο νέο χαρακτηριστικό στην ανθρώπινη εξέλιξη. Μπορεί να χρειαστούν 100.000 χρόνια για την εμφάνιση της μοναδικής αυτής δικλείδας ασφαλείας πέρα από τα σύνορα κατά τα άλλα οριοθετημένων πληθυσμών, χωριών, κοινοτήτων και εθνικών κρατών. Το γεγονός αυτό επικυρώνει την ιδέα ότι η φυσική επιλογή έχει επιλέξει σε αυτό το σύνολο συνθηκών οι οποίες επικρατούν στην μετα-Ολόκαινο εποχή μια χαρακτηριστική στάση κατακραυγής που αμφισβητεί ολοκληρωτικά κάθε γκραντουαλισμό μεταξύ των δεικτών απολίθωσης, επιστρατεύοντας αντ' αυτού μια εντελώς ξαφνική κατακραυγή που συνοδεύει τη γνώση μας για τους ανθρωπογενείς μαζικούς αφανισμούς, τη βιομηχανική γεωργία, την κλιματική αλλαγή, την παγκόσμια καταστροφή σε μια μόνο στιγμή στον γεωλογικό χρόνο.

Πόσο διαφορετική είναι η λατρεία της πέτρας από τον Χένρι Ντέιβιντ Θορώ, που έγραψε: "Θα σήμαινε την αναγέννηση της ανθρωπότητας, αν ανυψωνόταν αρκετά για να λατρέψει πραγματικά τα κλωνάρια και τις πέτρες… Ολόψυχα θα αυγάτιζα κάθε ευκαιρία να θαυμάσω και να λατρέψω, όπως ένα ηλιοτρόπιο καλωσορίζει το φως… μια πέτρα με ελκύει και με ανυψώνει, μου λέει πόσα μίλια έχω διανύσει, πόσα ακόμα απομένουν…Αν υπηρετούσε όλους το ίδιο, θα μπορούσε κάλλιστα να γίνει αντικείμενο δημόσιας αγαλλίασης" [14].

Η εικονογραφική εκδήλωση των συναισθημάτων που αποτελούν την ανθρώπινη ιστορία, ως συνάρτηση της χωροχρονικής μας απόστασης από τα εν λόγω γεγονότα, δεν θα πρέπει να περιπλέκει τους επίμονους ιδεασμούς που απομακρύνονται από το νόημα, ανεξάρτητα από το πόσο έχει εξασθενήσει το γεγονός ότι πέφτουν τα φύλλα το φθινόπωρο ή τα απομακρυσμένα έθιμα. Μπορεί οι χρόνοι αντίδρασης στην εποχή της αμεσότητας να έχουν περιοριστεί, ωστόσο τα βασικά ηθικά αντανακλαστικά έχουν καταπέσει, με υψηλά επίπεδα κατάθλιψης, πόνου και χαμηλή επίγνωση. Παρόλο που, ως είδος, έχουμε περάσει από τη Νεολιθική περίοδο διάφορα στάδια εξέλιξης από ήπειρο σε ήπειρο, τα αποτυπώματά μας, τα πόμολα, τα κρεβάτια και τα μαξιλάρια μας δεν έχουν αλλάξει ουσιαστικά. Και οι επικρατούσες νευροψυχολογικές μελέτες τείνουν προς την υπόθεση ότι "για να επιβιώσει κάποιος, πρέπει πρώτα να νοιαστεί για τον εαυτό του. Αυτό δεν σημαίνει ότι δεν πρέπει να νοιάζεσαι για τους άλλους, αλλά [να το επαναλάβω] πρέπει πρώτα να επιβιώσεις", γράφει ο επιστήμονας Ζαν Ντεσετί.[15] Η ρήση αυτή, που χρησιμοποιείται κάθε φορά στις οδηγίες χρήσης των ζωνών ασφαλείας στα εμπορικά αεροσκάφη, συμφωνεί επίσης με την εμφάνιση της "λειτουργικής βιοποικιλότητας", η οποία παραθέτει τρεις νέες, ιδιαιτέρως εξέχουσες, έννοιες ταξινομικής κατάταξης: "Ενώσεις", "Ομάδες Λειτουργικής Απόκρισης" και "Ομάδες Λειτουργικής Επίδρασης". [16] Η ταξινόμηση σε αυτές τις κατηγορίες καθιστά την ποιοτική ανάλυση των σχέσεων των οργανισμών πιθανώς πιο άμεση και αποκαλυπτική από την άποψη των συνιστωσών που συνεισφέρουν ή της βασικής παράπλευρης εμπλοκής στις μήτρες που χρησιμοποιούνται για τη θεία ανθρωπογενή έκλυση. Με άλλα λόγια, σε πολλούς τομείς της ανθρώπινης εμπειρίας, οι αντιδράσεις και η φύση της απόκρισης μπορούν να ποικίλλουν, να αναπτύσσονται και να μαθαίνουν. Ιδιαίτερη σημασία έχει το ακόνισμα των αντανακλαστικών και των μηχανισμών μνήμης του πόνου. [17]

Εάν η αιτία και το αποτέλεσμα της χωρικής και χρονικής ποσοτικοποίησης μπορούν να εκφραστούν ως λειτουργίες νευροεξελικτικών περιορισμών, πιέσεων και αλλαγών, τότε η ίδια η νευρική υποδομή είναι συνυπεύθυνη με κριτήρια που δεν έχουν γίνει ακόμα κατανοητά σε όλους τους οργανισμούς, με τις γνωστές εξαιρέσεις των σπόγγων και των Trichoplax (αρχέγονα πολυκύτταρα πλακόζωα), που πιστεύεται ότι *δεν* έχουν νευρώνες, σε αντίθεση με την όρκα, η οποία διαθέτει πάνω από 43 δισεκατομμύρια, ή τον διπλάσιο αριθμό νευρώνων που βρίσκονται στους ανθρώπους, με σημαντικό ποσοστό αυτών στον εγκεφαλικό φλοιό. Ωστόσο, δεν πρέπει να παραβλέπεται η σχεσιακή σπουδαιότητα της συνομοταξίας των πλακόζωων, η οποία δε διαθέτει νευρικές συνάψεις, και της οποίας τα

11.514 γονίδια είναι παρόμοια με πολλών άλλων ζώων.[18] Επιπλέον, επικοινωνούν από κύτταρο σε κύτταρο μέσω του μηχανισμού των πεπτιδίων ή ομάδων έως 20 αμινοξέων. Η αγαπημένη τους τροφή είναι τα χλωροφύκη. Ένα από τα Trichoplax, το είδος *adhaerens* - ένα από τα τρία είδη πλακόζωων- έχει προταθεί ως πρότυπος γήινος οργανισμός όσον αφορά την αινιγματική του πολυπλοκότητα σε σχέση με το γεγονός ότι σχετίζεται μεμονωμένα αλλά και συνολικά με άλλους σαν αυτόν. Για την ακρίβεια, η ικανότητά τους να μεταδίδουν σήματα έχει χαρακτηρισθεί "τεράστια". [19]

Ως εκ τούτου, πρόκειται για μια θαυμάσια σύγχυση που συμβάλει στην άρση των εμποδίων για την προβολή των ομοιοτήτων και των διαφορών μεταξύ των οργανισμών, και η οποία φέρνει πάλι στο προσκήνιο τη σκιά του χάσματος αυτού των αντανακλαστικών τα οποία κινητοποιούν τους μηχανισμούς υποστήριξης για ορισμένους οργανισμούς, αλλά όχι για κάποιους άλλους. Ο εντοπισμός και η αναγνώριση της έγκυρης ένδειξης αυτών των ρυθμίσεων γίνεται άμεσα. Όπως για παράδειγμα όταν ο Μπιρόν, ακόλουθος του Βασιλιά Φερδινάνδου της Ναβάρας στο έργο του Σαίξπηρ "Αγάπης Αγώνας Άγονος" (1595/1596), δηλώνει, "Για χάρη της σοφίας, μιας λέξης που όλοι οι άνθρωποι αγαπούν, ή για χάρη της αγάπης, μιας λέξης που αγαπά όλους τους ανθρώπους…" [20] Ή, σε αυτή την απόλυτη ομοιότητα που σπάει κάθε διαφορά για να υπαινιχθεί κάποια άλλη οντότητα ολόκληρη, ένα είδος ή μεμονωμένο υπερoργανισμό του οποίου η πρόθεση και ο τρόπος ύπαρξης και διαβίωσης υπερβαίνει τις μάταιες διαφοροποιήσεις μας. Αντ 'αυτού, ένα ον λογικής, μεταβολισμού, μόχθου και εσώτερης γνώσης, που δεν υπολογίζει κανέναν από τους προφανείς παράγοντες της ζωής, μια δύναμη της φύσης που δεν έχουμε ακόμη εντοπίσει, η οποία προέρχεται ακόμη βαθύτερα και από τον ίδιο τον πυρήνα, ή οι ατομικές συνδέσεις, η τριβή, η μάζα και η ενέργεια. Και πάλι, μια νέα εξ ολοκλήρου μορφή ζωής. Όπως έγραψε ο αείμνηστος ποιητής Τσαρλς Τόμλινσον σε κάποιο από τα πρώιμα έργα του, "Για ένα δευτερόλεπτο νομίζεις ότι είναι ολωσδιόλου νεκρό, μέχρι που βγαίνει από μέσα του ο άηχος βρυχηθμός του στο καλαμπόκι παραπέρα"[21].

Όλη αυτή η συζήτηση έχει να κάνει με τις *σχέσεις*. Είτε αυτή αφορά τα μέσα με τα οποία η ανθρώπινη ιστορία καταγράφει τις ανακεφαλαιώσεις μας, είτε τον τρόπο με τον οποίο οι ερευνητές καταγράφουν τα απολιθώματα. Και στις δυο περιπτώσεις ξεκινάμε από μια ιδέα σχετικά με την εικόνα μας για τη φύση. Αυτή μπορεί να συνδέεται με κάποια προσωπικά αντικείμενα ή κάποιες εκκρεμότητες από το παρελθόν που έχουν έρθει να μας υπενθυμίσουν το παρόν. Στο κεφάλαιό του με τίτλο "Πολιτική αυτοβιογραφία ενός νέου", ο Ίταλο Καλβίνο γράφει "…αυτός που, στη γενική καταστροφή των ανθρώπινων υποθέσεων όταν αφήνονται σε βιολογικές παρορμήσεις, ξέρει πώς να επιλέγει τι πρέπει να σωθεί, να προστατευθεί, να αναπτυχθεί και προσεχθεί ώστε να δώσει καρπούς"[22]. Οι ιδέες, οι ερμηνείες, οι μέθοδοι και οι εμπνεύσεις που μας βοηθούν να προσανατολίζουμε τη ζωή μας μέσα στη βιόσφαιρα μεταβάλλονται συνεχώς, παρά την περιοδική σταθερότητα ορισμένων σημείων μιας φαινομενικής μονιμότητας, όπως οι νόμοι της φύσης, της φυσικής, των μαθηματικών και ολόκληρων οικοσυστημάτων. Ωστόσο, μόλις πρόσφατα – πολύ καιρό μετά τη Θεωρία της Σχετικότητας, τη Θεωρία Παιγνίων, τη Θεωρία

Πληροφορίας και την Κβαντική Θεωρία, τον Ηλιοκεντρισμό ή τις αποκαλύψεις σχετικά με την "ψυχοφυσιολογία της όρασης"[23]- οι οικολόγοι κατάλαβαν ότι η διαδοχή των οικοσυστημάτων και των κοινοτήτων εκτυλίσσεται με τρόπους που υπερβαίνουν κατά πολύ την εξέλιξη των μεμονωμένων όντων, παρακάμπτοντας τους κανόνες που υπαγορεύουν τις βασικές έννοιες των γονιδίων, επικαλούμενοι, αντ' αυτού, την "εξέλιξη του οικοσυστήματος"[24]. Και παρά όλες αυτές τις σχέσεις που έχουν ιστορικά προωθήσει τη δημιουργία των επιστημονικών θεωριών, ο φιλοσοφικός *χαρακτήρας της Ύπαρξης* είναι αυτός που αντιτίθεται στη στασιμότητα, όχι σε κάποιο σταθερό επίπεδο αρχών, αλλά με απόλυτη ορμή, την αεικινησία που ενυπάρχει στη ζωή και τις παρορμήσεις της για αμοιβαιότητα.

Τι είναι αυτή η νομαδική στροφή στη συμπεριφορά, την προσωπικότητα, τη βούληση; Όλα όσα έχουμε μάθει μέχρι στιγμής για τη βιοποικιλότητα μας καθοδηγούν στη ρευστότητα, χωρίς να προδίδουν την πραγματική πλοκή, παρά μόνο ότι η αλλαγή αποτελεί προϋπόθεση της εξέλιξης. Πέραν όμως αυτού, με μια απλή μόνο εκπαίδευση σχετικά με τις πιο βραχυπρόθεσμες εξισώσεις, την επιβίωση δηλαδή του ατόμου, γνωρίζουμε ενστικτωδώς ότι οι πληθυσμοί μας εξελίσσονται, οι πρόγονοι ανακυκλώνονται, αυτοί που έζησαν πριν από εμάς, μέλη και μη της οικογένειάς μας, προδιαγράφουν κάποιας μορφής πορεία συνέχειας, ακόμη και με τον κίνδυνο των σχετικών αφανισμών, οι οποίοι γνωρίζουμε ότι αποτελούν έναν σπάνιο κανόνα. Τότε όμως, τι θα γίνει; Πώς θα ξεπεράσουμε το αμήχανο, ματαιόδοξο, *αλλά τι θα γίνει με εμάς*; Γνωρίζουμε με βεβαιότητα ότι ο προβληματισμός αυτός είναι σημαντικός για το παρόν, όμως οι συνέπειες για το κοντινό μέλλον πιθανώς είναι από μικρές έως μηδενικές.

ΣΗΜΕΙΩΣΕΙΣ

[1] *The Oxford English Dictionary, XIII Volumes*, Volume VIII, Oxford: At the Clarendon Press, UK, 1961, σελ. 242-243.

[2] Βλ. "Pity and Sympathy: Aristotle Versus Plato And Smith Versus Hume," του Christos Grigoriou, *The Journal of Scottish Philosophy* 16.1 (2018): 63-68. doi: 10.3366/jsp.2018.0183, Edinburgh University Press, https://www.researchgate.net/ publication/323491569_Pity_and_Sympathy_Aristotle_versus_Plato_and_Smith_versus_Hume

[3] Από το *A Philosophical Enquiry Into The Origin Of Our Ideas Of The Sublime And Beautiful*, στο *The Works Of The Right Honourable Edmund Burke*, Collected In Three Volumes, Vol. 1, J. Dodsley, London, 1792, p. 107. Πρώτη δημοσίευση το 1757.

[4] ό.π., *The Works…* in *A Vindication of Natural Society*, σελ. 23.

[5] ό.π, *A Vindication…*σελ. 25-26.

[6] *Essays Civil And Moral*, XIII Of Goodness And Goodness Of Nature, στο *The Works Of Francis Bacon, Lord Chancellor of England, A New Edition, With A Life Of The Author*, του Basil Montagu, σε Τρεις Τόμους, Τόμ. 1, A. Hart, Philadelphia, 1852,σελ. 21.

[7] ό.π., στο *Meditationes Sacrae*, σελ. 68.

[8] Titchener E.B. (1909/2014) Introspection and empathy *Dialogues in Philosophy, Mental and Neuro Sciences* 2014; 7: 25–30, Βλ. επίσης: https://plato.stanford.edu/ entries/ empathy/#HisInt

[9] Sir Thomas Browne, *Religio Medici*, Sixth Edition, With Annotations, Also Observations, του Sir Kenelm Digby, London, 1669, σελ. 142-144.

[10] Rizzolatti, G. and C. Sinigaglia, 2008. *Mirrors in the Brain: How our Minds Share Actions and Emotions*, Oxford: Oxford University Press.

[11] "Theodor Lipps and the Concept of Empathy: 1851-1914," to Dr. Montag, Images in Psychiatry, *Am J Psychiatry* 165:10, October 2008, σελ. 1261, ajp.psychiatryonline.org. Βλ. Lipps T: Ästhetik: Psychologie des Schönen und der Kunst: Grundlegung der Ästhetik, Erster Teil. Hamburg, Germany, L Voss, 1903 13. Preston SD, DeWaal FBM: Empathy: its ultimate and proximate bases. *Behav Brain Sci* 2002; 25:1–72

[12] Karsten Stueber, Stanford Center for the Study of Language and Information, The Metaphysics Research Lab, "Empathy," *Stanford Encyclopedia of Philosophy*, 27 Ιουνίου 2019, https://plato.stanford.edu/entries/empathy/#HisInt

[13] Edward Gibbon, *The History Of The Decline And Fall Of The Roman Empire*, in Twelve Volumes, Vol. XII, London, 1821, σελ. 420.

[14] *The Journal Of Henry D. Thoreau, In Fourteen Volumes*, Bound as Two, Volumes VIII-XIV 9 November, 1855-1861), Edited by Bradford Torrey and Francis H. Allen, With a Foreword By Walter Harding, Dover Publications, Inc., New York, 1962, σελ. 1063-1064.

[15] Βλ., "How do we balance self-interest with fairness for others?"του Jack Wang, 25 Αυγούστου 2020, *University of Chicago news*, https://news.uchicago.edu/story/how-do-we-balance-self-interest-fairness-others;

[16] Βλ. "Functional Biodiversity and the Concept of Ecological Function," του Antoine C. Dussault, στο *From Assessing to Conserving Biodiversity -Conceptual and Practical Challenges*, επιμ. Elena Casetta, Jorge Marques da Silva and Davide Vecchi, Springer Open Access Book, History, Philosophy and Theory of the Life Sciences book series, Vol. 24, https://link.springer.com/book/10.1007/978-3-030-10991-2

[17] Βλ. "Commonalities between pain and memory mechanisms and their meaning for understanding chronic pain," των Theodore J. Price και Kufreobong E. Inyang, Prog Mol Biol Transl Sci. Author manuscript; διαθέσιμο στο PMC 2015 Nov 30. *Prog Mol Biol Transl Sci.* 2015; 131: 409–434. Δημοσιεύθηκε διαδικτυακά στις 30 Ιαν. 2015. doi: 10.1016/bs.pmbts.2014.11.010, PMCID: PMC4664460 NIHMSID: NIHMS727455, PMID: 25744681, https://www.ncbi.nlm.nih.gov/pmc/articles/PMC4664460/

[18] Eitel, Michael; Osigus, Hans-Jürgen; DeSalle, Rob; Schierwater, Bernd (2 Απριλίου 2013). "Global Diversity of the Placozoa". *PLOS ONE.* **8** (4): e57131. Bibcode:2013PLoSO...857131E. doi:10.1371/journal.pone.0057131. PMC 3614897. PMID 23565136. Βλ. και Knowlton N. (February 2000). "Molecular genetic analyses of species boundaries in the sea". *Hydrobiologia.* 420 (1): 73–90. doi:10.1023/A:1003933603879. ISSN 0018-8158. S2CID 40909997.

[19] Βλ."Before nerves, there were peptides," του Nick Carne, 18 Οκτωβρίου 2018, *Cosmos*, https://cosmosmagazine.com/biology/before-nerves-there-were-peptides

[20] *The Complete Works of William Shakespeare,* Edited with a Glossary by W. J. Craig, Oxford University Press, Oxford, UK, 1947, "Love's Labour's Lost," ll. 358-359, p. 159.

[21] Charles Tomlinson, από το "An Insufficiency of Earth," στο *A Peopled Landscape -Poems,* Oxford University Press, London, 1963, σελ. 21.

[22] *Hermit In Paris -Autobiographical Writings*, του Italo Calvino, Μετάφραση από την ιταλική του Martin McLaughlin, Pantheon Books, New York, 2003, σελ. 146.

[23] Βλ. *Objectivity*, των Lorraine Daston και Peter Galison, Zone Books, New York, 2007, σελ. 380.

[24] Βλ. "The creation of 'Ecosystem Core' hypothesis to explain ecosystem evolution," των Kun Wang και Xiajie Zhai, *BMC Ecology*, 19, Αριθμός άρθρου: 33 (2019), BMC Ecology, doi: https://doi.org/10.1186/s12898-019-0251-y, https://bmcecol.biomedcentral.com/articles/ 10.1186/s12898-019-0251-y

Δύο φίλοι σε ναό του Rajasthani. Φωτογραφία © J.G. Morrison.

Κεφάλαιο 3

ΦΙΛΟΚΟΙΝΩΝΙΚΗ ΣΥΜΠΕΡΙΦΟΡΑ

Την τελευταία δεκαετία[1], η φιλοκοινωνική συμπεριφορά όπως αυτή εκδηλώνεται σε όλα τα είδη, το να κάνουμε δηλαδή τους άλλους χαρούμενους ή οι ευγενικές χειρονομίες σε κάθε περίσταση της ζωής, υφίσταται, ως αντικείμενο έρευνας, μια κοινωνική επιστημονική αναγέννηση. Η έρευνα στον τομέα της ψυχικής υγείας δείχνει ξεκάθαρα μια "θετική επίδραση" και τη βελτίωση διαφόρων νευρώσεων, η οποία μπορεί να συνδεθεί άμεσα με θετικές κοινωνικές πράξεις μεμονωμένων ατόμων.[2] Η αναπτυξιακή παιδοψυχολογία έχει λάβει ιδιαίτερη ώθηση από αυτή την ευρετική έρευνα.[3] Από τους βιώσιμους γάμους και την υποστήριξη στους ηλικιωμένους έως τις γενικές παγκόσμιες μελέτες σχετικά με την ανθρώπινη ευτυχία, κάθε συσχετισμός δικτύων κοινωνικής υποστήριξης αποκαλύπτει καλύτερη συνοχή, ψυχική υγεία και γενική ευημερία μέσω της επίδειξης φιλοκοινωνικής συμπεριφοράς.[4]

Παρατηρείται μια υφιστάμενη προκατάληψη των ερευνών που επικεντρώνονται σε μια υποτιθέμενη υψηλότερη εξελικτική φιλοκοινωνικότητα του *H. sapiens* σε σχέση με όλα τα άλλα πρωτεύοντα θηλαστικά, και, για να παραφράσω τον Martin Luther King, Jr., ότι *ναι, το μίσος είναι πολύ μεγάλο βάρος για να το σηκώνει κανείς*.[5] Όντως, υφίστανται στοιχεία σχετικά με φόνους μεταξύ χιμπατζήδων και για περιστασιακές συμπλοκές μεταξύ άλλων πρωτευόντων ειδών.[6] Ή ότι ένα σταβλοχελίδονο "μπορεί να καταναλώσει 60 έντομα ανά ώρα ή τον τεράστιο αριθμό των 850 εντόμων τη μέρα"[7]. Τυχόν ωστόσο συγκρίσεις, οι οποίες αφορούν την τάση τους είδους μας προς τη βία σε σύγκριση με άλλα είδη, τελειώνουν πριν καν η επιχειρηματολογία ξεκινήσει.

Ωστόσο, η συντριπτική καθολικότητα των διαφόρων επιπέδων συνεργασίας, όπως αυτή γίνεται αντιληπτή μέσω της ιδιαίτερα προκατειλημμένης οπτικής μας, αναντίρρητα μας οδηγεί στην κεντρικότητα τόσο της θετικής επίδρασης όσο και του αποτελέσματος, μέσω συνδέσεων που είναι εξαιρετικά χρήσιμες, είτε μεταξύ κυττάρων, είτε σε ολόκληρο το φυτικό και ζωικό βασίλειο. Και, ασφαλώς, μεταξύ ημών, των ανθρώπων, οι οποίοι, από πολύ παλιά, κατηχηθήκαμε έντονα με το χωρίο που βρίσκουμε στο απόσπασμα 23:2 του βιβλίου της *Εξόδου, να μην ακολουθούμε τους πολλούς στο κακό.*

Ένα προφανές εμπειρικό *πλαίσιο* περιπτωσιολογικής μελέτης της ορνιθοπανίδας του βόρειου Νέου Μεξικού στα τέλη Οκτωβρίου 2020 αμφισβητεί τις εύκολες γενικεύσεις περί των 500 εκατομμυρίων τόνων (τετράκις εκατομμυρίων) εντόμων που καταβροχθίζουν συνολικά ετησίως σε όλη τη γη τα περισσότερα από 18.000 είδη πτηνών. Δεν πρόκειται για κάποια ριζοσπαστική εξίσωση που αναζητά ψευδείς βιολογικές ισοδυναμίες μεταξύ των εντομοφάγων πτηνών και της ανθρώπινης συμπεριφοράς, στην οποία άλλα σύνολα δεδομένων αποδίδουν εμφατικά τη θανάτωση από δύο έως πέντε τρισεκατομμυρίων σπονδυλωτών ετησίως. Επιπλέον, τουλάχιστον δύο δισεκατομμύρια είναι οι άνθρωποι οι οποίοι σκοτώνουν εκατοντάδες τρισεκατομμύρια ασπόνδυλα, μεταξύ άλλων κεφαλόποδα (χταπόδια, καλαμάρια κ.λπ.), θαλάσσια καρκινοειδή (π.χ. αστακούς) και αρθρόποδα, μια συνομοταξία στην οποία περιλαμβάνονται οι μεγαλύτερες ομάδες οργανισμών στον πλανήτη (εκτός από το φυτοπλαγκτόν, το οποίο μοιάζει με φυτό), ιδίως τα έντομα και οι αράχνες. Ένα φθινοπωρινό μεσημέρι, η Jane και εγώ παρακολουθήσαμε πάνω από 50 διαφορετικά πουλιά από τουλάχιστον 16 διαφορετικά είδη, από κοκκινολαίμηδες και καρακάξες μέχρι τρυγόνια, τρυποφράχτες, σπίνους, λούγαρα, αιγίθαλους και διάφορα είδη σπουργιτιών, τα οποία, μαζί με διάφορα είδη σκίουρων, καθώς επίσης και τα ησύχως εργαζόμενα σκαθάρια Alniphagus aspericollis, μοιράζονται όλα μαζί μια μεγάλη ποικιλία από τροφές, χωρίς ανταγωνισμό και χωρίς συγκρούσεις. Ένα κοπάδι ελαφιών περιπλανιέται στο τοπίο, βοσκώντας κλαδιά αρκεύθου, βελανίδια, άγρια σμέουρα, κέδρους, ιτιές και κρανιές. Η παραδεισένια αυτή σκηνή προκαλεί το δέος των συγκρίσεων: μπορεί μια ανθρώπινη ιδέα, καθηλωμένη σε ένα ιδανικό, να είναι ευάλωτη, με την βέλτιστη σημασία της λέξης, στη φυσική επιλογή;

Μπορεί η επιλογή να κινηθεί αρκετά γρήγορα ώστε να επανεξελίσσεται σε πραγματικό χρόνο, περνώντας από την ιδέα στην εφαρμοσμένη μηχανική της; Τα παραδείγματα των αδερφών Ράιτ ή της ατμομηχανής του Τόμας Σέιβερι θα έδιναν θετική απάντηση. Τι γίνεται όμως με τη μετάβαση από την ελπίδα στην υλική εκδήλωση; Από την πείνα στην παράδοση τροφής; Από την ασθένεια στη θεραπεία; Από τη μη βία στην πραγματική μη βία; Από ένα σημείο που μεταμορφώνεται σε μοτίβο, από το ένα στα πολλά;

Για αυτά τα άγρια πλάσματα της γειτονιάς, φαίνεται να υπάρχει, τουλάχιστον προς το παρόν, ένα ποσοστό επιβίωσης 100% κατά τη διάρκεια της αιφνίδιας καθόδου της νυχτερινής θερμοκρασίας στους -10 βαθμούς Κελσίου περίπου, και με μισό μέτρο χιόνι, μετά από τρεις μήνες καλοκαιρινής ξηρασίας. Αυτό το φαινομενικά χαμηλού στρες, συνεργατικό περιβάλλον υπάρχει την ίδια στιγμή που εκατοντάδες χιλιάδες, πιθανώς εκατομμύρια αποδημητικά πουλιά – ειδικότερα, "ωδικά πτηνά, σπουργίτια και μυγοχάφτες", τα οποία συνήθως πετούν νότια από το Νέο Μεξικό τον χειμώνα, έχουν πεθάνει, πιθανώς λόγω βλάβης στους πνεύμονες από τον καπνό που δημιουργήθηκε από τις μαινόμενες πυρκαγιές στη Δυτική Ακτή.[8] Σε αυτό το πλαίσιο μεγέθους τοπίου, τα δεδομένα που μπορεί κανείς να παρεμβάλει υπογραμμίζουν μόνο μία από τις αμέτρητες συνέπειες του Ανθρωπόκαινου. Ωστόσο, αυτή η ποικιλία ειδών αναμιγνύεται σήμερα χωρίς τυμπανοκρουσίες ή οποιαδήποτε ένδειξη εδαφικής αναστάτωσης. Με πολλή ευκολία και εμφατική φιλία, μοιράζονται την τροφή τους απολαμβάνοντας πλήρως τη συνεργατική αυτή σχέση τους. Δεν τρώνε έντομα και αγνοούν τα σκουλήκια στο έδαφος, γεγονός το οποίο θα μπορούσε θεωρητικά να υποστηρίξει κάποιο ιδανικό διατροφικής απελευθέρωσης. Την ευρεία δηλαδή διάδοση γηγενών τύπων τροφής, που κατά τα άλλα σπανίζει, σε όλο τον κόσμο, επιδιώκοντας ένα αποτέλεσμα που θα άλλαζε το βιολογικό μωσαϊκό, δίνοντας βάρος στην υπόσχεση ενός ιδανικού ανθρώπινου λυτρωτικού περάσματος από τη μη συμμετοχή στη συμμετοχή, μέσω συνεργασιών που βασίζονται στην παρατήρηση. Μια συνεργατική παροχή βοήθειας που δεν διαφέρει από οποιαδήποτε άλλη παροχή επισιτιστικής βοήθειας σε καιρούς εκτάκτων κρίσεων. Διάφορα είδη σε όλο τον κόσμο λιμοκτονούν. Δεν είναι καθήκον μας να τα βοηθήσουμε; Οι περισσότεροι κρατικοί νόμοι απαγορεύουν τη σίτιση των άγριων ζώων.[9]

Εξίσου γρήγορα επισημαίνουμε τις διφορούμενες αδυναμίες κάθε παρόμοιας παρέμβασης στο οικολογικό καθεστώς (οι ελεγχόμενες έναντι των φυσικών πυρκαγιών αποτελούν άλλη μια ιστορική περίπτωση). Η δύναμη, ωστόσο, της λογικής είναι μεγαλύτερη όταν η καρδιά αρνείται να συνεργαστεί λόγω ουσιαστικής αδιαφορίας. Σε πολλές πόλεις (μεταξύ των οποίων το Φοίνιξ και το Λας Βέγκας), το τάισμα των περιστεριών είναι παράνομο. Το ίδιο συμβαίνει και σε πολλά μέρη του Λονδίνου (π.χ. στην πλατεία Τραφάλγκαρ). Στο Παρίσι, για την πράξη αυτή επιβάλλεται πρόστιμο 450 ευρώ. Στην πολιτεία της Γιούτα, η κλοπή οποιουδήποτε ζώου το οποίο εκτρέφεται για εμπορικούς σκοπούς, είτε πρόκειται για σπουργίτι είτε για χοίρο που πρόκειται να σφαχτεί, θεωρείται κακούργημα. Πολλοί είναι εκείνοι οι οποίοι έχουν καταδικαστεί για κακουργήματα σχετικά με τα δικαιώματα και την απελευθέρωση ζώων και μαραζώνουν στη φυλακή σε πολλές πολιτείες και κράτη.[10]

Σήμερα όμως, εδώ, στο βόρειο Νέο Μεξικό, η συντροφικότητα μεταξύ των ειδών υπογραμμίζει έντονα και το άγχος του να είσαι άνθρωπος και να παρακολουθείς με έχθρα αυτή τη γιορτή μέσα από μια φούσκα. Οι θερμοκρασίες του περιβάλλοντος αέρα στις οποίες όλα αυτά τα πουλιά πρέπει να αντιμετωπίσουν, από άποψη φυσιολογίας, τις επίμονες ψυχρές θερμοκρασίες που τα έπληξαν αιφνίδια. Ο υψηλός μεταβολισμός, η αποθήκευση λίπους, οι σμηγματογόνοι αδένες, ο καλλωπισμός του φτερώματος, οι διάφορες τεχνικές μόνωσης για την αποθήκευση αέρα, η μεγάλη ημερήσια κατανάλωση τροφής, άλλα ανατομικά χαρακτηριστικά τα οποία όλα μαζί λειτουργούν για την αποτροπή των τεράστιων ποσοστών θνησιμότητας. Οι κίνδυνοι που προκύπτουν από τη μετανάστευση του 75% περίπου όλων των πτηνών της Βορείου Αμερικής, σε σχέση με τους κινδύνους από την παραμονή στον Βορρά κατά τους χειμερινούς μήνες[11]. Οι διάφοροι τρόποι αναπαραγωγής και τυχαίας περιπλάνησης και παρορμήσεων και ο συνδυασμός των ανθρωπογενών κινδύνων.[12] Πρόκειται για πάρα πολλά στοιχεία τα οποία είναι δύσκολο να αφομοιώσουμε.

Ωστόσο, μέσα σε όλη αυτή την ταλαιπωρία μέσα από την οποία εδώ και αιώνες και εκατομμύρια χρόνια τα πουλιά απέκτησαν τις στρατηγικές τους εσωτερικές παραδόσεις και χάρες, ιδέες και έννοιες, και τη σοφία που γεννήθηκε από τους μηχανισμούς ανατομίας στους οποίους εμπλέκεται κάθε μορφή μορφολογικής εφευρετικότητας ώστε να αντεπεξέλθουν, αυτό που ξεχωρίζει, εκτός από τον αέρα μέσα στα χνουδωτά φτερά τους, είναι επίσης και η τάση τους για άνευ όρων συνεργασία, παρά τους απρόβλεπτους, αντιδραστικούς νομοθέτες. Ο συγχρονισμός των διαδρομών της καλοσύνης με τις διαπιστώσεις της ανθρώπινης αντίληψης ξεκινά με την υπόθεση ότι "Η συνεργασία και η φιλοκοινωνική συμπεριφορά είναι χαρακτηριστικά γνωρίσματα της ανθρώπινης συμπεριφοράς, και γενικά πιστεύεται ότι η συνεργασίας μας σε αξιοσημείωτα επίπεδα έχει συμβάλει στην επιτυχία του είδους μας"[13]. Ο στόχος των μελετών τέτοιου είδους συνήθως βασίζεται στην επιθυμία να βρεθούν σταθερά ή φυλογενετικά συγκεκριμένα χαρακτηριστικά για τα οποία μπορεί να ειπωθεί ότι συγκλίνουν μεταξύ των ορίων των ειδών καθόλη τη διάρκεια της εξέλιξης, ειδικότερα μεταξύ των πρωτευόντων θηλαστικών και των πτηνών. Το 2016, ερευνητές του Πανεπιστημίου της Βιέννης δημοσίευσαν δεδομένα που υποδηλώνουν ότι οι κυανόφτερες κίσσες "επιδεικνύουν ανθρώπινη γενναιοδωρία".[14] Στο Loro Parque Fundación, στο Πουέρτο ντε λα Κρουθ της Τενερίφης, διαπιστώθηκε ότι ένας αφρικανικός γκρίζος παπαγάλος με το όνομα Bella "βοηθά τους άλλους να βρίσκουν τροφή" και "δεν ζηλεύει αν οι άλλοι παίρνουν καλύτερες λιχουδιές". Τα ευρήματα θεωρήθηκαν ισοδύναμα της θεωρίας ότι το "επίπεδο νοημοσύνης [των αφρικανικών γκρίζων παπαγάλων] είναι συγκρίσιμο με εκείνο των ανθρωπίδων, των κουρούνων και των δελφινιών".[15]

Προσπαθώντας να απομονώσει έναν μηχανισμό ο οποίος πιθανώς προωθεί την τάση και την εκδήλωση φιλοκοινωνικής συμπεριφοράς, μια ομάδα ερευνητών αναζήτησε το ισοδύναμο του "ορμονικού μηχανισμού του νευροπεπτιδίου της ωκυτοκίνης" των ανθρώπων, το αντίστοιχο του οποίου στα πτηνά μάλλον είναι η "μεσοτοκίνη". Αυτή φαίνεται σίγουρα ότι ενθαρρύνει φιλοκοινωνική συμπεριφορά που ομοιάζει με τη συμπεριφορά των θηλαστικών στις κίσσες, "γεγονός που δείχνει έναν εξελικτικά διατηρημένο ορμονικό μηχανισμό φιλοκοινωνικής συμπεριφοράς"[16]. Η συμπεριφορά αυτή μπορεί να ερμηνευθεί ως μια δέσμη χαρακτηριστικών ειλικρινών σημάτων, και, ως εκ τούτου, σταθερών, εξελικτικών χαρακτηριστικών. Και επειδή πρόκειται – όπως έχει ερμηνευθεί μέχρι στιγμής – για μια καθολική μορφή σηματοδότησης που είναι εγγενής σε κάθε μορφή συγγένειας και φροντίδας, η πιο ενδιαφέρουσα πτυχή αυτής της συμπεριφοράς είναι η εμφάνισή της σε ετεροειδικά τοπία – *μεταξύ των ειδών*.

Σημειωσεις

[1] Βλ. *In the Light of Evolution V: Cooperation*, Οργ. από Joan E. Strassmann, David C. Queller, John C. Avise, και Francisco J. Ayala, 7-8 Ιανουαρίου 2011, Irvine, California, National Academy of Sciences Colloquia, www.nasonline.org/SACKLER_cooperation.

[2] Snippe, E.; et al. (2017). "The Reciprocity of Prosocial Behavior and Positive Affect in Daily Life". *Journal of Personality*. 86 (2): 139–146. doi:10.1111/jopy.12299. PMID 28093772. Βλ. και Schwartz, C. E., Meisenhelder, J. B.; Yusheng, A., Reed, G. (2003). "Altruistic social interest behaviors are associated with better mental health". *Psychosomatic Medicine*. 65(5): 778–785. doi:10.1097/01.psy.0000079378.39062.d4. PMID 14508020.

[3] Για εκατοντάδες ερευνητικές μελέτες στο θέμα, βλ. https://psycnet.apa.org/ search/results?id=51e158b8-97f2-40db-f052-e436f82f393d. Βλ. Eisenberg, Nancy, Paul Henry Mussen (1989-08-25). *The Roots of Prosocial Behavior in Children*. Cambridge University Press. Βλ. και Eisenberg, Nancy; Fabes, Richard A; Spinrad, Tracy L (2007). "Prosocial Development". *Handbook of Child Psychology*. doi:10.1002/ 9780470147658.chpsy0311.

[4] Baumeister & Bushman (2007). *Social Psychology and Human Nature*. Cengage Learning. σελ. 254. Βλ. επίσης Helliwell, J. F.; Putnam, R. D. (2004). "The social context of well-being". *Philosophical Transactions of the Royal Society B: Biological Sciences*. 359 (1449): 1435–1446. doi:10.1098/rstb.2004.1522. PMC 1693420. PMID 15347534. Βλ. και Silk, J & House, B (2011). "Evolutionary foundations of human prosocial sentiments". *PNAS*. 108 Suppl 2: 10910–7. Bibcode:2011PNAS..10810910S. doi:10.1073/pnas.1100305108. PMC 3131813. PMID 21690372. Βλ. επίσης, Alden; Trew (2013). "If it makes you happy: Engaging in kind acts increases positive affect in socially anxious individuals". *Emotion*. 13 (1): 64–75. doi:10.1037/a0027761. PMID 22642341. Βλ. επίσης Penner, Louis A; Dovidio, John F; Piliavin, Jane A; Schroeder, David A (2005). "Prosocial Behavior: Multilevel Perspectives". *Annual Review of Psychology*. 56: 365–92. doi:10.1146/annurev.psych.56.091103.070141. PMID 15709940. Βλ. "I've Decided To Stick with Love," by Javon Collines, Covelli Enterprises, https://www.covelli.com/wp-content/uploads/2015/02/Covelli-MLK-1st-place-essay_w-covelli_logo.pdf

[5] *Philos Trans R Soc Lond B Biol Sci.* 12 Σεπ 2010, 365(1553): 2723–2735. doi: 10.1098/rstb.2010.0118, PMCID: PMC2936168, PMID: 20679115 "On the psychology of cooperation in humans and other primates: combining the natural history and experimental evidence of prosociality," Adrian V. Jaeggi, Judith M. Burkart και Carel P. Van Schaik, The Royal Society Publishing, *Philosophical Transactions B*, https://www.ncbi.nlm.nih.gov/ pmc/articles/PMC2936168. Πρόσβαση 27 Οκτωβρίου 2020.

[6] "How Human Violence Stacks Up Against Other Killer Animals - Humans get some of their knack for killing each other from the primate family tree, a new study says—but that doesn't let us off the hook, Erika Engelhaupt, 28 Σεπτεμβρίου 2016, *National Geographic*, https://www.nationalgeographic.com/news/2016/09/human-violence-evolution-animals-nature-science/

[7] Βλ. "Birds That Eat Mosquitoes," Meg Campbell, 22 Νοεμβρίου 2019, *Sciencing*, https://sciencing.com/birds-eat-mosquitoes-6620996.html, όπου αναφέρεται η Linda deKort του Flathead Audubon Society.

[8] https://blogs.findlaw.com/law_and_life/2019/12/can-you-stop-neighbors-from-feeding-dangerous-wild-animals-.html

[9] Βλ. "In New Mexico, the Birds Are Dying at Alarming Rates," Claire Lampen, *The Cut*, 14 Σεπτεμβρίου 2020, https://www.thecut.com/2020/09/new-mexicos-migratory-bird-die-offs-are-unprecedented.html, πρόσβαση 27 Οκτωβρίου 2020.

[10] Βλ., για παράδειγμα, "Animal rights activists, facing 60 years in prison, to visit farm where they removed baby pig during 'open rescue'," Taylor Hartman, 6 Αυγούστου 2018, *Fox 13*, Salt Lake City, https://www.fox13now.com/2018/08/06/animal-rights-activists-facing-60-years-in-prison-visit-farm-where-they-removed-baby-pig-during-open-rescue/

[11] "How Many Birds Disappear Between Migration Seasons? We Now Have a Clue," της Pubrita Saha, *Audubon Magazine*, Winter 2018, https://www.audubon.org/magazine/winter-2018/how-many-birds-disappear-between-migration, Accessed October 27, 2020. Βλ. επίσης "Weather-related mass-mortality events in migrants," από I. Newton, IBIS, 15 Μαΐου 2007, https://doi.org/10.1111/j.1474-919X.2007.00704.x, https://onlinelibrary.wiley.com/doi/10.1111/j.1474-919X.2007.00704.x

[12] "A Summary and Comparison of Bird Mortality From Anthropogenic Causes with an Emphasis on Collisions," των Wallace P. Erickson, Gregory D. Johnson και David P. Young Jr., *USDA Forest Service Gen. Tech. Rep. PSW-GTR-191*. 2005, https://www.fs.fed.us/psw/ publications/documents/psw_gtr191/psw_gtr191_1029-1042_erickson.pdf

[13] "Studying the evolution of cooperation and prosciality in birds," του Jorg J M Massen, Ethology 126(2): 121-124, doi: 10.111/eth.13001, https://onlinelibrary.wiley.com/doi/10.1111/ eth.13001 https://www.researchgate.net/

publication/338555319_Studying_the_evolution _of_cooperation_and_prosociality_in_birds, Πρόσβαση στις 10 Δεκεμβρίου 2019, Ιανουάριος 2020.

[14] Lisa Horn, Clara Scheer, Thomas Bugnyar, Jorg J. M. Massen. Proactive prosociality in a cooperatively breeding corvid, the azure-winged magpie (Cyanopica cyana). *Biology Letters,* 2016; 12 (10): 20160649 doi: 10.1098/rsbl.2016.0649, αναφέρεται στο Πανεπιστήμιο της Βιέννης. "Azure-winged magpies show human-like generosity: Prosocial behavior due to cooperative breeding." *ScienceDaily. ScienceDaily*, 18 Οκτωβρίου 2016. <www.sciencedaily.com/releases/2016/10/161018193150.htm>. https://www.sciencedaily.com/releases/2016/10/161018193150.htm

[15] Βλ. "Prosocial and tolerant parrots help others to obtain food," Max-Planck-Gesellschaft, n.a., 9 Ιανουαρίου 2020, https://www.mpg.de/14328111/prosocial-grey-parrots, Πρόσβαση 28 Οκτωβρίου 2020.

[16] "Mesotocin influences pinyon jay prosociality," των J F Duque, W Leichner, A Ahmann, και J R Stevens, National Library of Medicine, *PubMed.gov*, PMID: 29643220, PMCID: PMC5938566, doi: 10.1098/rsbl.2018.0105, Biol Lett 2018 Apr;14(4):20180105. doi: 10.1098/rsbl.2018.0105. https://pubmed.ncbi.nlm.nih.gov/29643220/

Κεφάλαιο 4

Η ΑΝΑΛΟΓΙΑ ΤΩΝ ΧΕΙΡΑΨΙΩΝ

Αναφέρεται συχνά ότι ο Τόμας Χομπς (1588-1679) έγραψε στην αυτοβιογραφία του πως η μητέρα του γέννησε στις 5 Απριλίου 1588 δυο δίδυμα παιδιά: τον πρόωρο Τόμας και τον φόβο.[1] Το έργο του Χομπς "*Λεβιάθαν ή Η ύλη, μορφή και εξουσία μιας εκκλησιαστικής και λαϊκής πολιτικής κοινότητας*" (Λονδίνο, 1651) συνοψίζεται τελικά σε μισή πρόταση η οποία χαρακτηρίζει την τόσο δυϊστική κατάσταση της ανθρωπότητας: "συνεχής φόβος και κίνδυνος βίαιου θανάτου, και η ζωή του ανθρώπου, μοναχική, φτωχή, δυσάρεστη, βάναυση και σύντομη".[2] Γραμμένο δεκαεπτά χρόνια πριν το έργο του με τίτλο "*Behemoth: the history of the causes of the civil wars of England, and of the counsels and artifices by which they were carried on from the year 1640 to the year 1660" (γνωστό επίσης και ως "The Long Parliament*"), το *Λεβιάθαν* φέρει το βάρος ενός βιβλίου του οποίου η συγγραφή καθοδηγήθηκε από συμφορές. Τις φαινομενικά ατελείωτες προσβολές της αξιοπρέπειας χωρίς κανένα σημάδι επικείμενης βελτίωσης της κατάστασης. Την εμφάνιση της πραγματικής επιστήμης μέσα από τον τρόμο, τον πόνο και πάρα πολλές περιγραφές αυτοπτών μαρτύρων περί της επικρατούσας αναρχίας. Και την ίδια τη μοναρχία, που συγκρούεται καθόλη τη διάρκεια της ιστορίας καθοδηγούμενη σταθερά από όλο τον εγωισμό και τη σκληρότητα που μπορεί να υπάρχει μεταξύ αντιπάλων, από το αιματοβαμμένο πεδίο της μάχης του Έτζχιλ τον Οκτώβριο του 1642, μέχρι τις πολιορκίες, τους αποκεφαλισμούς, τις προδοσίες και τις αποκρουστικές συμπλοκές που ακολούθησαν. Εκστρατείες, εξεγέρσεις, προδοτική στάση και ατελείωτες συγκρούσεις. Μόλις δύο χρόνια πριν τη δημοσίευση του *Λεβιάθαν*, στο μυαλό οποιουδήποτε αναφερόταν στην ανθρώπινη φύση των Βρετανών ήταν χαραγμένο το γεγονός της θανάτωσης των Ιρλανδών Βασιλικών από τον Όλιβερ Κρόμγουελ, κατά τη διάρκεια της οποίας περίπου 3.500 άνθρωποι σφαγιάστηκαν στην πολιορκία της Ντρογκέντα.[3] Λες και αυτή ήταν η μοναδική εναλλακτική λύση για την αντιμετώπιση της *ευγενούς αγριότητας* που προτίμησαν να φανταστούν με νοσταλγία και να συλλάβουν ως ιδέα αυτοί που αντιτάχθηκαν στο έντονο γεωπολιτικό όραμα του Χομπς.

Η αστική αυτή εναλλακτική λύση είναι κρίσιμη για την αναπαράσταση του ιδανικού που παραμένει ακλόνητο σε αυτό που στις μέρες μας υπονοείται ως η πραγματικά τελευταία μας ελπίδα για τη σύλληψη ενός αντίχειρα της φιλοσοφίας και της ηθικής. Όχι την αγριότητα, αλλά την ευγένεια κάποιας πρωταρχικής αλήθειας: τι σημαίνει να είναι κανείς μέλος του ανθρώπινου είδους, μαζί με πιθανώς 100 εκατομμύρια άλλα είδη, ούτε περισσότερα, ούτε λιγότερα. Ο Χομπς υπήρξε ακριβής. Το αντίδοτο στις κατηγορίες του δεν έχει δοκιμαστεί ακόμα για λόγους που περιλαμβάνουν τις *αναμφισβήτητες* προκαταλήψεις, την απατηλή αυτοεκτίμηση και το καταστρεπτικό ωφέλιμο φορτίο της ανθρώπινης ψυχιατρικής, της δημογραφίας, του υπερεγώ και της οδυνηρής οικολογικής φέρουσας ικανότητάς του.

Αν επιλέξουμε να παρακάμψουμε με κάποιον μεταφυσικό τρόπο τους Εμφύλιους Πολέμους της Αγγλίας και της Αμερικής, τη Γαλλική Επανάσταση, χιλιάδες άλλες αιματοκυλισμένες εμφύλιες διαμάχες -από τους Αζτέκους μέχρι το Αφγανιστάν- θα μπορούσαμε να βρεθούμε τυχαία στο άκρο της ίδιας λεύκας την οποία η κίτρινη πεταλούδα Μαχάων - η οποία αναφέρθηκε για πρώτη φορά από το είδος μας στην *Ιλιάδα* του Ομήρου (και για αυτό ονομάστηκε *Papilio homerus*) - αναγνωρίζει ως το κατάλληλο μέρος για να γεννήσει τα αυγά της. Στον κόσμο αυτό πρέπει να συνεισφέρουμε, όχι να τον πολεμούμε. Στον βαθμό που γνωρίζουμε, 550 χαρισματικά είδη της οικογένειας των Παπιλιονιδών βρίσκονται σε όλες τις ηπείρους εκτός από την Ανταρκτική. Η αξιοθαύμαστη αυτή πεταλούδα (η λατινική της ονομασία ορίζει την υπόσταση όλων των λεπιδόπτερων) βασίζεται για την επικονίαση και τη διατροφή της κυρίως σε μια εξίσου τεράστια οικογένεια φυτών, ειδικότερα τα Rutaceae (Ρουτίδες), με περίπου 160 γένη σε δύο τεράστιες υποοικογένειες, τα Rutoideae και τα Cneroideae[4]. Το κοσμοπολίτικο αυτό σύνολο συμβιωτικών σχέσεων μεταξύ των πεταλούδων και των φυτών χρονολογείται βιογεωγραφικά στη Βερίγγεια γέφυρα, στο Βόρειο Ημισφαίριο, τουλάχιστον, με μια γενική εκτίμηση, από την εποχή του Ηώκαινου, περίπου 52 εκατομμύρια χρόνια πριν.[5]

Αξίζει να αξιοποιήσουμε ως βάση τα δεδομένα που έχουμε σήμερα για τις πεταλούδες για την εξαγωγή συμπερασμάτων για τα περισσότερο ορατά δίποδα σπονδυλωτά. Αν κάνουμε τις ίδιες ερωτήσεις και στις δυο ομάδες, πρέπει να αναμένουμε ότι οι απαντήσεις θα διαφέρουν μόνο ανάλογα με την προϋπάρχουσα λογική μας βάση, τα σενάρια που έχουμε αποδεχτεί ως βασισμένα σε συγκεκριμένες μεταφορές *στο μυαλό* μας, όπως η εξέλιξη, ή κάποιος Χομπς. Στον βαθμό που είμαστε προσηλωμένοι σε αυτά τα μοτίβα της φύσης όπως τα έχουμε περιγράψει, ορίσει, περιγράψει και τονίσει, παραμένουμε αιχμάλωτοι του συμβολισμού αυτού ο οποίος μετατράπηκε σε αρχή, των Νόμων δηλαδή της Φύσης τους οποίους επινοήσαμε. Επιβεβαιώνοντας την ανθρώπινη ταυτότητα ως το κέντρο της γήινης αυτής μαύρης τρύπας, ο πολλαπλασιασμός των εικασιών μας δεν προσφέρει παρά απειροελάχιστες ενδείξεις σχετικά με τον τρόπο με τον οποίο θα μπορούσαμε να δραπετεύσουμε από την εγγενή αυτοαναφορικότητά μας μπροστά στον καθρέφτη, σχεδόν όπως στο "πλέον διάσημο παράδοξο της φυσικής"[6]. Εάν οι συνεργατικές παρορμήσεις που βλέπουμε να επιδεικνύονται φυλογενετικά σε κάθε οικοσύστημα όντως μας εντυπωσιάζουν, αλλά μας κρατούν επίσης και ζωντανούς εκ των έσω (ενδοσυμβίωση), τότε γιατί, λες και έχουμε στιγματιστεί από τον Χομπς, εξακολουθούμε να εκδηλώνουμε συμπεριφορικά στοιχεία ισοδύναμα με αντισώματα ενάντια στα βασικά ζωτικά στοιχεία της βιολογίας μας; Σημειώνουμε ότι η βιολογία *μας* είναι η βιολογία όλων - η "δασκάλα χταπόδι", ο Μπάμπι, μια αράχνη που ονομάζεται Σάρλοτ. Ενώ η επιθετικότητα είναι εμφανής, η "αποφυγή του ανταγωνισμού" και μια πιο γενναιόδωρη και σύνθετη τοπολογία από αυτή που αποδίδεται περιστασιακά στα Σερενγκέτι του κόσμου, φαίνεται πιο πιθανό να είναι κεντρικός παράγοντας για την κατανόηση μιας "σταθερής" συνεργατικής θεωρίας της εξέλιξης, μεταξύ φυτών και ζώων ή μικροβίων.[7,8,9]

Αναλογιζόμενοι την "υπαρξιακή καταστροφή"[10], οι ερευνητές τείνουν να επικεντρώνονται στη νευρολογική αρχιτεκτονική που προσανατολίζεται στην ελαχιστοποίηση των απειλών και στο σχετικό στρες.[11] Ωστόσο, οι πραγματικές γνωστικές κατηγορίες είναι περιορισμένες. Τα γεγονότα που πρόκειται να εξελιχθούν σε βάθος αιώνα δεν μας ανησυχούν ιδιαίτερα, πόσω μάλλον σε βάθος ενός δισεκατομμυρίου ετών από τώρα. Δεν γνωρίζουμε καμία μορφή ζωής που να την ανησυχούν, όσο και αν θαυμάζουμε τα μακρόβια είδη όπως η Πεύκη η μακραίωνη (Pinus longaeva) ή ο κόκκινος αχινός (Mesocentrotus franciscanus).

Όμως, οι μηχανισμοί της ομαλής συμβίωσης, του συντονισμού της παρόρμησης πέρα από την αντίδραση φόβου ή φυγής, την απληστία ή την *απλή* προσωπική ανάγκη σε ευρύτερες συγκινησιακές αρένες που συνορεύουν τουλάχιστον με τη συνεργασία, αποτελούν μαγευτικές παγίδες αντίληψης που συνιστούν προμηνύματα τα οποία είναι τόσο δυσοίωνα για το μέλλον όσο και ευχάριστα για το παρόν. Το αντιληπτικό αυτό στάδιο της εξέλιξής μας εκτείνεται πέρα από το πλήθος των αμοιβαίων εκδηλώσεων αλτρουισμού που παρατηρούνται, από τις συμβιωτικές σχέσεις στα θαλάσσια περιβάλλοντα[12] έως τις προειδοποιήσεις μεταξύ συμπατρικών και αλλοπατρικών ειδών[13]. Στο είδος μας, αυτή η εννοιολογική μετάβαση σηματοδοτεί την αλλαγή της αναλογίας των αποτυπωμάτων της παλάμης σε μια πολύ μεγαλύτερη αναλογία χειραψιών.

ΣΗΜΕΙΩΣΕΙΣ

[1] Βλ. "Thomas Hobbes: A grim portrait of human nature," Lou Marinoff, *The Times Literary Supplement*, 5, 2020, https://www.the-tls.co.uk/articles/thomas-hobbes-footnotes-to-plato/ Πρόσβαση 29 Οκτωβρίου 2020.

[2] Oxford University Press, *Clarendon Edition of the Works of Thomas Hobbes*, Oxford University Press, 2012, XIII.9.

[3] Leniham, Pádraig (2008), *Consolidating Conquest: Ireland 1603–1727*, Pearson Education Ltd., Harlow, UK, σελ. 128.

[4] *Rutaceae (Οικογένεια Κίτρου) – 245 φωτογραφίες σχετικά με τη φυλογονία και την ονοματολογία του είδους στο PlantSystematics.org*. plantsystematics.org.

[5] "Global biogeographical pattern of swallowtail diversification demonstrates alternative colonization routes in the Northern and Southern hemispheres," των Fabien L. Condamine, Felix A. H. Sperling και Gael J. Kergoat, *Journal or Biogeography*, Wiley Online Library, 19 Οκτωβρίου 2012, https://doi.org/10.1111/j.1365-

2699.2012.02787.x, https://onlinelibrary.wiley.com/doi/10.1111/j.1365-2699.2012.02787.x. Ασφαλώς, δεν πρέπει να αγνοούνται οι αυτοπροστατευτικοί μηχανισμοί άμυνας. Ένα αξιοσημείωτο παράδειγμα βοτανολογίας προέρχεται από το Ηώκαινο, "Urticaceae leaves with stinging trichomes were already present in latest early Eocene Okanogan Highlands, British Columbia, Canada," των Melanie L. DeVore, Alphonse Nyandwi, Winnie Eckardt, Elias Bizuru, Myriam Mujawamariya, και Kathleen B. Pigg, *American Journal of Botany*, 22 Οκτωβρίου 2020, https://doi.org/10.1002/ajb2.1548, https://bsapubs.onlinelibrary.wiley.com/doi/abs/10.1002/ajb2.1548

[6] Βλ. "The Most Famous Paradox in Physics Nears Its End," του George Musser, *Quanta Magazine*, 29 Οκτωβρίου 2020, https://www.quantamagazine.org/the-black-hole-information-paradox-comes-to-an-end-20201029/?utm_source=pocket-newtab, Πρόσβαση 31 Οκτωβρίου 2020.

[7] "The evolution of cooperation," των R. Axelrod και W D Hamilton, *Science* 27 Μαρ 1981: Vol. 211, Issue 4489, σελ. 1390-1396, doi: 10.1126/science.7466396, https://science.sciencemag.org/content/211/4489/1390

[8] Βλ. Roberto Cazzolla Gatti. A conceptual model of new hypothesis on the evolution of biodiversity. *Biologia*, 2016; 71 (3) doi: 10.1515/biolog-2016-0032

[9] Βλ. "Symbiosis as a General Principle in Eukaryotic Evolution," της Angela E. Douglas, *Biology, CSH Perspectives*, 2014, http://m.cshperspectives.cshlp.org/content/6/2/a016113.short. Για μια σύντομη βιβλιογραφική σύνοψη πολλών ερευνητικών άρθρων για τις συμβιωτικές σχέσεις των μικροβίων, βλ.:https://scholar.google.com/scholar?q=symbiosis+ among+eukaryotes&hl=en&as_sdt=0&as_vis=1&oi=scholart#d=gs_qabs&u=%23p%3DTobRGg1fj2YJ

[10] "The Survival of Humanity," του Lawrence Rifkin, *Scientific American*, 13 Σεπτεμβρίου 2013, https://blogs.scientificamerican.com/guest-blog/the-survival-of-humanity/

[11] Βλ. "The ecology of human fear: survival optimization and the nervous system, των Dean Mobbs, Cindy C. Hagan, Tim Dalgleish, Brian Silston και Charlotte Prévost, PMC, *Frontiers in Neuroscience*, 2015; 9: 55. <arcj 19. 2015, doi: 10.3389/finis.2015.00055, PMCID: PMC4364301, PMID: 25852451, https://www.ncbi.nlm.nih.gov/pmc/articles/PMC4364301/

[12] Feder, H.M. (1966). "Cleaning symbioses in the marine environment". Στο έργο του Henry, S.M. *Symbiosis. 1.* Academic Press. σελ. 327–380. OCLC 841865823.

[13] Olendorf, Robert; Getty, Thomas; Scribner, Kim (22 Ιανουαρίου 2004). "Cooperative Nest Defence in Red–winged Blackbirds: Reciprocal Altruism, Kinship or By–product Mutualism?". *Proc. R. Soc. Lond. B.* 271 (1535): 177–182.

Κεφάλαιο 5

ΒΙΑ ΚΑΙ ΟΥΤΟΠΙΑ

Οι άνω των 560 πολυμερείς συνθήκες (εκ των οποίων οι 500 αφορούν το περιβάλλον, πλέον των άνω των 3.716 άλλων περιβαλλοντικών συμβάσεων που βρίσκονται εν ισχύ στα διάφορα έθνη) οι οποίες βρίσκονται στις μέρες μας εν ισχύ εντός των παγκόσμιων πολιτικών συνόρων[1] έλκουν την καταγωγή τους από την παλαιότερη γνωστή, χαραγμένη σε ιερογλυφικά, ειρηνευτική συμφωνία μεταξύ ανθρώπων, την "Αιώνια Συνθήκη", η οποία καταρτίστηκε στα μέσα του 13ου αιώνα π.Κ.Ε. από τους Αιγυπτίους και τους Χετταίους κατά τη διάρκεια της βασιλείας των Ραμσή Β' και Χατουσίλις Γ'. Δεν προκαλεί έκπληξη το γεγονός ότι μέρος της συμφωνίας αυτής αφορούσε την επί μακρόν παρατεταμένη κατάπαυση του πυρός στη Συρία.[2] Οι απρόθυμες συναινέσεις, οι παρατεταμένες διαπραγματεύσεις, η εκτόνωση των εντάσεων και όλα τα πρωτόκολλα που θεσπίστηκαν από τα (επί του παρόντος) 116 κράτη-μέλη της Σύμβασης της Βιέννης για το Δίκαιο των Συνθηκών (ΣΒΔΣ)[3] πιθανώς υποδηλώνουν ότι, εάν δεν υπήρχε η αντισταθμιστική αυτή βιολογική τάση προς την επίλυση συγκρούσεων, ο αριθμός των 250-500 εκατομμυρίων ανθρώπων που έχουν δολοφονηθεί από την εποχή της Ευρωπαϊκής Αναγέννησης θα ήταν πολύ μεγαλύτερος. Και, πράγματι, τα ποσοστά των ανθρωποκτονιών στην Ευρώπη κατά τη διάρκεια της τελευταίας χιλιετίας έχουν όντως μειωθεί. Ωστόσο, σε άλλα μέρη του κόσμου, μεταξύ άλλων και στις ΗΠΑ, τα ποσοστά παρουσιάζουν διακύμανση, καθώς ο μέσος όρος των θανάτων παραμένει στους 5 θανάτους ανά 100.000 ανθρώπους στην Αμερική και 6,2 θανάτους ανά 100.000 ανθρώπους παγκοσμίως.[4,5]

Οι πάνω από 240 γνωστοί στην καταγεγραμμένη ανθρώπινη ιστορία πόλεμοι δεν προηγούνται χρονικά των περσικών συγκρούσεων του 6ου αιώνα π.Κ.Ε. Για παράδειγμα, είναι γνωστό ότι, όπως αποκαλύφθηκε σε μια τοποθεσία στην Κένυα γνωστή ως Ναταρούκ, τουλάχιστον 27 άνθρωποι σκοτώθηκαν στο τέλος της Λίθινης Εποχής, πριν από περίπου 10.000 χρόνια. Ούτε οι αριθμοί των μαχών, των εκκαθαρίσεων, των γενοκτονιών ή της από ανθρωποκεντρική προσέγγιση σταδιακά αυξανόμενης προκαλούμενης και μη θνησιμότητας περιλαμβάνουν "τις εκτιμήσεις με τους μεγαλύτερους αριθμούς”. Για παράδειγμα, ορισμένοι ερευνητές έχουν υποθέσει ότι ο συνολικός αριθμός των θανάτων που προκλήθηκαν από τις εισβολές/αποικισμούς των Ευρωπαίων στο δυτικό ημισφαίριο μέσα σε διάστημα περίπου πέντε αιώνων προσεγγίζει τα 140 εκατομμύρια.[6] Εκτιμάται ότι, για κάθε άνθρωπο ο οποίος βρίσκεται σήμερα εν ζωή, έχουν πεθάνει πριν από αυτόν περίπου 15 άτομα, δηλαδή πάνω από 107 δισεκατομμύρια άνθρωποι τα τελευταία 50.000 χρόνια.[7] Εάν υπολογίσουμε με βάση αυτά τα δεδομένα, θα βρούμε ότι από τις απαρχές της ύπαρξής μας, περίπου 330.000 χρόνια πριν, έχουν γεννηθεί και πεθάνει συνολικά μερικά ακόμα δισεκατομμύρια άνθρωποι. Η ανάλυση της τρέχουσας βιομάζας του πλανήτη αποκαλύπτει ότι η ζωντανή παρουσία της ανθρωπότητας αποτελεί ένα μικρό κλάσμα του αριθμού των Άλλων. Για παράδειγμα, σήμερα μπορεί να υπάρχουν πάνω από δέκα τρισεκατομμύρια ζωντανά μυρμήγκια, και να έχουν περάσει πάνω από δέκα εκατομμύρια προηγούμενες γενιές μυρμηγκιών.[8]

Και παρόλο που υπολογίζεται ότι υπάρχουν σχεδόν 200 ακόμα εκατομμύρια είδη τα οποία δεν έχουν ακόμη ανακαλυφθεί και γύρω στα 750 εκατομμύρια άλλα είδη τα οποία έχουν εξαφανιστεί από τότε που άρχισε η ζωή (και οι δυο αριθμοί στα ανώτερα όρια των αξιόπιστων εκτιμήσεων)[9], ο λόγος της έμβιας ανθρώπινης βιομάζας προς όλα τα άλλα άγρια σπονδυλωτά και τα σπονδυλωτά που βρίσκονται σε αιχμαλωσία (σε αιχμαλωσία = “εξημερωμένα” ζώα για ανθρώπινη εκμετάλλευση) αποδίδει εκπληκτικά δυσανάλογους συντελεστές οικειοποίησης. Από το 1970, οι άνθρωποι έχουν εξοντώσει τουλάχιστον το 60% όλης της άγριας ζωής του πλανήτη, εξαιρουμένων των ασπόνδυλων και του καθαρού ρυθμού πρωτογενούς παραγωγής μας η οποία, με τη σειρά της, έχει μετατρέψει τουλάχιστον το “44 τοις εκατό” του αρόσιμου πλανήτη σε αγροτεμάχια για τους ανθρώπους.[10] Πριν από χιλιάδες χρόνια, ο ανθρώπινος πληθυσμός παγκοσμίως δεν ξεπερνούσε πιθανώς τα 5 εκατομμύρια άτομα, και δεν αντιπροσώπευε ούτε το 1% του συνολικού βάρους των σπονδυλωτών. Το 2020, το είδος μας αντιστοιχεί περίπου στο 32% του συνολικού βάρους των σπονδυλωτών, ενώ ο αριθμός των ζώων σε αιχμαλωσία αντιστοιχεί συνολικά στο 67% ή και παραπάνω. Κατά συνέπεια, ο αριθμός των υπόλοιπων ειδών της άγριας ζωής αντιστοιχεί σε κάτι λιγότερο από το 1% του συνόλου.[11] Από τα είδη που παραμένουν, σε κάθε ανθοφόρο φυτό του είδους “Penstemon haydenii”, κάθε σεκόγια ή κάθε γαλάζια φάλαινα αντιστοιχούν περίπου 400.000 άνθρωποι. Σε κάθε Πάνθηρα της Φλόριντα ή καριμπού αντιστοιχούν 156 εκατομμύρια άνθρωποι.[12] “Επειδή αυτός αγνόησε τους εφιάλτες και καταπάτησε τα λιβάδια τους”, έγραψε ο Ουίσταν Χιου Ώντεν στο ποίημά του “The Dirge”.[13]

Πρόκειται για μια άνευ προηγουμένου προβληματική παραχάραξη των στοιχείων καταγραφής των δεδομένων της οικολογίας. Το “doom scrolling” (η κατανάλωση θλιβερών και καταστροφικών ειδήσεων) παίζει σημαντικό ρόλο στην αριθμητική περιγραφή των προβλημάτων που προκαλεί στα υπόλοιπα σπονδυλωτά, τα ασπόνδυλα και τους βιότοπους η οικειοποίηση των πάντων από τον άνθρωπο. Δεν μας προσφέρει κανένα στοιχείο σχετικά με τους βιότοπους τους οποίους φιλοξενούμε εμείς οι ίδιοι, βιότοποι οι οποίοι αποτελούνται από εκατοντάδες δισεκατομμύρια συμβιωτικά πλάσματα, κυρίως γύρω στα 35 με 100 τρισεκατομμύρια βακτηριακά κύτταρα.[14,15] Μας προσφέρει, ωστόσο, με μαθηματική ακρίβεια, πολλές θλιβερές πληροφορίες σχετικά με την κυριαρχία μας επί του βιοχημικού χάους μέσω της ταξινόμησης των πάντων σε βαθμίδες και κατηγορίες. Στην υπερβολικά καταστροφική αυτή μέτρηση αναλογιών και ποσοστών οδήγησαν αθροιστικά οι πολλές βιομηχανικές επαναστάσεις. Ξεκίνησε σε μέρη όπως η χαλδαϊκή Ουρ κατά τη διάρκεια της τρίτης χιλιετίας π.Κ.Ε. Εκεί σφαγιάστηκαν συστηματικά χιλιάδες βοοειδή, *δαμάστηκε* το άλογο, υποδουλώθηκαν αμέτρητοι εργάτες και άρχισε να χαράσσεται στην πέτρα η σχολαστική καταμέτρηση αποθεμάτων και πλεονασμάτων. Πολεμικά εξαρτήματα στόλιζαν δεκάδες χιλιάδες στρατιώτες οι οποίοι, συγκεντρωμένοι σε φάλαγγες, φορώντας χιτώνες και θώρακες με ατσάλινες ραβδώσεις, προέλαυναν προς τη μάχη. Οι πιο γνωστές προσωπικότητες που επιδόθηκαν σε τέτοιου είδους αιματοχυσίες χάριν του υπερεγώ τους ήταν ο Σαργών (περίπου 24ος-23ος αιώνας π.Κ.Ε.) και ο Ναράμ-Σιν από τη βόρεια Ακκαδία (2254-2218 π.Κ.Ε.). Η επινίκια στήλη του Σιν στο Λούβρο παρουσιάζει τον πολεμιστή να κατακτά ένα βουνό, με τον στρατό του να τον επευφημεί.

Στη συνέχεια είχαμε τον Καρτέσιο (1596-1650), ο οποίος βασάνιζε με αυταρέσκεια ζώα προς τέρψη των συναδέλφων του, οι οποίοι είχαν διαμορφώσει μια ευρέως διαδεδομένη πεποίθηση ότι τα πλάσματα αυτά τα οποία σφάδαζαν αντιδρούσαν στον πόνο μηχανικά, χωρίς πραγματικά συναισθήματα.[17] Η διαμόρφωση των πεποιθήσεων αυτών συνέπεσε με τη μαζική αποστράγγιση των ελών σε όλη την Ευρώπη, την εφεύρεση των υψικαμίνων, των τεράστιων υφαντουργικών μύλων με τις ρυπογόνες βαφές και την επιμονή του Μακιαβέλι στη χρήση πρόσφορων μέσων και αυταρχικής εποπτείας.[18] Ακόμη και στην *Ουτοπία* του Τόμας Μορ (1516), για την εκτέλεση των εργασιών που δεν καταδέχονταν να κάνουν οι 80.000 κάτοικοι της καθεμιάς από τις 45 πόλεις εισάγεται η χρήση σκλάβων. Ταυτόχρονα, ο Μορ τόνιζε την ιδέα ότι οι κάτοικοι της Ουτοπίας του, στον βωμό της φυσικής θρησκείας και της αποχής από την υλική απληστία, ήταν σε θέση να προσφέρουν τις ειρηνευτικές υπηρεσίες τους σε όλο τον κόσμο. Όταν ο σερ Τόμας Μορ ανήλθε στην εξουσία ως Λόρδος Καγκελάριος, δεν έδειξε να ενστερνίζεται τη φιλοσοφία επιείκειας και καλοσύνης που προσπαθούσε να μεταδώσει με τη σάτιρά του, και αδιαφόρησε για τις διάφορες θανατικές ποινές που επιβλήθηκαν κατά την ταραχώδη βασιλεία του Ερρίκου Η' και της Αικατερίνης της Αραγωνίας, μέχρι τη στιγμή που ο ίδιος κατέληξε στον Πύργο του Λονδίνου και αποκεφαλίστηκε με την κατηγορία της προδοσίας το 1535.

Το 1829, μια ατμομηχανή ονόματι *Rocket*, σχεδιασμένη στο Νιούκασλ από κάποιον Ρόμπερτ Στίβενσον, διένυσε δώδεκα μίλια σε 53 λεπτά. Με αυτόν τον τρόπο, έθεσε ένα ανόητο ιστορικό προηγούμενο, τόσο στην Αγγλία όσο και σε όλη τη γη. Ένας από τους πατέρες των αρχών της θερμοδυναμικής, ο Σαντί Καρνό (1796-1832) έγραφε στα απομνημονεύματά του ότι "Ήδη με τη θερμική μηχανή λειτουργούν τα ορυχεία μας, κινούνται τα πλοία μας, εκτελούμε έργα εκβάθυνσης στα λιμάνια και τα ποτάμια μας, σφυρηλατούμε τον σίδηρο, διαμορφώνουμε το ξύλο, αλέθουμε τα σιτηρά μας, γνέθουμε και υφαίνουμε τα υφάσματά μας, έλκουμε τα πλέον βαρέα φορτία..."[19] Μέχρι τα μέσα του

19ου αιώνα, όπως έχει συνοψίσει ο ιστορικός Πωλ Τζόνσον, "δεν υπήρχε ορατή καμία έξοδος διαφυγής…[και] καθώς οι δυτικές ιδέες εξαπλώνονταν, όλα και μεγαλύτερα τμήματα της επιφάνειας της γης απαιτούσαν διεθνή αστυνόμευση… Θα έχανε ο αστυνόμος του νέου κόσμου τον έλεγχο στο δικό του σπίτι; Θα βαστούσε το κέντρο;".[20] Αντίθετα με την κατάσταση ευφορίας του Καρνό, ο Τζον Στιούαρτ Μιλ σκεφτόταν με περισσότερη νηφαλιότητα έναν κόσμο του οποίου η λειτουργία καθοδηγείτο από τον άνθρωπο, ο οποίος είχε καθυποτάξει κάθε δάσος και κάθε ανθοφόρο βασίλειο, ενώ "όλα τα τετράποδα ή τα πουλιά που δεν έχουν εξημερωθεί για ανθρώπινη χρήση έχουν εξοντωθεί ως αντίπαλοί του στην αναζήτηση τροφής… και μόλις και μετά βίας έχει απομείνει κάποιο μέρος όπου θα μπορούσε να αναπτυχθεί ένας άγριος θάμνος ή ένα λουλούδι χωρίς να εξαλειφθεί ως ζιζάνιο στο όνομα της βελτίωσης των γεωργικών καλλιεργειών".[21]

Οι περισσότερες από 250.000 προστατευόμενες περιοχές παγκοσμίως, καθεμιά από τις οποίες αποτελεί αποτέλεσμα κάποιας κοινωνικής αξίας, συλλογικής υπόσχεσης, ηθικής παραδοχής και κοινοτικής κωδικοποίησης σε νόμο, μαρτυρούν μια εδραιωμένη παράδοση μακροχρόνιων, σταδιακά αυξανόμενων συμβιβασμών, που για πρώτη φορά αναφέρεται στο *ουτοπικό παράδοξο* του σερ Τόμας Μορ, δηλαδή, πώς θα παρέχουμε πραγματική προστασία επιτόπου;[22] Και τι σημαίνει προστασία στην πραγματικότητα;

Στον πρόλογο του έργου μας *The Metaphysics of Protection* ("Η μεταφυσική της προστασίας")[23], περιγράψαμε τον ταπεινό άθλο της απελευθέρωσης εκατοντάδων μυγών οι οποίες είχαν γεννηθεί σε μια καλοκαιρινή κατοικία στη νότια Καλιφόρνια. Ως αντικείμενο συζήτησης, η *Musca domestica* (οικιακή μύγα) μπορεί να μοιάζει ελάχιστα ενδιαφέρουσα σε σχέση με άλλα ζητήματα, αφού το βάρος της είναι κατά μέσο όρο μικρότερο από 12 χιλιοστογραμμάρια, ενώ η ζωή της διαρκεί από 2 έως 4 ανθρώπινες εβδομάδες. Ωστόσο, εάν εμείς οι οποίοι διατεινόμαστε ότι σεβόμαστε τον ελέφαντα και τον Αμαζόνιο δεν μπορούμε να βρούμε την απαιτούμενη θέληση και ηθική για την προστασία ακόμα και των φαινομενικά πλέον μικροσκοπικών συγγενών μας εδώ στη γη, τότε τι αξία έχουν όλα αυτά; Πέρα από τις ετυμολογικές αναλύσεις, η λέξη "προστασία" στους νομικούς κανόνες παραμένει μόνο εν μέρει κατανοητή από κουλτούρα σε κουλτούρα. Υπάρχουν αναφορές για διάφορες περιπτώσεις παράλογων αποφάσεων, όπως η διαταγή εκτέλεσης ενός γουρουνιού – το οποίο κρίθηκε από κάποιους ηλίθιους άτακτο- από δικαστήριο στη Γαλλία το 1266, ενώ, σχεδόν 300 χρόνια αργότερα, ένα άλλο δικαστήριο χάρισε τη ζωή σε ένα θηλυκό γαϊδούρι μετά από την κατάθεση μαρτύρων υπεράσπισης ως προς την αρετή του.[24]

Το βίαιο ιστορικό μας, η αναπτυξιακή τεχνολογία και η μηχανική και οι προσπάθειες μείωσης του φαινομενικά αναπόφευκτου αποτυπώματος της ανθρωπότητας μας οδηγούν σε συμπεράσματα που δεν είναι ούτε νέα ούτε καινοτόμα: Το πλήθος των αντιφατικών δεσμών της ανθρώπινης ψυχής, καθώς αυτή συγκρούεται με τις έννοιες. Στα εθνικά πάρκα και προστατεύουμε αλλά και κυνηγάμε τα ζώα (και τα φυτά). Λέμε ότι αγαπάμε τα σπάνια και χαρισματικά ζώα, ενώ σφάζουμε μαζικά όλα τα άλλα, τα οποία απορρίπτουμε ως στερούμενες συναισθήματος μονάδες παραγωγής. Πρόκειται για μια πολύ παλιά ιστορία στην οποία περιπλέκεται η απόλυτη γραμμή κάθε ψυχαναλυτικής θεωρίας η οποία ελπίζει στην εύρεση καθαρά προβλέψιμων αιτίων και αιτιατών ή στην εύρεση διαφαινόμενων προοπτικών για ένα τοπίο δικαιοσύνης και σεβασμού μεταξύ ίσων στη βάση οποιουδήποτε θείου νόμου.

Δρ. Biruté Mary Galdikas με διασωσμένο ουρακοτάγκο στο Βόρνεο. Φωτογραφία © J.G. Morrison.

Σημειωσεισ

[1] Βλ. United Nations Treaty Collection, https://treaties.un.org/; see also, Βλ. Donald K. Anton, "'Treaty Congestion' in International Environmental Law," *International Law Reporter*, 24 Ιανουαρίου, 2012, http://ilreports.blogspot.com/2012/01/anton-treaty-congestion-in.html. Βλ. επίσης https://blogs.un.org/blog/2012/09/24/most-ratified-international-treaties/ Πρόσβαση 2 Νοεμβρίου 2020.

[2] Langdon, Stephen H.; Gardiner, Alan H. (1920). "The Treaty of Alliance between Hattusili, King of the Hittites and the Pharaoh Ramesses II of Egypt". *Journal of Egyptian Archaeology*. 6 (3): 179–205. JSTOR 3853914.

[3] "Vienna Convention on the Law of Treaties (1969)". *opil.ouplaw.com.* doi:10.1093/law:epil/9780199231690/law-9780199231690-e1498.

[4] Βλ. "Rap Sheet: Why is American history so murderous?" της Jill Lepore, *The New Yorker*, 2 Νοεμβρίου 2009, https://www.newyorker.com/magazine/2009/11/09/rap-sheet# ixzz0ePRpHvGz, Πρόσβαση 31 Οκτωβρίου 2020.

[5] Αριθμός ανθρωποκτονιών εκ προθέσεων και αναλογία ανά 100.000 πληθυσμού ανά χώρα/περιοχή (2000–2012). Δεδομένα (σε μορφή λογιστικού φύλλου) για την έκθεση του UNODC "Παγκόσμια Μελέτη για την Ανθρωποκτονία 2013". Γραφείο των Ηνωμένων Εθνών για τα Ναρκωτικά και το Έγκλημα (UNODC). Σύνδεσμος από την Παγκόσμια Μελέτη για την Ανθρωποκτονία – Δεδομένα: *UNODC Homicide Statistics* 2013

[6] *American Philosophy: From Wounded Knee to the Present*, των Erin McKenna, Scott L. Pratt; Bloomsbury Publishing, New York, 2015, σελ. 375.

[7] Βλ. "Do the dead outnumber the living?" του Wesley Stephenson, *BBC News*, 4 Φεβρουαρίου 2012, https://www.bbc.com/news/magazine-16870579, Πρόσβαση 1 Νοεμβρίου 2020.

[8] Βλ. *Life Counts - Cataloguing Life on Earth*, των Michael Gleich, Dirk Maxeiner, Michael Miersch, και Fabian Nicolay, σε συνεργασία με UNEP, IUCN, WCMC και Aventis, Μεταφρ. Steven Rendall, Atlantic Monthly Press, New York, 2002, σ. 23.

[9] ό.π., σελ/ 24, 261.

[10] Βλ. "Humanity has wiped out 60% of animal populations since 1970, report finds," του Damian Carrington, *The Guardian*, 29 Οκτωβρίου 2018, https://www.theguardian.com/ environment/2018/oct/30/humanity-wiped-out-animals-since-1970-major-report-finds, Πρόσβαση 15 Φεβρουαρίου 2019. Βλ. επίσης "All life on Earth, in one staggering chart," των Brian Resnick και Javier Zarracina, 15 Αυγούστου 2019, VOX, https://www.vox.com/science-and-health/2018/5/29/17386112/all-life-on-earth-chart-weight-plants-animals-pnas, Πρόσβαση 27 Απριλίου 2020, και προέρχονται από το έγγραφο ποσοτικής αξιολόγησης της βιομάζας (βιοπροσδιορισμός και βιογεωχημικός κύκλος), συμπεριλαμβανομένων των προβληματισμών σχετικά με την αφθονία, τη διανομή, τα "ευρεία πρότυπα για τις ταξινομικές κατηγορίες" και τα "τροφικά μέσα", συμπεριλαμβανομένων των δεδομένων "παγκόσμιας πυραμίδας θαλάσσιας βιομάζας" που αντιπροσωπεύουν τους καταναλωτές έναντι των παραγωγών: "The biomass distribution on Earth," by Yinon M. Bar-On, Rob Phillips, and Ron Milo, *PNAS, Proceedings of the National Academy of Sciences of the United States of America*, PNAS 19 Ιουνίου 2018 115 (25) 6506-6511. Πρώτη δημοσίευση 19 Μαΐου 2018 https://doi.org/10.1073/pnas.1711842115, https://www.pnas.org/content/115/25/6506, Πρόσβαση 27 Απριλίου 2020. Για τα ποσοστά σφετερισμού για καθαρή πρωτογενή παραγωγή, βλ "Global human appropriation of net primary production doubled in the 20th century," Fridolin Krausmann, Karl-Heinz Erb, Simone Gingrich, Helmut Haberl, Alberte Bondeau, Veronika Gaube, Christian Lauk, Christoph Plutzar, και Timothy D. Searchinger. PNAS 18 Ιουνίου 2013 110 (25) 10324-10329; https://doi.org/10.1073/pnas.1211349110του Wolfgang Lucht, Potsdam Institute of Climate Impact Research, Potsdam, Γερμανία, και έγινε δεκτή από το Συμβούλιο Εκδοτών στις 5 Μαΐου 2013, υποβλήθηκε για αναθεώρηση στις 4 Ιουλίου 2012, *PNAS, Proceedings of the National Academy of Sciences of the United States of America*, https://www.pnas.org/ content/110/25/10324#:~:text=The%20human%20appropriation%20of%20net,managed%20lands%20with%20lower%20productivity.

[11] Βλ. *On The Nature of Ecological Paradox*, των M. C. Tobias και J. G. Morrison, Springer Nature, New York, 2021, Κεφ. 72.

[12] Βλ. "Rare – Portraits of America's Endangered Species," του Joel Sartore, Focal Point, *National Geographic*, Washington D.C., 2010.

[13] Από το *The Age Of Anxiety*, του W. H. Auden, A Random House Book, New York, 1947, σελ. 105.

[14] Βλ. *Microbe Magazine*, Φεβρουάριος 2014, https://www.asmscience.org/content/journal/ microbe/10.1128/microbe.9.47.2citing: *J. Fox, Microbe*, Δεκέμβριος 2014, σελ. 494. M. Pollan, *New York Times Magazine*, 15 Μαΐου 2013. Science 342:1440, 2013; Cell Metab. 12:111, 2010.

[15] ό.π., *Life Counts*, σελ. 28-29.

[16] Βλ. *After Eden: History, Ecology & Conscience*, του M. C. Tobias, Avant Books, San Diego, 1985, σελ. 36-37.

[17] Βλ. "'A Brute to the Brutes?': Descartes' Treatment of Animals," του John Cottingham, *Philosophy*, 53, 1978, http://people.whitman.edu/~herbrawt/classes/339/Descartes.pdf

[18] Βλ. *De Principatibus/Il Principe*, 1532.

[19] Παράθεση στο Wilson Clark, *Energy for Survival – The Alternative to Extinction*, Anchor Books, Garden City New Jersey, 1974, σελ. 10. Ό.π., After Eden, Κεφ .6, σελ. 91-114, "A Tale of Two Centuries".

[20] *The Birth of the Modern – World Society 1815-1830*, του Paul Johnson, Harper Collins Publishers, New York, 1991, σελ. 355.

[21] J. S. Mill, *Principles of Political Economy with some of their Applications to Social Philosophy, Part IV*, Κεφάλαιο VI, John W. Parker, London, 1848.

[22] Βλ. *2018 United Nations List of Protected Areas – Supplement on protected area management effectiveness*, United Nations Environment Programme (Eric Solheim, Executive Director) and UNEP-WCMC (Director Neville Ash), UNEP-World Conservation Monitoring Centre, Cambridge, UK 2018, in Collaboration with the CBD

(Executive Secretary, Cristiana Pasca Palmer), the IUCN (Director General, Inger Andersen), the WCPA (World Commission On Protected Areas, Chaired by Kathy MacKinnon) and UN environment, Editors: Marine Deguignet, Heather C. Bingham, Neil D. Burgess and Naomi Kingston, https://wdpa.s3.amazonaws.com/UN_List_2018/2018%20List%20of%20Protected%20Areas_EN.pdf.

[23] *The Metaphysics of Protection*, του M. C. Tobias και J. G. Morrison, A Dancing Star Foundation Book, Waterside Press, Cardiff by the Sea, California, 2014.

[24] Βλ. *The Criminal Prosecution and Capital Punishment of Animals*, των E. P. Evans, William Heinemann, London, 1906. Βλ. επίσης, *The Trial: Four Thousand Years of Courtroom Drama*, New Random House, New York, 2006.

Κεφάλαιο 6

Η ΥΠΟΘΕΣΗ ΤΟΥ ΠΟΥΦΕΝΤΟΡΦ

Εν μέσω μιας ιστορικής σειράς αντιφατικών ισχυρισμών προσώπων και κυβερνήσεων, ο φιλόσοφος/δικηγόρος Ούγκο Γκρότιους από το Ντελφτ της Ολλανδίας (1583-1645) έθεσε τα θεμέλια για μια ορθολογική και συνεπή κατανόηση του φυσικού δικαίου (του ελληνορωμαϊκού "ανωτέρου δικαίου"). Μεταξύ των σημαντικότερων συμπερασμάτων του, με τα οποία ζητούσε εμφατικά να μπει ένα τέλος στους πολέμους και όλες τις εχθροπραξίες, ο Γκρότιους κατέστησε σαφές ότι "το να στερούμε από τον άλλον αυτό που του ανήκει, απλώς για δικό μας όφελος, αντιβαίνει στο φυσικό δίκαιο".[1] Η θεμελιώδης αυτή έννοια δικαιοσύνης και ισονομίας τονίζει την έννοια της συμφιλίωσης σε όλους τους τομείς, ωστόσο δεν έχει ακόμη εφαρμοστεί καθολικά, ούτε έχει χρησιμοποιηθεί στο δικαστήριο των διαειδικών δικαιωμάτων στη ζωή, την ελευθερία και τη δικαιοσύνη. Οι λέξεις οι ίδιες, όταν αντιπαραβάλλονται, καταντούν σχεδόν στερεότυπες, δεδομένου του χλευασμού που απειλεί σε μεγάλο βαθμό την εύθραυστη διατήρησή τους ως έννοιες ενσωματωμένες σε οποιαδήποτε καθιερωμένη *ανθρώπινη φύση*. Ή, όπως δήλωσε ο Ρουσσώ στο έργο του *Κοινωνικό Συμβόλαιο (Du contrat social; ou Principes du droit politique*, 1762), "Ακριβώς επειδή η ισχύς των περιστάσεων τείνει συνεχώς να καταστρέφει την ισότητα, η ισχύς της νομοθεσίας πρέπει πάντα να μεριμνά για τη διατήρησή της". Αναφέρει, όμως, και ότι "αυτός που τολμά να αναλάβει τη δημιουργία των θεσμών ενός λαού οφείλει να νιώθει ικανός, τρόπος του λέγειν, να αλλάξει την ανθρώπινη φύση…" [2]. Στη μελέτη του για *τον Μαρξ και τον Βιτγκενστάιν*, ο Ντέιβιντ Ρουμπινστάιν υποστήριξε ότι ο Μαρξ έδωσε μεγάλη έμφαση στη διαλεκτική μεταξύ της φύσης και του Εαυτού στο σημείο διακλάδωσης όπου "ο άνθρωπος είναι, με την πιο κυριολεκτική έννοια της λέξης, όχι μόνο ένα κοινωνικό ζώο, αλλά ένα ζώο που μπορεί να εξελιχθεί σε άτομο μόνο εντός της κοινωνίας".[3]

Κάθε μία από αυτές τις σκέψεις προκαλεί μια ατελείωτη ταραχή έναντι της *πραγματικής φύσης* που τελικά θα προδώσει κάθε φιλοσοφία και θα απομυζήσει την αισθητική αγαλλίαση που προκύπτει στο ίδιο σώμα του οποίου το σφρίγος τελικά θα φθίνει. Ό,τι κι αν πιστεύει ο καθένας μας, ακόμα και *μέχρι τον τελευταίο*, όπως διαμαρτύρεται ο Τζον Ράσκιν (1819-1900), τίποτα δεν μένει στον κόσμο, παρόλη τη δύναμη της αποφασιστικότητάς του. Ωστόσο, παρά τα όλα αυτά καλέσματα στην άγρια φύση, την ανθρώπινη αγριότητα και την απείρως μεγαλύτερη άγρια φύση, ο Ράσκιν ικανοποίησε κάθε ταξινομικό όριο με ένα καταληκτικό επιχείρημα που υποστήριζε ότι "όλη η αποτελεσματική πρόοδος προς την αληθινή ευτυχία της ανθρώπινης φυλής πρέπει να προέρχεται από ατομική και όχι δημόσια προσπάθεια".[4] Πρόκειται ακριβώς για αυτή την *προσπάθεια* που ο Αριστοτέλης παρερμήνευσε εντελώς, συγκρίνοντάς την με μια "Μικρή Δημοκρατία" εντός μιας Ολιγαρχίας. Χαιρετίζοντας τη θεαματικότητα και την εθιμοτυπία, σαν άλλος Νέρων, ως το καλύτερο μέσο για τη διασφάλιση της εμπιστοσύνης του κοινού σε αυτό που θεώρησε εσφαλμένα ότι σημαίνει "η σταθερότητα της διακυβέρνησης".[5] Ο σύγχρονος του Ράσκιν, αισθητικός φιλόσοφος Γκίλμπερτ Χάμερτον (1834-1894), συνέβαλε στην εδραίωση των πεποιθήσεων αυτών υποστηρίζοντας ότι είναι ευκολότερο να γνωρίζει κανείς τις σκέψεις ενός αγρότη παρά ενός αλόγου, και ότι "η απιθανότητα αυτή να γνωρίζουμε τα πραγματικά αισθήματα των ζώων" μπορεί εν μέρει να διορθωθεί "με προσεκτική παρατήρηση και στοχασμό, υποβοηθούμενη από καλοσυνάτη συμπάθεια και επιείκεια…"[6]

Αυτή είναι, λοιπόν, η διαλεκτική την οποία τείναμε πάντα να χρησιμοποιούμε ως βάση για την εμπλοκή μας σε συζητήσεις, διαφωνίες και περισπασμούς: το δημόσιο αίσθημα, η ατομική κλίση. Εάν δηλαδή είναι οι κοινωνικές δυνάμεις που μας φτιάχνουν τον καφέ και τα ζεστά μας ντους, τα βιβλία μας και τις λαϊκές αγορές, και όχι η εσωτερική μας φωνή, οι καθοριστικοί παράγοντες που διαμορφώνουν αυτό που τελικά πιστεύουμε, υπερασπιζόμαστε και προσευχόμαστε.

Γενικές λέξεις και σκέψεις, συντονισμένες σύμφωνα με την πιο συγκεκριμένη κατάταξη μιας διαρκούς διαλεκτικής. Δεν πρόκειται για απλό εξοστρακισμό της λογικής, αλλά για σταθερό διαλογισμό στη ρίζα όλης της φυσικής επιστήμης. Εξ ου και η σύσταση του Γκούο Σι (1020-1090) ότι "αν κάποιος μπορεί να αναπτύξει μια φυσική, ειλικρινή, ευγενική και έντιμη καρδιά, θα μπορεί αμέσως να κατανοήσει τις πτυχές των δακρύων και των χαμόγελων και των αντικειμένων…"[7]. Ο Λεονάρντο εξέφρασε μια διαφορετική θέση σχετικά με αυτό το κάλεσμα να γίνουμε μέρος του φυσικού κόσμου, όταν συμβούλευσε ότι "η λαμπρότητα της απεικόνισης ενός τοπίου" (σε πίνακα) δεν θα φτάσει ποτέ αυτή του πραγματικού τοπίου, όπως αυτό φωτίζεται από τον Ήλιο, παρά μόνο αν ο ζωγράφος τοποθετήσει τον καμβά απευθείας στο φως του ήλιου, ωσάν ένας παιάνας σε ανοιχτό χώρο.[8] Οι διαφορετικοί αυτοί υπαινιγμοί αναφέρονται στην εγγύτητα κατά την πορεία της μετατροπής των ιδεών σε κάποιο είδος πραγματικότητας που εναρμονίζεται πιο σταθερά με τις εντυπώσεις και τη θέση των αισθήσεων στην κοινωνική, την αισθητική ή την ψυχολογική σφαίρα. Πώς να υπερβούμε το σημείο αναφοράς προς τα πίσω και ακόμα πιο πίσω στο λογικό προηγούμενο ή εμπρός σε κάποια αδιαμφισβήτητη διαφαινόμενη αλήθεια. Ο Ιμμάνουελ Καντ το οραματίστηκε ως την προϋπόθεση της *Αιώνιας Ειρήνης*[9] στην οποία κυριαρχούσε μια "οικουμενική φιλοξενία". Τα επιχειρήματά του ακολούθησαν πιστά αυτά του Σάμουελ φον Πούφεντορφ, ο οποίος, στην πραγματεία του το 1672 με τίτλο *Ο Νόμος της Φύσης και των Εθνών*[10], διακήρυξε ότι η φυσική κατάσταση της ανθρωπότητας ήταν η ειρήνη και όχι ο πόλεμος, όπως είχε υποστηρίξει ο Χομπς. Ότι το γεγονός αυτό και μόνο μας έκανε ανθρώπους και όλοι οι νόμοι ακολουθούσαν μια ελεύθερα υιοθετημένη προτίμηση προς την αρετή και την καλοσύνη. Η *υπόθεση του Πούφεντορφ* επιλύει τη διαφωνία που επικρατούσε στον διάλογο της μετα-Ολόκαινου εποχής σχετικά με την ισχύ των ανθρώπινων κοινωνικών προτιμήσεων έναντι του εσωτερικού πυρήνα της ίδιας της ανθρωπότητας.

Η ανθρώπινη επιθετικότητα έχει διαγνωστεί με άφθονες μεθόδους, από προϊστορικούς τόπους ταφής και δολοφονίες (π.χ., ο Ότσι, ο "άνθρωπος των πάγων" των αρχών της 4ης χιλιετίας π.Κ.Ε. που θανατώθηκε είτε με βέλος στην πλάτη είτε κατά τη διάρκεια ανθρωποθυσίας), τις ψυχοπαθολογίες της γενοκτονίας μέχρι τη φρίκη, την αστυνομική εργασία και τον ανώνυμο πόνο που προκαλεί καθημερινά το είδος μας. Άπειρες παρορμήσεις και θύματα. Η βία, εξίσου ή περισσότερο από την αγάπη και την καλοσύνη, καταβροχθίζει τις ανήσυχες και απερίσκεπτες ψυχές της ανθρωπότητας.[11,12] Εκατοντάδες εκατομμύρια λέξεις έχουν γραφτεί στην προσπάθειά μας να αναλύσουμε τη γνωσιακή μας βάση, η οποία, από μόνη της, είναι ο προπομπός της ηγεμονίας της φαντασίας και πολύ εύκολα μεταφράζεται ως καταπάτηση των Άλλων. Το αρχικό αυτό αμάρτημα αναλύθηκε όσο πιο διορατικά γίνεται στην αξέχαστη έρευνα του Ελίας Κανέτι στο βιβλίο του *Μάζα και Εξουσία* (1962), όπου έγραψε: "Αυτοί που αφιερώνονται στον φόνο είναι και αυτοί που έχουν την εξουσία".[13] Ο Εμίλ Σιοράν σχεδόν συνδύασε τις αποκαλύψεις του Αριστοτέλη για τη διακυβέρνηση και του Κανέτι για την εξουσία, όταν αναγνώρισε την ακόλουθη εξελικτική υπόθεση: "Ένα μεγάλο βήμα προς τα εμπρός έγινε την ημέρα που οι άνθρωποι κατάλαβαν ότι, αν ήθελαν να βασανίζουν ο ένας τον άλλο πιο αποτελεσματικά, θα έπρεπε να συγκεντρωθούν, να οργανωθούν σε μια κοινωνία".[14]

Οι συλλογικότητες αυτές – η συμμετοχή στις δραστηριότητες της κοινότητας, το εθνικό κράτος – είναι τα ίδια τα όργανα όλων αυτών των εισβολών που συνιστούν τους ατελείωτους ψυχοφθόρους πολέμους, ενός ψυχοκοινωνικού αδιεξόδου μοναδικού στο είδος μας. Όλη αυτή η ανθρώπινη κακία και εχθρότητα έχει αξιολογηθεί διεξοδικά σε όλη την τζαϊνιστική παράδοση, τη διαλεκτική της "hinsa/ahin(m)sa" (βία/μη βία).[15] Την περίοδο της έξαρσης των συζητήσεων σχετικά με τα δικαιώματα των ζώων κατά τη διάρκεια του 18ου και 19ου αιώνα, η κυβέρνηση και οι θεοδικίες της, ως βασικός φορέας ο οποίος επιτρέπει σε κάποιο βαθμό την ανθρώπινη επιθετικότητα προς τους άλλους, αμφισβητήθηκε για πρώτη φορά προκλητικά το 1742 από τον Τζον Χίλντροπ (1682-1756). Συχνά υποθέτουμε ότι ο Γάλλος φυσιοδίφης Ζωρζ Κυβιέ (1769-1832) ήταν εκείνος που πρότεινε για πρώτη φορά την έννοια της εξάλειψης των ειδών, με βάση τις μελέτες του σε απολιθώματα. Στην πραγματικότητα, όμως, ο Χίλντροπ, ο οποίος έθιξε την ανθρώπινη ύβρη και τη συλλογική αλαζονεία της ηδονιστικής ερμηνείας της Γραφής από την ανθρωπότητα, ήταν εκείνος που έκανε για πρώτη φορά αναφορά σε "μια ολοσχερή Εξάλειψη της Ύπαρξης" ως άμεσο αποτέλεσμα της ανθρώπινης ανοησίας και της διαειδικής επιβολής.[16]

Ο Χίλντροπ επικεντρώθηκε στη Θεϊκή Δημιουργία, αμφισβητώντας τις υποθέσεις που επικρατούσαν στα μέσα του 18ου αιώνα στην Ευρώπη ότι η θεϊκή δημιουργία ήταν πολύ μεγαλύτερης σπουδαιότητας (προφανώς) από οτιδήποτε θα μπορούσαν να δημιουργήσουν οι ίδιοι οι άνθρωποι και ότι η συμπόνια μας προς τους Άλλους πρέπει επομένως να εντάσσεται στο ίδιο σχετικό πλαίσιο ταπεινότητας. Το 2021, η διαμάχη αυτή μεταξύ ύβρης και ταπεινότητας συνεχίζεται. Οι μελέτες της σημειολογίας της ηθικής κάνουν για πρώτη φορά λόγο για "απεριόριστη ευθύνη" καθώς και για μια "βαθιά και διαπεραστική" "διαφορά" μεταξύ της "μονολογικής" και της "διαλογικής" επικοινωνίας. Η τελευταία σηματοδοτεί όλες τις σημαντικές "σχέσεις μεταξύ του εαυτού και του άλλου"[17], οι οποίες επιδιώκουν να καταστήσουν αποδείξιμη τη θεωρία ότι η ανθρωπιά αποτελεί ένα σημειολογικό *γεγονός* και ότι η ηθική της προέκταση είναι εγγενής σε οποιαδήποτε επικοινωνιακή παρόρμηση.

Έτσι, η πλήρης αυτή εξειδίκευση θα ωθήσει τον εναγκαλισμό των *άλλων ατόμων του ιδίου ζωικού είδους.* Το ερώτημα δε μετατίθεται στην αβεβαιότητα των αμαλγαμάτων παρά στα επίπεδα συμφωνίας τους σε επίπεδο πολιτικών και βιολογικών κοινοτήτων. Πρόκειται για μια ηθική συνοχή που βασίζεται στη συμπεριφορά της αγέλης και τις αυτοοργανωτικές στιγμέργειες[18], ή αλλιώς τις διάφορες μορφές άμεσου και έμμεσου συντονισμού, μια έννοια που διατυπώθηκε για πρώτη φορά στα τέλη της δεκαετίας του 1950 από τον Γάλλο ζωολόγο Πιερ Πωλ Γκρασσέ, τον λεγόμενο και "Εντουαρντ Όσμπορν Γουίλσον των τερμιτών". Οι αποικίες, τα σμήνη, η αρχιτεκτονική των φερομονών, η επισήμανση συγκεκριμένων περιοχών, οι μονομαχίες των τρυποκάρυδων Colaptes auratus που ορμούν ο ένα στον άλλον σχηματίζοντας τον αριθμό 8 ή άλλες ξεχωριστές χορευτικές επιδείξεις που παρατηρούμε σε είδη όπως τα παραδείσια πουλιά ή το παγόνι, καταδεικνύουν όλα μια διαλογική γενετική προδιάθεση προς την επίτευξη της επικοινωνίας η οποία συμβάλλει στη διατήρηση της ζωής και η οποία καθοδηγείται από πάνω από δύο άτομα και όχι μόνο από ένα.

Παρά το γεγονός ότι οι φυσικοί θεωρούν ότι όλοι αυτοί οι υποδεέστεροι οργανισμοί - από τα πυκνά πλήθη του ζωοπλαγκτόν και των κριλ μέχρι τις νυχτερίδες που βγαίνουν από κάποια σπηλιά το σούρουπο- βρίσκονται όπως υποδηλώνει η εξωτερική θερμοδυναμική ισορροπία εν κινήσει, στην πραγματικότητα, μέσα στη βιόσφαιρα δεν θα μπορούσε να υπάρξει πιο κομψή απόδειξη της σπουδαιότητας της δύναμης των επιχειρημάτων μεταξύ των σωματιδίων, τα οποία τυχαίνει να είναι μεμονωμένα άτομα, τοπικά και μη, γεγονός το οποίο μας φέρνει στο πεδίο της οικολογικής φυσικής, υπερτονίζοντας τη βιολογία των συνεργατικών σχέσεων.

ΣΗΜΕΙΩΣΕΙΣ

[1] Βλ. "Natural Law and Peace -A Biography of Hugo Grotius," *Libertarianism.org*, του Jim Powell, 4 Ιουλίου 2000, https://www.libertarianism.org/publications/essays/natural-law-peace-biography-hugo-grotius, Βλ. επίσης το έργο του Γκρότιους' *De iure belli ac pacis libri tres*, Buon, Παρίσι, 1625.

[2] Όπως αναλύθηκε στο έργο του T. V. Smith *Beyond Conscience*, Whittlesey House, New York, 1934, σελ. 328-329. Βλ. *An Inquiry into the Nature of the Social Contract: Or Principles Of Political Right*, του Ζαν Ζακ Ρουσσώ, μεταφρασμένο από τα γαλλικά από τους G. G. J.και J. Robinson, Λονδίνο, 1791.

[3] Από το έργο του Μαρξ *The Grundrisse: 17, Marx and Wittgenstein – Social Praxis and Social Explanation*, Routledge & Kegan Paul Ltd., Λονδίνο, 1981, σελ. 153.

[4] *"Unto This Last" – Four Essays on the First Principles of Political Economy*, του Τζον Ράσκιν, Λονδίνο, 1862, σελ. 168.

[5] Βλ. τη μετάφραση του έργου του Αριστοτέλη A Treatise On Government, μετάφραση από τα Ελληνικά των William Ellis, T. Payne, Λονδίνο, 1776, σελ. 332.

[6] *Chapters on Animals*, Philip Gilbert Hamerton, Βοστώνη, 1882, σελ. 12-14.

[7] *An Essay on Landscape Painting*, του Kuo Hsi, μετάφραση από τα Κινεζικά των Shio Sakanashi, John Murray, Λονδίνο, 1959, σελ. 52.

[8] *Treatise on Painting*, του Leonardo Da Vinci, George Bell & Sons, Λονδίνο, 1897, σελ. 145.

[9] *Zum ewigen Frieden. Ein philosophischer Entwurf* [*To eternal peace. A philosophical draft*], F. Nicolovius, Konigsburg, Γερμανία, 1795.

[10] *Ή, A General System Of The Most Important Principles Of Morality, Jurisprudence, And Politics*, 5η έκδοση στα Αγγλικά, Λονδίνο, 1749.

[11] Βλ. Flannery, D.J., Vazsonyi, A.T. & Waldman, I.D., *The Cambridge handbook of violent behavior and aggression.* Cambridge University Press, Cambridge UK, 2007.

[12] Βλ. τη βιβλιογραφία στο *Psychology Of Peace And Mass Violence – War, Ethnopolitical Conflict, And Terrorism: Informational Resources*, των Linda M. Woolf και Michael R. Hulsizer, Webster University, 2004, OTRP Online, Office of Teaching Resources in Psychology, http://teachpsych.org/resources/ Documents/otrp/resources/woolf04war.pdf

[13] Μετάφραση από τα Γερμανικά της Carol Stewart [Translated from the German by Carol Stewart], The Viking Press, New York, p. 219.

[14] *History And Utopia*, by E. M. Cioran, Μετάφραση από τα Γαλλικά του Richard Howard, Seaver Books, Νέα Υόρκη, 1987, σελ. 92.

[15] Βλ. *That Which Is – Tattvartha Sutra – A Classic Jain Manual For Understanding The True Nature of Reality*, μετάφραση Nathmal Tatia, Με σχόλια των: Umasvati/Umasvami, Pujyapada και Siddhasenagani, πρόλογο του L.M.Singhvi και εισαγωγή από Padmanabsh S. Jaini, The Institute of Jainology, HarperCollins Publishers, San Francisco, CA, 1994. 145 / 5000. Επιπλέον, βλ. τα τρία σχόλια για το Tattvartha Sutra "που χρησιμοποιήθηκαν ως βάση του σχολιασμού" στη μετάφραση του Nathmal Tatia, δηλαδή: *Svopajna Bhasya*, του Umasvati που αναγνωρίζεται ως αυθεντικό από την αίρεση Svetambara, εκδ. Pandit Khubchandji Siddhantashstri, Shrimad Rajchandra Jain Shastramala, Bombay, 1932; *Svopajna Bhasya Tika*, του Siddhasenagani, σέκτα Svetambara, 2 τόμοι., Seth Devchand Lalbhai Jain Pustakoddhara Fund, σειρές 67 & 76, και *Sarvarthasiddhi*, του Pujyapada Devanandi, της σέκτας Digambara, 4η έκδοση,Bharatiya Jnanapitha, Δελχί, 1989. Βλ. *Uttaradhyayana by Bhadrabahu*, Αγγλική μετάφραση του μεγάλου Γερμανού Ινδολόγου, Hermann Jacobi, στο *Sacred Books of the East*, Volume XLV, Oxford University Press, Oxford, UK, 1895, Επανατύπωση 1980, *Jaina Sutras*, Part II, από τους Motilal Banarsidass Publishers PVT. LTD., Delhi Lecture XXIII, σελ. 128. Βλ. επίσης "Ahimsa-NonViolence," ταινία στο PBS του M.C.Tobias, KRMA/1987; και τη συνέχειά της, *Life Force – The World of Jainism*, του M.C.Tobias, Asian Humanities Press, Freemont, CA, 1991.

[16] Βλ. *Free Thoughts Upon The Brute-Creation: Or, An Examination Of Father Bougeant's Philosophical Amusement, &c*, R. Minors, Λονδίνο, 1742, σελ. 44.

[17] *The responsibility of power and the power of responsibility: From the "semiotic" to the "semioethic" animal,"* της Susan Petrilli, Università di Bari, http://www.susanpetrilli.com/ files/The-responsibility-of-power-and-the-power-of-responsibility.pdf, n.d. Βλ. επίσης "Global communication, proximity and responsibility: beyond the logic of identity," των Augusto Ponzio και Susan Petrilli, http://www.susanpetrilli.com/files/Global-communication.pdf, n.d.

[18] Theraulaz, Guy (1999). "A Brief History of Stigmergy". *Artificial Life.* **5** (2): 97–116. doi:10.1162/106454699568700. PMID 10633572. Βλ. επίσης *Pages Choisies*, του Pierre-Paul Grassé, Masson Et Cie, Editeurs, 1967.

Κεφάλαιο 7

Η ΣΤΟΙΧΕΙΩΔΗΣ ΗΘΙΚΗ ΤΟΥ ΠΑΡΜΕΝΙΔΗ

Υπό το πρίσμα των λεπτών και οξειών εκφράσεων των ατόμων και των ειδών σε ολόκληρο τον πλανήτη, το ζήτημα της εξαρμογής – των αλλαγών δηλαδή των εξελικτικών χαρακτηριστικών – καθίσταται υψίστης σημασίας, κυρίως ως παράγοντας κρίσιμος για τη βιώσιμη αλλαγή εννοιολογικής σύλληψης και την οποιαδήποτε δυνατότητα επίδειξης ηθικής συμπεριφοράς. Σε άλλο σημείο εξετάσαμε ήδη διάφορες μορφές της εξέλιξης σε πραγματικό χρόνο και της επανεξέλιξης, όπως για παράδειγμα στην περίπτωση των μυγών, των κοχυλιών και των πεταλίδων, της σεξουαλικότητας των ακάρεων και της ιστορίας της ζωής των σαλαμάνδρων. Όπως επίσης και τις ανθρώπινες τροποποιήσεις σε ανθρώπινα κύτταρα, γενετικά γεγονότα που σηματοδοτούν την επιστροφή σε κάποιο προγονικό ή κλαδιστικό χαρακτηριστικό ή κάποια πτυχή του, στην οποία τα 6 δισεκατομμύρια bit δεδομένων που βρίσκονται μέσα στο ανθρώπινο γονιδίωμα μετατοπίζονται ελάχιστα, σκόπιμα ή μη.[1]

Ο Γερμανός πολυμαθής, εξερευνητής/βιολόγος/φιλόσοφος Ερνστ Χέκελ (1834-1919), ο οποίος επικρίθηκε από πολλούς για την εμμονή του με την ευγονική, ενστερνιζόταν επίσης και τη δαρβινική ιδέα του ανθρώπινου πλουραλισμού από την περίοδο της αρχικής διατύπωσης του και δήλωνε ότι "οι φυσικές εντολές της συμπόνοιας και του αλτρουισμού όχι μόνο προέκυψαν στην ανθρώπινη κοινωνία χιλιετίες προ Χριστού, αλλά έχει βρεθεί και ότι χαρακτηρίζουν τα ανώτερα ζώα που ζουν σε κοπάδια και κοινωνικές ομάδες. Τα χαρακτηριστικά αυτά φέρουν τις αρχαιότερες φυλογενετικές ρίζες τους στο σχηματισμό των φύλων στα κατώτερα ζώα και στον σεξουαλικό έρωτα και τη γονική φροντίδα, στα οποία στηρίζεται η διατήρηση του είδους".[2] Το πρόβλημα με τις θεωρίες γενετικής οι οποίες προκύπτουν από τις πεποιθήσεις ενός Χέκελ, ή ακόμα και του Ζαν Μπατίστ Λαμάρκ (1744-1829) αφορά την έννοια του ναζισμού, την εμφυτευμένη ιδέα ότι ο εκφυλισμός των ανθρώπων μπορεί να ξεριζωθεί με τον ίδιο τρόπο που θα μπορούσε να εξαλειφθεί οποιοδήποτε ανεπιθύμητο χαρακτηριστικό που υποτίθεται ότι βασίζεται στην κληρονομικότητα. Η υπέρτατη ρατσιστική φαντασίωση, όπως τη συναντάμε σε κάποιο παραλήρημα περί της υπέρτατης Άριας φυλής στον Übermensch (Υπεράνθρωπό) του. Ωστόσο, εάν αφαιρέσει κανείς αυτό το ιστορικό φορτίο, βρίσκει στη σκέψη του Λαμάρκ μια ισχυρή, γενναιόδωρη και σημαντική όψη της αντίληψής μας σχετικά με την εξέλιξη σε σχέση με το Ανθρωπόκαινο. Πρόκειται για την ιδέα ότι μπορούμε, θεωρητικά, να αλλάξουμε αυτά τα κληρονομικά χαρακτηριστικά του εαυτού μας κατά τη διάρκεια της ζωής μας, η οποία αποτέλεσε τη βάση πολλών ερευνών για το ανασυνδυασμένο DNA και το αγγελιοφόρο RNA. Το αν η συμπεριφορά μας είναι κληρονομική μπορεί να έχει σημασία, σίγουρα όμως δεν αποτελεί το κλειδί για την κατανόηση σημαντικών απαντήσεων οι οποίες σώζουν ζωές. Μπορεί ως γεγονότα να φέρουν αμέτρητα άλλα ονόματα, ωστόσο το ταξίδι του ανθρώπου στον Άρη, η κλωνοποίηση ενός προβάτου και – αναπόφευκτα, ενός ανθρώπου – αποτελούν όλα διαφορετικές εκδοχές της ευγονικής. Πράγματι, όλα τα στοιχεία της περιόδου του Ανθρωπόκαινου καταγγέλλουν το σύνολο των καταστροφικών παραλογισμών που έχουν διαπραχθεί στο όνομα της υποτιθέμενης βελτίωσης της ανθρωπότητας.

Εναλλακτικά, αυτό που φαίνεται ηθικά αληθοφανές, δεδομένων των μηχανισμών πλουραλισμού οι οποίοι είναι εγγενείς στη φυσική επιλογή (στο βαθμό που μπορούμε να τους μετρήσουμε ή να τους φανταστούμε) είναι κάποια εκδοχή ενός βιοσημειοσφαιρικού πλαισίου στήριξης[3] ως της κυρίαρχης αρχιτεκτονικής των συστημάτων για την υποστήριξη ενός οράματος ζωής που περιλαμβάνει κάθε διαστρωμάτωση πραγματικής ή/και δυνητικής ανάδρασης. Ο Τίμο Μαράν έχει αναφερθεί στις "οντολογικές, εξελικτικές, φυσιολογικές και γνωσιακές" πτυχές αυτού του επικοινωνιακού πεδίου των βιολογικών σχέσεων.[4] Ο εορτασμός του θριάμβου της "συμπόνιας" επί ενός δικτάτορα από εκατομμύρια ανθρώπους στον δρόμο εξασφαλίζει μια θέση στους χρόνους βιολογικής απόκρισης για την υπομονή, την επίλυση συγκρούσεων και την αγαλλίαση.[5] Η ελαστικότητα των ουραίων νεύρων που επιτρέπουν στα κολιμπρί να μεταναστεύουν μόνα τους (μέχρι και πάνω από 1400 μίλια)[6] βρίσκεται επίσης σε συμφωνία με άλλες δημιουργικές επικοινωνιακές ικανότητες μεταξύ των πτηνών και σχεδόν κάθε άλλου σπονδυλωτού που έχει μελετηθεί, με απεριόριστη μουσικότητα και βαθιά δομημένες εκφορές ήχων που αναπαριστούν μια σύνθετη υποδομή πρόθεσης, αμοιβαίας ανταλλαγής μηνυμάτων και σηματοδότησης η οποία είναι τόσο εννοιολογική όσο και ρεαλιστική.[7,8]

Πώς ξέρουμε ότι είναι εννοιολογική; Η προειδοποίηση για πιθανούς θηρευτές ή η αιφνίδια επισήμανση νέων πληροφοριών αποτελεί τόσο μια προγονική αλυσιδωτή αντίδραση όσο και μια "πειρατεία" του νηματοσκώληκα, του οποίου το "εσωτερικό πλήθος μιτοχονδρίων", όπως περιγράφει ο Ντέιβιντ Τζορτζ Χάσκελ, "τροφοδοτείται από τη συνεργασία".[9] Αυτές οι αρχικές κοινότητες των κυτταρικών οργανιδίων και των μοριακών ενώσεων ενέργειάς τους (τριφωσφορική αδενοσίνη, ή ATP) αποτελούσαν αυτό που πιθανώς ήταν η παλαιότερη "κοινότητα επαρκώς εξελιγμένων οργανισμών με βάση τον άργιλο που ζουν σε ρηχά νερά εκτεθειμένα στο φως του ήλιου", "τη λιμνούλα του Δαρβίνου",[10] δηλαδή τον τόπο της πρώτης μεταβολικής συμβιωτικής σχέσης και, θα μπορούσαμε να συμπεράνουμε, κοινοτικής ηθικής.

Αυτή η ιστορία ζωής και θανάτου, πόνου και χαράς, αφορά σχεδόν εξ ολοκλήρου άτομα που *εργάζονται με τα χέρια τους*, τα σκουλήκια και τα σαλιγκάρια και τα κύτταρα του σώματος ή κάθε άλλο ζωντανό πράγμα. Το γεγονός ότι έχουμε κατοχυρώσει τη χρήση της λέξης *πράγμα* στον ενικό, σε αντίθεση με την κατηγορία των (υλικών) *πραγμάτων,* υπογραμμίζει την πολιτική και την οικολογία του ενός και των πολλών. Ο Παρμενίδης (περί τα τέλη του 5[ου] αιώνα π.Κ.Ε.) υποστήριξε ότι "τα *είδη* δεν εξαρτώνται από τον Νου του Ανθρώπου· αντίθετα, είναι άγνωστα στην ανθρώπινη Φύση ή την Ανθρωπότητα. Το Γένος και η Ουσία κάθε πράγματος είναι έννοιες αυθύπαρκτες, δεν υπάρχουν στα Όντα αλλά στην υποστήριξη και τη βάση της ύπαρξης των Όντων". Και προσθέτει τα εξής: "τα άλλα (δηλ. τα Όντα) προέρχονται από το Ένα, αλλά το Ένα, το οποίο παρατηρείται στους Άλλους, έχει πάρει την Ουσία του από κάτι τρίτο· δηλαδή τα Άτομα υπάρχουν από αυτές τις δευτερεύουσες Ιδέες, αλλά έτσι ώστε οι δευτερεύουσες Ιδέες να έχουν από την πρωταρχική Ιδέα την Ουσία τους και τη δύναμη με την οποία δίνουν στα Όντα κάποια Ουσία".[11] Παρά τη δύσκολη μετάφραση ενός τρομακτικά διφορούμενου κειμένου, κατά την εξέταση της "Αρχέτυπης Ιδέας", ο Παρμενίδης[12] αναλύει τις Πλατωνικές Μορφές, στις οποίες η ικανότητα του ενός και των πολλών, των ανθρώπων και των θεών, να κατανοούν ο ένας τον άλλον επιβεβαιώνεται και ταυτόχρονα αμφισβητείται.[13]

Ο ίδιος ο Παρμενίδης, παρά ταύτα, θέτει τα θεμέλια για μια τελική -και κριτική- διόρθωση, επικαλούμενος "πινακίδες" που προσφέρουν πολλαπλές "διαδρομές" που θα έχουν περιλάβει την "εμφάνιση" και την "αποδοχή". [14] Οι στοιχειώδεις φιλοσοφικές αρχές δεν αφήνουν καμία αμφιβολία ότι οι αρχαίες ελληνικές επιφοιτήσεις, οι χρησμοί, οι αξιολογήσεις της τραγικής κατάστασης του νου, εμποτίζονταν εξίσου με ωδές χαράς και έναν πανθεϊστικό σεβασμό για τη φύση. Ήταν η ποιότητα αυτή της κατανόησης των άλλων ειδών που καθοδήγησε την ελληνική διάνοια, της οποίας το πιο ευαπόδεικτο στοιχείο ήταν η επένδυσή της σε όλα αυτά τα Άλλα. Οι δυνάμεις της φύσης και η κοσμική τάση για μεγαλειώδεις μεταμορφώσεις όλων των Όντων σε όλα τα Όντα παραμένουν η πιο αποκαλυπτική ιστορία αυτού του συνεχούς θαύματος.[15]

ΣΗΜΕΙΩΣΕΙΣ

[1] Βλ. *The Hypothetical Species: Variables of Human Evolution*, των M. C. Tobias και J. G. Morrison, Springer Nature, Νέα Υόρκη, 2019, σελ. 103-106.

[2] των ιδίων, σελ. 296

[3] Βλ. "Semiotic scaffolding: a unitary principle gluing life and culture together," του Jesper Hoffmeyer, *Studies in Ecocriticism*, Volume 19, 2015, Issue 3, "Biosemiotics and Culture," 18 Ιουνίου 2015, https://doi.org/10.1080/14688417.2015.1058175, https://www.tandfonline.com/doi/abs/10.1080/14688417.2015.1058175

[4] Βλ "Scaffolding and Mimicry: A Semiotic View of the Evolutionary Dynamics of Mimicry Systems," by Timo Maran, *Biosemiotics* 8 (2): 211-222 (2015), PhilPapers, https://philpapers.org/rec/MARSAM-2

[5] 11.25 π.μ., EST, November 7, 2020 in America; also the day the first woman – Jeannette Rankin – was elected to the U. S. Congress.

[6] Βλ. *Hummingbirds Of North America: Attracting, Feeding, and Photographing*, του Dan True, University of New Mexico Press, Albuquerque, NM, 1993, σελ. 18.

[7] Βλ *Theory and Natural Order from the Renaissance to the Early Twentieth Century*, της Suzannah Clark, Cambridge University Press, Cambridge, UK, 2001.

[8] "Humans recognize emotional arousal in vocalizations across all classes of terrestrial vertebrates: evidence for acoustic universals," Piera Filippi, Jenna V. Congdon, John Hoang, Daniel L. Bowling, Stephan A. Reber, Andrius Pašukonis, Marisa Hoeschele, Sebastian Ocklenburg, Bart de Boer, Christopher B. Sturdy, Albert Newen and Onur Güntürkün, Ημερομηνία έκδοσης:26 Ιουλίου 2017 https://doi.org/10.1098/rspb.2017.0990, *Proceedings of the Royal Society B, Biological Sciences*, https://royalsocietypublishing.org/doi/full/10.1098/rspb.2017.0990?etoc=&, Πρόσβαση στις 7 Νοεμβρίου 2020.

[9] *The Forest – A Year's Watch in Nature*, Viking Press, New York, 2012, σελ. 6.

[10] *Genetic takeover and the mineral origins of life*, του A. G. Cairns-Smith, Cambridge University Press, Cambridge, UK, 1982, σελ. 351.

[11] "Parmenides," in *The History Of Philosophy*, του Thomas Stanley, Εκτύπωση από Thomas Baffett, Λονδίνο, 1687, σελ. 748-749.

[12] Όπως αποδόθηκε από τον Πλάτωνα στους *Διαλόγους* του μια φανταστική συνομιλία, στην οποία ο Ζήνων και ο Σωκράτης σχολιάζουν τον Παρμενίδη, όπως τη θυμάται ένας άλλος ήρωας, ο Πυθόδωρος.

[13] S. Peterson, "The Greatest Difficulty for Plato's Theory of Forms: The Unknowability Argument of Parmenides 133c-134c," Archiv für Geschichte der Philosophie, 63: 1–16, 1981. Βλ. επίσης *The Hypothetical Species*, σελ. 27.

[14] *The Route Of Parmenides – A Study Of Word, Image, And Argument In The Fragments*, του Alexander P. D. Mourelatos, Yale University Press, New Haven, Connecticut, 1970, σελ. 197.

[15] Βλ. *Old Greek Nature Stories*, των F. A. Farrar, George G. Harrap & Company, Λονδίνο, 1914.

Ο Δρ Russell A. Mittermeier στην κορυφή του φυσικού καταφυγίου στο Βόλτσμπεργκ, στο Κεντρικό Σουρινάμ. Φωτογραφία © M. C. Tobias.

Κεφάλαιο 8

ΟΝ-ΣΤΟΝ-ΚΟΣΜΟ: *DAS-IN-DER-WELT-SEIN*

Από τις 176 ανατομικές μεταμορφώσεις που πραγματοποιήθηκαν μεταξύ διαφόρων οργανισμών στις *Μεταμορφώσεις,* τα δεκαπέντε, γραμμένα το 8ο έτος Κ.Ε σε εξάμετρο, αξιόλογα βιβλία του Πόπλιου Οβίδιου Νάσωνος (43 π.Κ.Ε - 17/18 Κ.Ε) – από τον Ακταίωνα, ο οποίος μεταμορφώνεται σε ελάφι με τη μεσολάβηση της Αρτέμιδος, έως τον Κύηκα, ο οποίος γίνεται πουλί, τα μανιτάρια που γίνονται άνθρωποι, τη βασίλισσα των Πυγμαίων που μεταμορφώνεται σε γερανό, τη Σεμίραμι που αποκτά φτερά για να έρθει σε πλήρη επαφή με τα περιστέρια – το ερώτημα που στοιχειώνει και παίζει με την ανθρώπινη ιστορία είναι το εξής: Ποιος ήταν ο στόχος συγγραφής αυτού του έπους; Τη χρονιά της απελευθέρωσής του, ο αυτοκράτορας Αύγουστος εξόρισε τον διάσημο ερωτικό ποιητή στην Τόμιδα (την αρχαία ρουμανική πόλη της Κωνστάντζας στη Μαύρη Θάλασσα). Η εξορία δεν ταίριαζε στον Οβίδιο, ο οποίος θρηνούσε για την κατάστασή του καθόλη τη διάρκεια της τελευταίας δεκαετίας της ζωής του μακριά από τη Ρώμη και τα δουλοπρεπή, κουρασμένα από τον πόλεμο πλήθη της. Είχε δει πάρα πολλά, είχε εμβαθύνει στα αντικρουόμενα σεξουαλικά και συναισθηματικά κοινοτικά πρότυπα της αυστηρά ιεραρχικής ρωμαϊκής πολιτικής σκηνής, και σαφώς αισθανόταν μια ανάμικτη περιφρόνηση προς το αρχαίο πρόσωπο του διαλόγου του Πλάτωνα, του Ευθύφρονος, στον οποίο ο Σωκράτης πασχίζει να κατανοήσει την ανεξαρτησία ή μη των ανθρώπων και των αξιών τους από εκείνες των Θεών.

Ο Οβίδιος ήταν σαφώς περισσότερο της νοοτροπίας που απλώθηκε ευφυέστατα στις σελίδες του έργου του Λουκρητίου (50 π.Κ.Ε.) *Περί της Φύσης των πραγμάτων*: ενός σύμπαντος που κυβερνάται από απίστευτη ροή και ατομικές μετενσαρκώσεις.[1] Ένα πολύ πιο νατουραλιστικό σύμπλεγμα άγριας φύσης από τον κόσμο που περιέγραψε ο πλούσιος αγρότης Ησίοδος (750-650 π.Κ.Ε.) στο *Έργα και Ημέραι* του ή στη *Θεογονία.* Παραθέτοντας ιστορικά σχόλια όπως αυτά του Σενέκα, του Βάρρωνος και του Δικαιάρχου, ο ιστορικός Κλάρενς Γκλάκεν ισχυρίζεται ότι η Χρυσή Εποχή του Οβίδιου ήταν ακριβώς εκείνη η εποχή που η γη που δεν είχε ακόμη αλλοιωθεί από την ανθρωπότητα. Ένας κόσμος στον οποίο υπήρχε παντού αφθονία, γονιμότητα, το "γλυκύ νέκταρ" και η "καταπράσινη βελανιδιά".[2] Οι μεταμορφικές μεταβάσεις ήταν όλες θεμελιώδεις για τις ομοιότητες που επέτρεπαν σε κάθε χαρακτήρα να συμμετέχει ή μη συμμετέχει, και να είναι ο εαυτός του ανεξάρτητα από την ανθρώπινη παρουσία και παρέμβαση. Η ανθρωπότητα δεν ήταν παρά ένας ακόμα ηθοποιός σε αυτή τη σκηνή, ούτε καλύτερος, ούτε χειρότερος όσον αφορά τη συμμετοχή της.

Παρά την εμφάνιση επιστημονικών αποδείξεων σε κάθε εργαστηριακό τραπέζι από το Πεκίνο έως το Λονδίνο, το εμφατικό νήμα βιολογικής αλλαγής του Οβίδιου, εμποτισμένο με αλχημιστικές δυνατότητες και ηθικές διακυμάνσεις, θα τροφοδοτούσε για πάντα την παμψυχική περιγραφή του κόσμου, γνωστού και άγνωστου: ενός νου/ψυχής που λειτουργεί σε όλο το σύμπαν. Ότι "τα πάντα είναι γεμάτα θεούς", [3] σύμφωνα με τον Αριστοτέλη, και ότι στη συζήτηση σχετικά με την "ουσία" περιλαμβάνονται πολλές και διάφορες δυνάμεις ώθησης όπως η "Φιλία", ο "Αέρας" και το "Άπειρο". [4]

Το αριστούργημα του Οβίδιου επέτασσε ένα πλήρως απελευθερωμένο περιβάλλον για τις αλληλεπιδράσεις όλων των Όντων, ημών και των Άλλων. Τα πάντα ήταν ένας σκεπτόμενος, αισθανόμενος οργανισμός ευάλωτος στη διαδραστικότητα. Αυτό θα μπορούσε να θεωρηθεί μόνο ως ένα εξαιρετικά περίπλοκο πεδίο ηθικής σκέψης και επισφαλούς ταυτότητας εν μέσω πληθώρας άλλων. Της ευθραυστότητας του ιδίου συμφέροντος και της παράλογης επιμονής ακόμα και στην λιγότερο μόνιμη υποτέλεια σε βιογραφικά στοιχεία και λεπτομέρειες.

Συνεπώς δεν προκάλεσε καμία, στα τέλη του 15ου αιώνα, ο καλλιτέχνης Μάρτιν Σόνγκαουερ, όταν έφτιαξε ένα απλό χαρακτικό σε ξύλο με τίτλο "The Inconvenient Petitioners", στο οποίο διάφορα τετράποδα, με τους κορμούς τους να φλέγονται, εξέφραζαν έντονα τα παράπονά τους ενώπιον ενός ανθρώπινου δικαστηρίου, ξεκάθαροι πρόδρομοι του αλόγου που σηκώνεται στα πίσω πόδια του στην "Γκέρνικα" του Πικάσο (1937). Το 1647, ο Ολλανδός ζωγράφος Πάουλους Πότερ θα επιστρέψει στον απολυταρχισμό των ζώων στις δεκατέσσερις εξαίσιες σκηνές του που απεικονίζουν την "Τιμωρία ενός Κυνηγού" (στο Ερμιτάζ). Δεκαπέντε χρόνια αργότερα, στην έκδοση των Βρυξελλών του *Δον Κιχώτη*, οι αναγνώστες είδαν τη βαθιά συναισθηματική αποθέωση του Σταχτή, του γαϊδάρου του Σάντσο Πάνσα, και του υπέροχου γέρικου αλόγου του Δον, του μεγαλειώδους Ροσινάντε, στις πρώτες εικονογραφήσεις της αρχικής ισπανικής έκδοσης.[5] Έως τότε, τα περίφημα δοκίμια του Πορφύριου που συνιστούσαν τη χορτοφαγική διατροφή (*Περί αποχής εμψύχων*), μαζί με εκείνο του Πλούταρχου (*Περί σαρκοφαγίας*) είχαν εκδοθεί πολλές φορές στην Ευρώπη. [6]

Το 1945 η *Φάρμα των Ζώων* του Τζορτζ Όργουελ έγινε τελικά αποδεκτή για δημοσίευση από τους Secker & Warburg (Λονδίνο), παρότι το μυθιστόρημα αποδοκίμαζε ξεκάθαρα τον σύμμαχο της Βρετανίας στον πόλεμο, τους Σοβιετικούς. Ήταν την ίδια χρονιά που ο Aleksandr Georgīevīch Romm έγραψε και δημοσίευσε *τη Διαμαρτυρία των Ζώων* (***протест зверей***), σε εικονογράφηση του Risunki Vladimir Gal 'ba που θυμίζει τον Ιβάν Κριλόφ ή τον Αίσωπο, κατά τη διάρκεια της πολιορκίας του Λένινγκραντ.[7] Μια σπάνια επίδειξη πολιτικής ανυπακοής στο όνομα των ζώων, ζωγραφισμένη σαν πάπυρος με κινούμενα σχέδια, στην οποία ξεκάθαρα βασίστηκε ο επιδέξια συνηγορών υπέρ των δικαιωμάτων των ζώων Πρόλογος του Όργουελ στην ουκρανική έκδοση της *Φάρμας των Ζώων* τον Μάρτιο του 1947.

Η πολιτική ενός πλούσιου φάσματος χαρακτήρων οικείων με εκείνους του Αισώπου αντηχεί εδώ και τουλάχιστον 2.500 χρόνια. Οι περίεργα διχασμένες σχέσεις τους καθοδηγούν έναν φιλοσοφικό χώρο του οποίου τα περιεχόμενα περιγράφουν μια βιόσφαιρα που επεκτείνεται διαρκώς από την ίδια την αυτοαναφορική συνείδηση των αισθητικών, φιλοσοφικών και επιστημονικών περιπλανήσεων τις οποίες οι άνθρωποι μπορούν τόσο εύκολα να ταυτιστούν. Υφίσταται διάλογος σε κάθε συλληπτό και ασύλληπτο επίπεδο στοιχειώδους ανταλλαγής. Η ίδια η ανταλλαγή γίνεται πρόθεση, η πρόθεση γεννά μέτρα συνεργασίας που στοχεύουν τόσο στην ύπαρξη όσο και στην αποδοχή, ή στην καταδίωξη και την καταδίκη. Στην αποδοχή βρίσκεται η ταπεινότητα, με την ταπεινότητα βρίσκεται κάποιο ξεκάθαρο μονοπάτι προς την κατανόηση. Αυτό που ξεπερνά όλα αυτά τα βήματα και τις συνέπειές τους είναι το ίδιο με αυτό που μας χαρίζει τη μέρα και τη νύχτα, τις εποχές και τις αισθήσεις τους. Γίνεται σαφές από αυτά τα σημεία καμπής ότι δεν υπάρχει καμία αξία, κανένας λόγος για να ζει κανείς μέσα στην επιθετικότητα, την καταστροφή και την άρνηση συνεργασίας, Μόνο σε ειρηνικές συνθήκες μπορεί κάποια ιδεολογία να εξασφαλίζει μια βιώσιμη διαδραστικότητα για την οποία μπορεί να προβλεφθεί ότι θα ευημερήσει.

Αυτή είναι η ουσία του βιολογισμού, η nibbana (νιρβάνα), η παύση της dukkha, δηλαδή του πόνου, και – στον Βουδισμό – της Ευγενούς Οκταπλής Ατραπού, *ή ariya aṭṭhaṅgika magga*, στην αρχαία Πάλι. Αυτά είναι τα μέσα με τα οποία κανείς μπορεί να επιτύχει (φτάσει σε) εκείνο τον τόπο, την κατάσταση, το πλαίσιο της ειλικρινούς αντίληψης. "Ορθή κατανόηση, ορθή σκέψη, ορθός λόγος, ορθή πράξη, ορθός βιοπορισμός, ορθή προσπάθεια, ορθή επίγνωση και ορθό *samadhi* ("διαλογιστική απορρόφηση ή ένωση"). [8] Ο Κικέρωνας και ο Κουιντιλιανός μίλησαν για ορθότητα, συμφωνώντας σε μεγάλο βαθμό με τις συζητήσεις του Αριστοτέλη και του μαθητή του Θεόφραστου για τη ρητορική θεωρία. [9] Όμως, μια πολύ πιο τεκμηριωμένη αντίληψη της συμπεριφοράς μπορεί να καθοριστεί ηθικά in situ (στον κόσμο). Είναι εκείνος ο σκοπός, η οντολογία, όλοι εκείνοι οι ουσιαστικοί στόχοι των οποίων οι ευρείες σειρές ενεργειών απαραίτητα περιλαμβάνουν σε κάποιο σημείο ένα ή περισσότερα από τα τέσσερα αίτια (*αιτίαι*) που περιγράφονται από τον Αριστοτέλη, δηλαδή την ύλη, το είδος (μορφή), το ποιητικό αίτιο και το τελικό αίτιο. Αυτά παρέχουν τη βάση για την εξέταση κάθε *αποτελεσματικού* στοχοθετημένου προσανατολισμού, όπως ο Αριστοτέλης έβλεπε τη λειτουργία της Φύσης. Με μια απώτερη κατάσταση ευημερίας και ευτυχίας, της λεγόμενης ευδαιμονίας. [10] Τόσο ο Αριστοτέλης όσο και ο Αίσωπος προσέγγισαν τη βιολογία της καθημερινότητας σύμφωνα με παρόμοια βασικά στοιχεία: τις οικείες συναντήσεις και τα αντικείμενα απασχόλησης που συνθέτουν μια ανθρώπινη ζωή σε σύμπραξη με όλους τους Άλλους γύρω μας.

Ο Γκέοργκ Βίλχελμ Φρίντριχ Χέγκελ) (1770-1831) χαρακτήρισε αυτόν τον συλλογισμό (από εξαγωγή συμπερασμάτων σχετικά με το άγριο και το εξημερωμένο) ως "ανακλαστικό συμβολισμό": "μια μορφή τέχνης στην οποία η ιδέα όχι μόνο εμπεριέχεται σε αυτή, αλλά επίσης τοποθετείται ρητά ως διακριτή από την αισθητική μορφή με την οποία αναπαρίσταται." [11] Θα μπορούσε κανείς να συνεχίσει με αυτό προς πολλές κατευθύνσεις. Ο Ιμμάνουελ Καντ (1724-1804) έγραψε: "Έχουμε *a priori* επίγνωση της δικής μας πλήρους ταυτότητας σε σχέση με όλες τις αναπαραστάσεις που μπορούν ποτέ να αποτελούν μέρος της γνώσης μας, και θεωρούμε το γεγονός αυτό απαραίτητη προϋπόθεση για τη δυνατότητα όλων των αναπαραστάσεων".[12] Ενώ ο Άρτουρ Σοπενχάουερ στο έργο του Ο κόσμος σαν βούληση και σαν παράσταση (1819) υποστήριξε ότι ο κόσμος δεν είναι ανεξάρτητος από ένα γνωσιακό πλήθος αντίληψης, δεν υπάρχει αντικειμενική ή υποκειμενική συνωνυμία που να μην ενσαρκώνει το ταυτολογικό όλον. Δεν είναι απλώς θέμα μαθηματικών, φυσικής ή γραμματικής. Ούτε η χημεία ούτε η συνάρτηση πιθανοτήτων υποστηρίζουν την ψευδαίσθηση. Η κατανομή των ατόμων, των πληθυσμών, των συνόλων μεγάλων βιολογικών κοινοτήτων και των οικοσυστημάτων μπορεί να ποικίλλει στο έπακρο, ωστόσο οι οικολογικές αρχές, από τις οποίες γεννιέται κάθε φιλοσοφία, θεολογία και αισθητική ευφορία, είναι ενδεικτικές μιας πραγματικότητας που είναι η βαθιά κρυμμένη *στον Εαυτό της* οικουμενική πίστη. Πώς το γνωρίζουμε αυτό; Δεν μπορούμε να γνωρίζουμε, ούτε όμως πρόκειται για φαινόμενο λήθης. Κάθε άτομο έχει ένα *προαίσθημα* για τη φύση της ζωής, της δικής του ζωής, το οποίο ακροβατεί μεταξύ της εμπειρικής εικασίας και της επιφοίτησης. Ή μαραζώνει στον τρομακτικό πόνο της αίσθησης των τόσο πολλών συναισθημάτων για την τόσο ευάλωτη και εκτεθειμένη κατάσταση στην οποία βρίσκεται ο γνωστός κόσμος. Η θέση μας είναι η εξής: Όσο και αν η δυτική φιλοσοφία έχει προσπαθήσει για την γενικότερη νοητική αναπαράσταση του κόσμου, η διαφαινόμενη συνειδητοποίηση ότι *είμαστε* αυτός ο κόσμος γενικότερα, οδηγεί στην παθητική μας *συμμετοχή σε αυτόν·* μεταμορφώνοντάς τον σε αληθινό Ον, όχι σε κάποιο ηδονιστικό Dasein, αλλά σε ένα *das-in-der-Welt-sein,* ένα Ον-στον-Κόσμο.

Οι καταπράσινες ζώνες ενός αειθαλούς ονείρου – η απαρέγκλιτη αποθέωση της φυσικής ιστορίας – αντέχουν πέρα από τους όποιους φόβους ή υπολογισμούς, πέρα από την εξουσία των "Θεών" επί των Εαυτών μας, ή τυχόν περίτεχνα ερωτήματα περί πεπρωμένου μέσα στα σύδεντρα και τη λάσπη της αρχέγονης εξέλιξης. Από τις εκατοντάδες εκδόσεις των *Μεταμορφώσεων*, όλες καταλήγουν καταφανώς σε ένα τελικό αδιέξοδο στο Βιβλίο XIV, στο οποίο η φημισμένη ιστορία του Ετρούσκου Βερτούμνου και της παραπλάνησης της Πομόνα από αυτόν διεγείρει το αποκαλυπτικό πλαίσιο των μεταβαλλόμενων εποχών και το απόλυτο δράμα της βιοποικιλότητας.

Η τελική μαρτυρία για τις μεταμορφωτικές δυνάμεις των Θεών και των Θνητών σε συνεργασία, όπως αποτυπώθηκε στο συγκλονιστικό πορτρέτο του Αυτοκράτορα της Αγίας Ρωμαϊκής Αυτοκρατορίας Ροδόλφου Β '(1552-1612) ως Βερτούμνου από τον Τζουζέπε Αρτσιμπόλντο (1590 περίπου). Από την έδρα της αυλής του στην Πράγα, ο Ροδόλφος συγκέντρωσε την πιο εντυπωσιακή ιδιωτική συλλογή φυσικής ιστορίας στον κόσμο, έγινε δηλαδή ο Γουόλτερ Ρότσιλντ της Αναγέννησης. Η υποστήριξή του ως πάτρωνα στους Ρούλαντ Σάβεραϊ, Εχίντιους Σάντελερ, Γιαν Μπρίγκελ τον πρεσβύτερο, καθώς και στους Γιοχάνες Κέπλερ και Τύχο Μπράχε, μεταξύ δεκάδων άλλων, εδραίωσε σταθερά την ευρωπαϊκή εμμονή με την αλλαγή της Φύσης, τα ταξινομικά συστήματα και τον ρόλο της ανθρωπότητας στη συνειδητή συγχώνευση της εξαίρετης ομορφιάς με τις εμπειρικές μελέτες.

Οι συλλογές του Ροδόλφου – το Kunstkammer – που απογράφηκαν για πρώτη φορά το 1607- διασκορπίστηκαν τελικά σε όλη την Ευρώπη, με την αρχή να γίνεται το 1612, σε μια σειρά μοναρχικών επιδρομών και ιδιωτικών και δημόσιων δημοπρασιών.[14] Μεταξύ των χιλιάδων πινάκων, γλυπτών, ορυκτών, νεκρών και ζωντανών ζώων, τεχνουργημάτων που ξεδιαλέχτηκαν από μοναστήρια, άλλους βασιλικούς και από τους αιώνες, υπήρχαν τα οδυνηρά υπολείμματα μιας εμμονής που περιελάμβανε τις τελευταίες και μοναδικές γνωστές εικόνες του ντόντο, όπως ζωγραφίστηκε ζωντανό, και αυτού που φημολογείτο ότι ήταν δράκος. Αλλά ήταν η απόλυτη υπερβολή της φυσικής ιστορίας που ξεχύθηκε σε όλη την αυτοκρατορική επικράτεια ενός πειρατή αυτό που ξεκάθαρα υπέδειξε στον Arcimboldo την πλήρως ειρωνική και φρικιαστική προσωπογραφία της ανθρώπινης κυριαρχίας που είχε κυριευθεί από αυτό που είχε αντίστοιχα κατακτήσει, τον Αμαζόνιο που καταβροχθίζει το πρόσωπο του ανθρώπου. Πρόκειται για μια διεισδυτική κατανόηση του πολύ αγνότερου τοπίου των αλληλεξαρτώμενων πλασμάτων και φυσικών παραλλαγών του Οβίδιου. Ο Ροδόλφος, μετά τον Αίσωπο και τον Οβίδιο, έθεσε τις βάσεις για τους ζωολογικούς κήπους και τις λοιπές απρέπειες. Για τις μελλοντικές γενιές που είχαν χάσει τα ίχνη των φυσικών φραγμών, σφετεριζόμενες την παραδοσιακή πυξίδα με ένα νέο όραμα ανθρώπινης κυριαρχίας στο όνομα της αριστοτελικής επιστήμης.

ΣΗΜΕΙΩΣΕΙΣ

[1] Βλ. “Transformations in Ovid’s Metamorphoses,” του Ian Johnston, Μετάφραση Ian Johnston, Vancouver Island University, British Columbia, 2012 (Νέα έκδοση το 2017), http://johnstoniatexts. x10host.com/ovid/transformationshtml.html. *Ovid’s Metamorphoses in Latin and English*, Translated by the Most Eminent Hands, With Historical Explications of the Fables by the Abbot Banier, Μετάφραση John Dryden, Alexander Pope, John Addison, John Gay, William Congreve, Εξώφυλλο Bernard Picart, Σημειώσεις των Antoine Banier, Wetsteins και Smith, Amsterdam, 1732. Βλ. G. B. Riddehough, “Man-into-Best Changes in Ovid,” Phoenix, Vol. 13, No. 4 (Winter, 1959) https://doi.org/10.2307/1086030, https://www.jstor.org/stable/1086030, Εκδότης: Classical Association of Canada, https://www.jstor.org/stable/ 1086030?seq=1, σελ. 201-209. Βλ. “Ionian Enchantment: A Brief History of Scientific Naturalism,” του Ignacio Prado, Tufts University, Ιούνιος 2006, στο Naturalism, https://www.naturalism.org/worldview-naturalism/history-of-naturalism. Πρόσβαση στις 16 Νοεμβρίου 2020. Βλ. επίσης *Animal Acts: Configuring the Human in Western History,* τωνJennifer Ham και Matthew Senior, Routledge, New York/London, 1997.

[2] Στο *Traces on the Rhodian Shore – Nature And Culture In Western Thought from Ancient Times to the End of the Eighteenth Century*, University of California Press-Berkeley and Los Angeles, 1967, σελ.133.

[3] Αριστοτέλους Περί Ψυχής, 411a7–8. Βλ. “Aristotle’s Psychology,” Christopher Shields, *Stanford Encyclopedia of Philosophy*, 12 Οκτωβρίου 2020, https://plato.stanford.edu/entries/aristotle-psychology/# TextTranCommDeAnim

[4] *Aristotle's Metaphysics*, Μετάφραση με σχόλια και γλωσσάρι από τον Hippocrates G. Apostle, The Peripatetic Press, Grinnell, Iowa,1979, σελ. 163.

[5] *Don Quixote*, House of Juan Mommarte, Brussels, 1662.

[6] *Porphyrii De absteinentia ab esu animalium, libri quatuor* [*The porphyrii absteinentia from eating animals in four books*], Ioanne Bernardo Feliciano interprete, Ion. Gryphium, Venice 1547; και πάλι όπως δημοσιεύτηκε από τον Claudii Morillon στο Λούγδουνο (Λυών) το 1620.

[7] *Protest Zverei* [*Protest of Animals*], του G. Romm, Εικονογράφηση του Vladimir Gal'ba (Galberstad) του The Militant Pencil, Poligrafmasterskaya LSSKh, Leningrad 1945.

[8] Grzegorz Polak, *Reexamining Jhana: Towards a Critical Reconstruction of Early Buddhist Soteriology*, UMCS, Lublin, Poland, 2011, σελ.153-156, 196–197. Βλ. επίσης *Noble Eightfold Path and Its Factors Explained (Magganga-Dipani)*, του Ledi Sayadaw, Buddhist Publication Society, Sri Lanka, 1998.

[9] Βλ. "Cicero as a Reporter of Aristotelian and Theophrastean Rhetorical Doctrine," του William W. Fortenbaugh, *Rhetorica: A Journal of the History of Rhetoric*, Vol. 23, No. 1 (Winter 2005), σελ. 37-64, University of California Press. https://doi.org/10.1525/rh.2005.23.1.37, https://www.jstor.org/stable/10.1525/rh.2005.23.1.37, https://www.jstor.org/stable/10.1525/ rh.2005.23.1.37?seq=1, Πρόσβαση στις 16 Νοεμβρίου 2020.

[10] Βλ. Aristotle's *The Nicomachean Ethics*, 1095a15-22. David Ross, Μετάφραση με εισαγωγή. Oxford: Oxford University Press, Oxford, UK, 1925.

[11] *The Philosophy of Art: Being The Second Part of Hegel's Aesthetik In Which Are Unfolded Historically the Three Great Fundamental Phases of the Art-Activity of the World,* των Wm. M. Bryant, D. Appleton & Co., Νέα Υόρκη, 1879, σελ. 29.

[12] *Kant's Prolegomena To Any Future Metaphysic,* των John P. Mahaffy και John H. Bernard, Vol. II, The Prolegomena Translated, With Notes And Appendices, Macmillan And Co., London, 1889, σελ. 171.

[13] "The Story of Vertumnus, and Pomona," στο *Metamorphoses In Fifteen Books – Translated by the most Eminent Hands*, Εκτύπωση Jacob Tonson, Λονδίνο, 1717, σελ. 500.

[14] Βλ. "Rudolph II's art cabinet," *Codex Gigas*, των Michael Gullick, Ivan Hlaváček, Jan Svanberg, Anna Wolodarski, Μεταφραστές: Göran Bäärnhielm, Bengt Ellenberger, Jana Holá, Janis Kreslins, Christina Manetti, Roger Tanner, http://193.10.12.180/codex-gigas/eng/Long/handskriftens/rudolf-ii/#Rudolph%20II, Πρόσβαση στις 17 Νοεμβρίου 2020.

Κεφάλαιο 9

Η ΜΕΤΑΦΥΣΙΚΗ ΤΟΥ "ΕΙΡΗΝΙΚΟΥ ΒΑΣΙΛΕΙΟΥ"

Το ότι ο Κουάκερος ιεροκήρυκας Έντουαρντ Χικς (1780-1849) μπόρεσε να πάρει τους στίχους του προφήτη Ησαΐα "καὶ συμβοσκηθήσεται λύκος μετ᾽ ἀρνός, καὶ πάρδαλις συναναπαύσεται ἐρίφω, καὶ μοσχάριον καὶ ταῦρος καὶ λέων ἅμα βοσκηθήσονται, καὶ παιδίον μικρὸν ἄξει αὐτούς..." (Κεφάλαιο 11: 6-9) και, αρχίζοντας το 1820, να ζωγραφίσει τουλάχιστον 62 εκδοχές του "Ειρηνικού Βασιλείου" του πρέπει να αποτελεί το πυρήνα αυτής της πίστης σε κάποια εντελώς θαυμάσια, επαναστατική δυνατότητα για την ανθρώπινη κατάσταση.[1] Η βιογραφική ιστορία του Χικς περιγράφει την τέλεια κατάλληλη ηθική στάση για κάθε στιγμή. Γεννημένος έναν αιώνα μετά την υπογραφή από τον Κουάκερο Γουίλιαμ Πεν της συνθήκης ειρήνης με τους αυτόχθονες Ινδιάνους της φυλής των Λενάπε, έζησε την Αμερικανική Επανάσταση, τον Πόλεμο του 1812 και την αυξανόμενη κοινωνική, φυλετική, οικονομική και οικολογική αναστάτωση που σηματοδοτήθηκε από την ταχεία αύξηση του πληθυσμού έξι κοινοπολιτειών του Αλεγκένι που είχαν μόλις προστεθεί στην Ένωση μέχρι το 1830, και με αυτές, πάνω από "3.600.000" ανθρώπων. [2] Μυημένος στις φιλοσοφίες των Κουάκερων περί μη-βίας και "του φωτός εντός", κάνοντας διαρκείς αναφορές στη δύναμη του Χριστού να επηρεάσει τις προσωπικές σχέσεις με έναν κόσμο που κατοικείται από τον Μεσσία[3], ο Χικς μεταμόρφωσε τις μεγαλειώδεις επικλήσεις των πρώτων τοπιακών παραδόσεων στην Αμερική στην αμιγή εκπλήρωση των διαειδικών επιθυμιών των Κουάκερων.

Παρόλο που η συντριπτική πλειοψηφία των πινάκων με ζώα που είχαν δημιουργηθεί από Άγγλους και Γάλλους την εποχή του Χικς βρίσκονταν σε μεγάλο βαθμό σε προθήκες επάθλων ή αγροτικές εκθέσεις και απεικόνιζαν δαμάλες, μοσχάρια και άλλα τυπικά παχύσαρκα είδη του "ζωικού κεφαλαίου" και υπάκουα ιπποειδή – με απεικονίσεις από καλλιτέχνες όπως οι Τόμας Ουίβερ, Τζον Σέριφ, Χένρι Σαλόν, Τζέιμς Γουόρντ και Τζορτζ Σταμπς[4] – ο Χικς αποφάσισε να συμφιλιώσει τις αντιθέσεις μεταξύ των διάφορων κουακερικών (της Αγγλικής Εκκλησίας των Κουάκερων και της Εταιρείας σε μια αποικισμένη Αμερική) και άλλων χριστιανικών μεταφυσικών με μια εντελώς διαφορετική, καθημερινή ερμηνεία της Δημιουργίας. Να "ζωγραφίσει ένα κήρυγμα", όπως το χαρακτήρισε η Άλις Φορντ στην υπέροχη βιογραφία που συνέγραψε για τον πλέον προφητικό ηθολογικό ζωγράφο της Αμερικής. [5] Είναι αλήθεια ότι βιοποριζόταν ως ζωγράφος και ότι ασχολήθηκε με πολλές σκηνές "ζωικού κεφαλαίου", όπως στον πίνακα του 1849 με τίτλο "A May morning view of the Farm and Stock of David Leedom" (Abby Aldrich Rockefeller Folk Art Center).[6] Ωστόσο, το κατά πάσα πιθανότητα κύκνειο άσμα του Χικς υπήρξε το πιο λεπτό και ολοκληρωμένο όραμά του για το τι θα μπορούσε να γίνει η Βόρεια Αμερική και ο κόσμος: "The Peaceable Kingdom" (το Ειρηνικό Βασίλειο) του 1849 (βρίσκεται στο μουσείο Nelson-Atkins Museum, στο Κάνσας Σίτυ, του Μισούρι, δανεισμένος από τον τρισέγγονο του καλλιτέχνη).[7]

Αυτό που ήταν εκπληκτικά αξιοσημείωτο στην απάντηση του Χικς σε αυτή την ατέλειωτη τάση προς τη βιοφιλία ήταν η αναπαράσταση τριών παιδιών στο προσκήνιο και ενός κύκλου σχηματισμένου από αποικιστές της Πενσιλβάνια και ντόπιους Ινδιάνους που κρατιούνται χέρι-χέρι σε μια όχθη - η αναπαράσταση της υπογραφής της φημισμένης συνθήκης ειρήνης. Το ίδιο μοτίβο χρησιμοποιήθηκε σε δεκάδες πίνακες του Βασιλείου του Χικς. Θα δυσκολευόταν να βρει κάποιος κάτι τέτοιο σε οποιαδήποτε καταγεγραμμένη παράδοση τέχνης προηγουμένως. Οι πλησιέστεροι πρόδρομοι όσον αφορά την ανθρώπινη εγγύτητα/σχέση αγάπης με άλλα είδη προέρχονται από την παράδοση mille-fleurs, ιδίως το έργο "Η Κυρά και ο Μονόκερως", ένα έργο του 1500 περίπου (στο Παρίσι, Μουσείο Κλυνύ) και οι αμέτρητες σκηνές του έργου "Η Γέννηση" - ένα είδος που εμφανίζεται το 1223 με τον Άγιο Φραγκίσκο της Ασίζης, καθώς επίσης και η "Ανάπαυση κατά τη Φυγή στην Αίγυπτο", η οποία αναπαραστάθηκε με την ίδια μεγαλοπρέπεια ήδη από το 1304 στο έργο "Η Φυγή στην Αίγυπτο" του Τζόττο ντι Μποντόνε στο Παρεκκλήσι Σκροβένι στην Πάντοβα. Για έργα πριν από εκείνη την εποχή, κάποιος θα πρέπει να ερευνήσει όλα τα μεσαιωνικά βιβλία με μύθους ζώων[8], μερικές από τις χιλιάδες εικονογραφήσεις που βρίσκονται στα πολλά Βιβλία των Ωρών[9], σε πολυάριθμα ψαλτήρια και τις πρώιμες αποδόσεις των *Μύθων* του Αισώπου (οι οποίες, παρά τα κίνητρα πίσω από φαινομενικά καλοπροαίρετες συνδέσεις, όπως η εικονογράφηση του "The Parliament of Birds", περιέχουν στην πραγματικότητα άφθονες εικόνες βίας)· ή, το γνωστότερο έργο, η εικονογράφηση του Θεού καθισμένου με άφθονα ζώα στην εικονογραφημένη Βίβλο του Holkham, γυρω στο 500.[10] Πριν από αυτό το είδος, μπορεί να βρει κανείς εντελώς αντίθετους κυλινδρικούς σφραγιδόλιθους των Αχαιμενιδών που απεικονίζουν, για παράδειγμα, ήρωες που παλεύουν με λιοντάρια και ταύρους. [11]

Στο πυκνό και θεολογικά δυσνόητο κήρυγμά του στο Goose Creek, το οποίο εκφωνήθηκε σε ένα εντευκτήριο στην Κομητεία Λάουντον της Βιρτζίνια στις 22 Φεβρουαρίου 1837, ο Χικς επανέλαβε τα αισθήματα για τα ζώα όπως είχε κάνει και ο εκλιπών και σημαίνων εξάδελφός του, ο Κουάκερος ιερέας Ελίας Χικς, ισχυριζόμενος ότι όντως υπήρχε ένα Εσωτερικό Φως σε όλη τη φύση, το οποίο εκδηλώθηκε στους ανθρώπους και το καθήκον αυτών στη ζωή πρέπει να μεταφράζεται σε ειρήνη, σε αντίθεση σε κάθε καταπίεση, στο κυνήγι και σε κάθε μορφή κακοποίησης των ζώων. [12] Το πνευματικό δόγμα που συνοδεύει τις θεωρίες των Κουάκερων για την αλυσίδα της ύπαρξης έκλινε προς μια πίστη στην ανθρωπότητα ως την πλέον εμψυχωμένη (ανώτατη) μορφή πανίδας. Τα σαρκοφάγα – τόσο σημαντικά στους διάσημους στίχους του Προφήτη Ησαΐα που ενέπνευσε τους πίνακες του Ειρηνικού Βασιλείου του Χικς - πράγματι θεωρούνταν σαρκοφάγα, έτσι απλά, στο πρόσωπο των οποίων συνοψίστηκαν οι δυσάρεστες ιδιότητες της Φύσης με την υπέρβαση των οποίων είχαν επιφορτιστεί οι πεφωτισμένοι άνθρωποι. Αυτό υποδηλώνει μια δύσκολη αντίφαση όταν κάποιος αξιολογεί το έργο του Χικς, υπογραμμίζοντας μια αδιαμφισβήτητη αλήθεια για τη διττή προσωπικότητα ενός ανθρώπου που κήρυττε δυσνόητα αλλά ζωγράφιζε με εκπληκτική σαφήνεια. Δεν υπάρχει καμία απολύτως ασάφεια στις ζωηρές απεικονίσεις ενός κόσμου ικανού να επιτύχει εκ νέου, με αποφασιστικότητα, να ξαναγίνει ο απόλυτος παράδεισος που ήταν.[13] Μια παρόμοια διαλεκτική μπορεί να εντοπισθεί στον Κουάκερο φυσιοδίφη – κηπουρό και εξερευνητή - Γουίλιαμ Μπάρτραμ[14] – ο οποίος έχει χαρακτηρίστηκε "ο Θορώ του δέκατου όγδοου αιώνα…"[15] Στις πρώτες σειρές του αριστουργήματός του *Travels*…(1791), ο Μπάρτραμ συμφιλιώνει αυτή τη θεμελιωμένη στη Βίβλο διαφορά: "…οτιδήποτε μπορεί να συμβάλει στο μεγαλείο μας είναι εξίσου σημαντικό, είτε βρίσκεται στο ζωικό είτε στο φυτικό βασίλειο… είναι προφανές ότι ο Δημιουργός έχει κατανείμει αμερόληπτα τις χάρες του στα πλάσματά του, έτσι ώστε τα χαρακτηριστικά του καθενός να φαίνονται επαρκώς σημαντικά για να εκδηλώσουν τη θεία και αμίμητη δεξιοτεχνία". [16]

Ο Χικς προεικονίζει τους εμβληματικούς πίνακες της σχολής του ποταμού Χάντσον και της σχολής του Λουμινισμού που φιλοτεχνήθηκαν αργότερα, τον 19° αιώνα, όπως τα πολυάριθμα "Indian Encampments" του Άλμπερτ Μπίρσταντ και το επιδραστικό του έργο "The Rocky Mountains, Lander's Peak," (1863, Μητροπολιτικό Μουσείο Τέχνης, Νέα Υόρκη). Αυτό που ο Χικς δεν θα μπορούσε ποτέ να προσυπογράψει ήταν ένας πίνακας όπως ο "The Last Of The Buffalo" (Ο Τελευταίος Βούβαλος) του Μπίρσταντ (1889, The Corcoran Gallery of Art, Ουάσιγκτον), στον οποίο μια ομάδα Ινδιάνων σκοτώνει το εμβληματικό εθνικό ζώο. Ο ίδιος ο Μπίρσταντ δεν ενέκρινε τέτοιους είδους συμπεριφορές και, όσον αφορά τον πίνακά του, έγραψε ότι είχε "προσπαθήσει να δείξει τον βούβαλο από κάθε πλευρά και να απεικονίσει τη βάναυση σφαγή ενός ευγενούς ζώου που τώρα έχει σχεδόν εξαφανιστεί."[17] Η πραγματικότητα των πορτρέτων της Αμερικής μετά τον Έντουαρντ Χικς μπορεί να περιγραφεί εν γένει ως ευρισκόμενη μεταξύ της ρομαντικής νοσταλγίας για μια Χρυσή Εποχή, στο ίδιο επίπεδο με παρόμοιες παραπομπές σε όλη την ιστορία της ευρωπαϊκής τέχνης, και της απεικόνισης της καταστροφής σε διάφορες μορφές, όπως αυτή επιβλήθηκε στις αμέτρητες εκδοχές του έκδηλου πεπρωμένου.[18] Αυτό το αναπτυξιακό παράδειγμα, μια μάστιγα για το σπίτι μας, ξεκίνησε από την απεικόνιση του μεγαλείου του Γιοσέμιτι και έφτασε στις σκηνές της απεικόνισης της λεηλασίας και της αποψίλωσης δασών, της εξόρυξης ορυκτών, της κατασκευής δρόμων και της ανέγερσης πόλεων που φιγουράρουν στο δεκάτομο αριστούργημα που εξέδωσε ο Τζον

Μιούιρ το 1888 και το οποίο "είναι εικονογραφημένο με γκραβούρες, φωτογκραβούρες, χαρακτική σε ξύλο, κ.λπ. από διαπρεπείς Αμερικανούς καλλιτέχνες". [19]

Οι καλλιτέχνες οποιασδήποτε τάσης δεν κατάφεραν να αγνοήσουν και να μην αναλογιστούν τη διαρκή κακοποίηση των ανθρώπων, των άλλων ζώων και των τοπίων της Αμερικής, παρά την ψήφιση της 14ης Τροπολογίας των ΗΠΑ για την κατάργηση της δουλείας (κυρώθηκε στις 9 Ιουλίου 1868) και του Νόμου περί Πολιτικών Δικαιωμάτων του 1866. Μέχρι την εποχή της μετεμφυλιακής ανοικοδόμησης, ο "συλλογικός κόσμος των Ινδιάνων ήταν στα πρόθυρα της εξαφάνισης". [20] Το εικοσάτομο έργο του Έντουαρντ Σ. Κέρτις *The North American Indian*, το οποίο εκδόθηκε μεταξύ 1907 και 1930, επρόκειτο αρχικά να ονομαστεί "The Vanishing Race" ("Η φυλή που αφανίζεται"). [21] Στο μεταξύ, σε όλη την Ευρώπη, δεκάδες εκατομμύρια ανθρώπων σφαγιάζονταν, καθώς πόλεμοι και γενοκτονίες, η συγκέντρωση των ραγδαίως αναπτυσσόμενων ανθρώπινων πληθυσμών και των ενεργοβόρων αυτοκινήτων στις πόλεις, οι ανεξέλεγκτες χημικές νοθευτικές ουσίες και η ραγδαία άνοδος της βιομηχανοποιημένης γεωργίας και των σφαγείων που δημιουργήθηκαν ως αποτέλεσμα οδήγησαν στην ανάπτυξη της ατομικής βόμβας, από κοινού με τα πρώτα τραγικά δεδομένα που σκιαγραφούν την έναρξη του Ανθρωπόκαινου. Όλα αυτά κατέστησαν γρήγορα την κοσμοθεωρία ενός ανθρώπου όπως ο Έντουαρντ Χικς τραγικά μακρινή σε πολύ σύντομο χρονικό διάστημα.

Αυτό, ωστόσο, που απομένει από το όραμα του Χικς είναι το αντανακλαστικό του· ένα αμάλγαμα εμπειρικών αντιδράσεων ενός ατόμου σε *κάτι* που, εάν βρίσκεται σε κάποιον, πιθανώς να κατοικεί κάπου, με κάποιο τρόπο, σε όλους.

Σημειώσεις

[1] Βλ. *Notes, Critical, Explanatory, and Practical on the Book of the Prophet Isaiah*, του Albert Barnes, Leavitt & Allen, New York, 1860, Volume 1, σελ. 243-246.

[2] "The Colonization of the West," του Frederick J. Turner, *The American Historical Review*, Jan. 1906, Vol. 11, No. Oxford Journals, Oxford University Press, σελ. 304, https://www.jstor.org/stable/ 1834646, https://www.jstor.org/stable/pdf/1834646.pdf

[3] Βλ. "The Peaceable Kingdoms of Edward Hicks," της Victoria Emily Jones, *Art & Theology*, 6 Δεκεμβρίου 2016, https://artandtheology.org/2016/12/06/the-peaceable-kingdoms-of-edward-hicks/

[4] Βλ. *Farm Animal Portraits*, των Elspeth Moncrieff και Stephen και Iona Joseph, Antique Collectors' Club, Suffolk, UK, 1996.

[5] *Edward Hicks – His Life and Art*, της Alice Ford, Abbeville Press, New York, 1985, σελ. 46.

[6] *The Kingdoms of Edward Hicks*, της Carolyn J. Weekley, με τη βοήθεια των Laura Pass Barry, Harry N. Abrams, Inc., Publishers, New York, 1999, 179.

[7] Του ιδίου έργου της Alice Ford, σελ. 247.

[8] Βλ. για παράδειγμα The Sherborne Missal, Alnwick Castle, στο *Birds in Medieval Manuscripts* του Brunsdon Yapp, Schocken Books, New York 1981, σελ. 153.

[9] Βλ. για παράδειγμα *The Master of Mary of Burgundy – A Book of Hours for Engelbert of Nassau*, The Bodleian Library, Oxford, Introduction and Legends by J. J. G. Alexander, George Braziller, New York, 1970.

[10] Βλ. http://www.bl.uk/manuscripts/FullDisplay.aspx?ref=Add_MS_47682

[11] Βλ. *Animals and Men – Their relationship as reflected in Western art from prehistory to the present day*, William Morrow and Company, Inc., New York, by Kenneth Clark, 1977, σελ. 23.

[12] Ομοίως, *Weekley*, σελ.223-232, και σελ. 51-64.

[13] Βλ. "Friends and Nature in America: Towards An Eighteenth-Century Quaker Ecology," του Donald Brooks Kelley, *Pennsylvania History: A Journal of Mid-Atlantic Studies*, Vol. 53, No. 4 (October, 1986), σελ. 257-272, Penn State University Press, https://www.jstor.org/stable/27773139, https://www.jstor.org/stable/27773139?seq=1.

[14] Βλ/ "The Quaker Background of William Bartram's View of Nature," του Larry R. Clarke, *Journal of the History of Ideas*, Vol. 46, No. 3 (Jul.-Sep. 1985), σελ. 435-448, University of Pennsylvania Press, JSTOR, https://doi.org/10.2307/2709478, https://www.jstor.org/stable/2709478, https://www.jstor.org/stable/2709478?seq=1, Πρόσβαση στις 19 Νοεμβρίου 2020.

[15] Βλ. "The Nature of William Bartram," του Thomas P. Slaughter, *Pennsylvania History: A Journal of Mid-Atlantic Studies*, Vol. 62, No. 4 (Fall 1995), σελ. 429-451, Penn State University Press, https://www.jstor.org/stable/27773844, https://www.jstor.org/stable/27773844?seq=1, Πρόσβαση στις 19 Νοεμβρίου 2020.

[16] *Travels Through North and South Carolina, Georgia, East and West Florida*… του William Bartram, 3η έκδοση, 1793, σελ. ix-xii.

[17] Βλ. *Landscape Painting – A History*, του Nils Büttner, Abbeville Press Publishers, New York, σελ. 290.

[18] Βλ. *Extended Travels in Romantic America*, Chosen and Displayed by Joseph Jobé, Εκδόσεις Edita, Λωζάννη, Ελβετία, 1966.

[19] *Picturesque California: The Rocky Mountains and the Pacific Slope – California, Oregon, Nevada, Washington, Alaska, Montana, Idaho, Arizona, Colorado, Utah, Wyoming, Etc.*, J. Dewing Publishing Company, New York and San Francisco.

[20] Βλ. "Curtis And America's Environmental Social Justice Movement," του M. C. Tobias, στο *Edward S. Curtis – One Hundred Masterworks*, του Christopher Cardozo, με τη συνεισφορά των A. D. Coleman, Louise Erdrich, Eric J. Jolly και Michael Charles Tobias, Delmonico Books/Prestel Verlag Munich, Random House New York 2015, σελ. 141.

[21] Βλ. http://curtis.library.northwestern.edu/

Κεφάλαιο 10

ΕΞΙΣΩΣΕΙΣ ΔΥΝΑΜΙΚΟΤΗΤΑΣ

Ο εφοδιασμός της βιολογικής ιστορίας με τον στόχο να κατανοήσουμε τι συμβαίνει με την παρατήρηση του Χέγκελ, σύμφωνα με την οποία ένα βασικό γεγονός που μαθαίνουμε από την ιστορία είναι η (πρωτεύουσα) αλήθεια ότι δεν μαθαίνουμε τίποτα από αυτή, τονίζει αυτό που πολλοί ερευνητές έχουν ονομάσει δεύτερη και τρίτη τάξη της κυβερνητικής.[1] Συστήματα παρατήρησης τα οποία, με τη σειρά τους, επιτάσσουν την παρατήρηση της αυτοαναφορικότητας. Οι εσωτερικοί αυτοί γρίφοι δημιουργούν επίσης δεύτερης και τρίτης τάξης αμοιβαιότητες, προβλεπόμενα κέρδη, ζημίες και μια πλήρη περιφέρεια ηθικής της αρετής όπου η ίδια η φύση του αιτίου και του αιτιατού πρέπει να υποστηρίζει πιο λεπτούς μηχανισμούς για την κατανόηση του πώς το άτομο στέκεται στο ύψος των περιστάσεων. Υπάρχουν τόσο φιλοσοφικές όσο και πρακτικές δυσκολίες σε σχέση με το αίτιο που προκαλεί τις αντιδράσεις. Επιπλέον, κάθε ειδική γνώση που θα μπορούσε να μεταβάλει τον βαθμό απόκρισης ενός μέρους στη συναλλαγή αποδυναμώνει την ιδέα του αντικειμενικού παρατηρητή ή της συμμετοχής. Αυτό έχει αποδειχθεί επαρκώς από τον Ρόμπερτ Γκούντιν στην ανάλυσή του για τις διάφορες κατηγορίες ευπάθειας/ ευαλωτότητας.[2]

Το σύστημα διακοπτών και κυκλωμάτων που συντηρεί ένα καλολαδωμένο μηχανισμό – ανεξαρτήτως αν πρόκειται για βιολογικό μηχανισμό ή τεχνολογικό μηχάνημα - χαλάει και παύει να λειτουργεί ανεξαρτήτως των όποιων δικλείδων ασφαλείας, είτε μιλάμε για το ανοσοποιητικό σύστημα και τα αντισώματα, είτε για κάποιο σύστημα εφεδρείας, κάποια αδιάλειπτη παροχή ισχύος, κάποιο σύστημα ασφαλείας με ενσωματωμένες εφεδρικές σφραγίδες και δακτυλίους, είτε ακόμα και σε συστήματα με πιο ολοκληρωμένες δικλείδες ασφαλείας. Στην πλειοψηφία των αεροσκαφών, η δικλείδα αυτή μπορεί να είναι μια τριπλή δομοστοιχειωτή εφεδρεία. Στους υπολογιστές, ο (πολλαπλός λειτουργικά ισοδύναμος) προγραμματισμός Ν εκδόσεων. Η μηχανική αξιοπιστίας, η οποία λαμβάνει υπόψη στοχαστικές (τυχαίες) παραμέτρους, όπως εφαρμόζεται στον κίνδυνο ολόκληρων συστημάτων, συνυπολογίζεται σε μέτρα εμπιστοσύνης, περιθώρια σφαλμάτων συν ή πλην σε όλες τις δημοσκοπήσεις και τις απογραφές. Ο υπολογισμός αυτός συμπεριλαμβάνει τα όρια ανοχής, την οριακή αντοχή σε εφελκυσμό, τους διακόπτες φορτίου, τους θερμοστάτες και τα σήματα απενεργοποίησης ασφαλείας, τις παρεμβολές σύνθετων και Μπεϋζιανών πιθανοτήτων, τα συστήματα διαχείρισης ποιότητας ISO 9000 και άλλους διαδικαστικούς μετρητές ασφάλειας, καθώς και τα ίδια τα όρια καταπόνησης σε οποιοδήποτε τριβολογικό σύστημα στο οποίο οι εισροές επηρεάζουν τις επαληθεύσεις συνολικής προκύπτουσας απόδοσης. Δεν υπάρχει εγγύηση προϊόντος χωρίς κάποιο μέσο ταχείας ανίχνευσης των κύκλων ζωής, διαχείριση των προϊόντων και εύλογη ευθύνης του κατασκευαστή στην παγκόσμια εφοδιαστική αλυσίδα.[3] Αυτό που διακρίνει όλες αυτές τις ανησυχίες στο σύνολο των μηχανορραφιών του *Homo sapiens* είναι η πραγματικότητα ότι μόνο το δικό μας είδος, μεταξύ όλων των άλλων, εγκαταλείπει χωματερές, εξαλείφει τους βιότοπους και προκαλεί επιδημιολογικά κατάλοιπα. Πιστεύουμε ότι η θέση βρίσκεται κάπως *εκτός της Φύσης*, την ίδια ώρα που την καταστρέφουμε - έχοντας αυτοαπομονωθεί, πεπεισμένοι για την σουρεαλιστική μας ασυλία. Τα μέρη μας δεν είναι συνυπαίτια με την ιδέα ενός συνόλου. Το γεγονός ότι το βιολογικό μας ρολόι, οι απόλυτες φυσιολογικές απαιτήσεις και οι κιρκαδιανοί ρυθμοί μας δεν συσχετίζονται καθόλου με τις νοητικές μας κατασκευές σημαίνει ότι όχι μόνο φαίνεται ότι είμαστε σκλάβοι της πρόβλεψης του Χέγκελ, αλλά και ότι, υπηρετώντας τα συμφέροντα του εγώ, της απληστίας και της στιγμιαίας ικανοποίησης, δεν μπορούμε να δούμε μπροστά μας το άμεσο αίτιο και αιτιατό. Με άλλα λόγια, τα κακόβουλα πάθη μας – που χρονολογούνται τουλάχιστον 45.000 χρόνια πριν, κατά τη διάρκεια των οποίων οι πρόγονοί μας εκμεταλλεύτηκαν τα τροπικά δάση[4] – δεν αντιμετωπίζουν την καταστροφή που σπέρνουμε στο πέρασμά μας με συγκράτηση ή ουσιαστικό μετριασμό. Ή τουλάχιστον αυτό δεν γίνεται σε παρατηρήσιμη κρίσιμη μάζα, μέχρι σήμερα.

Αν αναλογιστούμε έστω και λίγο όλες τις μορφές βίας και βλάβης που προκαλούμε στους άλλους και στους εαυτούς μας, το ψυχαναλυτικό μονοπάτι των δακρύων αυξάνεται εκθετικά. Μεταξύ της ζωής, του θανάτου και όλων των προοπτικών μετενσάρκωσης – για τις οποίες η γνώση μας εξασθενεί – που καθορίζουν ομοιότητες μεταξύ, για παράδειγμα, του Μανχάταν και μιας σεκόγιας, υπάρχουν ευφάνταστα σκαλιά, η δημιουργία ποιητικών μεταφορών και παρομοιώσεων που δεν στηρίζονται πουθενά παρά μόνο στον νου μας. Αν με κάποιο τρόπο μετρήσουμε το βάρος, τον ρυθμό της πρόκλησης αντιδράσεων, το φορτίο του ιδεασμού σε κάποιο ασαφές ποσοστό *όλης* της νευρικής δραστηριότητας, των ιδεών και των ιδανικών μας, γνωρίζουμε ότι υπάρχει ένα τεράστιο χάσμα, το οποίο περιγράφεται παραδοσιακά ως η διάσταση μεταξύ του νου και του σώματος. Ακόμα, όμως, και αν ανάγουμε αυτή τη διαφοροποίηση στη δεύτερη βαθμίδα, δηλαδή στα διανοητικά παράγωγα αλληλεπίδρασης με τον φυσικό κόσμο σε πραγματικό χρόνο και τόπο, πιθανότατα θα εκπλαγούμε από την πλήρη έκταση της αποσύνδεσής μας.

Από το πλήρες φάσμα των σχεδόν ανεξέλεγκτων περαστικών γεγονότων στο μυαλό μας, μέχρι τις λεπτομέρειες που πρέπει να υπολογίσουμε ώστε να προετοιμάσουμε και να καταναλώσουμε ένα γεύμα, να κοιμηθούμε και μετά πάλι να σηκωθούμε και να κινηθούμε, μπορούμε να πούμε ότι η ζωή μας είναι ελεγχόμενη - από εσωτερικούς βιοχημικούς διακόπτες, αισθητηριακά ερεθίσματα και διάφορους περιβαλλοντικούς. στρεσογόνους παράγοντες και εξωτερικούς περιορισμούς - ακριβώς όπως οι διαθέσεις, οι παρορμήσεις, οι ανάγκες και οι επιθυμίες μάς ωθούν να κάνουμε πράγματα. Ποιο όμως είναι το μέρος αυτού του συνόλου δραστηριοτήτων μας το οποίο αδυνατούμε να κατανοήσουμε; Οι αποδείξεις, όπως γίνεται και κατά τη διαδικασία της μαθηματικής ή επιστημονικής επαλήθευσης, χάνουν γρήγορα τη συνάφειά τους, ανάλογα με το ερώτημα. Για παράδειγμα, έχει η κίνηση ενός ακάρεος στην Ουρουγουάη ως αποτέλεσμα τη μεταβολή της περιοδικότητας της σεισμικής δραστηριότητας στην Ιαπωνία; Εάν υποβιβάσουμε αυτό το είδος ερώτησης (το κλασικό "φαινόμενο της πεταλούδας") σε μια συγκεκριμένη *οικολογική κατηγορία*, μπορούμε να διαμορφώσουμε μια βαθμίδα άλλων, αμέτρητων κατηγοριών, ώστε να συμφωνούν με αυτό που θεωρούμε σχετικό. Εάν η ατομική συνάφεια θεωρείται η νόρμα, τότε η εξέλιξη μοιάζει, σε αυτό το νοητικό παιχνίδι, να είναι τόσο απίθανα ευρεία και ασήμαντη όσο θα μπορούσε να είναι οτιδήποτε. *Η προσωπική εξέλιξη* αφορά περισσότερο κάποιο αναγνωρίσιμο χαρακτηριστικό, υποδηλώνοντας κάτι πολύ πιο ατομικό από, για παράδειγμα, την εξέλιξη των ριγών στις ζέβρες (τις οποίες τα ζώα αυτά ανέπτυξαν με σκοπό τον έλεγχο των σμηνών των μυγών). Πρόκειται για μια γενική αίσθηση της *δικής* μας εσωτερικής ανάπτυξης που, ωστόσο, δεν θα μπορέσει ποτέ να επιβεβαιωθεί πλήρως. Ακόμη και στο αίτιο και αιτιατό, εάν αυτό δεν δηλώνεται με τη σαφήνεια κάποιας εξίσωσης της φυσικής για κάτι τόσο (φαινομενικά) βασικό όσο το μπιλιάρδο, η οποία περιλαμβάνει τη "διατήρηση της ορμής", τη "στατική τριβή" και την "τριβή ολίσθησης", την "εφαπτομενική ταχύτητα", τη "σχετική ολίσθηση" και τις "γωνίες πρόσκρουσης" κ.λπ.,[5] η αιτιότητα μετατίθεται σε ένα ευρύ φάσμα *δυνητικών εξισώσεων* και θεωριών: της κατανόησης της κίνησης των ατομικών σωματιδίων το 1928 από τον Πολ Ντιράκ στο πλαίσιο της ταχύτητας του φωτός, της πλαισίωσης των συνθηκών σε μια λήψη απόφασης με στάθμιση κινδύνου βάσει της αθροιστικής θεωρίας προοπτικών, του εξελικτικού χρονισμού για να καθορίσουμε την πιο πρόσφατη προγονική σύγκλιση μας, κ.λπ. Κάθε μία από αυτές τις θεωρίες είναι σε μεγάλο βαθμό άσχετη με την καθημερινή συνειδητή ζωή μας.

Με άλλα λόγια, οι θεωρίες της εξέλιξης, της γενικής συνάφειας και σχετικότητας, της εφεύρεσης της κατηγοριοποίησης, της καταπάτησης, της απόλυσης και της αντιπαράθεσης χάνουν τη βάση θεμελίωσής τους όσο απομακρυνόμαστε από αυτά που θεωρούμε απαραίτητα: τον αέρα, το νερό, την τροφή και τον ύπνο. Οποιαδήποτε προτεινόμενη βάση για την κατανόηση των αμοιβαίων λειτουργιών της ζωής είναι ήδη πολύ απομακρυσμένη από τις πιο απαιτητικές και προφανείς επιταγές. Αυτό σημαίνει ότι ολόκληρη η μελέτη της αμοιβαιότητας είναι εσωτεριστική; Ούτε κατά διάνοια. Ας δούμε ένα, ίσως παράξενο, παράδειγμα. Όταν διψάω και πίνω νερό, ενεργώ αμοιβαία, προσφέροντας στην αίσθηση της δίψας μου το απαιτούμενο διάλυμα για την πραγματοποίησης της σύνθεσης αφυδάτωσης. Αυτό προϋποθέτει, λοιπόν, ότι υπάρχουν *αρκετοί εαυτοί* μέσα μου, οι οποίοι φυσικά και υπάρχουν. Δεκάδες τρισεκατομμύρια κύτταρα. Οι οργανισμοί των οποίων είμαστε ξενιστές είναι τόσο πολλοί, όσες είναι και οι εικοτολογίες και οι κατηγορίες υποθέσεων: διμερείς, εμφανείς, δυσδιάκριτοι, ανταγωνιστικοί, μη ανταγωνιστικοί, πολυπαραγοντικοί, εσωστρεφείς, συνειδητοί, υποκειμενικοί, ακούσιοι και ούτω καθεξής. [6] Στο επίπεδο των νευρώνων "η κατανόηση του συνόλου των μετασχηματισμών με τους οποίους το εισερχόμενο αισθητηριακό ερέθισμα οδηγεί σε συμπεριφορικές αποκρίσεις (δηλαδή, ο νευρικός κώδικας) παραμένει κεντρικό πρόβλημα στη νευροεπιστήμη" εν μέρει λόγω των "ισχυρών ετερογενειών στους νευρικούς πληθυσμούς" "ακόμη και εντός του ίδιου κυτταρικού είδους". [7]

Οι κυτταρικές μεταβολικές αποκρίσεις, κυρίως η κατάσταση της "σηματοδότησης της αδρεναλίνης στα μυϊκά κύτταρα"[8], αλλά και κάθε πτυχή της γονιδιακής έκφρασης, προϋποθέτουν ένα in situ βιολογικό σύνολο εξελικτικών συνθηκών που συγκεντρώνονται σε ένα άτομο · σε ετοιμότητα ενώπιον της πρόκλησης του είδους του, ομάδων, πληθυσμών και άμεσων συνδέσεων. Στο πλαίσιο των νευρωνικών συνδέσεων "φλοιωδών μικροκυκλωμάτων" σε όλους τους γνωστούς οργανισμούς (εκτός από τους σπόγγους, τα πλακόζωα και τα μεσόζωα) φαίνεται ότι υπάρχουν "δομές μεταξύ προσυναπτικών αξόνων και μετασυναπτικών δενδριτών" των οποίων οι πολλαπλότητες εφεδρικής επαφής δείχνουν ότι σε αυτό το πλέγμα πολλαπλασίων συντελείται βελτιστοποίηση της πληροφορίας και άλλα είδη επεξεργασίας. [9] Έχει υποστηριχθεί ότι η συσσώρευση διακριτών, μικροσκοπικών μορφών ζωής σε όντα που τους παρέχουν υποστήριξη συμβαίνει επειδή έχουν ξεπεραστεί οι άμεσες εξαρτήσεις μεταξύ του κεντρικού νευρικού τους συστήματος και των εξωτερικών συνθηκών. Η γλώσσα, μια εντολή μεταφορικής έκφρασης, ο συμβολισμός, η τελετουργία, η αυτογνωσία, ο φουτουρισμός, ο πολιτισμός έχουν μετακινηθεί από τα μικρότερα βιολογικά κβάντα εντός ενός πεπερασμένου συνόλου κυττάρων και νεύρων σε μια πυκνότητα νευρικής χωρητικότητας που είναι απαλλαγμένη από "αντίστοιχες περιβαλλοντικές σχέσεις".[10]

Μια τέτοια απελευθέρωση, αν και εννοιολογική, προσκαλεί την καθαρή αφαίρεση της σύνθετης λογικής: της αλληλουχίας των νοητικών γεγονότων που εκτυλίσσονται με ή χωρίς προηγούμενο – δεν υπάρχει τρόπος να γνωρίζουμε – αλλά που όλα οδηγούν, τελικά, σε ένα σκόπιμο, οικολογικό τέλος του Εαυτού. Ακριβώς όπως ο πληθυσμός (ένα ζευγάρι) προετοιμάζεται για την αρχή, το άτομο προετοιμάζεται για το τέλος. Ας σκεφτούμε όλους τους τρόπους. Καθένας μας είναι ένας έγκυρος οικολογικός συντελεστής, μέρος μιας εξίσωσης που είναι θεμελιώδης για τις βιολογικές επιστήμες. Σε αντιπαράθεση με τις επικλήσεις του ότι η επιστήμη, από μόνη της, φιλτράρει διεξοδικά, γίνεται, θαυμάζει και ενώνεται με την προσωπικότητα, το συνολικό άθροισμα της οποίας ορίζει τη δική της πλοκή και το δικό της τέλος. Στην ουσία, προσφέρει την απόλυτη αμοιβαιότητα, δίνοντας τα πάντα στον εαυτό του και μετά επιστρέφει στη Γη που τον γέννησε για πρώτη φορά. Ο κύκλος, απρόσβλητος από κάθε κρίση ή απόδειξη, ολοκληρώνεται.

Προς αυτόν τον νατουραλισμό, αυτή την αλληλουχία αναπόφευκτων γεγονότων, η ενεργοποίηση οποιουδήποτε συλληπτού και ασύλληπτου τρόπου σκέψης είναι φαινομενικά απεριόριστη. Το 40 Κ.Ε. ο Σενέκας ο Νεότερος (4 π.Κ.Ε. -65 Κ.Ε..) έγραψε το “De Consolatione ad Marciam”, έναν από τους πιο βαθιούς διαλογισμούς του σχετικά με τον θάνατο και το να ζει κανείς με τον θάνατο, ένα έργο που συνέλαβε πολύ πριν αυτοκτονήσει. Υποτίθεται ότι ο Σενέκας έκοψε τις φλέβες του και κατάπιε δηλητήριο, ωστόσο μάλλον πέθανε από βίαια πρόκληση ασφυξίας. Προφανώς ο θάνατος αυτός δεν ήταν και ο πιο λεπτός τρόπος για να ασπαστεί το τέλος. Τα ελαφρά φορητά όπλα - όπως το μπαρουτοπίστολο του αυτοκράτορα Κάρολου Έ (στη δεκαετία του 1540) και το περίστροφο του Georg von Reichweins (1597), τα οποία γρήγορα θα οδηγούσαν στο περίστροφο Colt (1836), ήταν προφανώς πιο ανώδυνες μέθοδοι για την επίτευξη της ακατανίκητης αυτής επιθυμίας.[11] Το 1805 ο Φρίντριχ Σέρτυρνερ κατάφερε να εντοπίσει τη βασική δραστική ουσία της Μήκωνος της υπνοφόρου (παπαρούνας), αν και θα χρειάζονταν άλλα 92 χρόνια για την ανάπτυξη του εμπορευματοποιημένου προϊόντος, της ηρωίνης, πάλι με προώθηση του σκοπού της αυτοθεραπείας και παρά τα εμπόδια της πολιτισμικής αποδοκιμασίας.[12] Την 1η Δεκεμβρίου 2020, η Υπηρεσία Τροφίμων της Σιγκαπούρης ενέκρινε την εμπορική πώληση εργαστηριακά καλλιεργημένων κυττάρων κοτόπουλου που θα πωλούνταν με τη μορφή μερίδων κοτομπουκιών στα εστιατόρια της Σιγκαπούρης, μια σημαντική εξέλιξη για την ταχέως αναπτυσσόμενη παγκόσμια αγορά χορτοφαγικών/βίγκαν προϊόντων.[13] Θα μπορούσε κανείς να αναφέρει και άλλες τέτοιες καινοτομίες (ξεκινώντας όχι από το μηδέν, απλώς από κάποια εξαιρετικά αρνητική αξία την οποία αυτή τη στιγμή στηρίζει ο υπερβάλλων ζήλος της ανθρωπότητας), όπως: ο Γκεϊτσιανός πράσινος χάλυβας, το πράσινο υδρογόνο, το πράσινο σκυρόδεμα, τα ψυκτικά υγρά νατρίου στους αντιδραστήρες, η μηδενική βία, οι μηδενικές εκπομπές, η μηδενική καταπάτηση σε κάθε ενέργεια, παραγωγή και – με τη σειρά της, πρόθεση.

Στο ανθρώπινο χρονοδιάγραμμα της απάλυνσης του πόνου, οι ακτιβιστικές τεχνολογικές προσπάθειες αυτού του είδους έχουν πρωταρχική σημασία. Υπογραμμίζουν την προσαρμοστικότητα των ανθρώπινων ιδεών και υποδηλώνουν ότι, εάν η φαντασία μας μπορεί πράγματι να καλλιεργήσει την ενάρετη συμπεριφορά, η εξέλιξη έχει κάθε λόγο να λαμβάνει υπόψη αυτό που σκεφτόμαστε. Σκέψεις χωρίς υλικά εμπόδια. Όπως είπε κάποτε ο γλύπτης Χένρυ Μουρ, αυτό είναι το πραγματικό σημείο εκκίνησης κάθε φορά που διασταυρώνονται ο ιδεαλισμός με την πράξη.

ΣΗΜΕΙΩΣΕΙΣ

[1] Βλ. Ranulph Glanville, (2003), “Second Order Cybenetics,” στο Systems Science and Cybernetics, Ed. Francisco Parra-Luna, στο *Encyclopedia of Life Support Systems (EOLSS)*, Υπό την αιγίδα της UNESCO, Eolss Publishers, Oxford, UK, http://www.eolss.net.

[2] *Protecting the Vulnerable – A Reanalysis of Our Social Responsibilities*, του Robert E. Goodin, University of Chicago Press, Chicago, Ill., 1985.

[3] Βλ. *Waste as Opportunity – A MAHB Dialogue with Materials Scientist Veena Sahajwalla*, του Geoffrey Holland, MAHB, Millennial Alliance for Humanity and the Biosphere, 12 Νοεμβρίου 2020, https://mahb.stanford.edu/blog/mahbdialogues/waste-as-opportunity-a-mahb-dialogue-with-materials-scientist-veena-sahajwalla/Accessed November 20, 2020.

[4] Βλ. "The deep human prehistory of global tropical forests and its relevance for modern conservation," των Patrick Roberts, Chris Hunt, Manuel Arroyo-Kalin, Damian Evans και Nicole Bolvin, *Nature Plants* 3, Article number: 17093 (2017), 3 Αυγούστου 2017, https://www.nature.com/articles/nplants201793,

[5] Βλ. *Real World Physics Problems, – The Physics of Billiards*, n.a., https://www.real-world-physics-problems.com/physics-of-billiards.html

[6] Βλ. "A Theory of Reciprocity," by Armin Falk and Urs Fischbacher, *Games and Economic Behavior*, Volume 54, Issue 2, February 2006, σελ. 293-315, ScienceDirect, Elsevier, https://doi.org/10.1016/j.geb.2005.03.001 https://www.sciencedirect.com/science/article/abs/pii/S0899825605000254

[7] From the Introduction to "Neural Heterogeneities Determine Response Characteristics to Second-, but Not First-Order Stimulus Features," των Michael G. Metzen και Maueice J. Chacron, *The Journal of Neuroscience, J Neurosci.* 2015 Feb 18; 35(7): 3124–3138. doi: 10.1523/JNEUROSCI.3946-14.2015, PMCID: PMC4529327, CAMSID: CAMS4969, PMID: 25698748, https://www.ncbi.nlm.nih.gov/pmc/articles/PMC4529327/, Πρόσβαση στις 21 Νοεμβρίου 2020.

[8] *Response to a signal*, Khan Academy, https://www.khanacademy.org/science/ap-biology/cell-communication-and-cell-cycle/signal-transduction/a/cellular-response

[9] Βλ. *Redundancy in synaptic connections enables neurons to learn optimally*, των Naoki Hiratani και TGomoki Fukai, Edited by Terrence J. Sejnowski, Salk Institute for Biological Studies, La Jolla, CA, PNAS July 17, 2018 115 (29) E6871-E6879; Ημερομηνία πρώτης δημοσίευσης 2 Ιουλίου 2018, https://doi.org/10.1073/pnas.1803274115, PNAS, https://www.pnas.org/content/115/29/E6871, Πρόσβαση στις 21 Νοεμβρίου 2020.

[10] Βλ. για πράδειγμα *Does Altruism Exist? Culture, Genes, and the Welfare of Others*, του David Sloan Wilson, Yale University Press, Templeton Press, New Haven Connecticut, 2015, σελ. 53.

[11] *9 Oldest Guns in the World*, n.a., n.d., https://www.oldest.org/technology/guns/ Πρόσβαση στις 2 Δεκεμβρίου 2020.

[12] Βλ. "Defining a Good Death (Succerssful Dying): Literature Review and a Call for Research and Public Dialogue," των Emily A Meier, Jarred V. Gallego, Lori P. Montross Thomas, Colin A. Depp, Scott A. Irwin και Delip V. Jeste, *The American Journal of Geriatric Psychiatry*, Volume 24, Issue 4, April 2016, σελ.261-271, Elsevier, https://www.sciencedirect.com/science/article/abs/pii/S106474811600138X, Πρόσβαση στις 4 Δεκεμβρίου 2020.

[13] "Singapore becomes the first country to approve lab-grown meat," by Amy Woodyatt and Danielle Wiener-Bronner, *CNN Business*, December 2, 2020, https://www.cnn.com/2020/12/02/business/lab-grown-chicken-intl-scli-scn/index.html, Accessed Δεκεμβρίου 2, 2020.

Three Bhutanese Monks, Rinpung Dzong, Paro, Bhutan. Φωτογραφία © J.G. Morrison.

Κεφάλαιο 11

ΙΘΑΓΕΝΗΣ ΚΡΙΣΙΜΗ ΜΑΖΑ: ΝΕΥΡΩΝΙΚΗ ΚΑΙ ΒΙΟΠΟΛΙΤΙΣΜΙΚΗ

Από το κλασικό μοντέλο της μπάλας μπιλιάρδου που χρησιμοποιείται στη φυσική για την περιγραφή των σημείων "ελαστικής σύγκρουσης" στο σύστημα κινητικής ενέργειας[1] έως τη χρήση πειραματικών παιχνιδιών και της θεωρίας παιγνίων για την καλύτερη κατανόηση των θεμελίων της ανθρώπινης ανταλλαγής, της καλοσύνης και της δικαιοσύνης – των συστατικών δηλαδή στοιχείων των θεωριών γενικής αμοιβαιότητας,[2] το εγγενές χαρακτηριστικό κάθε περίπτωσης αφορά ένα αποτέλεσμα ενεργοποίησης κάποιου είδους και έκτασης. "Οι μετατοπίσεις καθεστώτος και οι κρίσιμες μεταβάσεις",[3] όπως διερευνήθηκαν το 2018 (Rocha et al.,), αποκάλυψαν ότι, από τα παραδείγματα μελέτης 30 γνωστών συμβάντων ενεργοποίησης μεταξύ οικολογικών στόχων, το "45%" αυτών των αλλαγών έδειξε τουλάχιστον "μία πιθανή δομική αλληλεξάρτηση". [4] Το κύριο δίλημμα που συναντούμε στην έρευνα της βιβλιογραφίας της ενεργοποίησης είναι ότι πρόκειται για ένα τόσο μεγάλο σώμα σχολιασμού και παραδειγμάτων που αυτό που μπορεί να κάνει κάποιος, στην καλύτερη περίπτωση, είναι να διαλέξει επιλεκτικά αιτίες και αποτελέσματα για να διατυπώσει κάποιο επιχείρημα. Ωστόσο, καμία τέτοια πρόταση, πόσο μάλλον υπόθεση, δεν μπορεί να τεκμηριωθεί ανεξάρτητα, επειδή όλες οι επιλογές είναι υποκειμενικές, ακόμη και σε ακραίες περιπτώσεις. Ακόμη και η στιγμιαία δυσλειτουργία του ωλένιου νεύρου δεν αποτελεί σε καμία περίπτωση μια καθολική αντίδραση του αγκώνα, όπως δεν είναι και εκατομμύρια άλλα φυσιολογικά εκούσια και ακούσια φαινόμενα αιτίας και αποτελέσματος. Για παράδειγμα, οι συνήθεις υποθέσεις σχετικά με τις "κινητικές διαταραχές" κάθε μορφής (δηλαδή, την "υπερκινητική", την "υποκινητική" ή τη "δυσκινησία") γενικά καταλήγουν στην εύρεση εξηγήσεων από το πεδίο των "γενετικών παθήσεων, των τραυματισμών, των ασθενειών του νευρικού συστήματος, των λοιμώξεων, των παρενεργειών φαρμακευτικής αγωγής και άλλων παραγόντων…" [5]

Άλλες κινήσεις, αντιδράσεις, αντισταθμίσεις ως γραμμική ή χρονική συνάρτηση, ποσότητα και ποιότητα της βιοχημείας απόκρισης – για παράδειγμα, ενός παγετώνα που προχωρά ή υποχωρεί ή της διέγερσης όταν οι νευρώνες στον εγκέφαλο εκτίθενται σε "σήματα αποστροφής"[6]- είναι απτές αλλά εννοιολογικά άυλες. Τέτοιες ασάφειες περιγράφηκαν στις θεωρίες του μαρξιστή κοινωνιολόγου Ανρί Λεφέβρ (1901-1991) σχετικά με την "τριαλεκτική" ή, τη "διαλεκτική της τριπλότητας" – μιας νέας γεωγραφίας των χωρικών πραγματικοτήτων για τις βιωθείσες ζωές που βρίσκονται διαρκώς σε ροή.[7] Είτε κάτι είναι νεκρό, είτε ζωντανό, είτε σε μεταβατικό στάδιο ανάμεσα στα δυο, η αλλαγή είναι διαρκής. Ακόμη και το όρος Έβερεστ εξακολουθεί να μεγαλώνει σε ύψος κατά περίπου ενάμιση εκατοστό τον χρόνο[8], ενώ και τα ατομικά ρολόγια, τα οποία βελτιώνονται συνεχώς, είναι πιθανό να πραγματοποιήσουν μια πιο ακριβή μέτρηση ακόμη και σε σχέση με το σημερινό ατομικό ρολόι καισίου που μπορεί να "μετρά τον χρόνο εντός ενός δευτερολέπτου για πάνω από 100 εκατομμύρια χρόνια".[9] Το αρχικό πρόβλημα με τα εκατοστά και τα δευτερόλεπτα είναι η φύση των ανθρώπινων μετρήσεων. Θα μπορούσε να ισχυριστεί κανείς ότι πρόκειται για αυθαίρετες επιδείξεις υποκειμενικής βούλησης που προϋποθέτουν αντικειμενικότητα. Το δευτερόλεπτο, για παράδειγμα, απέκτησε νέο ορισμό το 1967, και ήταν η πρώτη από τις επτά θεμελιώδεις μονάδες που θεσπίστηκαν από το Διεθνές Σύστημα Μονάδων: "η διάρκεια 9.192.631.770 ταλαντώσεων της μικροκυματικής ακτινοβολίας που απορροφάται ή εκπέμπεται όταν ένα άτομο καισίου μεταπηδά ανάμεσα σε δύο συγκεκριμένες ενεργειακές καταστάσεις". [10]

Η γαλλική μεταεπαναστατική αποφασιστικότητα (μετά το 1795) για τη δημιουργία προτύπων για τη μέτρηση του χρόνου, του μήκους, του βάρους, του ηλεκτρικού ρεύματος, της θερμοκρασίας, της μάζας και της φωτεινότητας προέκυψε λόγω του χάους που είχε επικρατήσει προηγουμένως στο εμπόριο και σε διάφορα μέρη της χώρας.[11] Μέχρι το 1837, οι Γάλλοι επέμεναν στο μετρικό σύστημα. Ωστόσο, οι επιπτώσεις της θέσπισης από τους ανθρώπους κανόνων και μέτρων και, ακόμα περισσότερο, νόμων, αρχών, θεωριών, σταθερών, δικαστικών και κοινοτικών προτύπων, πρέπει να έρχονται σε αντίθεση με μια πολύ βασική δήλωση του Έντμουντ Μπερκ, ο οποίος παρατήρησε το εξής: "Σκεφτείτε πώς θα πρέπει να κυβερνήσετε έναν λαό που πιστεύει ότι θα έπρεπε να είναι ελεύθερος και πιστεύει πως δεν είναι…" [12] Ο Πορφύριος είχε αναπτύξει διορατικά το δίλημμα που συναντά κάθε απόπειρα σφράγισης της αβεβαιότητας με συμμόρφωση όταν ανέλυε πόσο απίθανο είναι να αντιληφθούμε διορατικά τις "άυλες φύσεις", τις "ιδιαιτερότητες" στη φύση, και τις ίδιες "ιδιότητες [που] αφορούν τα σώματα…". Ο ίδιος κατέληξε στο συμπέρασμα ότι οι μόνοι πιθανοί λόγοι για την καθολικοποίηση μιας αρχής που αφορά τις σκέψεις και τις αντιλήψεις μας για τον φυσικό κόσμο πρέπει να διέπονται από μια πεποίθηση - γεννημένη από εκτενή έρευνα του κόσμου γύρω μας - ότι κάθε οργανισμός, πράγμα, δύναμη, ουσία απλώς υπάρχει σε μια κατάσταση και θέση "σε συμφωνία με τη φύση του…"[13]. Με άλλα λόγια, για κάθε νόμο και αρχή της φύσης υπάρχει ένα μοναδικό άτομο που είναι ελεύθερο να ζήσει έξω από τον κανόνα. Η έλλειψη οξυγόνου και νερού μπορεί να μας σκοτώσει, στο μεταξύ όμως, τίποτα δεν μπορεί να σταματήσει τη συγχώρεσή μας, τα δώρα μας, την αγάπη μας, την ανταλλαγή συναισθημάτων.

Για κάθε ιδανικό υπάρχει μια στραβή γραμμή· οι ιδανικοί κύκλοι και σφαίρες αποδεικνύουν εξίσου κάτι άλλο. Πριν από πολλά χρόνια, ερευνητές στο Ινστιτούτο Μαξ Πλανκ για τη Βιολογική Κυβερνητική στη Γερμανία έδειξαν ότι πολλοί άνθρωποι, όταν χάνονται αισθητηριακά, τείνουν να περπατούν λίγο-πολύ σε κύκλους.[14] Το αποτέλεσμα της σύγκρισης ευθειών με κύκλους δεν είναι μαύρο και άσπρο, αλλά αποτελεί θέμα ορισμών της μέτρησης και των εννοιολογικών ισοδυναμιών μεταξύ πολιτισμών και ατόμων. Υπάρχει ένας ορισμός της "καλής προσαρμογής" που αναλύεται στο επίπεδο, για παράδειγμα του δείκτη νοημοσύνης ή άλλων μεθοδολογιών βαθμολόγησης.[15] Είναι όμως πολύ περισσότερο ενδιαφέρον από αυτό. Στις ρίζες της κυβερνητικής θεωρίας, σε κάθε δράση/αντίδραση, στην επισκόπηση συστημάτων και την πρόβλεψη, από τους μηχανισμούς που καθοδηγούν ένα ζωντανό κύτταρο έως την τεχνητή νοημοσύνη, καθώς και σε κάθε πτυχή του κοινωνιολογικού χαρακτήρα ενός είδους ή πληθυσμού και της αλληλεπίδρασης των ατόμων μέσα στο μεγαοικοσύστημά τους, υπάρχει μια μεγάλη και ουσιαστική διακύμανση η οποία, τελικά, είναι γερά προσκολλημένη στην υποκειμενική εποχή και τις αόριστες τοπολογίες της διατύπωσής της.[16]

Αυτή η ατελείωτη κυμαινόμενη έμφαση στην οποία ζει το είδος μας, τόσο μέσω του περιορισμού όσο και της επιλογής, θα εμπλέκει συμπληρωματικές ιδέες, αποκαλύψεις και μετρήσεις. Τα παραδείγματα εξελίσσονται και επιστρέφουν σε προηγούμενο στάδιο ενώ το ίδιο διάστημα αυτοαναθεωρούνται διαρκώς. Από βιολογική άποψη, ο Λινναίος, ο Γουάλας και ο Δαρβίνος - αν τους διαβάσετε και τους ξαναδιαβάσετε - λένε ελάχιστα πράγματα που θα μπορούσε κανείς να περιγράψει ως κάτι καινούργιο ή επαναστατικό. Η φυσική επιλογή και η εξέλιξη, αν και αναπτύχθηκαν σε ένα εύρος διαφορετικών συνθηκών από τις θεωρίες που αναπτύχθηκαν στην Αγγλία στα τέλη του 19^{ου} αιώνα, δεν ήταν λιγότερο συναρπαστικές στο βιβλίο *Περί της Φύσης των πραγμάτων*[17] του Τίτου Λουκρητίου (περίπου 99-50 π.Κ.Ε), στο *Summa Theologiae*[18] του Θωμά Ακινάτη (1225-1274) ή στις πρωτο-επαναστατικές ιδέες του Ζορζ Λουί Λεκλέρκ ντε Μπιφόν (1707-1788) στο αποτελούμενο από 44 τόμους έργο του *Histoire Naturelle*. Κατά τη διάρκεια των περίπου δύο χιλιάδων ετών της πιο πρόσφατης ανθρώπινης σκέψης, η πάλη μας με την ξαφνική αλλαγή που επέφερε η καταστροφή έναντι της αργής ανάπτυξης, όπως αυτή γίνεται αντιληπτή με την πάροδο των ετών, των δεκαετιών και των αιώνων, έχει προφανώς εμπλουτίσει τον διάλογο με αξιοσημείωτα *γεγονότα* τα οποία είναι άξια των νόμων που έχουμε εφεύρει. Μπορούμε όλοι να αποδεχτούμε τη πραγματικότητα των αιτίων της πείνας, της δίψας, της βαρύτητας και άλλων παρόμοιων μισαλλόδοξων βεβαιοτήτων. Ωστόσο, τα απολιθώματα, η έννοια της εξαφάνισης, η μακροζωία ορισμένων ειδών και κυτταρικών σειρών έχουν προσθέσει κάποια ένταση σε αυτόν τον διάλογο, χλευάζοντας τη βεβαιότητα με διάφορες μορφές ευελιξίας, ακριβώς όπως οι μάχες που δίνουμε ενάντια σε κάθε μορφή ασθένειας τονίζονται με παρατηρήσιμα πρότυπα επιτυχίας και αποτυχίας.[19]

Όταν η καθηγήτρια δασικής οικολογίας Σουζάν Σιμάρ μεταδίδει την τεράστια γνώση της σχετικά με τη βιοχημεία του δασικού υπόγειου περιβάλλοντος, οι αναλύσεις της είναι γεμάτες ενθουσιασμό. Οι σχέσεις των υπερorganισμών σε εντόνως συνεργατικά μα και καιροσκοπικά "μυκορριζικά δίκτυα"[20] αποδεικνύουν έναν ακρογωνιαίο λίθο της ικανότητας συσχέτισης που διαθέτει κάθε άνθρωπος ο οποίος έχει θαυμάσει ένα παρθένο δάσος ή απλά έχει νιώσει ευδαιμονία κάτω από τη μεγάλη σκιά μιας κορφής[21]. Η αρχέγονη εμπειρία μας δίνει πληροφορίες χωρίς την ανάγκη οποιασδήποτε επιστήμης ή ανάλυσης. Η μεταλαμπάδευση εμπειρικοποιημένων αλληγοριών εμπλουτίζει, όμως δεν αποτελεί προαπαιτούμενο για την εύρεση παρηγοριάς. Ωστόσο, εάν η εστίαση της Δρος Σιμάρ στα "μητρικά δέντρα" μπορεί να εμπνεύσει δεόντως όλους εκείνους που εδώ και χιλιετίες κόβουν δέντρα με διάφορες δικαιολογίες να μετριάσουν την απληστία τους, τότε μπορούμε όλοι να συμφωνήσουμε ότι τα δίκτυα αμοιβαιότητας στα οποία αυτή και συνάδελφοί της σε όλο τον κόσμο έχουν επιστήσει την προσοχή είναι ζωτικής σημασίας. Κρίσιμα για κάθε βαθμό περιβαλλοντικού ακτιβισμού, για να μην αναφέρουμε την επαναφορά της γοητείας τους, την πολύ σαφή δηλαδή γνώση ότι αυτά τα μυθικά βασίλεια δάσους – τρισεκατομμύρια μεμονωμένα δέντρα – αποτελούν επικράτειες μέσα σε επικράτειες, το ήμισυ των οποίων έχει ήδη καταστραφεί από το είδος μας. Το να ξαναμεγαλώσουν, χωρίς τις αρχαίες πηγές σπόρων που περιγράφουν η Σιμάρ και οι συνάδελφοί της, απλώς δε γίνεται. Στην πραγματικότητα, στη Βρετανική Κολομβία όπου και διδάσκει η Σιμάρ, οι λεγόμενες "νεκρές ζώνες απομόνωσης" – 8.064.000 στρέμματα δάσους που έχουν αποψιλωθεί τα τελευταία δώδεκα χρόνια[22] – αποτελούν την οδυνηρή μαρτυρία μιας πραγματικότητας η οποία αμφισβητείται σθεναρά από τα συμφέροντα του τομέα υλοτομίας, με τον ισχυρισμό ότι "λιγότερο από το 3% του αρχικού δάσους [στη Βρετανική Κολομβία] μετατρέπεται για ανθρώπινη χρήση".[23]

Ο στόχος μας εδώ δεν είναι να βυθιστούμε στο τέλμα του συναγωνισμού μεταξύ αριθμητικών μετρήσεων αλλά να θυμόμαστε ότι για κάθε συνεχιζόμενη ανθρώπινη μορφή εκμετάλλευσης διακυβεύονται ζωές και ότι το ανθρώπινο είδος κυριαρχεί δυσανάλογα στη συζήτηση. Τα 86 δισεκατομμύρια περίπου νευρώνες που εργάζονται αδιάκοπα μέσα στον εγκέφαλό μας συνεχίζουν να αποτυγχάνουν στη συστηματική οικολογική αποκατάσταση όταν πολλαπλασιάζονται σε οποιαδήποτε μεγάλης κλίμακας κοινότητα ομοϊδεατών, ανεξάρτητα από το πόσο ενημερωμένοι, καλοπροαίρετοι ή αυτοχθόνως βιώσιμοι είναι. Σε μικρότερες κλίμακες και πολύ πιο λιτούς αριθμούς, οι οικισμοί μπορούν να λαμβάνουν αποφάσεις που να εφαρμόζονται. Ή αυτή ήταν η αρχική λογική που καθοδήγησε τους περισσότερους μικρούς πληθυσμούς πριν την παγκόσμια εξάπλωση του πληθυσμού της ανθρωπότητας.[24]

Πλέον, η καταστροφή ενός δάσους περιγράφεται όχι μόνο από την άποψη της αποδέσμευσης του άνθρακα στην ατμόσφαιρα ή την απώλεια δεκάδων χιλιάδων ειδών που κατοικούν σε αυτό, τα οποία μπορεί να είναι εκατοντάδες εκατομμύρια ανά στρέμμα. Όσον αφορά την πραγματική διατάραξη του εδάφους, έχει διατυπωθεί η άποψη ότι υπάρχουν πολλά δισεκατομμύρια οργανισμοί των οποίων η ποσότητα διαφέρει ανά περιοχή. Σε ένα τετραγωνικό μέτρο, ανάλογα με την τοποθεσία, μπορεί να βρίσκονται μέχρι και 300 γαιοσκώληκες.[25]

Ως ζήτημα *απλής* φιλοσοφίας ή προσανατολισμού, η μελέτη των "εξωομαδικών" "αμετάβλητων" δυνάμεων του παγκόσμιου "πλουραλισμού" αποκαλύπτει ότι η συνεξέλιξη της συνεργασίας και του ανταγωνισμού δεν έχει επιτύχει κάποια προβλέψιμη ισορροπία, αλλά, μάλλον, την εύκολη πιθανότητα προκατάληψης, παρά την παροχή εκπαίδευσης.[26] Οι διαφοροποιήσεις αυτού του είδους υποδηλώνουν ότι η κλίμακα αισιοδοξίας είναι πράγματι εύθραυστη, υποκείμενη τόσο στην τύχη στην ανθρώπινη συμπεριφορική σφαίρα όσο και σε εκείνες τις θεμελιώδεις αλήθειες που έχει ανακαλύψει η επιστήμη. Όπως και με κάθε οικοσύστημα, η ανθρωπότητα μπορεί να συλλάβει το μέγεθος της κρίσιμης μάζας των συνδυαστικών συμπεριφορικών εξελίξεων εγκαίρως ώστε να απομακρυνθεί από μια συνολική συντριβή των κομβικών οικολογικών σημείων καμπής, ή μπορεί και όχι. Δεν μπορούμε να χειριστούμε λογικά την τύχη, *μπορούμε* όμως να θέσουμε το υπόδειγμα μέσα σε ένα πεδίο πιθανοτήτων που θα προετοιμαστεί ως υποδοχέας για τη συλλογική νευροδιαβίβαση αυτού που ικανοποιεί τη θεμελιώδη απαίτηση για πραγματική αμοιβαιότητα σε ολόκληρο τον φυσικό κόσμο.

ΣΗΜΕΙΩΣΕΙΣ

[1] Βλ. "Physics Of Billiards – Ball Collision," στο *Real World Physics Problems*, n.a., n.d., https://www.real-world-physics-problems.com/physics-of-billiards.html#:~:text=The%20physics%20behind%20billiards%20(or,conserved%20before%20and%20after%20impact, Πρόσβαση στις 4 Δεκεμβρίου 2020.

[2] Βλ. "A Theory of Reciprocity," των Armin Falk και Urs Fischbacher, *Games and Economic Behavior*, Volume 54, Issue 2, February 2006, σελ. 293-315, https://doi.org/10.1016/j.geb.2005.03.001, Elsevier, https://www.sciencedirect.com/science/ article/abs/pii/S0899825605000254, Πρόσβαση στις 4 Δεκεμβρίου 2020.

[3] Βλ. "Cascading regime shifts within and across scales," των Juan C. Rocha, Garry Peterson, Örjan Bodin και Simon Levin, *Science* 21 Dec 2018:Vol. 362, Issue 6421, σελ. 1379-1383, doi: 10.1126/science.aat7850, https://science.sciencemag.org/content/362/6421/1379, Πρόσβαση στις 4 Δεκεμβρίου 2020.

[4] ό.π.

[5] Βλ "Movement Disorders," *RadiologyInfo.org*, 25 Νοεμβρίου 2019, n.a.

[6] Βλ. "Both Predator and Prey – Emotional Arousal in Threat and Reward," των Andreas Löw, Peter J. Lang, J. Carson Smith και Margaret M. Bradley, *Psychol Sci.* 2008 Sep: 19(9): 865-873. doi: 10.1111/j.1467-9280.2008.02170.x, PMCID: PMC3612950, NIHMSID: NIHMS451987, PMID: 18947351, HHS Public Access, https://www.ncbi.nlm.nih.gov/pmc/ articles/PMC3612950/ Πρόσβαση στις 4 Δεκεμβρίου 2020.

[7] "La production de l'espace," στο *Editions Anthropops, L'Homme et la Société, 1974, και The production of space,* του Henri Lefebvre, Μετάφραση Donald Nicholson-Smith, Basil Blackwell, Oxford UK, 1991

[8] "Giant Mount Everest Is Still Growing And Shaking," του Hill Williams, *The Seattle Times*, 17 Σεπτεμβρίου1990, https://archive.seattletimes.com/archive/?date=19900917&slug= 1093659, Πρόσβαση στις 4 Δεκεμβρίου 2020.

[9] "Precise atomic clock may redefine time," του Phillip Ball, 9 Ιουλίου 2013, *Nature*, doi:10.1038/nature.2013.13363, https://www.nature.com/news/precise-atomic-clock-may-redefine-time-1.13363

[10] ό.π.

[11] *La loi du 18 Germinal an 3; la mesure [républicaine] de superficie pour les terrains, égale à un carré de dix mètres de côté* [*The law of 18 Germinal year 3; the [Republican] measure of surface area for land, equal to a square with a side of ten meters.*]. Le CIV (Centre d'Instruction de Vilgénis) – Forum des Anciens.

[12] "Πρώτη ομιλία για τη συμφιλίωση με την Αμερικής [*First Speech on Conciliation with America*], "American Taxation," 19 Απριλίου 1774.

[13] "Auxillaries to the Perception of Intelligible Natures", του Πορφύριου, στο *Select Works Of Porphyry*, Μετάφραση από τα ελληνικά του Thomas Taylor, Εκτύπωση Thomas Rodd, Λονδίνο 1823, σελ. 223.

[14] "Scientists show that people really walk in circles when lost," *PhysOrg*, 20 Αυγούστου 2009,n.a., https://phys.org/news/2009-08-scientists-people-circles-lost-video.html

[15] Βλ. "Examining Equivalence of Concepts and Measures in Diverse Samples," των Tracy W. Harachi, Yoonsun Choi, Robert D. Abbott, Richard F. Catalano και Siri L. Bliesner, *Prevention Science*. 2002 Dec: 7(4): 359-368. doi: 10.1007/s11121-006-0039-0, PMCID: PMC3293252, NIHMSID: NIHMS293849, PMID: 16845592, HHS Public Access, https://www.ncbi.nlm.nih.gov/pmc/articles/PMC3293252/#R37, Πρόσβαση στις 4 Δεκεμβρίου 2020.

[16] Charles François, "Systemics and cybernetics in a historical perspective". In: *Systems Research and Behavioral Science*. Vol 16, σελ. 203–219, 1999.

[17] Βλ. *De rerum natura, Venice έκδοση* του 1515. Βλ. επίσης *Traces on the Rhodian Shore-Nature and Culture in Western Thought from Ancient Times to the End of the Eighteenth Century*, του Clarence J. Glacken, University of California Press, Berkeley and Los Angeles, 1997, σελ. 29-30.

[18] Βλ. *Summa Theologiae*. 61 τόμοι, Cambridge University Press, Cambridge, UK, 2006.

[19] Για μια πιο γενική επισκόπηση, βλ. *Successes and Failures in Fighting Disease*, n.a.. Watchtower Online Library, https://wol.jw.org/en/wol/d/r1/lp-e/102004362, Πρόσβαση στις 5 Δεκεμβρίου 2020.

[20] Βλ. "The Social Life of Forests," του Ferris Jabr, Φωτογραφία Brendan George Ko, *The New York Times Magazine*, 2 Δεκεμβρίου 2020, https://www.nytimes.com/interactive/ 2020/12/02/magazine/tree-communication-mycorrhiza.html, Πρόσβαση στις 5 Δεκεμβρίου 2020.

[21] Τον Ιούνιο του 2016 πάνω από 4 εκατομμύρια θεατές παρακολούθησαν την ομιλία της Σιμάρ στο TED με θέμα: "How trees talk to each other," https://www.ted.com/talks/suzanne_simard_how_trees_talk_to_each_other?language=en.

[22] Βλ. *Clearcut Carbon – A Sierra Club BC report on the future of forests in British Columbia*, του Jens Wietimg και ανάλυση χαρτογράφησης του David Leversee, Δεκέμβριος 2019, https://sierraclub.bc.ca/wp-content/uploads/2019-Clearcut-Carbon-report.pdf

[23] "Logging Statistics from the Council of Forest Industry (CoFI)," *Protect Glade Watershed*, https://www.protectgladewatershed.com/forestry-stats-cofi.html

[24] Βλ. *Hope, Risk and Reality: Perspectives from Indigenous Voices on our Relationship with Mother Earth, Συνέντευξη με τον Shanny Spang Gion, Καλλιτεχνική επιμέλεια Brenda Mallory*, 9 Φεβρουαρίου 2021, MAHB (Millennial Alliance for Humanity and the Biosphere), Stanford University, https://mahb.stanford.edu/blog/introduction-to-a-four-interview-cycle-an-interview-with-shanny-spang-gion/

[25] Βλ. *Soil Biota*, n.a., https://nematode.unl.edu/gramsoil.htm, Πρόσβαση στις 5 Δεκεμβρίου 2020.

[26] Βλ. "Indirect Reciprocity and the Evolution of Prejudicial Groups," των Roger M. Whitaker, Gualtiero B. Colombo και David G. Rand, *Scientific Reports 8*, Article number: 13247 (2018), 5 Σεπτεμβρίου 2018, https://www.nature.com/articles/s41598-018-31363-z, Πρόσβαση στις 5 Δεκεμβρίου 2020.

Κεφάλαιο 12

ΕΞΑΡΤΩΜΕΝΕΣ ΜΕΤΑΒΑΤΙΚΕΣ ΚΑΤΑΣΤΑΣΕΙΣ

Στην παραδοσιακή δημογραφική ανάλυση, οι οικολόγοι έχουν κάνει αναφορές στους πλέον βασικούς παράγοντες "της γέννησης, της μετανάστευσης, του θανάτου και του εκπατρισμού".[1] Αυτό δεν σχετίζεται καθόλου μα καθόλου με τη συνείδηση. Με άλλα λόγια, ένα γράφημα, μια μέτρηση, μια αλλαγή που αποκαλύπτει εν συντομία κάποιο είδος αλλαγής, μια έννοια τόσο απλή όπως ο παραδοσιακός ορισμός της "περιοχής ενός είδους", ήτοι "η μικρότερη περιοχή μέσα στην οποία το είδος εμφανίζεται σε κατάλληλα περιβάλλοντα και έξω από την οποία απουσιάζει εντελώς".[2] Η πρώτη εξαίρεση σε έναν τέτοιο κανόνα προέρχεται από τον ουρανό. Στην περίπτωση του *Calidris canutus*, της βορειοαμερικανικής κοκκινοσκαλίδρας, μεταξύ των πρώτων καταγεγραμμένων επισκεπτών στα δυτικά τμήματα της λίμνης Ήρι βρέθηκαν "τυχαίοι καλοκαιρινοί επισκέπτες που δεν αναπαράγονταν", όπως σημειώθηκε την τελευταία εβδομάδα του Ιουνίου (1954) και την πρώτη εβδομάδα του Ιουλίου (1985).[3] Ο λαμπρός μελετητής των κολιμπρί, Αλεξάντερ Σκουτς, παρατήρησε εκατό σκιουροπίθηκους της νότιας Κόστα Ρίκα να στέκονται δίπλα σε πιθήκους καπουτσίνους και να τρώνε μαζί τους από κοινού διάφορα "έντομα, μικρά ώριμα φρούτα, όπως σύκα και νέκταρ" σε όλο και μεγαλύτερες αποστάσεις μέσα από το γύρω δάσος. Όχι μόνο οι περιοχές τους ήταν ευέλικτες, αλλά - σημείωσε ο Σκουτς – παρά τις "ευρέως χωριστές καταγωγές" επέδειξαν πραγματική αμοιβαία "οικολογική σύγκλιση" με "μάρμοζετ, ταμαρίνους, πιθήκους τίτι και πουλιά", κάτι που υποδηλώνει "αποκλειστικότητα" μόνο όταν έρχονται αντιμέτωποι με "γειτονικές οικογένειες του ίδιου είδους στα όρια της περιοχής" που οριοθετούνται από "φωνητικά" σημάδια.[4]

Συνδυάζοντας την παλαιότητα των οικοπεριοχών της οροσειράς Ταλαμάνκα και των τροπικών υγρών δασών της Κόστα Ρίκα -η Βόρεια και η Νότια Αμερική ενώθηκαν πριν από τρία έως πέντε εκατομμύρια χρόνια [5] (και δεκάδες χιλιάδες παρόμοια είδη δασών βροχής βρίσκονται ακριβώς στα νότια, στον 55 εκατομμυρίων ετών Αμαζόνιο) με τουλάχιστον 12.000 ανθοφόρα φυτά, εκ των οποίων σχεδόν το 7% είναι ενδημικά, ο οικότοπος φυτοφάγων και εντομοφάγων είναι ιδιαίτερα εκτεταμένος. Ως εκ τούτου, ένας κανόνας που εδραιώνει την αλληλεπικάλυψη οικότοπων. Από τη σκοπιά της φυσικής, το άθροισμα αυτών των επικαλύψεων αντιπροσωπεύει μια κβαντική υπέρθεση. Τα πιο βασικά σημειωτικά σήματα εντός τέτοιων τοπολογιών ή άλλα είδη σηματοδότησης θέτουν τα θεμέλια για ένα πρότυπο που είναι ανοικτό σε οποιαδήποτε ερμηνεία επιλέξουμε: ακουστικά, γλωσσικά, εικονογραφικά ή οποιουδήποτε είδους αισθητηριακοί, ακόμη και μετεμψυχολογικοί φορείς πληροφοριών.

Το ζήτημα των βιολογικών επιπέδων αναφοράς έχει μεγάλο ενδιαφέρον, δεδομένου ότι σταματήσαμε να ξεκινάμε από το μηδέν αιώνες πριν. Οι ισοπεδωτικές καμπύλες εξαφάνισης έχουν ξεπεραστεί από τις έννοιες της ανθεκτικότητας, του μετριασμού και των ποσοτικοποιήσεων της IUCN (της Διεθνούς Ένωσης Προστασίας της Φύσης) για τα είδη που βρίσκονται σε κίνδυνο. Η υποκείμενη υπόθεση υποδηλώνει μια κατάσταση μη επιστροφής.

Από την άλλη, η απόδοση νοήματος και πλαισίου σε είδη που *δεν* απειλούνται από τον άνθρωπο απαιτεί όχι μόνο εργαλεία, αλλά υποθέσεις που είναι νέες για τον 21^{ο} αιώνα. Όταν ο Τζορτζ Χάντσον (1867-1946) περιέγραψε τις μεταναστεύσεις των μυρμηγκιών *Atta antarctica* στη Νέα Ζηλανδία το 1892, η ιδιαίτερή του άποψη ήταν ότι “η ετήσια μετανάστευση των φτερωτών αρσενικών και θηλυκών αυτού του είδους συνήθως λαμβάνει χώρα μια ζεστή ημέρα κατά την τελευταία εβδομάδα του Μαρτίου…” [6]. Παρότι ο Χάντσον ήταν ένα παιδί θαύμα που μεγάλωσε σε αγρόκτημα, και μάλιστα δημοσίευσε τις πρώτες παρατηρήσεις του για τα έντομα σε ένα περιοδικό εντομολογίας όταν ήταν μόλις δεκατεσσάρων και αργότερα επινόησε την ιδέα του θερινού χρόνου, μεταξύ των 48 γνωστών ειδών μυρμηγκιών στη Νέα Ζηλανδία, δεν υπάρχει κανένα μέλος του γένους *Atta* (φυλλοκόπτης), όπως κατέστη σαφές από τους Bert Hölldolbler και Έντουαρντ Όσμπορν Γουίλσον έναν αιώνα μετά το έργο του. [7] Είναι πιθανό, φυσικά, κάποια στιγμή να έφτασαν φυλλοκόπτες στη Νέα Ζηλανδία με κάποιο πλοίο από τη Νότια Αμερική και στη συνέχεια να εξαφανίστηκαν από τη Ζηλανδία. Είναι αδύνατον να μελετηθεί η κατανομή των εντόμων σε ένα σύστημα συστημάτων με παγκόσμια διακύμανση, και όλοι οι κανόνες δίνονται στη μετάβαση.

Είναι, για παράδειγμα, γεγονός ότι σε ολόκληρο το βόρειο ημισφαίριο “η φωτοσύνθεση των φυτών διακόπτεται σε μεγάλο βαθμό” τον χειμώνα,[8] κάτι που με τη σειρά του μας βοηθά να εξηγήσουμε τον λόγο για τον οποίο πάνω από το 70% όλων των γνωστών πτηνών της Βόρειας Αμερικής μεταναστεύουν. Ως εκ τούτου, ο εφοδιασμός με τρόφιμα, τα όντα που τα καταναλώνουν και η γεωγραφία των μέσων επιβίωσής τους μεταβάλλονται συνεχώς. Όπως ακριβώς κάθε ζωή και τοπίο. Η καλλιτέχνης, βιολόγος και παλαιοντολόγος Λόρα Κάνινγκχαμ έγραψε: “Ξεκίνησα την εξερεύνησή μου στην Παλιά Καλιφόρνια προσπαθώντας να φανταστώ αρκούδες γκρίζλι να μασουλάνε πεσμένα βελανίδια, να ξεθάβουν βολβούς βροδιαίας ανάμεσα σε θάμνους και να κοιμούνται σε πρεμνοφυή δάση ιτιάς της διάσπαρτης με βελανιδιές κοιλάδας του Ηστ Μπέι, αυτής που τώρα είναι το γεμάτο με αυτοκίνητα εμπορικό κέντρο El Cerrito Plaza”. Αργότερα, κοντά στο σπίτι της στην περιοχή της Κοιλάδας του Θανάτου, στις 15 Αυγούστου 2004, έζησε μια αξιοσημείωτη αστραπιαία πλημμύρα στην έρημο και έγραψε: "Αυτές οι πλημμύρες πάντα συνέβαιναν και πάντα θα συμβαίνουν, και κάποια ζώα βρήκαν νέες ευκαιρίες σε αυτές. Μπορούμε κι εμείς". [9] Οι αστικές αντιλήψεις μας για το πώς ζουν τα ζώα θολώνουν σε μεγάλο βαθμό την αντίληψή μας σχετικά με το πού ζουν *στην πραγματικότητα* και από πού έχουν προέλθει. Στην περίπτωση της μαυροσταχτάρας, της γκρίζας κουκουβάγιας και άλλων πτηνών, όπως "κοτσύφια, κοκκινολαίμηδες και θαμνοψάλτες", πολλοί φτερωτοί κάτοικοι πόλεων που υποτίθεται ότι προέρχονται από την Ευρώπη χτίζουν τις φωλιές τους στα πλησιέστερα φυσικά δάση που μπορούν να βρουν, μέσα σε κούτσουρα ή κοιλότητες δέντρων. Με τον δικό τους τρόπο, είναι απομεινάρια, τουλάχιστον σαν τον Ευρωπαίο βίσονα, μιας αρχέγονης εποχής που βλέπουμε να εξαφανίζεται.[10] Στη Βόρεια Αμερική, ένα άλλο μέλος της οικογένειας των βοοειδών σχεδόν αφανίστηκε εντελώς μπροστά στα μάτια μας. Όπως γράφουν ο Τιμ Φλάνερι και ο Πίτερ Σχάουτεν, "60 εκατομμύρια βούβαλοι περιφέρονταν στις πεδιάδες σε κοπάδια ενός ή περισσότερων εκατομμυρίων ζώων". Τον δέκατο ένατο αιώνα, όταν περνούσε ένα τέτοιο κοπάδι, περνούσαν μπροστά στα μάτια κάποιου τα περισσότερα μαζεμένα κιλά ενός μόνο είδους που θα μπορούσε να δει οπουδήποτε αλλού στον πλανήτη."[11] Μέχρι το τέλος του έτους, το 1883, μια ομάδα ηττημένων Ινδιάνων, σε συνεργασία με λευκούς κυνηγούς βουβαλιών, κατάφερε να σφάξει τον τελευταίο από τους βόρειους βίσωνες στην περιοχή του ποταμού Κάνονμπολ, στη Βόρεια Ντακότα. Όπως έγραψε ο Πίτερ Ματιέσεν στο αιχμηρό κλασικό του έργο *Wildlife In America*, "περιορίζοντας την πρόσβαση στο νερό, οι κυνηγοί κατάφεραν να καταστρέψουν ολόκληρο το κοπάδι. Μέχρι το τέλος του έτους, εξαιρουμένων των αδέσποτων ζώων, ο βούβαλος (πλέον εθνικό ζώο των ΗΠΑ, μαζί με τον φαλακρό αετό) είχαν εξαφανιστεί από τη Βόρεια Αμερική."[12]

Όλες αυτές οι αλλαγές, είτε πρόκειται για ένα εφήμερο καιρικό φαινόμενο είτε για μια πραγματική βιολογική εξαφάνιση, έχουν αμφισβητηθεί από την ανθρώπινη αυτοαναφορά, τις πεποιθήσεις που έχουμε σχετικά με τον αντίκτυπό μας στους Άλλους και τη συνύπαρξή μας με αυτούς. Ο Roger Caras (1928-2001), δέκατος όγδοος πρόεδρος της ASPCA (της Αμερικανικής Ένωσης για την Πρόληψη της Βίας κατά των Ζώων), ήταν πολύ ευθύς όσον αφορά το ρόλο των εξημερωμένων ζώων, δηλώνοντας εμφατικά ότι, εάν οι άνθρωποι δεν είχαν ταΐσει και καλωσορίσει τη συντροφιά τους, αλλά τελικά καταπίεζαν και/ή θανάτωναν όπως ήθελαν "υποζύγια" όπως η κατσίκα, το πρόβατο, ο τάρανδος, η άλκη, ο σκύλος, τα βοοειδή, ο χοίρος, ο ελέφαντας, οι καμηλίδες, τα ιπποειδή, η γάτα, το κουνέλι, το κοτόπουλο, η γαλοπούλα και αμέτρητα άλλα, δε θα είχαμε κατοικήσει ποτέ τον μισό σχεδόν πλανήτη, θα εξακολουθούσαμε να ζούμε εξ ολοκλήρου στην ύπαιθρο, χωρίς πόλεις και δε θα είχαμε επιδοθεί σε μεγάλους πολέμους ή κατακτήσεις. Οι πολιτισμοί μας, οι οικισμοί μας, πράγματι το μεγαλύτερο μέρος της ανθρώπινης προόδου, θα είχαν σταματήσει.[13] Αυτή η σχέση σίγουρα υπογραμμίστηκε από τους πρώτους αποικισμούς στην ανατολική ακτή των ΗΠΑ. Για παράδειγμα, τα στοιχεία δείχνουν ότι η Βιρτζίνια μπορεί να μην είχε επιβιώσει αν οι άποικοι δεν είχαν καταφέρει, μετά την πτώχευση της

Εταιρείας της Βιρτζίνια το 1624, να διατηρήσουν τις απαγορεύσεις κατά της βλάβης ζώων που μπορούσαν να αναπαραχθούν, κάτι που το 1629 οδήγησε σε μια απογραφή που περιελάμβανε "20.000 βοοειδή, 200 άλογα, 50 γαϊδούρια, 3.000 πρόβατα, 5.000 κατσίκες, αναρίθμητα γουρούνια και πουλερικά." Τέτοιοι αριθμοί αντανακλούσαν τις εξαρτώμενες εξισώσεις της πρόωρης σφαγής στη Βόρεια Αμερική, την επιμονή των νέων αποίκων σε μια επιβλητική ήπειρο.[14]

Η ανθρώπινη ιστορία έχει παγιδεύσει τις συνδέσεις της σε τέτοιο βαθμό που είμαστε *σχεδόν* ανίκανοι να αντιληφθούμε, να αισθανθούμε ή ακόμα και να θυμηθούμε πώς θα μπορούσε να ήταν. Ωστόσο, η μεταβατική φάση - στη χημεία, το αέριο, το υγρό ή το στερεό – εξαρτάται πλήρως από "διαμοριακές δυνάμεις"[15] οι οποίες, με τη σειρά τους, στηρίζονται στη σταθερότητα της θερμοδυναμικής. Οι οικολογικές μεταβάσεις, ωστόσο, δεν εξαρτώνται μόνο από τη θερμοκρασία, το έργο, την ενέργεια και την εντροπία, αν και θα μπορούσε τελικά κανείς να εντοπίσει τα πάντα σε αυτό το μεγάλο άγνωστο πριν από τη θέρμανση/διαστολή και την επακόλουθη ψύξη των υποατομικών σωματιδίων πριν και μετά από θεωρητικά κοσμολογικά γεγονότα, όπως η Μεγάλη Έκρηξη. Πιο κοντά στην εποχή μας, εδώ, σε έναν μικρό βράχο στο διάστημα, παρατηρούμε μια σχεδόν άπειρη σειρά σημείων ενεργοποίησης και ορίων. Η τεράστια έκτασή τους περιλαμβάνει όλο τον οικολογικό χρόνο, τις τάσεις και τα σημεία στίξης. Αυτά περιλαμβάνουν: "τροφικούς καταρράκτες" που θα μπορούσαν να προκληθούν από "μολυσματικούς παράγοντες" ή "χωροκατακτητικά είδη"·[16] στατιστικές διακυμάνσεις του πλούτου των ειδών σε οποιαδήποτε δεδομένη περιοχή και τα πλήθη των προτεινόμενων ποσοτικών και ποιοτικών αιτίων της αφθονίας ή των μειούμενων αριθμών·[17] "το φαινόμενο της επιβράδυνσης" – που θεωρείται σε πολλές περιπτώσεις "έγκαιρη προειδοποίηση" επερχόμενης μετάβασης ή "διακλάδωσης" (ο χώρος/χρόνος μεταξύ σταθερότητας και διατάραξης).[18]

Άλλα γεγονότα ενεργοποίησης θα μπορούσαν να περιλαμβάνουν τις τεράστιες απώλειες βιοποικιλότητας όταν μετρώνται σε δασικά σύνολα όπου δεν έχουν υπολογιστεί οι μεταβάσεις των "νέων δασών",[19] διαρκώς μεταβαλλόμενες όψεις "οικολογικής ακεραιότητας, ποικιλομορφίας και ανθεκτικότητας"[20] καθώς και τα μέτρα της "προόδου της επαναφοράς στη φυσική κατάσταση"[21], υπό το πρίσμα τόσο των σημερινών αλλαγών, όσο και μετρήσεων του παρελθόντος (οικολογικών μετρήσεων), όπως στην περίπτωση των κλιματικών χαρακτηριστικών που ευνόησαν τα "λιβάδια" ως κυρίαρχη χλωρίδα στις Μεγάλες Πεδιάδες κατά τη διάρκεια του Μειόκαινου της Βόρειας Αμερικής.[22] Άλλες ποικίλες μεταβατικές μετρήσεις περιλαμβάνουν "μεταβολικούς ρυθμούς σε συνδυασμό με εξειδικευμένες μεταβάσεις" (όπως στην περίπτωση των υπόγειων εκσκαφών και της νυχτερινής δραστηριοποίησης των σπονδυλωτών εξώθερμων" (π.χ. διαφόρων σαυρών),[23] τις "τοξικές επιπτώσεις των μικροπλαστικών" σε οποιοδήποτε περιβάλλον,[24] τις εξαιρετικά ποικιλόμορφες εκκινήσεις της διαδικασίας της "προσαρμοστικής διασποράς" σε είδη όπως τα ποντικόφιδα του Νέου Κόσμου, τα οποία οι εξελικτικοί βιολόγοι, χρησιμοποιώντας το εργαλείο των "βαθμονομημένων χρονικά μοριακών φυλογονιών", μπόρεσαν να εξετάσουν από την άποψη των "οικολογικών ευκαιριών" που επιτρέπουν "γρήγορη ειδογένεση", όχι μόνο από αρχεία απολιθωμάτων, αλλά και με ζωντανούς οργανισμούς[25].

Οι εξαρτώμενες μεταβάσεις αυτού του είδους λειτουργούν είτε στο μέγεθος του μηδενός, είτε, εάν κανείς το προτιμά, του συμβόλου "ίσον" (=). Είναι ευκαιρίες, φέρουν όμως από κοινού μια ενιαία σταθερά: είναι ευάλωτα στις ανατροπές, όπως όταν τα "πολλαπλασιαστικά αποτελέσματα" επεκτείνονται σε αλληλεξαρτώμενες λεπτομέρειες, η απώλεια ενός φυτού προκαλεί μια αλυσιδωτή επίδραση στα ζωικά είδη (ή άλλα φυτικά είδη) που ουσιαστικά το έχουν ανάγκη, μια προφανής αιτία τροφικών καταστροφών.[26] Μια από τις πιο τεράστιες τραγικές μεταμορφώσεις συμβαίνει τώρα στον Αμαζόνιο, καθώς οι ανθρωπογενείς δυνάμεις που ασκούνται πάνω του σε κάθε πλευρά του και από τον αέρα, μέρα με τη μέρα υπονομεύουν σωρευτικά την ικανότητα κατακράτησης υγρασίας και δέσμευσης χημικών ουσιών του ποταμού, επιταχύνοντας έτσι την μετατροπή του σε ένα μεγαοικοσύστημα σαν της σαβάνας. [27]

ΣΗΜΕΙΩΣΕΙΣ

[1] Βλ. "Demographic Processes," Κεφ. 7 στο *Principles of Conservation Biology*, του Gary K. Meffe, C. Ronald Carroll, and Contributors, Sinauer Associates, Inc., Sunderland, Mass., 1994, σελ. 180.

[2] Βλ. *The Butterflies Of North America*, του William H. Howe, Coordinating Editor And Illustrator & Twenty Contributors, Doubleday & Company, Inc., Garden City, New York, 1975, σελ. 33.

[3] *The Birds of Ohio*, του Bruce G. Peterjohn, Original Paintings by William Zimmerman, Indiana University Press, Bloomington, Indiana, 1989, σελ. 76.

[4] *Trogons, Laughing Falcons, and Other Neotropical Birds*, του Alexander F. Skutch, Εικονογράφηση: Dana Gardner, Texas A&M University Press, College Station, TX, 1999, σελ. 169.

[5] Βλ. *Costa Rica – Wildlife of the National Parks and Reserves*, των Michael και Patricia Fogden, Fundacion Neotropica Editorial Heliconia, San José, Costa Rica, 2005, σελ. 16.

[6] *An Elementary Manual Of New Zealand Entomology*, του G. V. Hudson, West, Newman, & Co., London, 1892, σελ. 37.

[7] *The Ants*, των Bert Hölldolbler και Edward O. Wilson. The Belknap Press of Harvard University Press, Cambridge, Massachusetts, 1990, σελ. 600.

[8] "State Of The States: Geographic Patterns Of Diversity, Rarity, And Endemism", των Bruce A. Stein, Lynn S. Kutner, Geoffrey A. Hammerson, Lawrence L. Master and Larry E. Morse, στο *Precious Heritage -The Status of Biodiversity in the United States*, των Bruce A. Stein, Lynn S. Kutner και Jonathan S. Adams, εικονογράφηση της Nicole S. Rousmaniere, Oxford University Press, New York, New York, 2000, σελ. 143.

[9] *A State of Change – Forgotten Landscapes of California*, Κείμενα και καλλιτεχνική επιμέλεια της Laura Cunningham, Heyday, Berkeley CA, 2010, σελ. 299-300.

[10] *The Afterlife Of A Tree*, εκδότης: Andrzej Bobiec, συγγραφείς: Jerzy M. Gutowski, William F. Laudenslayer, Pawel Pawlaczyk, Karol Zub, World Wildlife Fund Poland, Warsaw, Poland, 2005, σελ. 175.

[11] *A Gap in Nature – Discovering The World's Extinct Animals*, Text Publishing, Melbourne, Australia 2001, σελ. xx.

[12] *Wildlife In America*, Peter Matthiessen, Revised, Updated Edition, Drawings by Bob Hines, 1987, Viking Press, New York, σελ. 149-150.

[13] *A Perfect Harmony – The Intertwining Lives Of Animals And Humans Throughout History*, του Roger A. Caras, Simon & Schuster, New York, 1996.

[14] Βλ. *Creatures of Empire – How Domestic Animals Transformed Early America*, της Virginia DeJohn Anderson, Oxford University Press, New York, 2004, σελ. 103.

[15] *Chemistry*, *LibreTexts*, 15 Αυγούστου 2020, n.a. https://chem.libretexts.org/Bookshelves/Physical_and_Theoretical_Chemistry_Textbook_Maps/Supplemental_Modules_(Physical_and_Theoretical_Chemistry)/Physical_Properties_of_Matter/States_of_Matter/Phase_Transitions/Fundamentals_of_Phase_Transitions, Πρόσβαση στις 8 Δεκεμβρίου 2020.

[16] "Infectious Agents Trigger Trophic Cascades," των Julia C. Buck and William J. Ripple, Trends in Ecology & Evolution, Volume XX, 2266, αρ. σελ. 14, CellPress, 20 Ιουλίου 2017, doi: https://doi.org/10.1016/j.tree.2017.06.009, http://trophiccascades.forestry.oregonstate.edu/ sites/trophic/files/Buck2017_TREE.pdf; See also, "Invasive species triggers a massive loss of ecosystem services through a trophic cascade," των Jake R. Walsh, Stephen R. Carpenter και M. Jake Vander Zanden, PNAS 12 Απριλίου 2016 113 (15) 4081-4085; πρώτη έκδοση 21 Μαρτίου 2016; https://doi.org/10.1073/pnas.1600366113, Proceedings of the National Academy of Sciences of the United States of America, PNAS, https://www.pnas.org/content/113/15/4081

[17] "Measuring Ecological Biodiversity," του R. Hengeveld, Biodiversity Letters, Vol. 3, No. 2 (Mar., 1996), σελ. 58-65, https://doi.org/10.2307/2999770, https://www.jstor.org/stable/ 2999770, Wiley, https://www.jstor.org/stable/2999770?seq=1

[18] "Early Warning Signals of Ecological Transitions: Methods for Spatial Patterns," των Sonia Kéfi, Vishwesha Guttal, William A. Brock, Stephen R. Carpenter, Aaron M. Ellison, Valerie N. Livina, David A. Seekell, Marten Scheffer, Egbert H. van Nes, και Vasilis Dakos, *PLOS ONE*, 21 Μαρτίου 2014, https://doi.org/10.1371/journal.pone.0092097,
https://journals.plos.org/plosone/article?id=10.1371/journal.pone.0092097

[19] See "Forest ecosystem-service transitions: the ecological dimensions of the forest transition," των Sarah Jane Wilson, John Schelhas, Ricardo Grau, A. Sofia Nanni και Sean Sloan, *Ecology and Society*, 22(4): 38. https://doi.org/10.5751/ES-09615-220438, https://www.ecologyandsociety.org/vol22/iss4/art38/

[20] "Measuring ecological complexity," του Lael Parrott, Ecological Indicators 10(6): 1069-1076, doi: 10.1016/j.ecolind.2010.03.014, *ResearchGate*, https://www.researchgate.net/ publication/248412781_Measuring_ecological_complexity

[21] "Measuring rewilding progress," των Aurora Torres, Néstor Fernández, Sophus zu Ermgassen, Wouter Helmer, Eloy Revilla, Deli Saavedra, Andrea Perino, Anne Mimet, José M. Rey-Benayas, Nuria Selva, Frans Schepers, Jens-Christian Svenning, και Henrique M. Pereira, *Philosophical Transactions Of The Royal Society B, Biological Sciences*, The Royal Society Publishing, 22 Οκτωβρίου 2018, https://doi.org/ 10.1098/rstb.2017.0433, https://royalsocietypublishing.org/doi/10.1098/rstb.2017.0433

[22] Βλ. "Measuring Earth-Life Transitions: Functional Traits from Late Cenozoic Vertebrates," των P. David Polly και Jason J. Head, *The Paleontological Society Papers*, Volume 21: Earth-Life Transitions, 21 Ιουνίου 2017, https://www.cambridge.org/core/journals/the-paleontological-society-papers/article/abs/measuring-earthlife-transitions-ecometric-analysis-of-functional-traits-from-late-cenozoic-vertebrates/C0068D2B30ACCF76F415B2EB75F8BEB2

[23] "Shifts in space and time: ecological transitions affect the evolution of resting metabolic rates in microtelid lizards," των Melissa Bars-Closel, Agusti Camacho και Tiana Kohlsdorf, *Journal of Experimental Biology* 2018 221:jeb175661 doi:10.1242/jeb.175661, 30 Ιουλίου 2018, https://jeb.biologists.org/content/221/14/jeb175661

[24] "Microplastics in aquatic environments: Toxicity to trigger ecological consequences," των Hui Ma, Shengyan Pu, Shibin Liu, Yingchen Bai, Sandip Mandal and Baoshan Xing, *Envion Pollut.* Ιούνιος 2020, 261:114089. doi: 10.106/j.envpol.2020.114089. https://pubmed.ncbi.nlm.nih.gov/32062100/

[25] Βλ. "Ecological Opportunity: Trigger of Adaptive Radiation," των E. A. Myers και F. T. Burbrink, *Nature Education Knowledge* 3(10):23, 2012. https://www.nature.com/scitable/ knowledge/library/ecological-opportunity-trigger-of-adaptive-radiation-84160951/

[26] Βλ. "Domino effect – the loss of plant species triggers the extinction of animals," του Senckenberg Research Institute and Natural History Museum, 5 Ιανουαρίου 2017, *PHYS ORG*, https://phys.org/news/2017-01-domino-effectthe-loss-species-triggers.html: Βλ. Matthias Schleuning et al. "Ecological networks are more sensitive to plant than to animal extinction under climate change, *Nature Communications* (2016). doi: 10.1038/ncomms13965

[27] Βλ. "Amazon rainforests could transition to savannah-like states in response to climate change, new study predicts," του Shreya Dasgupta, Mongaby, 29 Δεκεμβρίου 2015, *PHYS ORG*, https://phys.org/news/2015-12-amazon-rainforests-transition-savannah-like-states.html, Πρόσβαση στις 9 Δεκεμβρίου, 2020.

"The Vision of Saint Eustace," Engraving by Albrecht Dürer, ca. 1501, Private Collection. Φωτογραφία © M.C.Tobias.

Κεφάλαιο 13

Ο Αλμπρεχτ Ντυρερ και η Φύση της Οικολογικής Μετατροπής

Είναι και γίνεσθαι, οι πηγές όλης της οντολογίας της ενδοσκόπησης, το αντικείμενο με οποίο ασχολήθηκε με τεράστια εμμονή όλη η δυτική λογοτεχνία και φιλοσοφία. Αναφορές σχετικά με την περισυλλογή έχει βρεθεί αρχικά στα ελάχιστα έργα του σαμάνου (Ιατρομάντη) Παρμενίδη (515 π.Κ.Ε. - περίπου 450 π.Κ.Ε.) και του λίγο παλαιότερου Ηράκλειτου (περίπου 535 π.Κ.Ε.- 475 π.Κ.Ε.), στα αντίστοιχα έργα τους "Περί Φύσεως", από τα οποία σώζονται μόνο "αποσπάσματα" και στις δύο περιπτώσεις. Ο Παρμενίδης μάς έδωσε ένα από τα πιο διορατικά και επαναστατικά επίπεδα αναφοράς όταν έγραψε ότι "*Τα είδη* δεν εξαρτώνται από τον Νου του Ανθρώπου, αντίθετα, είναι άγνωστα στην ανθρώπινη Φύση ή την Ανθρωπότητα".[1] Ο Ηράκλειτος υποθέτει ότι ο ανθρώπινος νους αντιλαμβάνεται τόσο την ύπαρξη όσο και την αλλαγή, λέγοντας ότι "Αυτό που ήταν διάσπαρτο συγκεντρώνεται. Αυτό που είχε συγκεντρωθεί καταστρέφεται".[2] Στις εισαγωγικές παρατηρήσεις του στον "Φαίδρο" του Πλάτωνα (περίπου 424 π.Κ.Ε. -347 π.Κ.Ε.), έναν διάλογο που συνετέθη τουλάχιστον έναν αιώνα μετά το θάνατο του Παρμενίδη και του Ηρακλείτου, ο Benjamin Jowett υποστήριξε ότι "η διαλεκτική μπορεί να οριστεί ποικιλοτρόπως, είτε ως η δύναμη της διαίρεσης ενός συνόλου σε μέρη και της ένωσης των μερών σε ένα σύνολο, είτε ως η διαδικασία του νου που μιλάει με τον εαυτό του".[3] Πρόκειται για μεταφυσικούς πυλώνες νόησης οι οποίοι δεν μπορούν να ειπωθούν, χωρίς μια τελικά προβληματική ύβρη, για να *αντικατοπτρίσουν* τη Φύση, τόσο πολύ ώστε να αποδώσουν σχολιασμό που είναι, όσο και αν είναι το βάθος του, μια μαθηματικά τυχαία επιλογή των πραγμάτων που θα μπορούσε κάποιος να πει ή να αισθανθεί. Αυτή είναι μια πολύ αδύναμη πρόβλεψη, χωρίς συνέπειες. Στην πραγματικότητα, δεν γνωρίζουμε τίποτα. Τα κομμάτια της ιστορίας που ενώσαμε παραμένουν τα ίδια. Εντυπώνοντάς τα με την τοποθέτησή τους σε ένα εννοιολογικό πλαίσιο, οπισθοχωρούμε από τα επίπεδα αναφοράς που θα χρειαζόμασταν σε άλλη περίπτωση για να αντιληφθούμε την κατάστασή μας.

Μόλις προσθέσουμε ένα συν ένα, το είδος μας είναι απίθανο να προσφέρει ποτέ αφαίρεση με καθαρή συνείδηση. Αυτό όμως δεν είναι λόγος για να μην προσπαθήσουμε.

Από αυτή τη βάση – την ολοκληρωτική άγνοια – όλες οι κατηγορίες ρέουν, σταματούν και ρέουν ξανά, σε αδιάλειπτα μεταβαλλόμενες αποδόσεις, οι οποίες αποτελούν την ουσία όλης της τέχνης και της συζήτησής της γύρω από την αρχέγονη φωτιά. Η διεκδίκηση της αιώνιας ανεξαρτησίας των άλλων ειδών, όλων των Άλλων, αποτελεί μια μικρή αντίθεση στην ιδέα της ενότητας μέσα στην ποικιλότητα. Εάν το ζήτημα της βιοποικιλότητας, συγκεκριμένα, απουσιάζει και εκείνο από τη *διαλεκτική*, τότε λογικά οι αυθαίρετες αναφορές μας - συναισθηματικές, πνευματικές ή κάπως φυσιολογικά υποχρεωτικές- δεν έχουν καμία απολύτως σχέση με οτιδήποτε άλλο εκτός από την ανθρώπινη επιλογή. Ούτε η ελεύθερη βούληση ούτε ο δημιουργικός έλεγχος βρίσκονται εντός των ορίων των δυνάμεών μας. Τι είδους επιλογές, λοιπόν, είμαστε ικανοί να κάνουμε; Αυτό είναι το μοναδικό ερώτημα που μπορεί να προκαλέσει τη δύναμη της στιγμιαίας μετα-σκέψης, το βασίλειο, με άλλα λόγια, της επιφοίτησης.

Η επιλογή καταφέρνει, στην πρώτη της αρχή, να συμπεριφέρεται σαν ολόκληρος ο κόσμος να είναι ακριβώς εκεί, μπροστά μας. Μια δραματική σκηνή επί της οποίας είναι δυνατές όλες οι επιλογές για μια πλήρη τάση αποκατάστασης που μπορεί να προκύψει από το άτομο ή/και την κοινότητά του. Μπορεί οι αντιστοιχίες με άλλα είδη να μοιάζουν κατάλληλες, ωστόσο - αν πιστέψουμε τον Παρμενίδη - ο πλήρης διαχωρισμός τους από το είδος μας θα πρέπει να δικαιολογεί τις μεταβάσεις για των οποίων την υλοποίηση είμαστε ικανοί, και όχι εκείνες για τις οποίες είμαστε ανίσχυροι. Η υπόθεση αυτή τοποθετεί την επιλογή στο επίκεντρο όλων.

Ας δούμε τον Βοραγινό, πρώην Αρχιεπίσκοπο Γένοβας, ο οποίος ήταν γνωστός και ως Ιάκωβος ο Βαραγινός (περίπου 1230-1298). Η συλλογή θείων επιφοιτήσεων που συνέγραψε με τίτλο *Legenda Aurea* αποτέλεσε ένα από τα πιο περιζήτητα βιβλία του τέλους του 15^{ου} αιώνα. Αρχικά μεταφρασμένο από τα λατινικά στα καταλανικά[4], σίγουρα αποτέλεσε ανάγνωσμα του Μιγκέλ ντε Θερβάντες, και, μεταξύ των εκπληκτικών ιστοριών του, περιλαμβάνει το *spiritualem conversionem* ενός Ρωμαίου στρατιώτη του 2^{ου} αιώνα, του Πλάκιδου. Το γεγονός μνημονεύεται σε όλη τη σειρά των χριστιανικών ονομασιών και αντιπροσωπεύει μια αρχετυπική στιγμή αποκάλυψης. Η μαχόμενη αφοσίωσή του στον αυτοκράτορα Τραϊανό, τότε Αδριανό, καθώς και στους θεούς της Ρώμης, μεταμορφώνεται σε μια στιγμή από ένα ιδιαίτερο ελάφι που στέκεται από πάνω του σε ένα δάσος. Το ελάφι δηλώνει: "εγώ είμαι ο Ιησούς, τον οποίο τιμάς χωρίς να γνωρίζεις." Το άλογο και τα πέντε κυνηγόσκυλα του Πλάκιδου μένουν εμβρόντητα, αποσβολωμένα στην ευδαιμονία της συνειδητοποίησης. Αργότερα, ο Πλάκιδος - που παίρνει το όνομα Ευστάθιος – δοκιμάζεται μαζί με τη σύζυγο και τα δυο παιδιά του. Εντέλει θανατώνεται μετά το θαύμα της φιλίας που επέδειξαν τα, σταλμένα να τον κατασπαράξουν, άγρια "θηρία" στο Κολοσσαίο.[5] Αντ 'αυτού, επικρατεί μια γενναιόδωρη βιοφιλία και η ψυχρή κτηνωδία της αρένας μετατρέπεται, ενώπιον όλων, σε αυτό το ειρηνικό βασίλειο. Και αυτό γιατί ο Πλάκιδος είχε επιλέξει να αφήσει κάτω το τόξο και το βέλος του, να γονατίσει στο έδαφος και να εκφράσει τη λατρεία του για το ελάφι σε μια σειρά προσευχών στην οποία συμμετείχαν όλοι με μια φωνή μέσα στο όλο βαρύ τοπίο του ρομαντικού έρωτα, της αγάπης και του ειλικρινούς σεβασμού στα είδη. Ο στρατιώτης έγινε χορτοφάγος και οι παραλλαγές της ιστορίας του τις επόμενες χιλιετίες συμβόλιζαν την ικανότητα των ανθρώπων και όλων των ζώων να αναζητούν την αρμονία και όχι τη διαμάχη.

Στο αποκορύφωμα της κραυγής συσπείρωσης του Αγίου Ευστάθιου, ο Ιταλός μαιτρ της ζωγραφικής Πιζανέλλο (περίπου 1395 - περίπου 1455) ζωγράφισε το "Όραμα του Αγίου Ευστάθιου" (Εθνική Πινακοθήκη του Λονδίνου) κάποια στιγμή μεταξύ του 1438 και του 1442. Υπάρχει ο χαρακτηριστικός Εσταυρωμένος, τοποθετημένος στο κέντρο ενός στηρίγματος που σχηματίζεται από τα πλατιά απλωμένα κέρατα του ελαφιού. Ο Πλάκιδος φοράει φορεσιά κυνηγού, και, παραδόξως, παραμένει στην αγιοποίησή του ως ο προστάτης άγιος όλων των κυνηγών, αλλά με τον ρητό όρο ότι είναι επίσης άγιος της προστασίας των αγρών και όλης της ζωής. Η κυρωμένη από το Βατικανό μέθοδολογία δράσης του παρομοιάζεται με του Αγίου Φραγκίσκου που κινητοποιείται, με γεωμετρική πρόοδο, να μεταλλαχτεί από κάποιον που ζει μια ζωή μέσα στη βία και την τυφλή υποταγή στην εξουσία σε έναν άνθρωπο του οποίου ο μοναδικός σκοπός ύπαρξης έχει μετατραπεί, σε διάστημα ίσο με ένα χτυποκάρδι του ελαφιού, στην άνευ όρων αγάπη και πίστη στον Θεό. Στην εξαίσια εκδοχή του Πιζανέλλο, ο Θεός είναι η Φύση, κατοικημένος με ένα πλούσιο βιολογικό πιστεύω, τόσο όσον αφορά την κυκλική κίνηση και την λεπτομέρεια όσο και την αιώνια επίκληση.

Για να τιμήσει τα 1300 χρόνια ταραχώδους ιστορίας μετά από αυτή την επική μεταστροφή στο δάσος, ο εικοσιεννιάχρονος τότε Άλμπρεχτ Ντύρερ (1471-1528), ιδιοφυΐα της Νυρεμβέργης και ήδη ένας από τους πιο διάσημους καλλιτέχνες σε όλη την Ευρώπη, δημιούργησε τη μεγαλύτερη, πλέον λεπτομερή και διάσημη χαλκογραφία του, αφιερωμένη στον Άγιο Ευστάθιο και το σατόρι του στην άγρια φύση. Αναγνωρίστηκε αμέσως ως το πιο σημαντικό χαρακτικό στην Αναγέννηση, διαστάσεων περίπου 35x25 εκατοστών. [6] Άλλοι Γερμανοί καλλιτέχνες, ανάμεσά τους ο Carl Wilhelm Kolbe (1759-1835) και ο Johann Heinrich Wilhelm Tischbein (1751-1829), απεικόνισαν ένα ρομαντικό τοπίο.[7] Έως τα τέλη του 18^{ου} αιώνα, ο εικοσιτετράχρονος Wilhelm Heinrich Wackenroder (1773-1798), τον οποίο πολλοί θυμούνται ως τον Ουόρντσγουορθ της Γερμανίας, έγραψε μια μικρή πραγματεία με περιγραφές ιστοριών των καλλιτεχνών της Αναγέννησης, όπως αυτές είχαν ερμηνευθεί από τη γερμανική ρομαντική σχολή των Ναζαρηνών και των προσπαθειών της να "επιδείξει την πνευματική φύση της αληθινής τέχνης." [8] Το πλέον έξοχο έμβλημα αυτού του ανακαλυφθέντος παραδείσου ήταν ο Ντύρερ. Όπως έχει αναφερθεί, το να κοιτά κανείς αυτό το χαρακτικό μέσα στο σπίτι του είναι σαν να γίνεται κατ΄ιδίαν κοινωνός ενός "θεϊκού γεγονότος". "Οι θεατές έχουν το προνόμιο να γίνονται μάρτυρες αυτής της εφήμερης στιγμής και καλούνται να μιμηθούν την πράξη λατρείας του Αγίου Ευσταθίου". [9] Για τον Άγιο, όπως έχει πειστικά συλλογιστεί ο Γάλλος φιλόσοφος Εμανυέλ Λεβινάς, "… με την πλήρη δύναμη της αναπόφευκτης ευθύνης τους. Να αναγνωρίσουν και να κατονομάσουν αυτές τις αδιάλυτες ουσίες και να τις εμποδίσουν να εκραγούν στη βία, τον δόλο ή την πολιτική, να επαγρυπνούν εκεί όπου τείνουν να ξεσπούν συγκρούσεις, μια νέα θρησκευτικότητα και αλληλεγγύη -αγαπώντας τον πλησίον μας... Είναι η ύπαρξη του άλλου η βάση στην οποία η δική μου η ύπαρξη θεωρείται ανθρώπινη".[10] Ή, όπως γράφει ο Τζωρτζ Στάινερ, είναι "αυτό που έρχεται να μας καλέσει...αυθόρμητη επίσκεψη και κάλεσμα..."[11]

Εκτός από την τεχνική του δεξιοτεχνία, το χαρακτικό "Όραμα του Αγίου Ευστάθιου" του Ντύρερ αναδύεται το ίδιο εξαγνιστικά στην καρδιά της Βόρειας Αναγέννησης με τη "Γέννηση της Αφροδίτης" του Σάντρο Μποτιτσέλι, 120 χρόνια νωρίτερα (στην Πινακοθήκη Ουφίτσι στη Φλωρεντία). Όπως και ο άμεσος προκάτοχός του, Μάρτιν Σόνγκαουερ) (1448-1491) του Κολμάρ, τόσο στο χαρακτικό του Αγίου Ευστάθιου καθώς και σε εκατοντάδες άλλα αριστουργήματά του, ο Ντύρερ κάνει μια αποφασιστική και ολοκληρωμένη επιλογή όσον αφορά τη δύναμη και την καλοσύνη της Φύσης, καθώς επίσης και τον ρόλο του ατόμου στην επανένωσή του με τη φύση εντός της εκστατικής ηθικής της, η οποία ομοιάζει με την επανένωση με τον Θεό, αυτό που οι οικολόγοι θα μπορούσαν να χαρακτηρίσουν ως μια "κρίσιμη μετάβαση σε μια εναλλακτική κατάσταση".[12] Μετά από πέντε ολόκληρους αιώνες ακαδημαϊκής μελέτης του ογκώδους έργου του Ντύρερ, ο Γάλλος ιστορικός Elie Faure ήταν εκείνος που χρησιμοποίησε με τη μεγαλύτερη ακρίβεια γλώσσα ανάλογη με μέγεθος του έργου του, ίσως την πιο αληθινή άποψη, όταν έγραψε: "Είναι σαν ο Χριστός να έγινε επιστήμονας και να επιδιώκει τη σωτηρία του κόσμου σε μια εντατική μελέτη των πτυχών του... έκοψε τον χαλκό με το χέρι ενός εργάτη, την καρδιά ενός ποιητή και τον νου ενός φιλοσόφου, [όμως] ποτέ δεν τακτοποίησε τη φύση. Την θεωρούσε "τον μοναδικό αφέντη" ... Τα τεράστια άλογα της Γερμανίας, τα μυώδη κυνηγόσκυλά της, τα ελάφια της, οι λαγοί της, οι αγελάδες της, τα γουρούνια της που σέρνονταν στη λάσπη των χωριών, όλα τα έντομά της και όλα τα πουλιά της συμμετείχαν, σχεδόν πάντα, στις περιπέτειες της αγάπης..." [13]

Πόσο όμοιοι, στην πραγματικότητα, είναι αυτοί οι αισθητικοί πανηγυρισμοί με τη φαινομενικά πολύ πιο πεζή συγκέντρωση των σημερινών ομοσπονδιακών υπηρεσιών διαχείρισης γης των ΗΠΑ που εργάζονται υπό το λάβαρο της "Εποπτείας Οικολογικού Μετασχηματισμού". [14] Αυτός είναι ένας μόνο τρόπος σκέψης για την εφαρμογή της εμπνευσμένης πνευματικής αποκάλυψης στην πραγματιστική συμπεριφορά. Ένας άλλος αφορά τη θεμελιώδη θεωρία αμοιβαιότητας, η οποία υποστηρίζει - όπως διατυπώθηκε από τον Gerald Carter – "ότι οι μύκητες, τα φυτά, τα ψάρια, τα πουλιά, οι αρουραίοι και τα πρωτεύοντα θηλαστικά ενισχύουν το αμοιβαίο όφελος μεταβάλλοντας ενδεχομένως τις επενδύσεις τους στη συνεργασία βάσει των αποδόσεων της συνεργασίας..." [15] Ο Carter καταλήγει επισημαίνοντας τρία, έως τώρα κλασικά, αν και ανεπαρκή, "πλαίσια που έχουν καθοδηγήσει εμπειρικές μελέτες της συνεργασίας" και είναι η "συμπεριληπτική ικανότητα/καταλληλότητα" η "θεωρία της αμοιβαιότητας" και η "θεωρία της βιολογικής αγοράς". [16] Είναι σαφές ότι το "Όραμα του Αγίου Ευστάθιου" δεν ταιριάζει σε καμία από αυτές τις κατηγορίες εμπειριών και, ως εκ τούτου, απαιτεί μια τελείως νέα προσέγγιση της αναγνώρισης των κρίσιμων θεμελίων αυτού που οι θεολόγοι ξέρουν ως *μεταστροφή*. Μιας εμπειρίας έξω από τα όρια του ίδιου συμφέροντος ή οποιασδήποτε προσδοκίας κέρδους.

ΣΗΜΕΙΩΣΕΙΣ

[1] "Parmenides," Chapter III, στο *The History of Philosophy*, του Thomas Stanley, Thomas Bassett, Λονδίνο, 1683, σελ. 748.

[2] #4 των *The Fragments, The Collected Wisdom of Heraclitus*, Μετάφραση Brooks Haxton, Εισαγωγή James Hillman, Viking Penguin Group, New York, 2001, σελ. 27.

[3] *Plato, Lysis, Or Friendship, The Symposium, Phaedrus*, Μετάφραση από τα ελληνικά με εισαγωγικές αναλύσεις, από τον Benjamin Jowett, Εισαγωγή από την Whitney J. Oates και εικονογράφηση του Eugene Karlin, The Easton Press Norwalk, Connecticut, 1979, σελ. 137.

[4] *Legenda aurea santorum sive Lombardica historia*, του Jacobus de Voragine, et al., Jordanus de Quedlinburg, Strassburg 1486. Μερικοί δαιτείνονται ότι μαρτύρησε το έτος 118 μ.Χ. υπό τον Πάπα Άγιο Σίξτο Α΄. Βλ. "St. Eustace the Hunter and the Magnificent 'Stag,'" του Larry Peterson, *Catholic365.com*, 29 Ιουνίου 2018, https://www.catholic365.com/article/9024/st-eustace-the-hunter-and-the-magnificent-stag.html

[5] *Martyrologium Romanum*, Libreria Editrice Vaticana, Ρώμη, 2001.

[6] 369 x 270 mm; Bartsch 57; Dogson 32; Hollstein 60 (Meder b); Rowlands 1971, 78; Bartrum 16; Schoch 32, as described by Giulia Bartrum in *Albrecht Dürer and his Legacy – The Graphic Work of a Renaissance Artist*, with contributions by Günter Grass, Joseph L. Koerner and Ute Kuhlemann, Princeton University Press, Princeton New Jersey, 2002, pp. 140-142. See also, *Albrecht Dürer – Paintings, Prints, Drawings*, by Peter Strieder, Translated by Nancy M. Gordon and Walter L. Strauss, Abaris Books, New York, 1982, σελ. 174-176.

[7] Βλ. *German romantic prints and drawings from an English private collection*, Εισαγωγή του Antony Griffiths, πρόλογος της Giulia Bartrum, The British Museum Press, London, 2011.

[8] *Herzensergießungen eines kunstliebenden Klosterbruders* [Heartfelt Effusions of an Art-Loving Monk], und L. Tieck, Berlin J F Unger, 1797, Βλ. Keith Moxey, "Motivating History," *Art Bulletin 77*, no. 3 (September 1995), σελ. 393–401. Βλ. *German Romantic Painting*, του William Vaughan, Yale University Press, New Haven and London, 1980, σελ. 163.

[9] Από την πύλη *Washington University Digital Gateway Image Collections & Exhibitions*, http://omeka.wustl.edu/omeka/exhibits/show/durernatureworks/ creatingmasterpieces/steustace

[10] Emmanuel Levinas, *Alterity & Transcendence*, Μετάφραση Michael B. Smith, Columbia University Press, New York, 1999, σελ. 87-88 και 167.

[11] George Steiner, *Real Presences*, The University of Chicago Press, Chicago, Ill., 1989, σελ. 179.

[12] Βλ. "Can we detect ecosystem transitions and signals of changing resilience from paleo-ecological records?" των Zofia E. Taranu, Stephen R. Carpenter, Victor Frossard, Jean-Philippe Jenny, Zoee Thomas, Jesse C. Vermaire, και Marie-Elodie Perga, *Ecosphere, Ecological Society of America*, 2018, 9 (10), ff10.1002/ecs2.2438ff. ffhal-01959637f, https://hal.archives-ouvertes.fr/hal-01959637/document

[13] Elie Faure, *History of Art – Renaissance Art*, Garden City Publishing, New York 1937, σελ. 356, 358.

[14] *National Adaptation Forum – Action today for a better tomorrow, – Linking the science of ecological transformation to RAD [resist, accept, direct] decisions*, http://www.nationaladaptationforum.org/archive/2019/wednesday/concurrent-sessions-4/linking-science-ecological-transformation-rad-decisions, 10-22 Μαΐου 2022.

[15] Βλ. "The Reciprocity Controversy," του Gerald Carter, *Animal Behavior and Cognition*, 1(3). 368-386. doi: 10.12966/abc.08.11.2014. http://animalbehaviorandcognition.org/uploads/journals/3/11.Carter_FINAL.pdf

[16] ό.π., σελ. 379.

Κεφάλαιο 14

Η ΔΙΑΛΕΚΤΙΚΗ ΤΗΣ ΕΠΙΦΟΙΤΗΣΗΣ

Στα διάφορα γραπτά του σχετικά με το θέμα της "μεταστροφής", ο Μαχάτμα Γκάντι διακήρυξε το πασίγνωστο "Δεν θεωρώ κανέναν άνθρωπο μολυσμένο/διεφθαρμένο επειδή εγκατέλειψε το κλαδί πάνω στο οποίο καθόταν και πήγε σε ένα άλλο του ίδιου δέντρου".[1] Σε αυτή την ηθική πυξίδα υπάρχει σιωπηρή η κυριολεκτική κατανόηση ότι, παρά το γεγονός ότι υπάρχουν περίπου 60.000 (γνωστά) είδη δέντρων στον κόσμο,[2] τελικά, *αυτά* και *εμείς* υπάρχουμε όλοι στο ίδιο δάσος, αν με τον όρο δάσος εννοήσουμε τη Φύση στο σύνολό της. Πρόκειται μια απλή, αν και λογική κατηγορία συμπερασμάτων, σε καμία περίπτωση η πλέον μελετημένη ή εξεζητημένη. Ωστόσο, παραμένει αληθινή. Ένα άτομο μπορεί να προβεί σε επιλογές εντός ενός συγκεκριμένου πλαισίου, οι οποίες αλλάζουν το συνολικό πλαίσιο. Ένα κομμάτι μπορεί να αλλάξει το σύνολο. Το σύνολο μπορεί να αμφισβητήσει τον εαυτό του μέσω οποιουδήποτε εκ των συστατικών, μελών, ή ατόμων του. Εάν το πλαίσιο εξετάζεται υπό το πρίσμα κάποιας εξίσωσης, τότε πάντα κάτι θα συμβαίνει, ακόμη και αν η ισορροπία αποτελεί τελικό αποτέλεσμα. Για το άτομο, το πλαίσιο *υποβάλλεται* σε μια κατάσταση αλλαγής.

Η δύναμη της αποκαλυπτικής εμπειρίας που προκύπτει από μια συγκλονιστική αίσθηση ή κάποια ιδέα που έχει γεννηθεί σε κάποιο φυσικό περιβάλλον υπογραμμίζει έναν διαρκή πνευματικό μεταβολισμό που μπορεί να εντοπισθεί, σε ένα πασίγνωστο παράδειγμα, στον Βούδα, ο οποίος κάθεται για σαράντα εννιά μέρες και νύχτες κάτω από τη σκιά ενός *Ficus religiosa*. Όμως, σχεδόν κάθε άγιος και διάσημος μοναχός μπορεί να ιδωθεί στο πλαίσιο των φυσικών του επιφοιτήσεων, των στιγμών αποκάλυψης στις οποίες συμμετέχει το πλαίσιο της άγριας φύσης και οι μεταβαλλόμενες διαστάσεις του.[3]

Υπάρχει ο Άγιος Αιγίδιος, με το μόνιμο τραύμα που απέκτησε καθώς έσωζε μια ελαφίνα στη σπηλιά που χρησιμοποιούσε ως ερημητήριό του. Ο Χριστόφορος, προστάτης όλων των ορειβατών και των δύσκολων ταξιδιών στα ποτάμια. Ο Άγιος Βενέδικτος συνδέεται με όλους τους αγρότες, αλλά και τους σπηλαιολόγους. Ο Άγιος Αντώνιος, προστάτης άγιος των κατοικίδιων ζώων. Ο πρώτος ερημίτης της ερήμου, Παύλος. Ο

Ιωάννης ο Βαπτιστής και το αρνί που τον συντροφεύει. Ο Άγιος Ιερώνυμος και το λιοντάρι που τον φύλαγε. Ο Άγιος Φραγκίσκος, προστάτης όλης της οικολογίας. Ο Αρχάγγελος Ραφαήλ, προστάτης όλων των ταξιδιωτών. [4]

Όλα αυτά τα παραδείγματα ζητούν τις αναλύσεις των ανθρώπινων γενετικών προδιαθέσεων εντός των εκτεταμένων τεταρτημορίων της βιοφιλίας, των συναισθηματικών και ψυχολογικών αναγκών μας, των επιπέδων ανοχής, των ορίων της νευρολογικής θεραπείας, του τρόμου, της ταραχής, του πάθους και της ηρεμίας. Της ποσότητας των πληροφοριών που μπορούν να επεξεργαστούν οι άνθρωποι, όπως αυτές ορίζονται από οποιοδήποτε οικοσύστημα στο οποίο έχουμε επιβάλει την ύπαρξή μας, και των σχετικών καθημερινών εργασιών μας, οι οποίες είτε ανατρέπονται είτε επιβεβαιώνονται σε συνεργασίες από τις διάφορες απόπειρές μας να επιβιώσουμε. Όμως οι άγιοι και οι σοφοί που θαυμάζουμε συστηματικά αποκαλύπτουν μια κρίσιμη οικολογική σταθερά, ένα έναυσμα εντός της ψυχής και των ενστίκτων της που υποστηρίζει την εγκυρότητα της πιο εκτεταμένης περιοχής τόσο της συνειδητής όσο και υποσυνείδητης πίστης, δηλαδή, αυτή την ικανότητα και την ανάγκη να "αγκαλιάσουμε πλάσματα που δεν μπορούν να μας ανταποδώσουν", όπως έγραψε ο David W. Orr, παραθέτοντας τον Λιούις Μάμφορντ.[5] Φυσικά, όσο περισσότερο εξασκούμαστε στην επίδειξη ταπεινοφροσύνης, τόσο μεγαλύτερα πλεονεκτήματα αρχίζουμε να αναγνωρίζουμε στη συντροφιά Άλλων που δεν χρειάζεται να μιλούν τη γλώσσα μας, να μοιράζονται τα ανατομικά μας χαρακτηριστικά ή να συγκατοικούν με εμάς σε τόπους όπως οι γειτονιές που συνήθως θεωρούμε μόνιμη βάση μας σε έναν πλανήτη μέσα στο άπειρο. Ενώ παραμένουμε ουσιαστικά περιορισμένοι στους εαυτούς μας και στις πολυπληθείς συμπάθειες/ προτιμήσεις μας - σε συμφωνία - με άλλα είδη, μας παραχωρούνται πρωτογενώς διάφορες εγγύτητες. Η αποδοχή αυτών των άλλων σπινθήρων είναι η βαθιά ανταμοιβή της ευκαιρίας μας και, δεδομένης μιας τόσο βίαιης ιστορίας, το ιερό καθήκον μας.

Ο οικονομολόγος, ιστορικός, κοινωνιολόγος Φράνσις Κλίνγκεντερ (1907-1955) έκανε έναν εκπληκτικό απολογισμό των επιλογών που έχουν πραγματοποιηθεί από πολιτισμό σε πολιτισμό στο μεγαλόπνοο έργο του *Animals in art and thought*[6]. Ξεκινά με έναν Πρόλογο στον οποίο κάνει αναφορές "στο φανταστικό παιδί των ονείρων του Λιούις Κάρολ σε φιλική κουβέντα με τα πουλιά και τα θηρία" και σε έναν αρκτικό μύθο δημιουργίας των Τσιπέουα, ο οποίος λέει: "Στην αρχή δεν υπήρχαν άνθρωποι, παρά μόνο ζώα· ωστόσο αυτά έμοιαζαν με ανθρώπινα όντα και μπορούσαν να μιλήσουν· όσο τα ζώα μπορούσαν να μιλήσουν ήταν καλοκαίρι, και όταν έχασαν τη δύναμη να μιλούν ήρθε ο χειμώνας".[7] Το τεράστιο εικονογραφημένο έπος του Κλίνγκεντερ είναι αυτή η αφήγηση των διπόλων που παρουσίαζαν την τάση της ανθρωπότητας να χαίρεται να ενδύεται τον μανδύα της ζωομορφικής ολομέλειας. Το ερώτημα που στηρίζει το τι θα μπορούσε να επιλέξει κάθε γραφέας, ιερέας, καλλιτέχνης, απλός άνθρωπος οποιασδήποτε ροπής να σκεφτεί, να κάνει, να εκπροσωπήσει στη συντροφιά τόσων Άλλων είναι ο άξονας πάνω στον οποίο ο ρεαλισμός είτε προχωρά προς το ιδανικό, είτε το αγνοεί. Στοιχειωμένο από την κακοφωνία των κυνηγετικών μύθων και τη σκληρότητα που χαρακτηρίζει τον έναν πολιτισμό μετά τον άλλο στον οδικό χάρτη που επιβάλει τη χρήση συμβόλων, την εξημέρωση, τη σφαγή και/ή τη λατρεία, είναι σαφές από την ασσυριακή θεώρηση του 6ου αιώνα π.Κ.Ε. ότι ο βασιλιάς διατηρεί την ψευδαίσθηση της εξουσίας του κρατώντας δύο λιοντάρια ανάποδα.[8] Ότι τα άλογα ήταν απλώς ένα μέσο για την επίτευξή κάποιου σκοπού όταν ο Αλέξανδρος ξεκίνησε να κατακτήσει τον κόσμο. Ή, όπως απεικονίζεται στη σκέψη του ανυπέρβλητου Εβραίου φιλοσόφου Φίλωνα (περίπου 20 π.Κ.Ε. - περίπου 50 Κ.Ε.), "τα φαινόμενα της φύσης, όπως τα αντιλαμβάνονται οι αισθήσεις, δεν εκτιμώνται, επομένως, ως αυτά καθαυτά, παρά μόνο ως ενδείξεις των βαθύτερων αληθειών στις οποίες πρέπει να διεισδύσει η ανθρώπινη σκέψη από τους παράλληλους δρόμους της ορθολογικής αφαίρεσης και της αλληγορικής ερμηνείας". [9] Από τους τοίχους των σπηλαίων μέχρι και τα εικονογραφημένα ψαλτήρια, χαραγμένα σε μαρμάρινες στήλες, εραλδικά κουζινικά σκεύη και τα γραπτά που βρέθηκαν σε κρύπτες, παραμύθια και βιβλία με μύθους ζώων, στα χειρόγραφα της Αποκάλυψης και κυριολεκτικά εκατομμύρια έντυπες σελίδες σε τόσα πολλά Βιβλία των Ωρών και Φλαμανδικούς τοιχοτάπητες, κάθε εποχή και ζώο και φυτό κοσμεί την ανθρώπινη ιστορία για τον ίδιο, θεμελιώδη για τη ζωή μας, λόγο που θέλουμε να αλλάξουμε, για να δούμε το πλαίσιο όχι αμαυρωμένο από την ανθρωπότητα, αλλά απαλλαγμένο από αυτή.

Σύμφωνα με την ερμηνεία του Κλίνγκεντερ, στην περίπτωση των εικονογραφήσεων μιας νορμανδικής κυνηγετικής ομάδας, "Με τη θανάτωση του ζώου-υποκατάστατου του συζύγου της κυράς του, ο νεαρός εραστής μπορεί να απελευθερώσει την αφόρητη ένταση που προκύπτει από την αυτοσυγκράτηση που του επιβάλλεται από την πραγματική ζωή και την πίστη του στον κύριό του".[10] Με διαφορετική περιγραφή αλλά την ίδια έμφαση στην προβληματική μετάβαση που ενέχει η μετάθεση της ακεραιότητας, της ανησυχίας και της προτίμησης, ο Φινλανδός ιστορικός Νιλς Έρικ Ένκβιστ υποστήριξε ότι "οι Χριστιανοί Αγγλοσάξονες ποιητές που πασχίζουν να ξεφύγουν από τους τρομερούς χειμώνες του ηρωισμού στο ιδανικό κλίμα του παραδείσου όπως αυτός μνημονεύεται επί μακρόν στα θρησκευτικά γραπτά της Ευρώπης" πιθανώς περιπλέκει τις τρέχουσες εμμονές μας για τη φύση. Ο Ένκβιστ θα μπορούσε με βεβαιότητα να δηλώσει ότι "μια ιδιοφυΐα όπως ο Τσώσερ μπορεί να εμποτίσει ακόμη και τις πιο φθαρμένες κοινοτοπίες με ένα νέο πνεύμα ιμπρεσιονιστικής φρεσκάδας", και ότι αυτή η δυνατότητα του ανθρώπινου πνεύματος πρέπει να εξευγενίζει ακόμη και τις πιο σκοτεινές μέρες.[11]

Μπορούμε να τελειοποιούμε και να συνοψίζουμε τις αναφορές περί της ανθρώπινης ανοησίας και οργής όπως αυτές αναφέρονται σε σχέση με τη ζωή των άλλων ζώων. Αυτό έγινε ιδίως έναν χρόνο μετά τη δημοσίευση της μελέτης του Ένκβιστ, από τον Αλμπέρ Καμύ. Όπως διηγήθηκε ο Τζον Μάξγουελ Κούτσι, ένα νεαρό αγόρι στην Αλγερία, ο Καμύ διατάχθηκε από τη γιαγιά του να της φέρει μια κότα στην κουζίνα, όπου την είδε έντρομος να της κόβει το κεφάλι. "Η κραυγή θανάτου αυτής της κότας", γράφει ο Κούτσι, "εντυπώθηκε στη μνήμη του αγοριού τόσο έντονα που το 1958 [ο Καμύ] έγραψε μια παθιασμένη πολεμική για τη λαιμητόμο. Ως αποτέλεσμα, εν μέρει, αυτής της πολεμικής, η θανατική ποινή στη Γαλλία καταργήθηκε". [12] Αυτή η βαθιά, μετεπαναστατική πολιτιστική κατάργηση ίσως ήρθε και ενίσχυσε τη φρίκη που έγινε αισθητή τις στιγμές που ακολούθησαν τον αποκεφαλισμό ενός σαρανταεπτάχρονου δασκάλου στο Παρίσι, του Σαμουέλ Πατί, από έναν τρομοκράτη τον Οκτώβριο του 2020.[13]

Επιπλέον, ο Γάλλος υπουργός Γεωργίας τιντιέ Γκιγιόμ ανακοίνωσε ότι, έως το τέλος του 2021, η Γαλλία θα απαγορεύσει τη θανάτωση "ανεπιθύμητων αρσενικών κοτόπουλων" (την άλεση δηλαδή των πτηνών αυτών όσο είναι ακόμα ζωντανά σε μηχανήματα τεμαχισμού), μια πρακτική με την οποία θανατώνονται τουλάχιστον επτά δισεκατομμύρια αρσενικά κοτόπουλα κάθε χρόνο. "Τίποτα δεν θα είναι όπως ήταν", είπε ο Γκιγιόμ. [14] Ωστόσο, λαμβάνοντας υπόψη ότι σχεδόν 67 εκατομμύρια Γάλλοι μεγαλώνουν σε ένα σχολικό σύστημα που απαιτεί το 40% όλων των γευμάτων που σερβίρονται στα σχολεία να περιέχει κρέας και ψάρι και ότι ένα σλόγκαν του σωματείου αγροτών "Coordination Rurale" είναι "Σώσε έναν αγρότη, φάε έναν βίγκαν",[15] η Γαλλία είναι απίθανο να βιώσει σύντομα μια επιστροφή στον κόσμο της, μεταφρασμένης από τον Γάλλο J. Corbechon, μεγάλης μεσαιωνικής εγκυκλοπαίδειας φυσικής ιστορίας *De Proprietabius rerum* ("Περί των ιδιοτήτων των πραγμάτων"). Σε αυτόν τον θησαυρό του πραγματικού και του φανταστικού του 13ου αιώνα, ο συγγραφέας του, Bartholomeus Anglicus (1203-1272), και στην έκδοση που γράφτηκε στην Μπρυζ από τον Jehan de Ries, υπάρχει μια κομψά εικονογραφημένη σελίδα στην οποία απεικονίζεται η καθημερινότητα μιας περιοχής, και στην οποία πέντε φίλοι στέκονται δίπλα σε ένα κανάλι που κυριαρχείται πλήρως από τα πουλιά: Αναγνωρίζει κανείς αμέσως "έναν λευκό πελαργό, μια καρακάξα, έναν κύκνο, έναν γερανό… έναν κόκορας και μια κότα, μια κουρούνα και μια μαυροκουρούνα ή ένα χαβαρόνι που την αντικρίζει. Κάποια άλλα ίσως να μπορούμε να τα μαντέψουμε. Ωστόσο ο θρύλος έχει παρεισφρήσει. Η κουκουβάγια στον γκρεμό είναι μυθολογικό πουλί, όχι φυσικό…"[16]

Το κλειδί σε αυτή την εικόνα της καθημερινότητας είναι η ευθυμία ανθρώπων και πτηνών. Δεν υπάρχει καμία ένδειξη εκείνης της αβάσταχτα πραγματικής στη Γαλλία έντασης, της ανικανότητας αντιπαραβολής αντιθέτων έξω από ένα ανήσυχο μυθιστόρημα ή ριζωμένης σε μια πίστη μακριά από τις πόλεις. Το Αβαείο του Κλαιρβώ, για παράδειγμα, το Κιστερκιανό μοναστήρι (σήμερα ερειπωμένο) που μάλλον βρισκόταν σε μια θέση στο Ville-sous-la-Ferté που "καταπραΰνει τα κουρασμένα μυαλά, ανακουφίζει από τις ανησυχίες και τις έγνοιες και βοηθά τις ψυχές που αναζητούν τον Κύριο…". [17] Σε αντίθεση με "ένα μεσαιωνικό δείπνο" που μπορεί να "περιέχει πάνω από μια ντουζίνα διαφορετικά είδη φαγητού στοιβαγμένα σε μεγάλες πιατέλες" και τα οποία περιλαμβάνουν "ωμοπλάτες από αγριογούρουνο βραστό και τουρσί, αποφλοιωμένο σιτάρι βρασμένο σε γάλα και κρέας ελαφιού, λαδερά στιφάδα, παστό ελάφι, φασιανό, κύκνο, καπόνια, λάμπρενα, πέρκα, κουνέλι, αρνί" κ.λπ. [18] Ένα χάσμα σε αυτό το λογοτεχνικό πάθος στο οποίο όλα έχουν να κάνουν με τον άνθρωπο και κυριαρχεί το παράδοξο. Αυτό για το οποίο, στο τέλος του μιούζικαλ του Άλαν Τζέι Λέρνερ, "Brigadoon", ο χαρακτήρας του Τόμμυ αντιδρά στον φίλο του, Τζεφ, και στην ευλογημένη ανάμνηση της Φιόνα στην σκωτσέζικη αθάνατη ουτοπία της: "Λοιπόν, κλείστε εισιτήρια για το αεροπλάνο αμέσως!.../Ξέρω ότι δεν είναι εκεί, αλλά θέλω να δω πού ήταν…/Ποιος νοιάζεται αν δεν βγάζει νόημα;...Θέλω να πάω…”.[19]

ΣΗΜΕΙΩΣΕΙΣ

[1] Βλ. *Gandhiji on Religious Conversion*, Επιλογή και σύνθεση με εισαγωγή των Sandhya Mehta, Mani Bhavan Gandhi Sangrahalaya, Meghshaym T. Ajgaonkar, Executive Secretary, Mumbai, India 2002, σελ. 28.

[2] Βλ. *The Tree – A Natural History of What Trees Are, How They Live, and Why They Matter*, του Colin Tudge, Crown Publishers, New York, New York, 2006, σελ. 20.

[3] Βλ. *The Paradise Or Garden of the Holy Fathers Being Histories of the Anchorites Recluses Monks Coenobites and Ascetic Fathers of the Deserts of Egypt…* Δύο τόμοι, μεταφρασμένο από το συριακό με σημειώσεις και εισαγωγή του Ernest A Wallis Budge, Chatto & Windus, London, 1907.

[4] Βλ. *Saints in Art*, της Rosa Giorgi, Επιμέλεια Stefano Zuffi, μετάφραση Thomas Michael Hartmann, The J. Paul Getty Museum, Los Angeles, CA 2003.

[5] “Love It or Lose It: The Coming Biophilia Revolution,” του David W. Orr, στο *The Biophilia Hypothesis*, Επιμέλεια Stephen R. Kellert και Edward O. Wilson, A Shearwater Book, Island Press, Washington D.C., σελ. 434. Ο Orr παραθέτει το βιβλίο του Lewis Mumford, *The Conduct of Life*, Harcourt Brace Jovanovich, New York, N.Y., 1970, σελ. 286

[6] *Animals in art and thought – to the end of the Middle Ages*, του Francis Klingender, Επιμέλεια Evelyn Antal και John Harthan, The M.I.T. Press, Cambridge, Mass., 1971.

[7] ό.π.., σελ. xxv.

[8] ό.π.., σελ. 43, Cylinder seal from the Pierpont Morgan Library, New York.

[9] ό.π.., σελ. 150.

[10] ό.π.., σελ. 472.

[11] *The Seasons of the Year – Chapters on a Motif from Beowulf to The Shepherd's Calendar*, του Nils Erik Enkvist, Societas Scientlarum Fennica, Commentationes Humanarum Litterarum. XXII. 4. Helsingfors, Ejnar Munksgaards Forlag, Kobenhavn, 1957, σελ.180.

[12] Από το *The Lives Of Animals* του J. M. Coetzee, (με σχόλια των) Marjorie Garber, Peter Singer, Wendy Doniger και Barbara Smuts, Επιμέλεια και εισαγωγή της Amy Gutmann, Princeton University Press, Princeton, New Jersey, 1999, σελ. 63. Ο Coetzee παραθέτει το έργο του Albert Camus, *The First Man*, μετάφραση του David Hapgood, Hamish Mailton, London, 1995, σελ. 181-183; και Réflexions sur la guillotine,” στο *Essais*, επιμέλεια R. Quilliot και L. Faucon, Gallimard, Paris, 1965, σελ. 1019-1064.

[13] Βλ. “France teacher attack: Seven charged over Samuel Paty's killing,” n.a., BBC News, 22 Οκτωβρίου 2020, https://www.bbc.com/news/world-europe-54632353, Πρόσβαση στις 13 Δεκεμβρίου 2020.

[14] “France says its poultry industry will stop shredding male chicks alive by 2022,” του Michael Brice-Saddler, 29 Ιανουαρίου 2020, *The Washington Post,* https://www.washingtonpost.com/ science/2020/01/29/shredding-chicks-ban/, Πρόσβαση στις 13 Δεκεμβρίου 2020.

[15] “Violent clashes amid the growing beef between vegans and farmers,” του Adam Sage, *The Times*, U.K., 9 Οκτωβρίου 2017, σελ. 33.

[16] *Le Livre des Propriétez des choses*. B.L. MS. Roy. 15 E.iii, f. llr. Flemish, 1482, στο *Birds in medieval manuscripts*, του Brunsdon Yapp, Schocken Books, New York, 1982, σελ. 116.

[17] Βλ. *Passing Strange and Wonderful – Aesthetics, Nature, and Culture*, του Yi-Fu Tuan, A Shearwater Book, Island Press, Washington D.C., 1993, σελ. 218.

[18] ό.π., σελ. 47-48.

[19] *Brigadoon – A Musical Play*, του Alan Jay Lerner, Coward-McCann, Inc., New York, 1947, *Act II*, σελ. 135.

Κεφάλαιο 15

ΕΠΑΝΑΔΙΑΠΡΑΓΜΑΤΕΥΣΗ ΤΟΥ ΛΑΒΥΡΙΝΘΟΥ

Οι νευροφυσιολόγοι έχουν μελετήσει τη βασική λήψη αποφάσεων σε διάφορα είδη ποντικών, μυγών και σκουληκιών, ερευνώντας τους "νευρώνες που ελέγχουν τις αποφάσεις σχετικά με την τροφή" και το "αποτέλεσμα της ανταμοιβής μετά την πρόσληψη τροφής", μεταξύ των πολλών άλλων αντιληπτών μηχανισμών τους οι οποίοι βρίσκονται σε λειτουργία.[1] Παρατηρούνται θεμελιώδεις δυσκολίες όσον αφορά την εξαγωγή συμπερασμάτων για τις διατροφικές επιλογές οι οποίες εξαρτώνται από την "αίσθηση των θρεπτικών συστατικών" στον εγκέφαλο (ή από κάποιο πιο σύγχρονο "περιφερειακό εντοπισμό" για την "εξισορρόπηση της μεταβολικής ομοιόστασης") στη *Drosophila melanogaster* · ή εκείνες τις χημικές ουσίες που *επιλέγονται* με κάποιο τρόπο από τα βακτήρια για τη διατήρηση της ανάπτυξης[2], σε αντίθεση με άλλους παράγοντες και οργανισμούς σε ένα πρωτεύον θηλαστικό όπως εμείς, όχι πως δεν μοιραζόμαστε όλοι μας πολλά από τα ίδια βακτηριολογικά δομικά συστατικά της ζωής.

Οι αναπτυξιακοί ψυχολόγοι συζητούν για την ελεύθερη βούληση, τη συμμετοχή της διχοτόμησης εγκεφάλου/νου στην απεικόνιση των μοτίβων λήψης αποφάσεων και συγκεκριμένων επιλογών και για τις συνακόλουθες ευθύνες σε επίπεδο ηθικής οι οποίες προκύπτουν στις κοινωνικές ιεραρχίες.[3] Ακόμη και οι πολλαπλές επιλογές οποιασδήποτε νοητικής ενέργειας υπόκεινται, όπως πιστεύουν ορισμένοι, σε αυτό που ονομάζεται "αρχιτεκτονική της επιλογής" και, από εξελικτική σκοπιά, "presence bias". [4] Το παράδοξο των επιλογών βρίσκεται στον πυρήνα της σκέψης του Γουώλτ Γουίτμαν που ξεκινά στα "Φύλλα Χλόης" με το περίφημο "Song of the Open Road" (Τραγούδι του Ανοιχτού Δρόμου" – "Πεζός και ξένοιαστος παίρνω τον ανοιχτό δρόμο, Είμαι υγιής και ελεύθερος, ο κόσμος απλώνεται μπροστά μου, ο μακρύς χωματόδρομος μπροστά μου όπου διαλέξω εγώ με πάει."[5] (μετάφραση Ελένη Ηλιοπούλου και Κατερίνα Ηλιοπούλου, https://www.o-klooun.com/anadimosiefseis/walt-whitman-to-tragoydi-tou-anoixtoy-dromou) Ο Αριστοτέλης, ο Θωμάς Ακινάτης και ο Σπινόζα αντικατοπτρίζουν ο καθένας τους χιλιάδες ανθρώπινα χρόνια σκέψης σχετικά με τη φύση της ελεύθερης βούλησης και των προϋποθέσεών της. Εάν οι λογικοί οργανισμοί δεν μπορούν να κάνουν διαφορετικές επιλογές και είναι ως εκ τούτου απαλλαγμένοι από τις συνέπειες, η υπό συνθήκη ανάλυση υποδηλώνει ότι ο Ρόμπερτ Φροστ, όπως ο Γουίτμαν, παίρνοντας "τον λιγότερο περπατημένο [από δύο δρόμους]", το έκανε υπό την αιγίδα κάποιας βιολογικής παρόρμησης που είναι τόσο μια καθυπόταξη της ελευθερίας του όσο και η απόλυτη ηθική εξαίρεση όταν ο ποιητής συνεχίζει "και αυτό έκανε όλη τη διαφορά". Ποια είναι η διαφορά, θα μπορούσαμε να αναρωτηθούμε, σε έναν κόσμο μυριάδων αιτίων και αποτελεσμάτων όπου, ανεξάρτητα από τον πολλαπλασιασμό τους, δεν υπήρχαν καν πραγματικές επιλογές; Ή, αντιστρόφως, πού έγκειται ο πανθεϊσμός μας των ασυγκράτητων ξαφνικών επιθυμιών, βουλήσεων και διαβεβαιώσεων, αν δεν υπάρχει πραγματικός ντετερμινισμός; Η τελευταία αυτή ελευθερία αποτελεί τον υπαρξιακό πυλώνα της αυθεντικότητας και της πραγματικής ελευθερίας που υπερβαίνει την απολυταρχία ενός Θεού ή ένα σενάριο όπως αυτό που τίθεται από τους "καλύτερους όλων των δυνατών κόσμων".[6] Οδηγεί παντού. Ο Αριστοτέλης αρχικά περιέγραψε το σενάριο (αργότερα γνωστό ως ο "όνος του Μπουριντάν"), παρουσιάζοντάς μας "έναν άνθρωπο, εξίσου πεινασμένο και διψασμένο, τοποθετημένο μεταξύ φαγητού και ποτού, [ο οποίος] πρέπει αναγκαστικά να παραμείνει εκεί που βρίσκεται και να λιμοκτονήσει μέχρι θανάτου".[7] Ο Σπινόζα είδε αυτή τη δύσκολη θέση ως ένα αδύνατο σενάριο λογικής, προάγγελο ενός καθολικού παραλογισμού[8] και αυτό που ένας ψυχολόγος έχει αναφέρει ως "το παράδοξο της επιλογής",[9] έναν υπολογισμό που έχει πάει στραβά και ο οποίος, σε οποιοδήποτε οικολογικό περιβάλλον, είναι συνώνυμος με την κατανάλωση και την "τραγωδία των κοινών". [10] Υπάρχουν εξαιρέσεις που σχετίζονται με την επιλογή και το συμφέρον του καταναλωτή στις κοινές καταστροφές των περισσότερων *κοινών*. Η "οικονομία του ντόνατ" στις Κάτω Χώρες, η ακαθάριστη εθνική ευτυχία στο Μπουτάν, οι σχετικά εξαιρετικές χρηματοοικονομικές επενδύσεις της Κόστα Ρίκα για την εκπαίδευση και το περιβάλλον, το εξαιρετικά υψηλό ποσοστό προστατευόμενων παρθένων δασών στο Σουρινάμ.[11]

Το περιβαλλοντικό ιστορικό της ασαφούς αιτιότητας η οποία δεσμεύεται από την ανθρώπινη επιλογή και τις υπονοούμενες, γεμάτες από αυτοκαταστροφή ανησυχίες της, διακρίνεται περισσότερο στις μεθόδους μέτρησης που χρησιμοποιούνται στη διασταύρωση τάσεων όπως η κλιματική αλλαγή και η μεγάλη έξοδος των δεκάδων, εκατοντάδων εκατομμυρίων οικολογικών μεταναστών/προσφύγων και οι επιλογές που γίνονται υπό το πρίσμα ενός ολοένα και περισσότερο μη κατοικήσιμου πλανήτη. Σε ένα σύνολο προβλέψεων με αξιοποίηση πολυδιάστατων μοντέλων κλίματος/μετανάστευσης, η οργάνωση ProPublica και οι New York Times, με την υποστήριξη του Pulitzer Center, βασιόμενοι σε πολλές προηγούμενες εκθέσεις, συμπεριλαμβανομένης της δημοσίευσης "Groundswell" της Παγκόσμιας Τράπεζας το 2018, "τροφοδότησαν το μοντέλο τους με περισσότερα από 10 δισεκατομμύρια σημεία δεδομένων." Υπέβαλαν υποθέσεις και προβλέψεις σε "προηγούμενες παρόμοιες καταστάσεις" (ο τρόπος με τον οποίο εξελίχθηκαν οι συνιστώσες των διεθνών μεταναστευτικών μοτίβων εμπειρικά εκ των υστέρων) και στη συνέχεια οι ερευνητές "μελέτησαν με ποιον τρόπο θα μετακινούνταν οι άνθρωποι στην περίπτωση της αύξησης του παγκόσμιου επιπέδου συγκεντρώσεως άνθρακα σε πέντε διαφορετικά σενάρια." Στο Σενάριο 2, του 2050, "υπάρχουν δεκάδες εκατομμύρια άνθρωποι πιο απελπισμένοι και με λιγότερες επιλογές. Η δυστυχία βασιλεύει και μεγάλοι πληθυσμοί παγιδεύονται." Αυτό, λίγο-πολύ, συνόψισε όλες τις προοπτικές, σε κάθε μορφή, της ανθρωπότητας.[12]

Θα μπορούσε κανείς να υπολογίσει τους 409.000 θανάτους από ελονοσία το 2019,[13] τους "επτά εκατομμύρια πρόωρους θανάτους κάθε χρόνο" από εσωτερική και εξωτερική ατμοσφαιρική ρύπανση[14] ή τους 4.400-200.000 θανάτους ετησίως από σχιστοσωμίαση.[15] Προσθέστε σε αυτούς τους αριθμούς τις "περίπου 121 εκατομμύρια ανεπιθύμητες εγκυμοσύνες" κάθε χρόνο μεταξύ 2015 και 2019, εκ των οποίων το "61% κατέληξε σε άμβλωση."[16] Και τέλος, τα "43 κιλά κρέατος" που καταναλώθηκαν κατά μέσο όρο ανά κάτοικο παγκοσμίως το 2014, με τους αριθμούς να αυξάνονται ραγδαία έκτοτε.[17] Τόσες επιλογές όσες είναι και οι ακούσιες, επώδυνες συνθήκες από τις οποίες ποτέ δεν επιβεβαιώθηκε, αποδόθηκε ή αποφεύχθηκε η ατομική συμμόρφωση με το συνολικό άθροισμα των πραγματικοτήτων. Δόλια καταστροφή, αθέλητη βλάβη, ελεύθερα καθορισμένες επιλογές, πολιτιστικές παρωπίδες, καταπίεση, πολύ κακή τύχη, όλα αναμεμιγμένα σε μια εξελικτική στιγμή, για κάθε άτομο, τα αλληλεξαρτώμενα είδη που πιάνονται στον ιστό των ανθρώπινων μηχανορραφιών.

Η ηθική των πιθανοτήτων, οι ηθικές αρχές των οικοσυστημάτων και οι βιοσυνθετικές καταστάσεις του νου συνθέτουν την ευαλωτότητα του ατόμου στη δύναμη της αλλαγής, την ελπίδα και την απελπισία της επιλογής και τους καταστροφικούς συνδυαστικούς αλγόριθμους της πλανητικής ροής. Αυτό είναι το πλαίσιό μας, η βιολογική γραμμή αναφοράς που πρέπει να αναγνωρίσουμε. Εάν το γνωρίζουμε αποσπασματικά, θα υποκινήσει το πλαίσιο την λυτρωτική επιλογή; Ή μήπως είμαστε γενετικά φτιαγμένοι να αγνοούμε αυτό που είναι το καλύτερο για εμάς; Η κλασική μυθολογία μας λέει ότι ο κατασκευαστής του μεγάλου κρητικού λαβύρινθου, ο Δαίδαλος, ήταν σχεδόν ανίκανος να βρει το δρόμο προς την έξοδο, και αυτό τον άφησε μόνο του με τον Μινώταυρο.[18] Ποιος ξέρει αν τα δύο *θηρία* θα μπορούσαν να γίνουν φίλοι ή αν, στην πραγματικότητα, η καταστροφή μας είναι ο ίδιος ο λαβύρινθος, όταν θα μπορούσαμε, ως είδος, να κάνουμε τα πράγματα τόσο απλά, οικουμενικά εξυπηρετικά, μια οικογένεια μελλοντικών πλαισίων που μοιάζουν περισσότερο με μια ευθεία γραμμή; Κάτι που όλοι μπορούμε να κατανοήσουμε, να κρατηθούμε από αυτό και να το τιμήσουμε με την καλοσύνη μας.

ΣΗΜΕΙΩΣΕΙΣ

[1] Βλ. "Cellular and molecular basis of decision-making," των Nilay Yapici, Manuel Zimmer και Ana I Domingos, *EMBO reports*, 2014 Oct: 15(10): 1023-1035, δημοσιεύτηκε online στις 14 Σεπτεμβρίου 2014. doi: 10.15252emr.201438993, PMCID: PMC24353843, PMID: 25239948, https://www.ncbi.nlm.nih.gov/pmc/articles/PMC4253843/

[2] "Predicting the decision-making chemicals used for bacterial growth," των Kazuha Ashino, Kenta Sugano, Toshiyuki Amagasa και Bei-Wen Ying, *Scientific Reports 9*, Article number: 7251 (2019) 10 Μαΐου 2019, https://www.nature.com/articles/s41598-019-43587-8

[3] Βλ. "Decision Making: Biology, Free-Will and Accountability," του "Spencer," UCLA doctoral student blog, *Psychology In Action*, 27 Νοεμβρίου 2011, https://www. psychologyinaction.org/psychology-in-action-1/2011/11/27/decision-making-biology-free-will-and-accountability

[4] Βλ. "Better decision-making for the planet," λεπτομερής ανάλυσης της έρευνας του καθηγητή του Princeton Elke Weber από τον Yasemin Saplakoglu, *Discovery: Research at Princeton*, January 4, 2018, https://www.princeton.edu/news/2018/01/04/better-decision-making-planet, Πρόσβαση στις 14 Δεκεμβρίου 2020.

[5] *Leaves Of Grass*, του Walt Whitman, Επιλογή και εισαγωγή του Christopher Morley, Εικονογράφηση Lewis C. Daniel, Doubleday, Doran & Co., Inc., New York, 1940, σελ. 3.

[6] Βλ. *Free Will as an Open Scientific Problem*, του Mark Balaguer, Cambridge, Mass., M.I.T. Press, 2010. Βλ. επίσης *Deep Control: Essays on Free Will and Value*, του John Martin Fischer, Oxford University Press, New York, 2012. Βλ. επίσης "What Happens When Someone Acts?" του David J. Velleman, Mind, 101: 461-81, 1992.

[7] *On the Heavens*, του Αριστοτέλη (στα αγγλικά), 295b. https://archive.org/details/ decaeloleofric00arisuoft/ page/n19/mode/2up

[8] *Ethics*, Part II, Proposition 49 Scholium. Βλ "Siris" -http://branemrys.blogspot.com/ 2005/10/spinoza-on-buridans-ass.html

[9] *The Paradox of Choice -Why More Is Less*, του Barry Schwartz, New York, Harper Perennial 2004. Βλ. επίσης Benjamin Scheibehenne, R. Greifendeder και P.M.Todd, "Can There Ever be Too Many Options? A Meta-Analytic Review of Choice Overload" (PDF). *Journal of Consumer Research. 37 (3): 409–425, 2010,* doi:10.1086/651235. Derek Thompson, "More Is More: Why the Paradox of Choice Might Be a Myth". *The Atlantic,* August 19, 2013; Βλ. επίσης Alexander Chernev, U. Böckenhold and J. K. Goodman, (2015). "Choice overload: A conceptual review and meta-analysis". *Journal of Consumer Psychology. 25 (2): 333–358.* doi:10.1016/j.jcps.2014.08.002.

[10] "The Tragedy of the Commons," του Garrett Hardin, *Science* 13 Dec 1968: Vol. 162, Issue 3859, σελ. 1243-1248 doi: 10.1126/science.162.3859.1243, https://science.sciencemag.org/ content/162/3859/1243

[11] Βλ. "Why These Countries Value Happiness Over Endless Economic Growth", της Megha Bahree, Εικονογράφηση Rebekka Dunlap, *FutureProof (HuffPost and Unearthed)*, https://www.huffpost.com/entry/economy-growth-gdp-happiness_n_5fad6f57c5b6d647a 39cf60c, Πρόσβαση στις 14 Δεκεμβρίου 2020.

[12] "Where Will Everyone Go?" του Abrahm Lustgarten, ProPublica, Φωτογραφές της Meridith Kohut για το *New York Times Magazine*, July 23, 2020, https://features.propublica.org/climate-migration/model-how-climate-refugees-move-across-continents/, Πρόσβαση στις 14 Δεκεμβρίου 2020.

[13] "Malaria," *World Health Organization*, 30 November 2020, https://www.who.int/news-room/fact-sheets/detail/malaria#:~:text=Disease%20burden,411%20000%20deaths%20in %202018.

[14] "Air Pollution," *World Health Organization*, https://www.who.int/health-topics/air-pollution#tab=tab_1, Πρόσβαση στις 14 Δεκεμβρίου 2020.

[15] Vos T, Allen C, Arora M, Barber RM, Bhutta ZA, Brown A, et al. (GBD 2015 Mortality and Causes of Death Collaborators) (October 2016). "Global, regional, and national life expectancy, all-cause mortality, and cause-specific mortality for 249 causes of death, 1980-2015: a systematic analysis for the Global Burden of Disease Study 2015". *Lancet*. 388 (10053): 1459–1544. doi:10.1016/s0140-6736(16)31012-1. PMC 5388903. PMID 27733281.

[16] "Unintended Pregnancy and Abortion Worldwide," Ιούλιος 2020, *Guttmacher Institute Fact Sheet*, https://www.guttmacher.org/fact-sheet/induced-abortion-worldwide#:~: text= Roughly%20121%20million%20unintended%20pregnancies%20occurred%20each%20year%20between%202015%20and%202019.&text=Of%20these%20unintended%20pregnancies%2C%2061,73%20million%20abortions%20per%20year.

[17] Βλ. "Meat and Dairy Production," των Hannah Ritchie και Max Roser, *Our World in Data*, Αύγουστος 2017, αναθεώρηση τον Νοέμβριο 2019, https://ourworldindata.org/meat-production

[18] Penelope Reed Doob, *The Idea of the Labyrinth: from Classical Antiquity through the Middle Ages*, Cornell University Press, Ithaca, New York, 1992, σελ. 36.

Οι Todas της Φυλής Kerrir, στα όρη Nilgiris, Tamil Nadu, Ινδία. Φωτογραφία © M. C. Tobias.

Κεφάλαιο 16

ΚΟΜΠΟΙ ΔΕΚΑ ΔΙΑΣΤΑΥΡΩΣΕΩΝ

Οι καταλύτες για την ατομική αλλαγή που θα μπορούσαν –μέσω της φυσικής επιλογής– να επηρεάσουν μια κρίσιμη μετασχηματιστική μάζα στον πληθυσμό αυτού του ατόμου, όπως ορίζεται από τις γεωπολιτικές παραμέτρους μιας πολιτιστικής κοινότητας, είναι τα σημεία καμπής που σχετίζονται περισσότερο με αυτήν τη συγκλίνουσα κρίση της Ανθρωπόκαινου Εποχής. Οι αλλαγές μας έχουν το πλεονέκτημα / μειονέκτημα ότι δοκιμάζονται όσον αφορά τη συνάφεια επιβίωσης, προφανώς μια πρωταρχική προϋπόθεση για την πρόσβαση στο μέλλον. Αυτή η δοκιμή είναι ο περιορισμός που διέπει όλα τα είδη. Μια τέτοια πρόκληση δεν είναι παρά το προκαταρκτικό ακρωτήριο: αυτό που, στις παραδόσεις που βασίζονται στην πίστη, θα ήταν σωτηριολογικό, δηλαδή προσανατολισμένο προς τη σωτηρία, αλλά από οικολογική άποψη σηματοδοτεί τόσο την εξοικείωση στις απαιτούμενες συνθήκες, όσο και την προσωπική υποδομή για τη διευκόλυνση των κατάλληλων επιλογών, της πληθώρας εθελοντικών ενεργειών νυχθημερόν. Οι πλήρεις διαστάσεις αυτής της φανταστικής οδύσσειας εντός του ανθρώπινου είδους παραμένουν άγνωστες, ειδικά όταν η συμπεριφορά μας είναι ορατή στον καθρέφτη της συνείδησής μας, της αυτοεπιβαλλόμενης απειλής μας ως ένα είδος που δεν λειτουργεί σωστά σε σχέση με αυτό που πραγματικά θέλουμε και ελπίζουμε.

Ο Πάουλ Έρλιχ γράφει: "Οι φύσεις μας είναι κατά κανόνα υποεξοπλισμένες για την καταγραφή τέτοιων απειλών, αλλά μπορούμε ωστόσο να εκπαιδευτούμε ώστε να έχουμε επίγνωση των βιολογικών εξελικτικών προεπιλογών που αναστέλλουν την αντίληψη ορισμένων βασικών προβλημάτων".[1] Ο Έρλιχ συνεχίζει ισχυριζόμενος ότι υπάρχει η ικανότητα για "αργά αντανακλαστικά" που, συνδυαζόμενα με γρήγορους χρόνους ανταπόκρισης, μπορούν να "ανιχνεύονται" και να ενσωματώνονται σε πιο αποτελεσματικές στρατηγικές αντίκρουσης των μακροπρόθεσμων απειλών που οι περισσότεροι είτε δεν βλέπουν ποτέ είτε επιλέγουν να αγνοήσουν.[2] Η πράξη της παρεμπόδισης της ίδιας της συμπεριφοράς κάποιου αντιπροσωπεύει ένα είδος κόμπου. Όμως, όπως αναγνώρισε ήδη από το 1876 ο Σκωτσέζος φυσικομαθηματικός Πίτερ Τέιτ

(1831-1901), οι ολόκληροι κοινωνικοί κόμποι παρουσιάζουν πολύ υψηλότερη πολυπλοκότητα. Μέσα σε μια δεκαετία, ο Τέιτ περιέγραψε "κόμπους 10 διασταυρώσεων" για τους οποίους έγραψε ότι "...είναι πολύ περίπλοκοι ώστε να μπορούν να είναι πάντα εντελώς ορατοί. Επομένως, δεν μπορούμε να είμαστε απολύτως βέβαιοι ότι μέσα από τέτοιες διαδικασίες έχουμε λάβει όλες τις ουσιαστικά διαφορετικές μορφές τις οποίες περιλαμβάνει ο ορισμός που χρησιμοποιούμε...".[3]

Υπάρχουν επίπεδες επιφάνειες, απομακρυσμένα χαρακτηριστικά και πολυώνυμα τμήματα με άπειρους στρόβιλους (οι αστρογεωφυσικές εκδοχές του λαβύρινθου). Οι συνδυασμοί τους στις δέκα κοσμικές διαστάσεις που διατυπώνει η θεωρία των χορδών, μέχρι σήμερα, θα σήμαινε, εάν μεταφράζονταν με γήινους όρους, ότι η οικολογική ολοκλήρωση, όπως στην περίπτωση ενός δάσους κλίμακος που υφίσταται ακόμη τις αλλαγές των συνεπειών ενός "νεκρού" δέντρου, δεν μπορεί να εξακριβωθεί, ακόμη και σε έναν τόσο φαινομενικά πεπερασμένο χώρο όπως μια επίπεδη επιφάνεια. Όπως ανέμενε ο Τέιτ, οι προκαταλήψεις μας είναι οι ίδιοι οι κόμποι που ξεδιπλώνονται τόσο ασαφώς όσο η συναρπαστική ανάλυση της Καρολίν Κοτσιάρντι σχετικά με τον δυνητικά διαφορετικό σχεδιασμό, λογική και προέλευση των κόμπων στον μπλε χιτώνα του Λεονάρντο ντα Βίντσι σε σχέση με τον κόκκινο χιτώνα στον πίνακά του "Salvator Mundi".[4]

Ίσης σημασίας είναι η αθωότητα κάθε δοκιμής από οικολογική άποψη. Απλώς δεν μπορούμε να πιαστούμε από καμία δύναμη συνήθειας εάν, πέρα από το ακούσιο νευρικό σύστημα και τις μυϊκές ανωμαλίες, οι φαντασίες μας εξουσιάζουν αυθαίρετα ή επίμονα πεπρωμένα, δηλαδή τα μέλλοντα που έχουμε εφεύρει.

Ως εκ τούτου, η πλήρης δύναμη της αμοιβαιότητας υπάρχει ως έμφυτη δυνατότητα, μια παρθένα σε κάθε συνάντηση κατάσταση με αρχικές προοπτικές. Εμείς (οι άνθρωποι) δεν έχουμε ιδέα τι αυτό σημαίνει, τι προμηνύει, τη δύναμη που είναι εγγενής ακόμη και σε μια ελάχιστη βιοχημική ακολουθία γεγονότων στο όνομα της sui generis πνευματικής, αισθητικής ή άλλης μη βίαιης ανταλλαγής. Η απόδοση ενός μαθηματικού στάτους στην ελεύθερη βούληση υπαινίσσεται επίσης τις διάφορες θεωρίες πιθανότητας που συνοδεύουν την επιλογή ή την απάρνηση στα αντίστοιχα ταξίδια τους προς το άπειρο. Τι σημαίνει αυτό για το άτομο; Για τη βιόσφαιρα;

Σύμφωνα με τον Άγιο Ευστάθιο ή τον τζαϊνιστή Samavasarana, το παν στην αρχή είναι η δοκιμή. Αυτό ενδέχεται να οδηγήσει κάποιον, πολιτιστικά μιλώντας, σε ένα Αμιρίμ -την πρώτη αποκλειστικά vegetarian κοινότητα του Ισραήλ (μαζί με μεγάλο μέρος της διατροφικής κουλτούρας του Τελ Αβίβ, καθώς και την Ντιμόνα, το "Χωριό της Ειρήνης". Μπορεί να οδηγήσει στις πολλές φυτοφαγικές μέρες την εβδομάδα στη Γάνδη, στους δύο φυτοφαγικούς μήνες το χρόνο στο Μπουτάν· στο Pushkar, μια ολόκληρη πόλη τέτοιων ηθικών διαστάσεων στο Ρατζαστάν, στους Todas, τους τζαϊνιστές, τους Bishnoi, στην Εκκλησία Αντβεντιστών της Εβδόμης Ημέρας, στα εκατομμύρια οπαδών της PETA, στους Inner Badui στην Ιάβα και ούτω καθεξής. Η φυτοφαγία δεν είναι παρά ένα έμβλημα του ανταλλακτικού συστήματος των συμβιωτικών αποκαλύψεων που μπορούν να προκύψουν από τα προαναφερθέντα *ταξίδια προς το άπειρο*. Γίνονται μεμιάς, από οικολογικής άποψης, ηθικά πλεονεκτήματα.

Άλλες τέτοιες στιγμές φέρνουν στη μνήμη την Τζέιν Γκούντολ με τους συντρόφους της τους χιμπαντζήδες στο Εθνικό Πάρκο Gombé Stream στην Τανζανία, τον σερ Ντέιβιντ Ατένμπορο να χαμογελά έμπλεως ευφορίας ανάμεσα σε μια οικογένεια γοριλών στα Όρη Βιρούνγκα στο Ηφαιστειακό Εθνικό Πάρκο της Ρουάντα, την Έλσα, τη λιονταρίνα (1956-1961) να φέρνει τα τρία λιονταράκια της από τις ερημιές του Εθνικού Πάρκου Meru στην Κένυα για να τα δείξει στους φίλους της τους ανθρώπους, την οικογένεια Άνταμσον (στην ταινία "Born Free"). Δέκα χιλιάδες βίντεο στο YouTube κάθε μέρα αποκαλύπτουν αυτές τις μαγικές, χαρούμενες συναντήσεις των ανθρώπων με άλλα είδη. Δεν υπάρχουν μαθηματικές θεωρίες πιθανοτήτων που να εξηγούν την άπειρη ποικιλότητα που ανά πάσα στιγμή αιωρείται μέσα στη βιόσφαιρα. Δεν υπάρχει ο διπλός κόμπος του ψαρά, που για πολύ καιρό θεωρείτο ως ένας από τους πιο επίμονα δισεπίλυτους κόμπους, όπως το "Πρόβλημα του Συνόλου των Τριών Κύβων" ή η ακόμα άλυτη "Υπόθεση του Ρίμαν".[5]

Όμως, ακόμη και τα "7 μεγαλύτερα αναπάντητα ερωτήματα στη Φυσική"[6], όπως το ερώτημα "Από τι αποτελείται η ύλη;" ή "Γιατί ο χρόνος φαίνεται να ρέει προς μία μόνο κατεύθυνση" μπορούν να λυθούν στη φαντασία μας, όπως και κάθε οικολογικό γεγονός, τόπος, γενικότερο πλαίσιο ή ελπίδα μπορούν επίσης να θεωρηθούν με ένα ηθικά πραγματιστικό, αν όχι ιδεαλιστικό τρόπο, με μαθηματική ακρίβεια. Στην πολλαπλή περιστροφική συμμετρία, "τις παρενθέσεις μέσα σε παρενθέσεις"[7] και στο (a) (n), δηλαδή το μέγεθος δείγματος από έναν δεδομένο πληθυσμό[8] κάθε αιτία και αποτέλεσμα, συμμετρία και ασυμμετρία υπόκεινται στην προσθήκη του ενός, ενός φυσικού αριθμού / ακέραιου αριθμού. Με άλλα λόγια, κάτι εντελώς μη δοκιμασμένο, εκείνος ο δρόμος που ποτέ δεν διαβήκαμε. Διαισθητικά, ο μέγας μαθηματικός Ντάβιντ Χίλμπερτ (1862-1943) διέβλεψε αυτό στο ανέκδοτο "Εικοστό Τέταρτο Πρόβλημά"[9] του, αναζητώντας το αριθμητικό αντίστοιχο του Ξυραφιού του Όκαμ, καθώς και μια εξίσωση για να το αποδείξει. Όμως, σε συνοικολογικούς όρους – ολόκληρες βιοτικές διαδικασίες και οι εμπλεκόμενες σε αυτές κοινότητες- κάθε άτομο αντιπροσωπεύει τη δυνατότητα μιας νέας, "επιπρόσθετης" εξίσωσης, και μαζί με αυτήν έναν χαρακτήρα που περιλαμβάνει μια βιογραφία, συναισθήματα, καθώς και περισσότερη φαντασία. Ακόμα και ένα κλάσμα αυτού του ατόμου κατέχει την ισότητα που πρέπει να προκύπτει από την αναγνώριση της ίδιας της ζωής.

Με όσο πιο απλά λόγια γίνεται, το νόημα της ύπαρξης κάθε οργανισμού είναι ανεξάρτητο από κάθε άλλου οργανισμού, όμως το κέρδος και για τους δυο τους στο πλαίσιο των βιολογικών συνεργασιών είναι ασύγκριτο, και οι συνεργασίες είναι ούτως ή άλλως αναπόφευκτες. Πρόκειται για το σιωπηρό, ανεξέλεγκτο, αναντίρρητο δόγμα που ενυπάρχει στην καρδιά κάθε διάταξης και κάθε φιλοσοφίας.

Εξού αυτό το πρελούδιο σε κάθε πειστική διατύπωση που οδηγεί στον κόσμο της αμοιβαιότητας.

Σημειώσεις

[1] *Human Natures – Genes, Cultures, and the Human Prospect*, του Paul R. Ehrlich, Shearwater Books, Island. Press, Washington D.C., 2000, σελ. 327.

[2] ό.π., σελ. 327.

[3] "On The Unknotting Problem," του Jorge Ramírez Alfonsín, *στο Combinatorics, Complexity, and Chance – A Tribute to Dominic Welsh*, Συγγραφείς Geoffrey Grimmett και Colin McDiarmid, Oxford Lecture Series In Mathematics and Its Applications – 34, Oxford University Press, New York, N.Y., 2007, σελ. 261.

[4] Βλ. *Leonardo's Knots*, της Caroline Cocciardi, Εκδόθηκε από Mona Lisa Knot, 2019. Αανασκόπηση του Scott Taylor, Mathematical Association Of America, 8 Δεκεμβρίου 2019, https://www.maa.org/press/maa-reviews/leonardos-knots, Πρόσβαση στις 15 Δεκεμβρίου 2020.

[5] Βλ. "The 10 Hardest Math Problems That Remain Unsolved," του Dave Linkletter, 26 Σεπτεμβρίου 2019, *Popular Mechanics*, https://www.popularmechanics.com/science/math/ g29251596/impossible-math-problems/. Πρόσβαση στις 15 Δεκεμβρίου 2020.

[6] Του Dan Falk, MACH, 8 Αυγούστου 2017, *NBC News*, https://www.nbcnews.com/mach/ science/7-biggest-unanswered-questions-physics-ncna789666, Πρόσβαση στις 15 Δεκεμβρίου 2020.

[7] Βλ. "Mathematical "N" Words," *themathlab.com*, https://themathlab.com/dictionary/ nwords/nwords.htm

[8] Βλ. "Sample sizes required," 7.2.2.2., *Engineering Statistics Handbook*, https://www.itl.nist.gov/ div898/handbook/prc/section2/prc222.htm

[9] "Mathematical Problems," του David Hilbert, *Bulletin (New Series) Of The American Mathematical Society*, Volume 37, Number 4, Pages 407{436S 0273-0979(00)00881-8Article electronically published on June 26, 2000,

https://web.archive.org/web/20090530182730/http://www.seas.harvard.edu/courses/cs121/handouts/Hilbert.pdf; Βλ. επίσης, "Hilbert's twenty-fourth problem," *του Rüdiger Thiele,* American Mathematical Monthly, *Ιανουάριος 2003.*

Κεφάλαιο 17

ΣΗΜΕΙΟΛΟΓΙΚΑ ΥΠΟΣΤΟΙΧΕΙΩΔΗ ΣΩΜΑΤΙΔΙΑ

Αρκούν λίγα παραδείγματα για να βάλουμε αποτελεσματικά τα θεμέλια για το απτό αφήγημα που υποδηλώνει αυτό που για τα ανθρώπινα πρότυπα αποτελεί ριζική αλλαγή. Όλο και περισσότεροι άνθρωποι αγκαλιάζουν αγελάδες και άλλα "εκτρεφόμενα" ζώα για θεραπευτικά οφέλη.[1] Στο Ofxordshire, η οργάνωση Kirtlington Wildlife & Conservation Society[2] έχει αξιοποιήσει στο έπακρο τη συμμετοχή 90.000 ατόμων σε ένα εθνικό πρόγραμμα γνωστό ως "Οδός Σκαντζόχοιρου"[3], στο οποίο ανοίγονται μικρές τρύπες, στο μέγεθος του σκαντζόχοιρου, σε παλιούς τοίχους από τούβλα αλλά και αλλού, προκειμένου οι σκαντζόχοιροι να μπορούν να μετακινούνται τη νύχτα κατά τη διάρκεια των ωρών τις οποίες συνήθως αναζητούν την τροφή τους. Όπως γίνεται σε περίπου το 25% όλης της άγριας ζωής αυτόχθονων θηλαστικών στο Ηνωμένο Βασίλειο -από τους κόκκινους σκίουρους ως τον βουνομυωξό και τους νεροαρουραίους- οι αριθμοί των σκαντζόχοιρων μειώνονται με γοργό ρυθμό και βρίσκονται στον κατάλογο των ειδών υπό εξαφάνιση της χώρας. Το κοινό τούς λατρεύει -όπως είναι εμφανές από τις δεκάδες χιλιάδες θεάσεις βιντεακού υλικού από κάμερες νυκτός με τις νυχτερινές τους αποδράσεις- και αυτό έχει αποδειχθεί ότι είναι ένα τρόπος για τη γρήγορη εξάπλωση μιας Pax Brittanica με τον πολυαγαπημένο, κοσμοπολίτικο *Erinaceus europaeus*. Σχετικά με το Ταμείο Ειδών υπό Εξαφάνιση του βρετανικού παραρτήματος του οργανισμού People's Trust και τους ντόπιους εν γένει, ο Chris Powles, δημιουργός της "εθνικής οδού" που μέχρι σήμερα διαθέτει πάνω από 12.000 τεχνητές τρύπες σκαντζόχοιρων, αναφέρει ότι "Πολύ λίγοι άνθρωποι λένε όχι…αυτή ακριβώς είναι η δύναμη των σκαντζόχοιρων… Νιώθεις ότι μπορείς να κάνεις κάτι θετικό και η ικανοποίηση που αντλείς από αυτό είναι μεγάλη..."[4]

Η ειλικρινής ταπεινοφροσύνη του Powles αποτελεί ξεκάθαρα στοιχείο αυτού που ονομάζουμε βιοφιλία, ή όπως αλλιώς θέλετε να την πούμε. Το βασικό στοιχείο εδώ είναι η απόφαση να κάνει κάποιος αυτήν την επιλογή. Όποιο και να είναι το νευρολογικό υπόβαθρο που στηρίζει τη σειρά των αισθητηριακών και γνωστικών αφομοιώσεων προκειμένου να φτάσουμε εκεί, η αντίδραση είναι το παν όταν συζητάμε για το μέλλον του *Homo sapiens*. Οι προαναφερθέντες στη βιβλιογραφία μαθηματικοί στρόβιλοι έχουν εξαιρετικό ενδιαφέρον. Στα μαθηματικά, το απόρρευμα ανατάραξης επηρεάζει με τρόπο που δημιουργεί ρευστότητα, ροή, ταχύτητα και μια γραμμή άξονα, ευθεία, καμπύλη ή πιο περίπλοκη από ό,τι γύρω από ποικίλες υπάρχουσες διάφορες ροές ή σε συνδυασμό με αυτές τις ροές, οι ασκούμενες δυνάμεις των οποίων μπορούν να περιγραφούν ως οικολογικές. Ο στροβιλισμός της περιστροφής ρευστού περιλαμβάνει σωματίδια. Όμως, στην κλίμακα της περιβαλλοντικής ροής, δεν υπάρχει αστρόβιλη (απούσα) διαδρομή, γιατί απλά υπάρχει πάντα. Τίποτα δεν είναι κενό και καμία φιλοσοφία δεν μπορεί να αποδείξει το αντίθετο. Υπάρχουν μη σταθερές ροές στη γεωμετρία του στροβίλου, και διάφορες διακυμάνσεις της πίεσης, της πυκνότητας, της ψύξης και θέρμανσης και της διάδοσης δονήσεων και κυμάτων, στο κέντρο όμως, όλων αυτών των μετρήσιμων φαινομένων βρίσκεται η ιδέα της ανατάραξης στη φύση, σε οποιοδήποτε επίπεδο, είτε διακριτό είτε όχι.[5]

Αυτό που τα παραπάνω συνεπάγονται για εμάς και για όλους τους Άλλους είναι η μεταναστευτική δύναμη των δεδομένων, της παρόρμησης και της προσωπικής ταυτοποίησης. Πρόκειται για μια οικολογία μεταφορών που έχουν έλξη στο σημείο της ατομικής αναγνώρισης και ανταπόκρισης. Η ποικιλία της τοπολογίας αυτών των πληροφοριών οικοδόμησης μεταβάσεων είναι ατελείωτη. Απλά σκεφτείτε τα εξής: θαλάσσια θερμικά κύματα[6]; το αναπόφευκτο της μετάβασης μεταξύ σταθερών καταστάσεων[7], η γεωοικοσυστημική ροή στο οροπέδιο του Θιβέτ[8], “σήματα” “από παλαιο-οικολογικά αρχεία”[9], “καμπύλη/ες μετάβασης υπηρεσίας οικοσυστήματος” δάσους[10], οι προκλήσεις για την “εννοιολογία” την επαύριο του θανάτου των δασών[11], η χαρτογράφηση της βλάστησης φυσικών και ημι-φυσικών συστημάτων[12], οι επιπτώσεις των κρίσιμων διαταραχών σε μια λίμνη εντός ενός οικοσυστήματος λόγω δασικών πυρκαγιών[13], τα δεδομένα του πλούτου των ειδών από (αμφιλεγόμενες) μελέτες καθεστώτος διαχείρισης θαλάσσιων οικοσυστημάτων της “περίπλοκα παράκτιας” Καλιφόρνιας[14], η αξιολόγηση των μεταβάσεων “εξανθράκωσης” σε τοπική, περιφερειακή και παγκόσμια κλίμακα, με διαρκώς μεταβαλλόμενες γραμμές βάσης[15]. Άλλες αξιοσημείωτες μεταφορές: το “υδρολογικό πλαίσιο” για “πρώτης τάξεως, αρχέγονους καθοριστικούς παράγοντες της τάσης των οικοσυστημάτων να υφίστανται μετάβαση”[16], η μέτρηση της έντασης των πολυπαραγοντικών διαταραχών στη λίμνη Michigan[17], η “δραματική” αντικατάσταση των “βυθισμένων μακρόφυτων” (υδρόβια φυτά) από “νηματώδη στρώματα φυκών” στις βόρειες πηγές της Φλόριντα[18], τα διάχυτα μονοπάτια ανάκαμψης των δασών, ένα έμβλημα των τόσο διαφοροποιημένων μεγαοικοσυστημάτων από εύκρατα, τροπικά και βόρεια δάση που περιλαμβάνουν πάνω από 3 τρισεκατομμύρια μεμονωμένα δέντρα[19], οι αλλαγές που επέρχονται ως αποτέλεσμα της “διαχείρισης βάσει του οικοσυστήματος” του Μεγάλου Κοραλλιογενούς Ύφαλου της Αυστραλίας[20], οι αριθμοί των κριλ και η Δεκαδική Ταλάντωση του Ειρηνικού[21], οι διάφοροι ορισμοί των “συστατικών, καταστάσεων, μεταβάσεων και ορίων” στη μοντελοποίηση των διαταραχών στις “φυτικές κοινότητες”[22]·, οι μεταμορφώσεις, σύμφωνα με αυτές τις τελευταίες γραμμές, συστημάτων στα οποία κυριαρχούν τα λιβάδια σε σαβάνα που αποτελείται κυρίως από "χόρτα και ξυλώδη φυτά"[23].

Κάθε μία από αυτές τις αντιπροσωπευτικές, μεταβατικές συναρτήσεις των βιοστροβίλων δηλώνει μια διαδικασία που περιλαμβάνουν οι οικολογικές διεπαφές και αναταράξεις οι οποίες εκφράζουν τις ζωές των πληθυσμών και των ατόμων. Οι επαναλαμβανόμενες ταξινομίες τους δεν συμβαδίζουν με την απόλυτη βιολογική αγαλλίαση που ενυπάρχει στη γνώση ότι από τα τέλη του 2016, για παράδειγμα, υπάρχουν τουλάχιστον 60.065 είδη δέντρων παγκοσμίως, ο μεγαλύτερος αριθμός των οποίων -8.715 είδη- βρίσκονται στη Βραζιλία, μετά στην Κολούμπια, και μετά στην Ινδονησία.[24] Τα ζωντανά αυτά δέντρα είναι ένας πρότυπος "ξενώνας" αναρίθμητων άλλων ζωνών επαφής, σκιάς και ηλιασμού, που είναι το άθροισμα των ακατέργαστων ξυλωδών συντριμμιών, ή νεκρό ξύλο που αλληλεπιδρά με τους κύκλους της μετάβασης που περιλαμβάνουν εκατομμύρια οργανισμούς, τρισεκατομμύρια ή τετράκις εκατομμύρια εμπλεκόμενων μορίων.[25] Υπάρχει ένα είδος ανακούφισης στη βιολογική ζώνωση, ακόμη και όταν δύσκολα μπορούμε να φανταστούμε και να αποδεχθούμε αδιανόητες πραγματικότητες. Αυτά τα μεγάλα σύννεφα άγνοιας που αποτελούν τη βιόσφαιρα μπορεί να είναι πιο ευπρόσδεκτα κάποιες μέρες από άλλες. Παρόλα αυτά, σε τελική ανάλυση, όλα τα σχέδια προσφέρουν κομψότητα, ομορφιά και ενδιαφέρον που διαρκεί όλο και περισσότερο όσο πιο μπερδεμένες μοιάζουν να είναι οι περιγραφές των ορίων τους και οι διασταυρώσεις τους. Οι εσωτερικοί αυτοί κόσμοι με τα δικά τους συνδετικά στοιχεία, περιθώρια, πυρήνες και εφεδρείες, συμπληρώνουν όλοι μαζί την αρωγή των απαντήσεων όπου συναντάμε τους Άλλους και έχουμε κάθε λόγο να πιστεύουμε στο θαύμα ότι αποτελούμε μέρος κάποιου πράγματος πέρα από την κατανόησή μας.

Τόσο πανταχού παρούσα είναι η ταυτοχρονία αυτού που έχει περιγραφεί ως "μεταβολική συνεργασία" ανάμεσα σε αυτές τις χωρίς όριο ανατροφοδοτήσεις[26], που οι παραδοσιακές παράμετροι που αποδίδονται σε σταθερές κοινότητες, ιδίως μεταξύ των πρωτευόντων[27] (με έμφαση στις εξελικτικές προϋποθέσεις για τον αμοιβαίο αλτρουισμό) αποκαλύπτουν ένα αδιαμφισβήτητο είδος απόδειξης για τη μελλοντική φύση των σχέσεων, κυρίως υπό το πρίσμα της ισορροπίας ισχύος που εξάρτάται από τις ανθρώπινες επιλογές. Εξού η αξέχαστη προσταγή του Τζον Ράσκιν: "Αν μπορείς να ζωγραφίσεις ένα φύλλο, μπορείς να ζωγραφίσεις τον κόσμο όλο".[28] Η Diana Beresford-Kroeger περιγράφει την οικογένεια των καρυδιών (Καρυοειδών), τα δέντρα της οποίας συνήθως αφήνουν να πέσει "ηρεμιστική γιουγλόνη στο νερό το φθινόπωρο... [που] επηρεάζει τη φυσική νάρκη των ψαριών και των πλασμάτων του νερού, συμβάλλοντας έτσι στη σταθεροποίηση και διατήρηση του μειωμένου βασικού μεταβολικού ρυθμού τους".[29] Οι περσικοί πολιτισμοί φημίζονται για τη χρήση του καρυδιού για τον έλεγχο ορισμένων τύπων διαβήτη.

Ο Γάλλος φιλόσοφος Ζακ Μαριτέν (1882-1973) πάλεψε με την οντολογία, όπως αυτή ήταν, του φύλλου του Ράσκιν, το δέντρο του, ολόκληρου του τοπίου του, επιχειρώντας να βυθοσκοπήσει την "κατανόηση", όπως αυτή αφορούσε την "προέλευση της επιστήμης" και τα σκοτεινά στοιχεία της ανθρώπινης φαντασίας, που περιορίζεται στην "πρώτη τάξη της αφαίρεσης".[31] Πέρα από τις αισθήσεις μας, τι είμαστε ικανοί να γνωρίζουμε, να ορίσουμε, να φανταστούμε; Και τι αξίζει; Οι βοτανολόγοι εξακολουθούν να είναι μπερδεμένοι από δύο μεγάλα μυστήρια της φυτικής επιστήμης: τη βιοσύνθεση του γιββερελικού οξέως, μιας από τις πιο σημαντικές και καθοριστικές ορμόνες για τις ρυθμιστικές λειτουργίες ενός φυτού, ιδιαίτερα στα θηλυκά, όπως η επιβίωση, η βλάστηση, η παραγωγή κάθε μεγαοικοσυστήματος που ανθίζει, σε σχεδόν 392.000 αγγειακά (ανθοφόρα) είδη φυτών στη γη.[32] Και το άλλο μεγάλο μυστήριο, δηλαδή, την αλλαγή των χρωμάτων των φύλλων, ιδίως το φθινόπωρο, την "αποκάλυψη των καροτενοειδών ή ανθοκυανών χρωστικών".[33]

Ρωτήσαμε τη Δρα. Melanie DeVore, μία από τις μεγαλύτερες παλαιοβοτανολόγους στον κόσμο, τι βρίσκει εκπληκτικό στα φύλλα και η απάντησή της ήταν εξαιρετικά εμπνευσμένη: "... Μπορείτε να συλλέξετε 30 διαφορετικά φύλλα δικοτυλήδονων, να προσδιορίσετε την αναλογία αυτών των φύλλων σε σχέση με ολόκληρα φύλλα και υπολογίσετε επακριβώς τη μέση ετήσια θερμοκρασία για οποιαδήποτε χλωρίδα από την Κρητιδική περίοδο έως σήμερα. Έστειλα μαθητές μου και μου έφεραν 30 φύλλα από δυο πανεπιστημιουπόλεις (μια στο Μίλετζβιλ της πολιτείας της Τζόρτζια και μια στην πόλη του Τζέρσεϊ στην πολιτεία του Νιου Τζέρσεϊ) και υπολόγισαν τη σωστή μέση ετήσια θερμοκρασία (ΜΑΤ) σε τρία σημαντικά ψηφία".[34] Η DeVore και η συνάδελφός της K.B. Pigg έχουν επανεξετάσει εκτενώς το καθεστώς ανακύκλωσης άνθρακα του θερμικού μέγιστου του Παλαιόκαινου-Ηώκαινου (PETM) πριν από 56 εκατομμύρια χρόνια, ιδιαίτερα σε μια περίοδο 200.000 ετών κατά την οποία η ανίχνευση της παρουσίας άνθρακα τα τελευταία 500 εκατομμύρια χρόνια στη γη φαίνεται να έχει φτάσει το υψηλότερο σημείο όλων των εποχών, τουλάχιστον μέχρι τη Βιομηχανική Επανάσταση και την επαύριόν της. Βασιζόμενες σε αρχεία από την ανάλυση "απολιθωμάτων φύλλων, γύρης και ξύλου", οι DeVore και Pigg συνόψισαν το μεθοδολογικό πλαίσιο που έχουν φτιάξει οι παλαιοβοτανολόγοι για να δουν με ποιον τρόπο οι φυτικές δομές, όπως τα φύλλα, "χρησιμεύουν ως μεσολαβητές για τις θερμοκρασίες, τις βροχοπτώσεις και τις ατμοσφαιρικές συγκεντρώσεις διοξειδίου του άνθρακα" κατά τη διάρκεια τεράστιων χρονικών διαστημάτων αξιοποιώντας μια "προσέγγιση, γνωστή ως ο πλησιέστερος ζων συγγενής (NLR)", καθώς και τη σημαντική "προσέγγιση συνύπαρξης".[35]

Επιπροσθέτως, οι DeVore και Pigg βασίζονται σε μια εξίσωση για τον υπολογισμό της "μέσης ετήσιας βροχόπτωσης... από ένα σύνολο απολιθωμένων φύλλων μέσω της ταξινόμησης μορφοτύπων σε κατηγορίες βάσει του σχεδίου ταξινόμησης μεγέθους Raukier-Webb" που τονίζει "ότι όσο ξηρότερο είναι το κλίμα, τόσο υψηλότερη είναι η αναλογία ξυλωδών δικοτυλήδονων με μικρά φύλλα στη χλωρίδα..."[36] Η παρακολούθηση των διαφόρων φυτικών ζωνών και των στρωματογραφικών μεταβατικών σημείων εδώ και εκατομμύρια χρόνια στο παρελθόν αποτελεί έναν ανακριβή αλλά παραταύτα εξαιρετικά αναλυτικό οδηγό για το βοτανικό αυτό καθεστώς που πιθανώς θα κυριαρχήσει στο Ανθρωπόκαινο τα επόμενα χρόνια και δεκαετίες και το φύλλο, σε όλη τη εκπληκτική πολυπλοκότητά του και ομορφιά αποτελεί έναν από τους καλύτερους οδικούς χάρτες που θα μπορούσε η ανθρωπότητα να κληρονομήσει από επιστήμονες όπως οι DeVore και Pigg. Η ανθρώπινη κουλτούρα αφοσίωσής τους στο φύλλο και όλα όσα αυτό αντιπροσωπεύει για τη γήινη ζωή θα αποβεί κρίσιμη για το κυρίαρχο ερώτημα σχετικά με το εάν η ανθρωπότητα μπορεί να συνεχίσει να συνυπάρχει με μονοκοτυλήδονα και δικοτυλήδονα στο μέλλον. Σύμφωνα με την DeVore, ο Leo Hickey (1940-2013) και ο Jack Albert Wolfe (1936-2005) ήταν τα λαμπρά μυαλά που πραγματικά "προλείαναν το έδαφος για να κάνουμε τα φύλλα τα εργαλεία που χρησιμοποιούμε σήμερα".[37]

Από την άλλη όμως, υπάρχει και το ίδιο το απόλυτο ανθρώπινο ζήτημα, η τάση μας για εισβολή -όσο καλές προθέσεις και να έχουμε- στο πλαίσιο του πεδίου εφαρμογής των εμμονικών ανθρωπικών μας ορισμών σε κάθε επίπεδο των οικολογικών επιστημών. Στη μελέτη τους για τα αυτο-επιβεβαιούμενα/ δυναμικά υποστοιχειώδη σωματίδια -τις βιολογικές οντότητες που γνωρίζουν (βιώνουν) τόσο τους εσωτερικούς όσο τους και εξωτερικούς ρόλους τους σε οποιοδήποτε οικοσύστημα- ως διακριτά από την αφηρημένη ταξινομική μονάδα (τάξον), οι Timothy F.H. Allen και Thomas B. Starr αμφισβήτησαν με δηκτική διάθεση την "επιβολή της κλίμακας του ατόμου επάνω στα εννοιολογικά μας μοντέλα της παρούσας κρίσης".[38] Το όλον παραμένει ένας ανθρώπινος ορισμός, ένας ορισμός ωστόσο που παρουσιάζει την ιδέα ότι πρόκειται για μια λέξη που αντιπροσωπεύει μια διαφάνεια που είναι πραγματική στον φυσικό κόσμο, δεδομένης της μετάδοσης υλικού και αισθητηριακών πληροφοριών μεταξύ του ίδιου του όλου και του περιβάλλοντος. Με άλλα λόγια, ένας Εαυτός, ένα μέλος των Άλλων. Μια κοινότητα που δεν πάσχει από αυταρχικά βιολογικά πρωτόκολλα επιβαλλόμενα εκ των άνω προς τα κάτω.

ΣΗΜΕΙΩΣΕΙΣ

[1] "Cow-hugging, an alleged wellness fad, has people cuddling farm animals to relieve stress," της Alexandra Deabler, *Fox News*, 17 Οκτωβρίου 2020, https://apple.news/AAABgH-QARZiP_bTCmRrtFA

[2] http://kirtlingtonvillage.co.uk/kwacs

[3] https://www.hedgehogstreet.org/

[4] *The age of extinction - Prickly business: the hedgehog highway that knits a village together*, της Phoebe Weston, 17 Οκτωβρίου 2020, https://www.theguardian.com/environment/ 2020/oct/17/prickly-business-hedgehog-highway-knits-a-village-together-kirtlington-oxfordshire-aoe

[5] Βλ. Rupert Klein και Lu Ting, *Viscous Vortical Flows*, *Lecture Notes in Physics*, SpringerLink, New York, 1991. Βλ. επίσης, *Quantum Ecodynamics*, αδημοσίευτο χειρόγραφο των Michael Charles Tobias, George Fitzgerald Smoot III, και Jane Gray Morrison, 2021.

[6] Βλ. T. Wernberg, S. Bennett, R. C. Babcock, T. de Bettignies, K. Cure, M. Depczynski, F. Dufois, J. Fromont, C. J. Fulton, R. K. Hovey, E. S. Harvey, T. H. Holmes, G. A. Kendrick, B. Radford, J. Santana-Garcon, B. J. Saunders, D. A. Smale, M. S. Thomsen, C. A. Tuckett, F. Tuya, M. A. Vanderklift, S. Wilson. Climate-driven regime shift of a temperate marine ecosystem. *Science*, 2016; 353 (6295): 169 doi: 10.1126/science.aad8745, American Association for the Advancement of Science. "Marine heatwave triggers dramatic ecosystem transition." *ScienceDaily*. *ScienceDaily*, 7 Ιουλίου 2016. www.sciencedaily.com/releases/2016/07/160707151113.htm, http://dx.doi.org/10.1126/ science.aad8745

[7] "Inferring critical thresholds of ecosystem transitions from spatial data," των Sabiha Majumder, Krishnapriya Tamma, Sriram Ramaswamy and Vishwesha Guttal, Πρώτη δημοσιευση: 03 Μαΐου 2019, https://doi.org/10.1002/ecy.2722, Corresponding Editor: Tom E. X. Miller., *ESA, Ecological Society of America*, Volume 100, Issue 7, July 2019, e02722, https://esajournals.onlinelibrary.wiley.com/doi/10.1002/ecy.2722

[8] *Geo-ecosystems in Transition on the Tibetan Plateau (TransTiP)*, Technische Universität Braunschsweig, https://www.tu-braunschweig.de/en/irtg-transtip, 10 Ιανουαρίου 2020.

[9] Zofia E. Taranu, Stephen R. Carpenter, Victor Frossard, Jean-Philippe Jenny, Zoe Thomas, et al. "Can we detect ecosystem critical transitions and signals of changing resilience from paleo-ecological records?". *Ecosphere, Ecological Society of America*, 2018, 9 (10), ff10.1002/ecs2.2438ff. ffhal-01959637, https://hal.archives-ouvertes.fr/hal-01959637/document

[10] "Viewing Forest Transition Outcomes through the Lens of Ecosystem Services," των Wilson, S.J., J. Schelhas, R. Grau, A. Sofia Nanni και S. Sloan 2017. Forest ecosystem-service transitions: the ecological dimensions of the forest transition. *Ecology and Society*, 22: 38. https://partners-rcn.org/viewing-forest-transition-outcomes-through-the-lens-of-ecosystem-services/Πρόσβαση στις 16 Δεκεμβρίου 2020

[11] *Ecosystem dynamics and management after forest die-off: a global synthesis with conceptual state-and-transition models*, Cobb R.C., Ruthrof K.X., Breshears D.D., Lloret F., Aakala T., Adams H.D., Anderegg W.R.L., Ewers B.E., Galiano L., Grünzweig J.M., Hartmann H., Huang C., Klein T., Kunert N., Kitzberger T., Landhäusser S.M., Levick S., Preisler Y., Suarez M.L., Trotsiuk V., και Zeppel M.J.B (2017), Critical Zone Observatory (Collaborative Network), Santa Catalina Mountains and Jemez River Basin, https://czo-archive.criticalzone.org/catalina-jemez/publications/pub/cobb-et-al-2017-ecosystem-dynamics-and-management-after-forest-die-off-a-gl/, Πρόσβαση στις 16 Δεκεμβρίου 2020

[12] *Monitoring Vegetation under Global Change*, Humboldt-Universität zu Berlin," 03.07.2020, https://www.enmap.org/data/projects/HU%20Berlin.pdf

[13] *Bushfires may have driven a critical ecosystem transition*, ANSTO Staff, http://onlinelibrary. wiley.com/ doi/10.1002/2017JG004135/full, https://www.ansto.gov.au/news/bushfires-may-have-driven-a-critical-ecosystem-transition

[14] *Bulletin of Marine Science*, 74(3): 693–708, 2004 Bulletin of Marine Science 693 © 2004 Rosenstiel School of Marine and Atmospheric Science of the University of Miami, *Transition From Low To High Data Richness: An Experiment In Ecosystem-Based Fishery Management From California*," των Les Kaufman, Burr Heneman, J. Thomas Barnes, και Rod Fujita, http://fisherysolutionscenter.edf.org/sites/catchshares.edf.org/files/ Transition%20from%20Low%20to%20High%20Data%20Richness%20-%20An%20Experiment%20in%20 Ecosystem-based%20Fishery%20Management%20from%20California.pdf

[15] "Keeping up with an evolving energy transition ecosystem," *ARDIAN Media Coverage*, 1 Απριλίου 2020, https://www.ardian.com/news-insights/keeping-evolving-energy-transition-ecosystem

[16] *Ecohydrology of ecosystem transitions: a meta-analysis*, by Viglizzo, Ernesto F.; Nosetto, Marcelo D.; Jobbágy, Esteban G.; Ricard, M. Florencia; Frank, Federico C., URI: http://hdl.handle.net/10625/54836, 2014, IDRC Digital Library, Canada, https://idl-bnc-idrc.dspacedirect.org/handle/10625/54836

[17] "Seasonal zooplankton dynamics in Lake Michigan: Disentangling impacts of resource limitation, ecosystem engineering, and predation during a critical ecosystem transition," των Henry A. Vanderploeg, Steven A. Pothoven, Gary L. Fahenstiel, Joann F. Cavaletto, James R. Liebig, Craig A. Stow, Thomas F. Nalepa, Charles P. Madenjian, και David B. Bunnell, *Journal of Great Lakes Research*, Volume 38, Issue 2, June 2012, σελίδες 336-352, Elsevier, *ScienceDirect*, https://www.sciencedirect.com/science/article/abs/pii/S0380133012000408

[18] "Flow reversals as a driver of ecosystem transition in Florida's springs," Robert T. Hensley, και Matthew J. Cohen, Freshwater Science, Volume 36, Number 1, 2017, *The University of Chicago Press Journals*, https://www.journals.uchicago.edu/doi/full/10.1086/690558? mobileUi=0

[19] "Forest ecosystem-service transitions: the ecological dimensions of the forest transition," των Sarah Jane Wilson; John Schelhas; Ricardo Grau; A. Sofía Nanni; Sean Sloan, 2017, *Ecology and Society Publication Series: Scientific Journal (JRNL)*, Southern Research Station, doi: https://doi.org/10.5751/ES-09615-220438, https://www.fs.usda.gov/treesearch/pubs/56948, USDA, U.S. Forest Service, Treesearch.

[20] "Navigating the transition to ecosystem-based management of the Great Barrier Reef, Australia," Per Olsson, Carl Folke, και Terry P. Hughes, PNAS Ιούλιος 15, 2008 105 (28) 9489-9494; https://doi.org/10.1073/pnas.0706905105, https://www.pnas.org/content/105/28/9489

[21] Βλ. "Analysis of Abrupt Transitions in Ecological Systems," Bestelmeyer, B. T., A. M. Ellison,W. R. Fraser, K. B. Gorman, S. J. Holbrook, C. M. Laney, M. D. Ohman, D. P. C. Peters, F. C. Pillsbury, A. Rassweiler, R. J. Schmitt, και S. Sharma. 2011. Analysis of abrupt transitions in ecological systems. *Ecosphere* 2(12):129. https://jornada.nmsu.edu/bibliography/11-036.pdf, The Jornada-Rangeland Research Programs, https://jornada.nmsu.edu/ltar/abrupt-transitions

[22] "State and transition modeling: An ecological process approach," των Tamzen K. Stringham, William C. Krueger, και Patrick L. Shaver, *J. Range Manage.* 56: 106-113 Μάρτιος 2003, https://www.nrcs.usda.gov/Internet/FSE_DOCUMENTS/16/nrcs143_023865.pdf

[23] "An Ecosystem in Transition: Causes and Consequences of the Conversion of Mesic Grassland to Shrubland," των John M. Briggs, Alan Knapp, John M. Blair και Jana L. Heisler, *ResearchGate*, March 2005; BioScience 55:243-254, doi: 10.1641/0006-3568(2005)005[0243: AEITCA] 2.0 CO;2, https://www.researchgate.net/publication/269038536_An_Ecosystem_in_Transition_Causes_and_Consequences_of_the_Conversion_of_Mesic_Grassland_to_Shrubland

[24] "There Are More Than 60,000 Tree Species Worldwide, Scientists Say," της Merrit Kennedy, 12 Απριλίου 2017, *NPR*, https://www.npr.org/sections/thetwo-way/2017/04/12/ 523616033/there-are-60-000-species-of-tree-worldwide-scientists-say, Πρόσβαση στις 16 Δεκεμβρίου 2020.

[25] Βλ. *The Afterlife Of A Tree*, των Jerzy M. Gutowski, Karol Zub, Pawel Pawlaczyk και William F. Laudenslayer, επιμέλεια Andrzej Bobiec, WWF Poland, Warszawa-Hajnowka 2005, σελ. 18-27.

[26] Βλ. "Reciprocal Fitness Feedbacks Promote the Evolution of Mutualistic Cooperation," των Daniel Preussger, Samir Giri, Linéa K. Muhsal, Leonardo Oña, και Christian Kost, *Current Biology*, 23 Ιουλίου 2020, doi: https://doi.org/10.1016/j.cub.2020.06.100, Article 1, Volume 30, Issue 18, P3580-3590.E7, 21 Σεπτεμβρίου 2020. https://www.cell.com/current-biology/pdfExtended/S0960-9822(20)30986-6#back-bib3

[27] Βλ. *The Symbolic Species – The Co-Evolution of Language and the Brain*, του Terrence W. Deacon, W. W. Norton & Co., New York and London, 1997, pp. 397-401.

[28] *Modern Painters*, του John Ruskin, Τόμος 5, Κεφ. 5, σελ. 61, *The Works of John Ruskin in 25 Volumes*, Merrill and Baker, New York, 1890.

[29] *The Global Forest*, της Diana Beresford-Kroeger, Viking Publishers, New York, 2010, σελ. 31.

[30] "Chemical Composition and the Effect of Walnut Hydrosol on Glycemic Control of Patients With Type 1 Diabetes," των Hossein Moravej, Alireza Salehi, Zahra Razavi, Mahmood Reza Moein, Hamed Etemadfard, Forough Karami και Forough Ghahremani, *International Journal of Endocrinology and Metabolism*, 14 (1): e34726, 30 Ιανουαρίου 2016, doi: 10.5812/ijem.34726, https://sites.kowsarpub.com/ijem/articles/17732.html, Πρόσβαση στις 18 Δεκεμβρίου 2020.

[31] *Philosophy of Nature, To which is added, Maritain's Philosophy Of The Sciences*, του Yves R. Simon, Philosophical Library, New York, 1951, σελ. 29.

[32] "Gibberellic acid in plant: still a mystery unresolved," των Ramwant Gupta και S K Chakrabarty, *Plant Signal Behav.* 2013 Sep:8 (9):e25504 doi: 10.4161/psb.25504. Epub 2013 Jun 28, PMID: 23857350 PMCID: PMC4002599, https://pubmed.ncbi.nlm.nih.gov/23857350/, Πρόσβαση στις 18 Δεκεμβρίου 2020.

[33] *The Biological Significance of Leaf Color Change*, του Harvard Forest, n.a., n.d., https://harvardforest.fas.harvard.edu/leaves/biological, Πρόσβαση στις 18 Δεκεμβρίου 2020.

[34] *Personal communication via email with Dr. DeVore*, 18 Δεκεμβρίου 2020.

[35] Βλ DeVore, Melanie L., και Kathleen B. Pigg, "The Paleocene-Eocene Thermal Maximum: Plants as Paleothermometers, Rain Gauges, and Monitors," των Melanie L. DeVore και Kathleen B. Pigg, στο *Nature Through Time – Virtual field trips through the Nature of the past*, Επιμέλεια Edoardo Martinetto, Emanuel Tschopp και Robert A. Gastaldo σελ. 109-128, Spring, Cham, Ελβετία, 2020, https://link.springer.com/book/10.1007/978-3-030-35058-1

[36] ό.π.., 4.3.

[37] Βλ. "East meets West: The contrasting contributions of David L. Dilcher and Jack A. Wolfe to Eocene systematic paleobotany in North America," Νοέμβριος 2007, CFS *Courier Forschungsinstitut Senckenberg* 258(258): 85-94, https://www.researchgate.net/publication/ 242342988_East_meets_West_The_contrasting_contributions_of_David_L_Dilcher_and_Jack_A_Wolfe_to_Eocene_systematic_paleobotany_in_North_America

[38] *Hierarchy – Perspectives for Ecological Complexity*, των Timothy F.H. Allen και Thomas B. Starr, The University of Chicago Press, Chicago, Ill. 1982, σελ. 217.

Ιερή λίμνη στη χορτοφαγική πόλη Pushkar, Rajashtan, Ινδία. Φωτογραφία © M. C. Tobias.

Κεφάλαιο 18

ΜΕΛΛΟΝΤΙΚΑ ΣΥΜΒΟΛΑ

Τα δομικά δίκτυα που υποστηρίζουν διάφορες μορφές αμοιβαιότητας χρονικά και γεωγραφικά υποδηλώνουν προφανείς συμβιώσεις: για παράδειγμα, ορισμένα είδη αμφιπόδων (*Podocerus jinbe)* στα στόματα των φαλαινοκαρχαριών[1], ταραντούλες και βάτραχοι, βραδύποδες και φύκη, ορισμένα έντομα (χρυσοπίδες) και λειχήνες[2], η μαγκούστα και ο βούκερως, οι τσικνιάδες των βοοειδών και οι πολυάριθμοι ξενιστές τους, από τα βοοειδή έως τους ιπποπόταμους, φυλλοκόπα μυρμήγκια και φύλλα, κάθε επικονιαστής φυτών από τις μύγες έως τα κολιμπρί, είναι ορισμένα προφανή παραδείγματα[3]. Αντίθετα, ο ανταγωνισμός μεταξύ των ειδών -π.χ. μεταξύ αρπακτικού και θηράματος- που ακολουθεί τις παραδοσιακές εξισώσεις Lotka-Volterra[4] είναι επίσης ένα γεγονός τεράστιας πλανητικής συνέπειας και τραύματος. Έχουμε την άποψη ότι η ανθρώπινη φαντασία είναι μια δύναμη που έχει ριζώσει επάνω σε μεταβατικά θεμέλια και που είναι απόλυτα ικανή να εφευρίσκει ένα ιδανικό για να υπερβαίνει την οικολογική θλίψη. Αυτή η ιδέα ωστόσο προτείνεται χωρίς την παρουσία αλγορίθμων που θα μπορούσαν να προβάλουν τις επιπτώσεις στον τρέχοντα φυσιολογικό μέσο όρο της βιοποικιλότητας γενικά. Αντί να προτείνουμε μια επίθεση σε έναν στατιστικά αμφισβητούμενο αστερισμό γενετικά τροποποιημένων οργανισμών και, κατ' επέκταση, ολόκληρων κοινοτήτων -μια μορφή επιθετικότητας που μοιάζει με έναν αστεροειδή που συγκρούεται- περιορίζουμε την ιδέα μας στις ανθρώπινες κοινότητες, κάτι που αποτελεί περισσότερο από πρόκληση.

Δεδομένου ότι ο *Podocerus* που καθαρίζει τα δόντια του φαλαινοκαρχαρία ζυγίζει ως και περίπου 6,389 γραμμάρια και ο φάλαινοκαρχαρίας έως και 9.072 κιλά, πρέπει να υποθέσουμε ότι η κλίμακα είναι άνευ σημασίας για την αξιολόγηση κάποιου ουσιαστικού μέτρου αίσθησης, συνείδησης, ευαλωτότητας, αθωότητας, ακόμη και αισθητηριακής εξουσίας. Είναι αναμφίβολο ότι και οι δύο οργανισμοί -ο καρχαρίας και τα αμφίποδα- αισθάνονται τον πόνο βαθιά συγκεντρωμένο στον εγκέφαλό τους και / ή σε νευρικές συστάδες σε άλλα μέρη του σώματός τους. Στην πραγματικότητα, οι γαρίδες και πολλές άλλες ομάδες καρκινοειδών αποτελούν το αντικείμενο εκτεταμένων δοκιμών σε αλγοϋποδοχείς.[5] Ο αριθμός των νευρώνων δεν σημαίνει τίποτα, όταν ασχολούμαστε με συγκρίσεις μεταξύ οργανισμών στο προαναφερθέν επίπεδο όλου.[6] Η ποσόστωση των νευρώνων ανά είδος έχει αποδειχθεί ότι αποτελεί ένα σύνολο αναντιστοιχιών που δεν βγάζει κανένα νόημα. Η βασική ομάδα του πολυκυτταρικού πλακόζωου τριχόπλαξ δεν εμφανίζει ούτε έναν νευρώνα. Με τις μηδενικές συνάψεις και "μόνο έξι τύπους κυττάρων" του, μπορεί ωστόσο να κατευθύνει τις εκκρίσεις του και να συντονίζει τη διατροφική συμπεριφορά στα φύκη, και αυτό υποδηλώνει "την παρουσία μηχανισμών για την κυτταρική επικοινωνία και ολοκλήρωση". Επιπλέον, αυτό το αξιοθαύμαστο πλάσμα διαθέτει "ομοιότητες με τους τοίχους των πεπτικών οδών των θηλαστικών...".[7]

Η ακτινοβολία των θηλαστικών, τα εξαφανισμένα και τα υπάρχοντα είδη, η εξέταση των μοριακών ρολογιών που παρέχουν πληροφορίες για τις "εκτιμήσεις απόκλισης-χρόνου", καθώς και η εφαρμογή των τελευταίων εξελίξεων στις "μοριακές μεθόδους" έχει επαναστατικοποιήσει ακόμη και αυτό το επίπεδο ταξινόμησης, στο οποίο όλοι ανήκουμε. Μόλις το 2018 τα θηλαστικά απέκτησαν τη δική τους "Βάση Δεδομένων Θηλαστικής Ποικιλότητας", πολύ καιρό μετά τη δημιουργία ψηφιακών αποθετηρίων για άλλες ομάδες οργανισμών. Δεν αποτελεί έκπληξη το γεγονός ότι την περίοδο 2005-2018 αναγνωρίστηκαν "1.251 νέα είδη (θηλαστικών)". Αυτή τη στιγμή, ο αριθμός έχει φτάσει στα "6.485" είδη θηλαστικών.[8] Ο συνολικός αριθμός των ειδών που βρίσκονται πάνω στη γη είναι εντελώς άγνωστος. Ωστόσο, με τη χρήση μιας ανάλυσης πολυμεταβλητής λογαριθμοκανονικής κατανομής[9] που συμπεριλαμβάνει τους κόσμους των μικροβίων, οι Kenneth J. Locey και Jay T. Lennon περιγράφουν "έναν ενοποιημένο νόμο κλιμάκωσης που προβλέπει την αφθονία των κυρίαρχων ειδών κατά μήκος 30 τάξεων μεγέθους στην κλίμακα όλων των μικροοργανισμών στη Γη. Χρησιμοποιώντας αυτόν τον νόμο κλιμάκωσης σε συνδυασμό με το λογαριθμοκανονικό μοντέλο βιοποικιλότητας, προβλέπουμε ότι η Γη φιλοξενεί μέχρι και 1 τρισεκατομμύριο (10^{12}) είδη μικροβίων".[10] Οι Locey και Lennon επισημαίνουν ότι όταν συζητάμε τους νόμους κλιμάκωσης στην οικολογία (που είναι λιγότερο κατανοητοί όταν μιλάμε για μικρόβια), "[…]Μεταξύ των πιο γνωστών είναι η κλιμάκωση του μεταβολικού ρυθμού... σύμφωνα με το μέγεθος του σώματος... και ο ρυθμός με τον οποίο ο πλούτος των ειδών... κλιμακώνεται ανάλογα με την περιοχή". Και ότι τέτοιοι τύποι "αποκαλύπτουν με ποιον τρόπο οι φυσιολογικοί, οικολογικοί και εξελικτικοί περιορισμοί ισχύουν στα γονιδιώματα, τα κύτταρα, τους οργανισμούς και τις κοινότητες με τεράστια ποικιλία στο μέγεθος".[11]

Ο αριθμός 1 τρισεκατομμύριο, που εξακολουθεί να μην εμπεριέχει όλους τους Άλλους, μας βάζει στον πειρασμό να συνδυάσουμε την έννοια του άπειρου με αυτήν της βιοποικιλότητας. Καμία από τις δύο δεν μπορεί να αντέξει την πολυτέλεια της αναγνωρίσιμης κλίμακας. Προσθέστε το *ένα* στην ιδέα και αυτό αποκτά έναν δελεαστικό, αν και ελάχιστα διακριτό αντίκτυπο. Προφέρετε αυτό το *ένα* δυνατά, και την ίδια στιγμή ένας άνθρωπος θα επινοήσει σύμβολα, επίθετα και άλλα χαρακτηριστικά για να το αναπαραστήσει. Ένα μεμονωμένο κύτταρο που μπορεί να διαιρείται, για παράδειγμα. Εάν αναγάγουμε αυτήν την αποδεικτική μοναδικότητα στο επίπεδο μιας ανθρώπινης κοινότητας, αμέσως η συζήτηση θα αφορά τις πολιτικές που πρέπει να αξίζουν σε αυτό *το ένα που είναι πέρα από το άπειρο*. Τα βλαστοκύτταρα; Πρέπει να κλωνοποιούνται; Ποια είναι η ταξινομική, κλαδιστική θέση τους σε κάποιο Δέντρο της Ζωής; Αν όμως αφήσετε κάτι άγραφο, χωρίς φωνή, και εντελώς έξω από τη σφαίρα της σκόπιμης ανθρώπινης ανακήρυξης, τότε, όπως λέει και ο Ρόμπερτ Κάπλαν, δεν είναι τίποτα περισσότερο ή λιγότερο από το "το τίποτα το οποίο υπάρχει".[12]

Ο Γερμανός μαθηματικός και φιλόσοφος Γκέοργκ Κάντορ (1845-1918), γνωστός κυρίως για τη θεωρία των συνόλων, επινόησε επίσης την έννοια των υπεραριθμήσιμων αριθμών, καθώς και το άπειρο των απείρων, στην παράδοση του τζαϊνιστή μαθηματικού Surya Prajnapti (περίπου 4ος-3ος αιώνας π.Χ.), που οραματίστηκε τρεις ιδιότητες του άπειρου: το σχεδόν, το αληθινά και το άπειρα άπειρο.[13] Στο Βιβλίο των Παροιμιών, μαθαίνουμε ότι "ο δρόμος ενός αετού" ξεπερνά κατά πολύ την ανθρώπινη κατανόηση. Δεν προαπαιτούμε την έννοια του απείρου για να το αποδείξουμε αυτό. Ούτε μπορούμε να δεχτούμε το απλό απόφθεγμα ότι η φτερούγα ενός πουλιού, όπως μπορούμε να την εξετάσουμε μορφολογικά, δεν αποτελεί τίποτα περισσότερο από την εξέλιξη της πτήσης που εγκολπώνεται από τον "αγώνα για ύπαρξη". Αυτή ήταν μια ιδέα που διατυπώθηκε μεταξύ των βρετανών ζωολόγων τέσσερα χρόνια μετά το τέλος του Α' Παγκοσμίου Πολέμου[14], κατά τη διάρκεια του οποίου καταρρίφθηκαν στις μάχες μεταξύ της Γερμανίας και των εθνών της Αντάντ 100.000 αεροπλάνα. Ήταν αυτός ο πόλεμος που προκάλεσε τον Ταξίαρχο Μπίλι Μίτσελ -ο οποίος είχε ηγηθεί των πρώτων αεροπορικών επιθέσεων στην ιστορία του πολέμου- να δηλώσει ότι "η κύρια δύναμη της άμυνας και η ισχύς της πρωτοβουλίας ενάντια σε έναν εχθρό έχει περάσει στον αέρα".[15] Αναγνωρίζουμε ότι τέτοιου είδους πολιτιστικά μιμίδια, καθώς η ανθρώπινη βία, μετατρέπονται εύκολα σε σύμβολα. Μπορεί να υλοποιηθούν με τη μορφή πτήσης, ή της ισχύος πυρός, ή των αυθαίρετων μαθημάτων ενός Πύργου της Βαβέλ ή κάποιας άλλης μηχανικής χίμαιρας. Εντούτοις, απομένει ακόμα προς ανακάλυψη ο βέλτιστος τρόπος με τον οποίο το είδος μας μπορεί να προετοιμάζεται -*ελλείψει* συμβολισμού- για αλληλεπίδραση με οποιοδήποτε *άλλο* είδος.

Μπορεί όλο αυτό να μοιάζει με άγονη γραμμή για συζήτηση, ωστόσο, θεωρούμε ότι αποτελεί τον κρίσιμο κοινωνικό, ψυχολογικό και υλικό άξονα από τον οποίο εξαρτάται ξεκάθαρα το μέλλον μας ως άτομα και ως είδος. Πώς μπορεί να διατηρηθεί η αλληλεξάρτηση ή η ανεξαρτησία της σκέψης και του συναισθήματος, και με τι σκοπό, μπροστά στη χρήση των επιφανειακών μας συμβολισμών για την κατάκτηση, ή τα παιδαριώδη μας παραδείγματα απεριόριστης ανάπτυξης, την αυτοϊκανοποίησή μας και την ανυπόφορη εκμετάλλευση των Άλλων? Η κατάκτηση της Σελήνης είχε γίνει ο απόλυτος ερευνητικός πόθος του είδους μας για πολλά χρόνια. Ύστερα, η ισχύς αυτής της επιθυμίας επισκιάστηκε από άλλες συγκλονιστικές επιθυμίες που έτρεφαν από κοινού τα αμέτρητα ήθη και όρια της μοναδικής ανθρώπινης εμπειρίας. Εάν κάποιο φωτεινό αντικείμενο στη σκοτεινή νύχτα, μια πανδημία ή η προσωπική φαντασίωση της απόκτησης μερικών εκατοντάδων δισεκατομμυρίων δολαρίων είναι απαραίτητα για την μερική τόνωση του ενδιαφέροντος του εγωιστικού είδους μας σε έναν κοινό σκοπό, τι άλλο θα μπορούσε να προσφερθεί από την πλευρά μιας εντελώς διαφορετικής κατηγορίας ηθικών μαγνητών; Βιολογικοί συντελεστές που μιλούν για τη μελλοντική διαιώνιση μιας σε μεγάλο βαθμό αδιατάρακτης βιόσφαιρας; Πόση ανακούφιση θα αντλούσατε ανακαλύπτοντας βαθιά στον πυθμένα του ωκεανού μια αόρατη, ακμάζουσα κοινότητα πυριτιόσπογγων της Ανταρκτικής? Θα δίνατε τη ζωή σας για αυτούς?

ΣΗΜΕΙΩΣΕΙΣ

[1] "A New Species of *Podocerus* (Crustacea: Amphipoda: Podoceridae) Associated with the Whale Shark *Rhincodon typus*," των Ko Tomikawa, Makio Yanagisawa, Takuo Higashiji, Nagisa Yano, Wim Vader, Species Diversity, Online ISSN: 2189-7301, Print ISSN: 1342-1670, 2019 Volume 24 Issue 2 Σελίδες 209-216, doi: https://doi.org/10.12782/specdiv.24.209, Βλ.: "[Research] New species of shrimp-like creatures discovered in a whale shark's mouth," Hiroshima University, Associate Professor Ko Tomikawa, https://www.hiroshima-u.ac.jp/en/news/55023, 10 Δεκεμβρίου 2019.

[2] Βλ. *Lichens Of North America*, των Irwin M. Brodo, Sylvia Duran Sharnoff, και Stephen Sharnoff, με επιλεγμένες εικόνες της Susan Laurie-Bourque, έκδοση σε συνεργασία με το Canadian Museum of Science, Yale University Press, New Haven, Connecticut, 2001, σελ. 52.

[3] Bascompte, J.; Jordano, P.; Melián, C. J.; Olesen, J. M. (2003). "The nested assembly of plant–animal mutualistic networks". *Proceedings of the National Academy of Sciences.* 100 (16): 9383–9387. Bibcode:2003PNAS..100.9383B. doi:10.1073/pnas.1633576100. PMC 170927. PMID 12881488. Βλ. επίσης, Ollerton, J. 2006. "Biological Barter": Interactions of Specialization Compared across Different Mutualisms. σελ. 411–435 στο: Waser, N.M. & Ollerton, J. (Eds) *Plant-Pollinator Interactions: From Specialization to Generalization.* University of Chicago Press. Βλ. επίσης Wright, David Hamilton (1989). "A Simple, Stable Model of Mutualism Incorporating Handling Time". *The American Naturalist.* 134 (4): 664–667. *doi:10.1086/285003.* Βλ. επίσης Bastolla, U.; Fortuna, M. A.; Pascual-García, A.; Ferrera, A.; Luque, B.; Bascompte, J. (2009). "The architecture of mutualistic networks minimizes competition and increases biodiversity". *Nature. 458 (7241): 1018–1020.* Bibcode:2009Natur.458.1018B. doi:10.1038/nature07950. PMID 19396144. Breton, Lorraine M.; Addicott, John F. (1992). "Density-Dependent Mutualism in an Aphid-Ant Interaction". *Ecology.* 73 (6): 2175–2180. doi:10.2307/1941465. JSTOR 1941465.

[4] "Evolution of predator-prey systems described by Lotka-Volterra equation under random environment," των Y. Takeuchi, N. H. Du, N. T. Hieu, και K. Sato, *Journal of Mathematical Analysis and Applications*, Volume 323, Issue 2, 15 Νοεμβρίου 2006, σελ. 938-957, https://doi.org/10.1016/j.jmaa.2005.11.009, Elsevier, *ScienceDirect*, https://www.sciencedirect.com/science/article/pii/S0022247X0501142X, https://www.jstage.jst.go.jp/article/specdiv/24/2/24_240213/_article/-char/en, Πρόσβαση στις 18 Δεκμβρίου 2020.

[5] Maldonado, H. & Miralto, A. (1982). "Effect of morphine and naloxone on a defensive response of the mantis shrimp (Squilla mantis)". *Journal of Comparative Physiology.* 147 (4): 455–459. doi:10.1007/bf00612010. S2CID 3013237. Βλ. επίσης, Hanke, J., Willig, A., Yinon, U. και Jaros, P.P. (1997). "Delta and kappa opioid receptors in eyestalk ganglia of a crustacean". *Brain Research.* 744 (2): 279–284. doi:10.1016/s0006-8993(96)01114-6. PMID 9027387. S2CID 26232087. Βλ. επίσης Casares, F.M., McElroy, A., Mantione, K.J., Baggermann, G., Zhu, W. and Stefano, G.B. (2005). "The American lobster, Homarus americanus, contains morphine that is coupled to nitric oxide release in its nervous and immune tissues: Evidence for neurotransmitter and hormonal signaling". *Neuro Endocrinol. Lett. 26 (2): 89–97.* PMID 15855878. Βλ. επίσης Lüschen, W., Buck, F., Willig, A. and Jaros, P.P. (1991). "Isolation, sequence analysis, and physiological properties of enkephalins in the nervous tissue of the shore crab Carcinus maenas L." *Proceedings of the National Academy*

of Sciences. 88 (19): 8671–8675. Bibcode:1991PNAS...88.8671L. doi:10.1073/pnas.88.19.8671. PMC 52571. PMID 11607217. Lozada, M., Romano, A. and Maldonado, H. (1988). "Effect of morphine and naloxone on a defensive response of the crab Chasmagnathus granulatus". *Pharmacology Biochemistry and Behavior. 30 (3): 635–640.* doi:10.1016/0091-3057(88)90076-7. PMID 3211972. S2CID 45083722. Βλ. επίσης Dyakonova, V.E. (2001). "Role of opioid peptides in behavior of invertebrates". *Journal of Evolutionary Biochemistry and Physiology.* 37 (4): 335–347. doi:10.1023/A:1012910525424. S2CID 23134631.

[6] Βλ. Calabrese, M. (2011). *Hierarchical-Granularity Holonic Modelling, Doctoral Thesis. University of Milan, Italy,* https://air.unimi.it/retrieve/handle/2434/155499/138859/ phd_unimi_R07647.pdf

[7] "Coordinated Feeding Behavior in *Trichoplax*, an Animal without Synapses," των Carolyn L. Smith, Natalia Pivovarova, και Thomas S. Reese, *PLOS ONE*, 2 Σεπτεμβρίου 2015, https://doi.org/10.1371/ journal.pone.0136098, https://journals.plos.org/plosone/article?id=10.1371/journal.pone.0136098, Πρόσβαση στις 19 Δεκεμβρίου 2020.

[8] Βλ. https://www.mammaldiversity.org/; *Illustrated Checklist of the Mammals of the World*, 2 Volumes, των Connor J. Burgin, Don E. Wilson, Russell A. Mittermeier, Anthony B. Rylands, Thomas E. Lacher, Wes Sechrest, και Albert Martinez-Vilalta, Marc Olivé, David M. Leslie, Jr., Εισαγωγή από Razan Al Mubarak, Texts based on the initial contributions from the authors of the *Handbook of the Mammals of the World* Volumes 1-9, Color plates by Toni Llobet (main artist), Ilian Velikov, Lluis Sogorb Faansie Peacock, Alex Mascarell, Francesc Jutglar, Blanca Martí, Stephen D. Nash and Jesus Rodriguez-Osorio, Lynx Edicions, Barcelona, 2020, Volume 1, σελ. 24.

[9] Johnson, Norman L.; Kotz, Samuel; Balakrishnan, N. (1994), 14: "Lognormal Distributions", Continuous univariate distributions. Vol. 1, Wiley Series στο *Probability and Mathematical Statistics: Applied Probability and Statistics* (2nd ed.), New York: John Wiley & Sons, ISBN 978-0-471-58495-7, MR 1299979

[10] "Scaling laws predict global microbial diversity," των Kenneth J. Locey και Jay T. Lennon, *PNAS* May 24, 2016, 113 (21) 5970-5975, Επιμέλεια David Michael Karl, University of Hawaii, https://www.pnas.org/content/ 113/21/5970, Πρόσβαση στις 19 Δεκεμβρίου 2020.

[11] ό.π.

[12] *The Nothing That Is – A Natural History of Zero*, των Robert Kaplan, Εικονογράφηση Ellen Kaplan, Oxford University Press, New York, NY, 2000.

[13] Ian Stewart, *Infinity: a Very Short Introduction*, Oxford University Press, New York, NY, 2017, σελ. 117.

[14] Βλ. *Birds In Flight*, του W. P. Pycraft, Εικονογράφηση Roland Green, Gay & Hancock Limited, London, 1922, σελ. 18.

[15] https://www.britannica.com/biography/William-Mitchell; Βλ. επίσης *Time Magazine*, 22 Ιουνίου 1942.

Κεφάλαιο 19

ΕΝΑΣ ΝΕΟΣ ΦΥΣΙΚΟΣ ΘΕΤΙΚΙΣΜΟΣ

Ο Δαρβίνος μίλησε για κάποιο "γενικό σχέδιο" που διέπει όλες τις συνήθειες και την οργάνωση της ζωής και επέμεινε ότι "τίποτα δεν μπορεί να είναι πιο απελπιστικό από το να επιχειρήσει κανείς να εξηγήσει αυτή την ομοιότητα του μοτίβου σε μέλη της ίδιας τάξης, χρησιμοποιώντας ως βάση εξήγησης τη χρησιμότητα ή το δόγμα των τελικών αιτιών". Η επιστημονική βάση πάνω στην οποία διαμόρφωσε την υπόθεσή του αφορούσε "τροποποιήσεις" και τροποποιητικές συσχετίσεις εντός του *οργανισμού*. Ενώ η εστίασή του σε αυτούς τους διαλογισμούς αφορούσε τη "μορφολογία", την "εμβρυολογία" και τα "υποτυπώδη όργανα", ποτέ δεν ήταν ανένδοτος στο θέμα της "ίδιας τάξης" και άφηνε εντελώς ανοιχτό τον τρόπο με τον οποίο έπρεπε να οριστεί ή να διακριθεί οποιαδήποτε οριστικότητα. Μάλιστα, το κεφάλαιο δεκατέσσερα στο βιβλίο του *Περί της προέλευσης των ειδών*[1] είχε τίτλο "Αμοιβαίες συγγένειες των οργανισμών".[2] Ο βιοφιλικός και φυσιολατρικός αυτός πυρήνας συνάδει επίσης σε μεγάλο βαθμό με αυτό που η Anne Buttimer έχει περιγράψει ως μια "αρχετυπική μεταφορά", τη φιλοσοφική ευελιξία που ενδυναμώνει οποιαδήποτε πιθανή ποικιλομορφία.[3] Η *οργάνωση*, όπως το ανέφερε ο Δαρβίνος, είναι αυτή η *σρχετυπική μεταφορά*. Ως εκ τούτου, μια κοινότητα θεωρητικά αντικατοπτρίζει το άτομο, στο συνδυασμένο φαινοτυπικό επίπεδο, εξακολουθεί όμως να μην είναι ο βαθμός της ζωής ενός ατόμου. Ως εκ τούτου, δεδομένου του απόλυτου πολλαπλασιασμού των διαφορετικών προοπτικών και των βιωμένων εμπειριών, ο John R. Stilgoe γράφει στο καταπληκτικό βιβλίο του *What Is Landscape?* ότι "Ομάδες ανθρώπων έχουν ορισμένες κοινές γενικές οπτικές γωνίες και συμφωνούν σε ορισμένες σκληρές απλότητες". Όλα τα άλλα όμως αποτελούν προσωπική υπόθεση: "Κάθε ερευνητής συνειδητοποιεί κάτι μοναδικό…" [4] Εκεί ενυπάρχει η φανταστική διαλεκτική, η ερμηνευτική των μυθοποιών, και τα σχόλια που βρίσκονται μέσα σε αυτά, αρθρώνουν τη γενετική προτίμηση, υποδηλώνουν την ολότητα και τη διάσπαση. Και τα δύο είναι απαραίτητα.

Οι μονοφυλετικές και απομορφικές διακρίσεις των κληρονομικών χαρακτηριστικών που προέρχονται από κοινούς προγόνους που χρονολογούνται από μία ή περισσότερες γενιές φέρουν διακριτά ταξινομικά όρια, ωστόσο το γένος και το είδος, ακόμη και ολόκληρες γενεαλογικές διαιρέσεις μπορεί να συγχωνευθούν. [5] Είναι ωστόσο πιο πιθανό να διαχωριστούν, ιδιαίτερα υπό το πρίσμα του καθαρού όγκου των φυλογενειών τους.

Ο παππούς του Δαρβίνου, Έρασμος Δαρβίνος (1731-1802), περιέγραψε άψογα τις διακριτότητες αυτές σε βαθμό που στον σχολιασμό του για το *Δοκίμιο προς μια νέα θεωρία της όρασης* ("Treatise on Vision, 1709) του επισκόπου Τζορτζ Μπέρκλεϊ (1685-1753) πρότεινε ότι "τα χρώματα που βλέπουμε, είναι μόνο μια γλώσσα..."[6] Έτσι ώστε να συμπεράνει ότι *τα πάντα* προσφέρουν κάποια γλώσσα, μοναδική στον εαυτό της. Η συνειδητοποίηση αυτή υπήρξε ο πυρήνας της αποκαλυπτικής μελέτης του Μωρίς Μαίτερλινκ (1862-1949) με τίτλο "Η Νοημοσύνη των λουλουδιών" όπου και έγραφε ότι "η ιδιοφυΐα της Γης…ψηλαφεί, διστάζει, διορθώνει τον εαυτό της κάθε φορά,… παλεύει ενάντια στην τεράστια, σκεπτική, σκοτεινή μάζα της ύπαρξής της. Δεν γνωρίζει περισσότερα από εμάς προς τα πού πηγαίνει, αναζητά και βρίσκει τον εαυτό της σταδιακά. Διαθέτει κάποιο ιδανικό…"[7]

Ένας εξίσου μεταφορικός συλλογισμός φτάνει στο επίπεδο της οικολογίας του πληθυσμού από αυτή τη συνεχιζόμενη αβεβαιότητα σε σχέση με το τι ακριβώς σημαίνει ή είναι ένα άτομο, σχετικά το τι αυτό προμηνύει. Η μορφολογία, τα μοτίβα, οι αριθμητικοί διαχωρισμοί, τα αναπαραγωγικά όρια, η δυναμική συνάθροισης, η βιοχημική κατασκευή και περιβάλλοντα παρουσιάζουν το καθένα τα δικά του ελαττώματα και ασάφειες εξεταζόμενα στη βάση του αυστηρού ορισμού. Το 1917 ο βοτανολόγος H.A. Gleason υποστήριξε "ότι τα είδη διέθεταν ατομικιστικά οικολογικά χαρακτηριστικά και συναθροίζονταν σε τοπικές κοινότητες με ένα μεγάλο στοχαστικό συστατικό σύμφωνα με τις ιδιομορφίες της διασποράς και τα διαθέσιμα περιβάλλοντα".[8] Στην περίπτωση των συμβατικών στοχαστικών διαδικασιών στις οποίες μια συνάρτηση περιλαμβάνει τουλάχιστον δύο μεταβλητές - ας πούμε, ένα άτομο και την κοινότητά του - η κατανομή οποιασδήποτε εκτιμώμενης σχέσης μεταξύ των δύο οντοτήτων, όταν μεταφραστεί στο επίπεδο του είδους, είναι αρκετή για να αναγνωρίσει κανείς μια μαθηματική αρχή αβεβαιότητας που είναι γεμάτη διακυμάνσεις. Το αν αυτό είναι κάτι τυχαίο ή αποτέλεσμα επιρροής κάποιου άλλου παράγοντα αποτελεί θέμα γραμματικής. Ο λευκός θόρυβος, οι συναρτήσεις κατανομής, οι ταλαντώσεις, οι αμοιβαίες πιθανότητες, οι "ιδιότητες διαδρομής", η σύγκλιση και η περιοδικότητα συνθέτουν το δίλημμα της εξακρίβωσης[9]. Οι μεταβλητές αυτές αντιπροσωπεύουν τις καμπύλες που βλέπουμε σε οποιονδήποτε τύπο γραφήματος. Η κάμψη είναι μια μετάβαση. Μπορεί να καθορίζεται γεωγραφικά για ένα συγκεκριμένο χρονικό διάστημα, ωστόσο η αλλαγή είναι θεμελιώδης. Κανείς δεν μπορεί να μπει στο ίδιο ποτάμι δύο φορές, κανένα σύννεφο δεν σχηματίζεται όπως τα προηγούμενα. Ο Ευκλείδης (μέσα 4ου αιώνα π.Χ.) θεωρούσε ότι η καμπύλη είναι το ίχνος που παραμένει μετά από κάποια φανταστική κίνηση, είτε καθορισμένη (πεπερασμένη) είτε απροσδιόριστη. Και στις δύο περιπτώσεις, η καμπύλη ορίζει κάποια επιφάνεια: την επιφάνεια, μια τοπολογία ή ένα τοπίο κινητών σημείων κατά μήκος των οποίων διέρχονται τα όντα, ή στο οποίο στέκονται αόριστα. Θα μπορούσε κανείς να φανταστεί μιγαδικά μηδενικά και πολυωνυμικές (πολλαπλές) ταυτότητες εντός του(ων) προτεινόμενου χώρου(-ων), αυτά όμως αποτελούν τα τεχνητά κλάσματα, εμβλήματα, αντιπροσωπευτικά στοιχεία και ταυτότητες στις οποίες εστιάζει η ανθρώπινη σκέψη, ανεξάρτητα από τη φευγαλέα ακεραιότητα.

Μπορούμε να τους εκχωρήσουμε ατομικούς ή ομαδικούς συσχετισμούς και ονόματα, τελικά όμως, όλο αυτό είναι κάτι αφηρημένο, είτε μιλάμε για μαθηματικά, φυσική, ή για κάποιον χώρο υπαρκτό ή ανύπαρκτο, εντός ή εκτός κάποιας διάστασης. Το θέμα είναι ότι, αν θέλουμε να *συσχετιστούμε* ως ζωντανά όντα με την εμπειρία αυτών των ταυτοτήτων, τότε θα πρέπει να διατυπώσουμε την αρχή της οικολογικής ανάκαμψης. Ένα βιολογικό δηλαδή σύνολο συντεταγμένων, σχέσεων και - το πιο σημαντικό – εννοιολογικών *δυνατοτήτων*.

Αυτό μπορεί να είναι μια θεμελιώδης ανταλλαγή εργασίας και αγαθών εντός ενός δεδομένου συστήματος.[10] Τελικά όμως, υπάρχει, όπως γράφει η Freya Mathews, "η μεγαλύτερη ηθική επείγουσα ανάγκη" στο ερώτημα του τι ακριβώς χαρακτηρίζει *τον εαυτό* και τους "εαυτούς";[11] Η απάντησή της, ή κάποια από τις πολλές απαντήσεις της, επιτυγχάνει, στο μυαλό μας, τεράστια ζωτικότητα, με έναν τρόπο όπως αυτός του Ανρί Μπεργκσόν ή του Άλμπερτ Σβάιτσερ. Γράφει σχετικά με τη "σημαντικότητα" ότι η "ενατένιση της υπέροχης Φύσης" ή όλες οι πολλές και διάφορες φιλότιμες προσπάθειές μας να "σώσουμε" τη Φύση, μπορεί να μην αποτελούν μια δύναμη τόσο ισχυρή όσο η βασική προσπάθεια που απαιτείται για "να διδάσκουμε τις καρδιές μας να εξασκούν την αυτοεπιβεβαίωση και αφυπνίζοντας την ικανότητά μας για την ενεργό, εξωστρεφή, παγκόσμια προσφορά αγάπης".[12]

Ο εαυτός, ο κόσμος, υποκειμενικός, αντικειμενικός: οι δυάδες είναι ευάλωτες σε κάθε χαρακτηρισμό, όμως αυτό δεν έχει καμία σημασία. Πλησιάζουμε όλο και πιο κοντά στην ουδετερότητα, ακόμα και όταν οι συνωμότες στο Παράδοξο του Ζήνωνα γνωρίζουν ότι δεν υπάρχει. Αυτό που αυτό σημαίνει για εμάς είναι ότι οι συνδέσεις μας δεν είναι, στην πραγματικότητα, *σταθερές*. Όλα έχουν να κάνουν με τον αριθμό, την πυκνότητα, την κατανομή, τη συνάρτηση πιθανοτήτων, τον χώρο στον οποίο οι ιδέες μας προσεγγίζουν αυτά τα μοναδικά ή πολλαπλά πεπρωμένα, τα χρονικά πλαίσια που εμπλέκουν αυτό που έχει ονομαστεί Κβαντικό Παράδοξο του Ζήνωνα.[13] Αφαιρούμε από αυτές τις ασάφειες την πραγματικότητα ότι η ανθρώπινη συμπεριφορά είναι άστατη, ολέθρια, αλλά οικολογικά ανοικτή, ανεξάρτητα από το πόσο τρομερά είναι τα γεγονότα του ορίζοντα που πλησιάζουν τη συλλογική μας ανοησία. Έτσι, η εναπομένουσα προοπτική, όπως περιγράφεται από τον Δρα Ράσελ Μιτερμάιερ και τους φίλους του, για τη "διατήρηση τουλάχιστον του 75% της χερσαίας βιοποικιλότητας του πλανήτη σε μόνο περίπου 8% της χερσαίας έκτασής του", η μέθοδος επιλογής βασίζεται στον προσδιορισμό των 36 γνωστών "κομβικών σημείων".[14]

Ανεξάρτητα από το πόσο εντατικές είναι οι οριοθετήσεις των ειδών και των οικοτόπων που κινδυνεύουν, το Ανθρωπικό Παράδοξο - αυτό που γνωρίζουμε μέσα από το δικό μας φακό, όχι αυτό ενός δέντρου – περιορίζει τα ερωτήματα και τις δεσμεύσεις μας στις ανησυχίες της ανθρωπότητας, οι οποίες φέρουν όλα τα ίδια αναπάντητα ερωτήματα που αποτελούν προϊόν των ατομικών/κοινοτικών αλληλεπιδράσεων, διασταυρώνοντας και δημιουργώντας κράματα σκέψης, γονιδίου, ενστίκτου και ψηφιδωτών της διαδικασίας. Η συνιστώσα της αρετής, αν και σύμμαχος του ανθρώπου και αυτοαναφορική, δεν αποτελεί ωστόσο κάποιο κλειστό οικοσύστημα στοχασμού. Το ξέρουμε γιατί ξέρουμε ότι δεν ξέρουμε τίποτα. Όλα τα γήινα και κοσμικά φαινόμενα τα οποία αποσπούν την προσοχή, με όποιον τρόπο και αν λαμβάνονται, δεν αγνοούνται. Για την επίτευξη του καλού δεν χρειάζεται να εξετάζουμε την κίνηση και τα κίνητρα υπό το πρίσμα κάποιου μαθηματικού παραδόξου. Μπορούμε και πρέπει να παρακάμπτουμε την επίπονη θεωρητική τοποθέτηση του τι είναι ή τι σημαίνει να είσαι άνθρωπος, και αντ 'αυτού, να επιδιώκουμε μια άγρια, συνεργατική πορεία της καρδιάς, μια αλλαγή πορείας, όσο μικρή και αν είναι η ικανότητα αντίληψης που κατευθύνεται προς αυτή την άγνωστη ποσότητα μηχανισμού, την ποιότητα μιας ψυχής.

Οι οικολόγοι κάνουν λόγο για σπάνια ή ελάχιστα γνωστά είδη στο πλαίσιο συστημάτων γνώσης που είναι γνωστό ότι έχουν κενά[15]. Οι κοινωνικές δομές, στα θηλαστικά ή σε οποιαδήποτε άλλη κατηγορία οργανισμών είναι τόσο μεταβλητές όσο το γεωγραφικό μέγεθος ή η διάνοια. Μπορεί να βρει κανείς συναισθηματικές ιδιοτροπίες στα Ιμαλάια, σε οποιαδήποτε ζούγκλα, θάλασσα, έρημο… παντού. Όπως έχει δηλωθεί σχ σχέση με τον Τζαϊνισμό, "Η οντολογία Jain Dharma δηλώνει ότι ψυχή μπορεί να έχουν όχι μόνο οι άνθρωποι αλλά οι ριπές του ανέμου, οι σπίθες της φωτιάς, τα τρεχούμενα νερά, τα φυτά, τα μικρόβια, οι βράχο, οι πέτρες και το χώμα, καθώς και τα έντομα, τα ερπετά, τα πουλιά, τα είδη που ζουν στη θάλασσα και τα θηλαστικά".[16] Ο πανθεϊστικός αυτός εναγκαλισμός αποτελεί τον πλέον πειστικό προσανατολισμό, ελλείψει σταθερής κατανόησης, ένα προσανατολισμό αντάξιο του φάσματος της βούλησής μας. Βρίσκεται στην καρδιά του αμοιβαίου αλτρουισμού σε πλανητική κλίμακα, εμποτίζοντας κάθε κλάδο με μια ανθρώπινη προβλεψιμότητα που μπορεί να κατευθύνεται προς την συντροφικότητα.

Αυτό που μας ανησυχεί είναι ο τρόπος με τον οποίο αυτή η *κατεύθυνση* θα μπορούσε πραγματικά να επηρεάσει τις ηθικές εικασίες στην καρδιά του φυσικού θετικισμού.

ΣΗΜΕΙΩΣΕΙΣ

[1] *On The Origin Of Species by Means of Natural Selection, or the Preservation of Favoured Races in the Struggle for Life*, With a New Preface by Charles G. Darwin, Illustrated with Wood Engravings by Paul Landacre, A Limited Edition. The Easton Press, Westport, Connecticut, σελ. 397.

[2] ό.π.., σελ. 376-418.

[3] *Geography and the Human Spirit*, της Anne Buttimer, Πρόλογος Yi-Fu Tuan, The Johns Hopkins University Press, Baltimore, MD, 1993, σελ. 85.

[4] *What Is Landscape?* του John R. Stilgoe, The MIT Press, Cambridge, Massachusetts, 2015, σελ. 17.

[5] Βλ *Podani, János (2010).* "Taxonomy in evolutionary perspective. An essay on the relationships between taxonomy and evolutionary theory". *Synbiologia Hungarica. 6: 1–42.*

[6] *The Botanic Garden Part II Containing The Loves Of The Plants, A Poem*, With Philosophical Notes, 3rd Edition, J. Johnson, London, 1791, σελ. 131.

[7] In *Hours Of Gladness*, του Maurice Maeterlinck, Μετάφραση Teixeira De Mattos, Εικονογράφηση E. J. Detmold, George Allen & Co., Λονδίνο, 1912, σελ. 94.

[8] Βλ. H. A. Gleason, "The structure and development of the plant association," Bulletin of the Torrey Botanical Club., 43, 463-81, ανάλυση του Robert P. McIntosh στο *The Background of Ecology – Concept and Theory*, Cambridge Studies In Ecology, Cambridge University Press, 1985, σελ. 265.

[9] "Stochastic Processes," στο http://physics.gu.se/~frtbm/joomla/media/mydocs/ LennartSjogren/kap4.pdf, του Lennart SJÖGREN, Associate professor (retired) of University of Gothenburg,Sweden; Βλ. επίσης "Non-equilibrium Stochastic Processes in Physics, *Chemistry, and Biology*, του T. Löfwander, 27 Μαΐου 2020.B. Mehlig

[10] Βλ. για παράδειγμα "Labour relationships and the role of reciprocity" από το "The Andean economic system and capitalism," του Rodrigo Sánchez, στο *Ecology and Exchange in the Andes*, Επιμέλεια David Lehmann, Cambridge University Press, Cambridge, UK, 1982, σελ. 167-183.

[11] *The Ecological Self*, της Freya Mathews, Barnes & Noble Books, Savage, Maryland, 1991, σελ. 129.

[12] ό.π. 160.

[13] W.M.Itano; D.J. Heinsen; J.J. Bokkinger; D.J. Wineland (1990). "Quantum Zeno effect," *Physical Review A*. 41 (5): 2295–2300. Bibcode: 1990PhRvA..41.2295I. doi:10.1103/ PhysRevA.41.2295. PMID 9903355.

[14] Βλ. "Forests And Primates Worldwide," Κεφάλαιο 9, στο *Saving Endangered Species – Lessons in Wildlife Conservation from Indianpolis Prize Winners*, Επιμέλεια Robert W. Shumaker, Εισαγωγή Harrison Ford, Johns Hopkins University Press, Baltimore, MD, 2020, σελ. 292. Βλ. επίσης www.hotspots-thefilm.com

[15] Βλ. "Social Considerations" του John D. Peine, Κεφάλαιο 9 στο *Conservation of Rare or Little-Known Species – Biological, Social, and Economic Considerations*, Επιμέλεια Martin G. Raphael και Randy Molina, Εισαγωγή Nancy Molina, Island Press, Washington-Covelo-London, 2007, σελ. 236.

[16] Σχόλιο του Sulekh Jain, Βλ. *Ecological Tragedy, Memory and the Birth of Hope – A Special MAHB Feature - An Ultimate Collision*, του M. C. Tobias, https://mahb.stanford.edu/library-item/ecological-tragedy-memory-and-the-birth-of-hope-a-special-mahb-feature-from-michael-charles-tobias/ , Πρόσβαση στις 19 Δεκεμβρίου 2020.

Άγνωστα άτομα με Σαμάνο, σε κανό, περίπου 4500 π.Χ., Tamaulipas, Μεξικό. Φωτογραφία © M. C. Tobias.

Κεφάλαιο 20

Μη Ασυμπτωτικες Λυσεις Οριων

Πρόκειται για μια στροφή η οποία θα μπορούσε να παρομοιαστεί με την υψίστης σημασίας κατηγορία της βιοκυβερνητικής, την αναγνώριση σχέσεων ανατροφοδότησης ("αντανακλαστικές", "συμμετρικές", "μεταβατικές" κ.λπ.)[1] που ενσωματώνουν συστήματα αντί να τα απομονώνουν μόνιμα. Αυτό θα μπορούσε να καθοδηγήσει την πιο αποκαλυπτική διαίρεση στην εξελικτική ανάλυση: την καλλιέργεια ορίων, διαιρέσεων, αντιθέσεων και δυάδων των οποίων οι συμπεριφορικές αντιδράσεις καλούνται να διαιωνίσουν τη διαφορά, να εγκαταλείψουν το διάλογο, να καταργήσουν τους συνασπισμούς. Αντίθετα σε αυτό το μεταβατικό κάλεσμα του να προαναγγείλουν όλες τις ενώσεις, τις συγχωνεύσεις, τις συλλογικότητες και τη συνεργασία. Η τελευταία συμφωνία καλεί για μια εκτίμηση της ελλειπτικής επιστροφής[2] (σε αντίθεση με μια αυστηρή αλλά απίθανη καθαρή κυκλικότητα) που έχει εισάγει πρωτόκολλα, προτεραιότητες, τη βαρύτητα των επιταγών στην καθημερινή μας ζωή.[3] Περιλαμβάνει μη ασυμπτωτικές λύσεις ορίων. Ως εκ τούτου, πρόκειται για ένα εξελικτικό μοντέλο του οποίου οι προγνωστικές συγκλίσεις προϋποθέτουν ένα όραμα που έχει τουλάχιστον στοιχειωδώς εντοπίσει ιδιαιτερότητες σε αυτήν την παγκόσμια συνένωση της αμοιβαίας αιτιότητας.

Εκτείνεται πέρα από τα πλανητικά όρια. Τα χιονογλάρονα, για παράδειγμα, ξοδεύουν μεγάλο μέρος της ενέργειάς τους μεταναστεύοντας για να επωφεληθούν από το καλοκαιρινό φως του ήλιου, από την Αρκτική στην Ανταρκτική.[4] Το άστρο αυτού του ηλιακού συστήματος τροφοδοτεί σχεδόν όλη την παραγωγή τροφής στη γη, υπενθυμίζοντάς μας ότι οι συνέπειες των ανθρώπινων πράξεών μας δεν είναι περισσότερο σημαντικές από τον πιο ταπεινό φωτοαυτότροφο οργανισμό που ασχολείται με την εξάπλωση του οξυγόνου, τη μετατροπή της χημικής ενέργειας σε υδατάνθρακες, την εκπομπή και τη διατήρηση των επιπέδων οξυγόνου, δημιουργώντας την πιο σημαντική διαδικασία στη γη: τη μετατροπή της ενέργειας του φωτός μέσω των πρωτεϊνών που βρίσκονται μέσα στους χλωροπλάστες, των οποίων βασικοί ξενιστές είναι όλα αυτά τα φύλλα, των οποίων η ηλικία χρονολογείται σε τουλάχιστον 1,6 δισεκατομμύρια χρόνια.[5] Δεν γράφουν επιτάφιους.

Αναγνωρίζουμε ότι κάθε αριθμός, κάθε υποτιθέμενο *αποδεικτικό στοιχείο*, κάθε προοπτική και τμήμα ανάλυσης, βασίζονται στην ανθρώπινη προκατάληψη και στις αυθαίρετες μεθόδους μετρήσεών μας. Η τεράστια μάζα των κριτηρίων τα οποία συγκεντρώνονται για τη διασφάλιση της εμπιστοσύνης σε όλα τα ισχυριζόμενα γεγονότα μπορεί να παραμερίζεται χωρίς αιτιολογία, ή να παραδίδονται στην αποθέωση της επιστημονικής νομιμότητας. Αυτοί οι καθοριστικοί παράγοντες της πραγματικότητας αποτελούν τους συντελεστές συναίνεσης ενός αμφίβολου κοινοβουλίου, ολοκληρωτικά πιστού στον εαυτό του και στην απλοϊκή ιδέα των συνολικών πολιτισμών του. Τα μαθηματικά παράδοξα είναι κόσμοι οι οποίοι βρίσκονται μέσα σε άλλους κόσμους οι οποίοι είναι παγιδευμένοι σε ακόμα περισσότερους κόσμους. Για παράδειγμα, θα μπορούσε κανείς να θέσει το εξής δίλημμα: Εάν ένας επιβάτης ο οποίος κάθεται στο πίσω μέρος ενός αεροσκάφους που βρίσκεται σε πτήση περπατήσει προς την μπροστινή περιοχή του αερομεταφορέα, πρόκειται για το ίδιο αεροπλάνο; Μπορούμε από περίσσεια πρακτικότητας να απορρίψουμε τέτοιου είδους ζηνωτικές λεπτολογίες, παρότι αφορούν έναν αριθμό διασταυρούμενων τομέων, ο καθένας με πιθανές συνθέσεις, αρχές ασάφειας. Μπορούμε ακόμη και να διαστρώσουμε τις αρχές αυτές με πλήρη εύρη ασάφειας και να επιδοκιμάσουμε τις αναλύσεις. Για λόγους απλότητας, ας χαρακτηρίσουμε αυτό το συγκεκριμένο είδος σεναρίου ως *παρομοίωση πρώτης τάξης*. Τέτοιες παραβολές βρίσκονται διάχυτες σε όλο το έργο του Αισώπου και των διαδόχων του. "Ο κόρακας που ήθελε να μοιάσει του αητού", του Λα Φονταίν. "Ο χωρικός και κάμποσοι αρουραίοι", του Πιλπάι. "Το τσακάλι, η λεοπάρδαλη και άλλα θηρία", του Τζον Γκέι. Οι διάφοροι παράδεισοι του Αγίου Αυγουστίνου", του Ισίδωρου, του Τζον Μάντεβιλ, του Αγίου Μπρένταν, του Ουόλτερ Ράλεϊ κ.λπ.[6]

Είναι όμως η *δεύτερης τάξης παρομοίωση* εκείνη που κλείνει τη μη ασυμπτωτική θεωρία και είναι αντιαναλυτική, ή, θα μπορούσε να προσθέσει κανείς, αντιαπαθής. Αυτό σημαίνει ότι, αντί να ζοριζόμαστε να φτάσουμε κούτσα-κούτσα στη γραμμή τερματισμού του Ζήνωνα, χωρίς ποτέ (θεωρητικά) να φτάνουμε εκεί, η λύση αυτού του σεναρίου της χελώνας και του λαγού, ενός από τα δέκα παράδοξα που πρότεινε ο μεγάλος μαθητής του Παρμενίδη, μπορεί να βρεθεί από ένα μάλλον μη επιστημονικό ερέθισμα που αφορά την ίδια την ιδέα ενός "υπεπεπερασμένου άπειρου αδιαίρετων μερών". Για το οποίο ο Αριστοτέλης διακήρυξε μια απλή λύση: "υπάρχουν μόνο δυνητικά άπειρα".[7] Αυτός ο ισχυρισμός απέναντι στο Παράδοξο του Αχιλλέα και το Παράδοξο της Διχοτομίας, έχει ανανεωθεί μετά από δύο χιλιετίες από την *Τυπική Λύση*, όπως περιγράφεται περιστασιακά, που σημαίνει ότι μπορεί κανείς πραγματικά να παρακάμπτει ακολουθίες που συνεχίζονται για πάντα (ένα άπειρο τμημάτων μήκους μερικών εκατοστών, για παράδειγμα) μέσα σε οποιοδήποτε πεπερασμένο χρονικό πλαίσιο.[8] Με άλλα λόγια, οι πρακτικές εκτιμήσεις είναι όντως σχετικές για την επίτευξη κάποιου αποτελέσματος, πολλώ δε μάλλον αν κάποιος θέλει να μετακινηθεί από τη μια μεριά του χαλιού στην άλλη ή να αγγίξει ένα δέντρο. Και με αυτό το απλό συμπέρασμα στον θεωρητικό πουαντιγισμό, μπορεί κανείς να συνεχίσει, χωρίς βροχές μετεωριτών ή ηλιακές κηλίδες λογικής.

Πρόκειται, χωρίς πολλές φανφάρες, για μια τεράστια καινοτομία. Σκεφτείτε τον αξιοσέβαστο Τσέζαρε Μπεκαρία (1738-1794), του οποίου η πραγματεία με τίτλο *Περί Εγκλημάτων και Ποινών* (1764) έθεσε στην ουσία τις βάσεις για τη διανεμητική δικαιοσύνη, στην οποία βασίστηκαν σε μεγάλο βαθμό οι ιδρυτές της Αμερικής.. Πρόκειται για ένα νομικό πρότυπο για τη συσχέτιση αιτίου και αποτελέσματος το οποίο απέδιδε απόλυτη προτίμηση στην αμοιβαιότητα και ξεκινούσε με μια προϋπόθεση: το τεκμήριο της αθωότητας. Με όρους του τομέα της οικολογίας, παραδέχεται ότι κάθε αθωότητα είναι δυνατή και ότι οι νόμοι θα πρέπει να χρησιμεύουν ως προοίμιο αυτής της γενναιόδωρης προσέγγισης της φύσης.

Αναφερθήκαμε στις παρομοιώσεις πρώτης και δεύτερης τάξης, στην αναγνώριση της διαίρεσης και της ενότητας, όπως εκδηλώνονται στον τρόπο που ερμηνεύουμε τη φυσική επιστήμη, τα γεγονότα, τις ιδιαιτερότητες. Υπάρχει όμως και μια εξαιρετικά σημαντική *παρομοίωση τρίτης τάξης* η οποία αντιστρέφει την υποτιθέμενη απροσδιοριστία ανάμεσα σε αντικείμενα, αισθητά, μη αισθητά, αφηρημένα, φανερά, ιδέα, ιδανικό. Δεν πρόκειται για Τάξη, Νόμο, Ισχύ ή άλλη επίφαση κανονικότητας. Αντίθετα, αυτή η παρομοίωση τρίτης τάξης προμηνύει την πραγματική οικολογική λύση και είναι εξαιρετικά σημαντική για την αποδοχή ορισμένων αρχών, έστω κι αν αυτές προτείνονται από συγκεκριμένο είδος. Αν μπορούμε να φανταστούμε μια *μετα-Ανθρωπική, διαειδική μεταφορά πρώτης τάξης* (PATSM, για λόγους απλότητας), τότε μπορούμε κάνουμε αποδεκτή την τυφλή αλλά καθόλα πρακτική εξίσωση που μοιάζει ως εξής: (patsm) x 1... = 1+?, όπου το "?" ισοδυναμεί με οποιοδήποτε αντικείμενο προτίμησης, στοργής, ενδιαφέροντος, διαδραστικότητας (οι λέξεις καθίστανται επουσιώδεις εντός ενός άνευ ορίων μεταφορικού χώρου). Αυτό θα περιλάμβανε, αναγκαστικά, παρατήρηση και σκέψη, με έμφαση στην υλοποιήσιμη ιδέα ότι αλληλεπιδρώντας με τον κόσμο, η αρχή "1+?" επιτρέπει οποιαδήποτε σύνδεση, υπερβαίνοντας την εξουθενωτική άπειρη σειρά με πραγματικές εκδηλωμένες σχέσεις, όπως και αν τις ορίσουμε: κοινωνικές, μοναχικές, ενάρετες ή άλλες.

Τουλάχιστον έχουμε κατοχυρώσει ότι το *patsm* επιτρέπει 1+ οποιοδήποτε άλλο Ον ή κοινότητα. Ομολογουμένως αδέξιο στην προσπάθειά του να περιγράψει αυτό που δίνει τη δυνατότητα στον *H. sapiens* να βγει από την απομόνωσή του, αποτελεί ένα χρήσιμο εργαλείο για να κατανοήσουμε πώς μπορούμε να αναγάγουμε μια ιδέα σε ιδανικό και να δράσουμε σύμφωνα με αυτό. Πρώτα ως γλωσσικό μοντέλο, μια βαθιά κατασκευή εκπλήρωσης επιθυμιών, ένα λεξικό ιδανικών, και στη συνέχεια ως η συμπιεσμένη ευθυγράμμιση βασικών εμπειρικών θεμελίων για την συμπλήρωση των κενών μεταξύ ιδανικών και παρατηρήσεων. Οι προκύπτουσες δυνατότητες δεν αποτελούν θέαμα, ούτε αυθόρμητες στιγμές επιφοίτησης, παρά μόνο τις σιωπηρές συνειδητοποιήσεις της δυνατότητάς μας να συνδεθούμε με το οτιδήποτε, σε οποιοδήποτε επίπεδο, ανά πάσα στιγμή, με έναν φυσικά θετικιστικό, ενάρετο τρόπο.

Κανείς δεν προτιμά τη χρήση απλών εξισώσεων ως αποδεκτή (ή κομψή) μεθοδολογία για την επεξήγηση συμπεριφορών. Αυτό είναι ακόμη πιο πολύπλοκο όταν υποθέτουμε ότι αποδεχόμαστε πολλαπλές μορφές ζωής, για τις οποίες δεν γνωρίζουμε σχεδόν τίποτα. Το τίποτα που προωθεί τον εαυτό του είναι, κυριολεκτικά, μια ανούσια άσκηση ματαιότητας. Αναγνωρίζοντας όμως αυτή τη ραγδαία αυξανόμενη ανοησία, μπορούμε να εισέλθουμε σε οποιοδήποτε πεδίο αριθμών θέλει κανείς να ορίσει. Δεν χρειάζεται να μπούμε σε αυτήν τη διαδικασία. Μπορούμε να επιλέξουμε να παρακάμψουμε την ασχήμια, την εχθρότητα, τη βλακεία και ολόκληρους πολιτισμούς του παρελθόντος, αν θέλουμε. Και συνήθως το κάνουμε. Υπάρχουν άπειρες θεωρίες που το εξηγούν και αυτό.

Όμως αυτό το *patsm* είναι ενδιαφέρον, όπως και η απλή προσθήκη του "ενός" στο άπειρο. Ο στόχος της βύθισής μας μέσα στη λάσπη τέτοιου είδους επιτηδευμένων περιφράσεων εξαρτάται από την αποθαρρυμένη επιθυμία να παρουσιάσουμε μια αλήθεια που ελλοχεύει μέσα στις μεταφορές και τις παρομοιώσεις της ανθρώπινης εμπειρίας· δεν πρόκειται για μοναδική αλήθεια παρά για μια κατηγορία αληθειών που είναι διαθέσιμες σε εκείνους που θα επωφεληθούν από μια τέτοια δυνατότητα. Την ηθολογική δηλαδή συνδεσιμότητα που μπορεί να φιλτραριστεί από όλα όσα συνθέτουν ένα άτομο. Αντί απλώς να επιδοκιμάζουμε το ηθικό, ή το πνευματικό, ή τη γνωστικό... ας δημιουργήσουμε κάποια συνδυαστική δύναμη η οποία, λόγω της ύπαρξής της σε οποιαδήποτε βαθμίδα στο νευρωνικό, υποσυνείδητο και συνειδητό οικοδόμημα που τυχαίνει να είναι η ατομική μας ζωή, δημιουργεί μια εξίσου συνδυαστική συμμαχία, συμμαχίες, μεταξύ Εαυτών και Κοινοτήτων κάθε κλαδιστικής, ταξινομικής, φυλογενετικών κληρονομιών και μελλοντολογίας. Αυτός ο γλωσσοδέτης έχει ως σκοπό την παράθεση του συνόλου πολύ πιο απλών ρητών που έχουν ενισχύσει τις προοπτικές της ανθρωπότητας από τις απαρχές της, παρά την τόση αρνητική εγκληματολογία.

Πιο πρόσφατα, βλέπουμε αυτή την αρχέγονη κοινοτοπία αραδιασμένη στις κολομβιανές ζούγκλες του Σέρο Ασούλ νότια της Μπογκοτά, ζωγραφισμένη σε οκτώ μίλια βράχου στην Σεράνια Λα Λιντόσα (γεωγραφικό διαμέρισμα Γκουαβιάρε) πριν από 12.600 έως 11.800 χρόνια. Σε αυτήν απεικονίζονται αρκετά από τα τελευταία είδη και διάφορα υβρίδια της Εποχής των Παγετώνων, καθώς και μια σκηνή που υποδηλώνει την ύπαρξη μιας ομάδας ανθρώπων που λάτρευαν τα μεγάλα ζώα.[9] Έχουμε ερευνήσει προσωπικά άλλες μεγάλες εκτάσεις με πετρογλυφικά και βραχογραφίες του Ύστερου Πλειστόκαινου και του Ολόκαινου στη νοτιοδυτική Αμερική, την Καλιφόρνια, την περιοχή Μπούργκος στο Ταμαουλίπας (Μεξικό), τη νοτιοανατολική Γαλλία και το νότιο Σουρινάμ. Όμως τα θεαματικά ευρήματα στο Σέρο Ασούλ, το Λιμοντσίλος και το Σέρο Μοντόγια φαίνονται χωρίς εξαίρεση να είναι τα πιο αποκαλυπτικά ηθολογικά έργα τέχνης που έχουν ανακαλυφθεί ποτέ, σίγουρα μετά από εκείνα του σπηλαίου Λασκώ (τα οποία βρέθηκαν το 1940) και του σπηλαίου Σοβέ (τα οποία ανακαλύφθηκαν το 1994).[10] Σχεδόν αμέσως η UNESCO πρόσθεσε αυτές τις τοποθεσίες – οι οποίες αποτελούν μέρος του Εθνικού Πάρκου Chiribiquete της Κολομβίας - στον κατάλογο των Μνημείων Παγκόσμιας Κληρονομιάς.[11] Δεν θα αποκρυπτογραφήσουμε ποτέ τους κώδικες που εκτίθενται τόσο πλούσια και εμφατικά στην επιφάνεια αυτών των πολύχρωμων λείων βράχων, για τον ίδιο ακριβώς λόγο, όπως περιγράφει η Melody Jue στο βιβλίο της *Wild Blue Media*, για τον οποίο είμαστε, στις μέρες μας, "ενσαρκωμένοι παρατηρητές".[12] Αυτός ο εγκλωβισμός στο *παρόν* μας αποδίδει μια εκπληκτικά σαφή κρίση της προκατάληψής μας. Ο Αντρέ Λερουά-Γκουράν το αποκαλεί "το παλαιολιθικό παράδοξο" [το οποίο] "συνίσταται σε σημαντικό βαθμό στο γεγονός ότι οι εικόνες αριστοτεχνικής τεχνικής και ελκυστικής μορφής μπορεί να φαίνεται ότι δεν αντιπροσωπεύουν τίποτα με καμία συνοχή.". Ο Λερουά-Γκουράν το αποκαλεί "το μη παραστατικό". Και διευκρινίζει: "Για το σύγχρονο μάτι, τα σμήνη των ζώων και των σημείων δεν εκφράζουν παρά αποσπασματική δράση, και δεν μπορεί να εντοπιστεί καμιά αφήγηση. Μπορούμε να αναρωτηθούμε αν αυτό συμβαίνει απλώς επειδή δεν μπορούμε να διαβάσουμε...".[13]

Η ιστορία της Σεράνια Λα Λιντόσα δεν είναι λιγότερο εμφατική από οποιαδήποτε παραδεισένια σκηνή του Ρούλαντ Σάβεραϊ ή του Γιαν Μπρίγκελ του Πρεσβύτερου, ούτε λιγότερο αφηγηματική από τη "Γκερνίκα" του Πικάσο. Είναι απίθανο να ανακαλύψουμε τι συνέβη πριν από πάνω από 12.000 χρόνια σε αυτές τις ζούγκλες της Κολομβίας. Μπορεί κάλλιστα να μην συνέβη *τίποτα* το ασυνήθιστο. Ίσως θα έπρεπε απλώς να αναρωτηθούμε: τι συνέβαινε από αιώνα σε αιώνα; Σε αυτή την εικοτολογία βλέπουμε τον κόσμο εκείνων των καλλιτεχνών που εμπνεύστηκαν να δημιουργήσουν αυτά τα υπαίθρια μουσεία. Δεδομένου του ύψους και της προβολής τους, όντως η λέξη *μουσείο* μπορεί να αποτελεί κατάλληλο χαρακτηρισμό. Αν ισχύει αυτό, τότε αισθανόμαστε αμέσως μια σχεσιακή απήχηση μεταξύ των ανθρώπων του 21^{ου} αιώνα και των προγόνων μας 480 γενιές πριν. Δεν πρόκειται για σχέση η οποία διακατέχεται από άγνοια. Θα μπορούσε κανείς να υποστηρίξει ότι παρατηρείται ένα αδιευκρίνιστο, ασαφές προφητικό νόημα, ας πούμε, στο έργο του Τζάκσον Πόλοκ "#1 Νο 5, 1948".[14] Υπάρχουν σύγχρονες ιστορίες στις οποίες διαβάζουμε αυτό που θέλουμε να αποκρυπτογραφήσουμε. Πρόκειται για έναν συνειδητό τρόπο λειτουργίας της φαντασίας που υπερβαίνει την αντικειμενική κατάληψη και είναι θεμελιώδης για κάθε τέχνη. Ακόμα συζητάμε για την σπασμένη κολόνα στην "La Tempesta" (1506/1508) του Τζορτζόνε ή για τους πίνακες του Πιτ Μοντριάν που είναι συνθέσεις από κόκκινο, κίτρινο και μπλε χρώμα.[15] Για τους περισσότερους μη-φυσικούς, το E=mc2 δεν μπορεί να *διαβαστεί*.[16] Αμφιβάλλουμε πολύ σοβαρά ότι το 0,00001 τοις εκατό όλων των ζωντανών ανθρώπων μπορεί να διαβάσει (να εξηγήσει) λεπτομερώς πώς ακριβώς λειτουργεί μια Κεντρική Μονάδα Επεξεργασίας, η τηλεόραση, ο Διεθνής Διαστημικός Σταθμός, το εμβόλιο κατά του Covid ή μια λάμπα.

Προσπαθώντας ωστόσο να *διαβάσουμε* ζωγραφισμένες ιστορίες 10.000 ή 14.000 χρόνων πριν, μπορούμε να είμαστε λογικά βέβαιοι ότι τα ένστικτα που προκαλούνται στα δικά μας πλαίσια λογικής, η αισθητηριακή δύναμη της πειθούς, ανεξάρτητα από τα πραγματικά σκαλοπάτια μιας σκάλας ή το μήκος του λαιμού ενός καμηλοειδούς που απεικονίζεται στους εξωτερικούς μουσειακούς τοίχους ενός βράχου της ζούγκλας, μπορούν να αφομοιωθούν στο σημερινό μας ψυχισμό. Διστάζουμε να ερμηνεύσουμε μόνο και μόνο επειδή γνωρίζουμε ότι βλέπουμε έναν πολιτισμό του παρελθόντος, ο οποίος λάτρευε άλλα είδη, ενώ ο δικός μας όχι. Αυτή η ιθαγενής σύγκρουση, όπως αντηχεί σήμερα σε ολόκληρο τον ανθρώπινο κόσμο, είναι σοκαριστική, επώδυνη και τόσο ανησυχητική ώστε να διαλύει την αυτοπεποίθησή μας στα πιο θεμελιώδη επίπεδα. Η πρόκληση που διατρέχει χιλιετίες *πρέπει* να απορροφηθεί. Όποιος έχει περάσει τη νύχτα αγκαλιά με ένα γιακ, κοιμισμένος κάτω από τον έναστρο ουρανό ή μια γιγαντιαία σεκόγια, γνωρίζει ότι αυτά τα κατά τα άλλα αμέτρητα περιθώρια λάθους μπορούν να περιοριστούν.

Marieta van der Merwe και φίλος της στο Καταφύγιό της στο Harnas στη Ναμίμπια. Φωτογραφίες © M. C. Tobias.

Σημειωσεις

[1] “Cause and Effect,” του John Denker, https://www.av8n.com/physics/causation.htm, 2003. Βλ. επίσης “Decision-Theoretic Foundations for Causal Reasoning,” των David Heckerman και Ross Shachter, *Journal of Artificial Intelligence Research* 3 (1995) 405-430, Δεκέμβριος 1995, https://arxiv.org/pdf/cs/9512104.pdf

[2] Jason Merchant, *The syntax of silence: Sluicing, islands, and the theory of ellipsis.* Oxford: Oxford University Press, Oxford UK 2001.

[3] “Systems biology ... is about putting together rather than taking apart, integration rather than reduction. It requires that we develop ways of thinking about integration that are as rigorous as our reductionist programmes, but different. ... It means changing our philosophy, in the full sense of the term.” Denis Noble, (2006). *The music of life: Biology beyond the genome*, Oxford University Press, Oxford, UK, σελ. 176.

[4] “To the Ends of the Earth,” των Tara Ramroop και Kara West, 16 Ιανουαρίου 2011, *National Geographic*, https://www.nationalgeographic.org/article/ends-earth/

[5] “Oldest plant-like fossils discovered are 1.6 billion years old,” του Mattias Karisson, *NewScientist*, 14 Μαρτίου 2017, https://www.newscientist.com/article/mg23331172-900-oldest-plantlike-fossils-discovered-are-16-billion-years-old/, Πρόσβαση στις 22 Δεκεμβρίου 2020.

[6] Βλ. *The Book of Legendary Lands*, του Umberto Eco, Μετάφραση του Alastair McEwen, Rizzoli ex libris, New York, 2013, σελ. 167-177.

[7] Βλ. “Zeno’s Paradoxes,” του Bradley Dowden, *Internet Encyclopedia of Philosophy*, https://iep.utm.edu/zeno-par/, Πρόσβαση στις 23 Δεκεμβρίου 2020. Βλ. επίσης “Why Mathematical Solutions Of Zeno’s Paradoxes Miss The Point: Zeno’s One And Many Relation And Parmenides’ Prohibition,” της Alba Papa-Grimaldi, http://philsci-archive.pitt.edu/2304/1/zeno_maths_review_metaphysics_alba_papa_grimaldi.pdf, *The Review of Metaphysics* 50 (December 1996): 299-314, Πρόσβαση στις 23 Δεκεμβρίου 2020

[8] ό.π., Dowden.

[9] “Tens of Thousands of 12,000-Year-Old Rock Paintings Found in Colombia,” Livia Gershon, *Smithsonianmag.com*, 1 Δεκ. 2020. https://www.smithsonianmag.com/smart-news/tens-thousands-12000-year-old-rock-paintings-found-colombia-180976427/, Πρόσβαση στις 23 Δεκεμβρίου 2020.

[10] "Colonisation and early peopling of the Colombian Amazon during the Late Pleistocene and the Early Holocene: New evidence from La Serranía La Lindosa," των Gaspar Morcote-Ríos[a]Francisco, Javier Aceituno, José Iriarte, Mark Robinson, Jeison L. Chaparro-Cárdenas, *Quaternary International*, 29 Απριλίου 2020, https://doi.org/10.1016/j.quaint.2020.04.026

[11] Βλ. *UN honors site of prehistoric Colombia cave paintings hidden for millennia*, του Jaime Ortega Carrascal, 3 Ιουλίου 2018, efe-epa, https://www.efe.com/efe/english/technology/un-honors-site-of-prehistoric-colombia-cave-paintings-hidden-for-millennia/50000267-3671693, Πρόσβαση στις 23 Δεκεμβρίου 2020.

[12] *Wild Blue Sea – Thinking Through Seawater*, της Melody Jue, Duke University Press, Durham, NC, 2020.

[13] *Gesture and Speech*, του André Leroi-Gourhan, Μετάφραση από τα γαλλικά της Anna Bostock Berger, εισαγωγή του Randall White, An October Book, The MIT Press, Cambridge, Massachusetts, 1993, σελ. 396.

[14] Βλ. https://learnodo-newtonic.com/jackson-pollock-famous-paintings, Πρόσβαση στις 23 Δεκεμβρίου 2020.

[15] Βλ. "How Mondrian Won the Art World with Just Three Colors," Jessica Stewart, *My Modern Met*, May 22 Μαΐου 2019, https://mymodernmet.com/piet-mondrian/, Πρόσβαση στις 23 Δεκεμβρίου 2020.

[16] Βλ. David Bodanis, *$E=mc^2$: A Biography of the World's Most Famous Equation*, Bloomsbury Publishing, New York, 2001.

Κεφάλαιο 21

PARASPAROPAGRAHO JIVANAM

Καμία παροιμία, προσευχή, σούτρα, θρησκευτικός στίχος, πνευματικό κείμενο, ηθική κοινή αρχή, θεολογική διαβεβαίωση, κανόνας του Ντάρμα ή ακρογωνιαίος λίθος συμπόνιας δεν έχει ποτέ αποκλίνει, ως πρώτη αρχή, από κάποια συνθετική εκδοχή του Χρυσού Κανόνα. Ωστόσο, η επιστήμη και η ψυχολογία αυτών των πεδίων μάχης στη Φύση που νομίζουμε ότι κατανοούμε, στην πραγματικότητα τεκμηριώνει μόνο εν μέρει την καταγραφή των ενάρετων σχέσεων. Υπερ-, υπο- και μεσοσαρκοφάγα, βλεννοφάγα παράσιτα, πτηνοφάγα που τρώνε άλλα πουλιά, δουροφάγα και παιδοφάγα ψάρια που τρώνε άλλα του είδους τους, μυρμηκοφάγα έντομα όπως οι τερμίτες και τα μυρμήγκια, και πληθώρα χερσαίων σπονδυλωτών, ωοφάγοι καρχαρίες που καταναλώνουν αυγά, ακόμη και εκείνα τα λεπιδοφάγα θαλάσσια είδη που ειδικεύονται στην κατανάλωση λεπιών ψαριών, είναι όλα ευρέως διαδεδομένα παραδείγματα κάθε άλλο παρά της διατροφικής συμπεριφοράς μιας Χρυσής Εποχής. Η επιστήμη εμπλέκει ως πρωταρχικό κίνητρο την ενδοφυλετική θήρευση - το να τρώει κανείς τους ανταγωνιστές του, κάτι που συχνά παρατηρείται σε οφιοφάγα, π.χ. μαγκούστες, φιδαετούς και γαλάζιους ερωδιούς, που κυνηγούν φίδια, όπως ακριβώς η κατά τα άλλα σχεδόν φυτοφάγα κεραμοχελώνα γίνεται βίαιη όταν έρχεται αντιμέτωπη με ένα θαλάσσιο σφουγγάρι- ή η κατά τα άλλα ευγενική σαλαμάνδρα ή ένα κουφαηδόνι θα σκότωναν για ένα σκουλήκι. Τουλάχιστον το 12% όλων των θηλαστικών και πάνω από το 90% όλων των πτηνών είναι αχόρταγοι καταναλωτές εντόμων. 270 γνωστά είδη θηλαστικών είναι αμιγώς σαρκοφάγα. Και από το 1500 π.Χ., στο 92% της ανθρώπινης ιστορίας είχαμε την ύπαρξη τουλάχιστον ενός πολέμου, κάποιοι από τους οποίους, όπως η Δεύτερη Σινοϊαπωνική διαμάχη, είχαν ως αποτέλεσμα μέχρι και 25 εκατομμύρια θύματα. Ο Πρώτος Παγκόσμιος Πόλεμος είχε διπλάσιο αριθμό θανάτων- ο Δεύτερος Παγκόσμιος Πόλεμος πάνω από τους τριπλάσιους, ή περί τα 85 εκατομμύρια πιθανούς θανάτους.

Τι θα πρέπει λοιπόν να αποκομίσουμε απ' όλο αυτό το μακελειό; Υπάρχουν πολλοί τρόποι να απορροφήσουμε τις έντονες αντιφάσεις του ιδανικού του Χρυσού Κανόνα. Ο

Φρίντριχ Νίτσε (1844-1900), που ποτέ δεν μάσησε τα λόγια του, δήλωσε: "Ζουν πάρα πολλοί άνθρωποι...". Ο Rüdiger Safranski αναλογίζεται αυτήν την σκοπιά του Νίτσε: "Για την αποτροπή της "τρέλας" των προηγούμενων γενεών από το να ξεσπάσει στις σημερινές και τις μελλοντικές γενεές και να προσγειώσει όλη την ιστορία σε μια φρικτή κατάσταση "εκφυλισμού", πρέπει να ληφθούν συγκεκριμένα μέτρα. Ποια;". Η απάντησή του: "Δώστε μεγάλη προσοχή, αδελφοί μου, σε κάθε ώρα που το πνεύμα σας θέλει να μιλάει με παραβολές: αυτή είναι η προέλευση της αρετής σας".[1]

Ο Μαχαβίρα (599-527 π.Χ.) ήταν πολύ πιο γενναιόδωρος στην ανάλυσή του σχετικά με τη φύση της πραγματικότητας, όταν περιέγραψε τις "είκοσι δύο δυσκολίες" που πρέπει να ανέχεται ένας μοναχός και, ως άμεση συνέπεια, πρέπει να απαρνηθεί τον "πόνο" όποια και αν είναι η αιτία του. "Υποφέροντας από έντομα ένας μεγάλος σοφός παραμένει ανενόχλητος".[2] Ο σοφός Umaswati (2ος αιώνας έως 5ος αιώνας μ.Χ.) στο περίφημο έργο του *Tattvartha Sutra* δήλωσε τα εξής: "Αποχή από τη βία, το ψέμα, την κλοπή, τη λαγνεία και την κτητικότητα - αυτοί είναι οι όρκοι"[3], μια φράση που διατυπώθηκε με το πλήρες υπόβαθρο και τη γνωστική φιλοσοφία του νου που θα μπορούσε επίσης να αντιπροσωπεύει με απόλυτη σιγουριά ότι: "Όλη η ζωή συγκροτείται από αμοιβαία υποστήριξη και αλληλεξάρτηση", "Parasparopagraho Jīvānām".[4]

Παρότι αυτό το δείγμα αληθειών αντιπροσωπεύει ένα ελάχιστο κλάσμα απόψεων, ορισμένες από τις οποίες έχουν τροποποιηθεί και περιέλθει στην αφάνεια με την πάροδο του χρόνου (η αδελφή του Νίτσε, Ελίζαμπεθ, παρανόησε και παραποίησε εντελώς το έργο του κατά την επιμέλεια του έργου του Νίτσε ως συντηρήτρια μετά το θάνατό του, όπως συνέβη και με τον Πέρσι Σέλλεϋ), παραμένουν πυλώνες μιας αντινομίας που παραμένει για πάντα άλυτη στα χρονικά της ζωής. Κοντά στον επίλογο του δικού του καθοριστικού βιβλίου, *Biophilia*, ο Έντουαρντ Όσμπορν Γουίλσον αποκαλύπτει με ειλικρίνεια την απροκάλυπτη αλήθεια για τον εαυτό μας. Στην αναφορά του στο κλασικό έργο του Christopher D. Stone με τίτλο *Should Trees Have Standing?*[5] του 1972, ο Γουίλσον αναφέρεται στην έννοια ως ένα πεδίο "διευρυμένης γενναιοδωρίας" και συνεχίζει: "Τα ανθρώπινα όντα, παρά το όποιο κατά δήλωση ήθος και την αδελφικότητά τους, κάνουν εύκολα διακρίσεις εις βάρος ξένων και αρκούνται στο να τους σκοτώνουν κατά τη διάρκεια πολέμων που έχουν κηρυχθεί για σχετικά επιπόλαιους λόγους. Είναι συνεπώς πολύ πιο εύκολο να βρεθεί δικαιολογία για την εξόντωση ενός άλλου είδους".[6] Και ο Γουίλσον καταλήγει ότι αυτό είναι "εκείνο το εκπληκτικό μείγμα προόδου και φάρσας που ανέκαθεν καθόριζε το είδος μας".[7]

Στις καμπύλες, τις γραμμές και τις παραβολικές εξισώσεις για τις οποίες έγινε αναφορά στο προηγούμενο κεφάλαιο, υπονοήσαμε πολυδιάστατο χώρο, στροβίλους και μαθηματικούς γρίφους εκατέρωθεν του μηδενός· με άλλα λόγια, άπειρα που είναι πάντα ικανά για την προσθήκη του "1=?". Αυτή η δίνη των Κοάν μας οδηγεί παντού και πουθενά. Το να αποδεχτούμε το τελευταίο σημαίνει ότι πιστεύουμε στο 1, ότι αναγνωρίζουμε τη δύναμη του 1+1, ότι ξέρουμε πως παρά την τόση οδύνη που είναι οικολογικά απλωμένη στο χώρο και στο χρόνο, η βιολογία ευνοεί την πρωτοτυπία, τον ενθουσιασμό όσο και την ήρεμη ενδοσκόπηση, την ομορφιά, την αυτοσυγκράτηση, την αγάπη. Ότι αυτές οι εποχές της αδαούς πίστης ανυψώνονται στο επίπεδο ακλόνητων ιδανικών, με τα οποία το βάρος των κολάσεων και της απελπισίας μπορεί να αναστηθεί όπως ο Φάουστ. Ότι ακούμε το τρυφερό κάλεσμα, όχι σε τυχαία αδιέξοδα τραγωδίας, αλλά με σκόπιμη πίστη που έχει γίνει ακτιβισμός, ως η αξιόπιστη αφήγηση που μας καθοδηγεί από εδώ και στο εξής: η αρχή "patsfom" που συζητήθηκε προηγουμένως, ή μετα-Ανθρωπική, διαειδική μεταφορά πρώτης τάξης που προκαλεί πραγματική πίστη συν δράση, διόλου εξωπραγματικά ανάμεσα σε Κογκρέσα και δημοτικά όργανα, όπου η αλλαγή έχει πάντα εφαρμογή στους νωθρούς και τους ασφαλείς. Καμία όμως στο άτομο, όπου ο θρίαμβος δε βρίσκει κανένα εμπόδιο.

Αν αυτό το "άτομο" πρόκειται να ξεφύγει από τα νύχια της δαρβινικής εξέλιξης και το Μεσαίωνα της μηχανιστικής σκέψης, να βγει έξω χωρίς εξάρτηση από τεχνολογικές αυταπάτες τέλειας αυτοδυναμίας και χρηματικές ορέξεις για ατελείωτο κέρδος, αλλά για να αναζητήσει νέες φιλίες και συμμαχίες που δεν εξαρτώνται ή απαιτούν κάποια απόκριση, συγκεκριμένη γλώσσα ή την επιδοκιμασία των προσπαθειών μας, τότε μπορούμε να είμαστε σίγουροι γι' αυτό που, αλλού, έχουμε περιγράψει ως "δυνατότητα αμοιβαιότητας".[8] RP = η τετραγωνική ρίζα της συμπόνιας, επί την τετραγωνική ρίζα του μέσου όρου του πληθυσμού διαιρούμενη με το άθροισμα ολόκληρης της ομάδας ή οποιουδήποτε πληθυσμού (n). Ο Μαχάτμα Γκάντι ονόμασε μια μορφή αυτού ως "το Ευαγγέλιο της Ανιδιοτελούς Υπηρεσίας". [9] Μήπως αυτό έρχεται σε αντίθεση με τη βιοχημεία και τη φυσική επιλογή; Ή φαντάζεστε να μπορεί να ταρακουνά ολόκληρα οικοσυστήματα, επιτιθέμενο σε ανεμόμυλους;

Δεν πιστεύουμε κάτι τέτοιο. Αντίθετα, είναι η μόνη ακλόνητη αλήθεια που μπορεί να σώσει το είδος μας και τα περισσότερα από τα υπόλοιπα από την απόλυτη καταστροφή. Όσο πολύπλοκα κι αν είναι τα συστήματα που αποκρυπτογραφούμε σε όλη τη Φύση, όταν μας καλοπιάνει η αδιάφορη, μαραμένη ταπεινότητα που το άτομο πρέπει να είναι, μπορούμε να αξιολογήσουμε κάθε αρχή του προσφέρω και λαμβάνω· αγαπώ και αγαπιέμαι. Κάποιοι μπορεί να ισχυριστούν ότι πρόκειται για θαύμα. Αλλά στην πραγματικότητα είναι απλώς αυτό το κλειδί για την εξάλειψη της πολυπλοκότητας.

ΣΗΜΕΙΩΣΕΙΣ

[1] *Nietzsche – A Philosophical Biography*, του Rüdiger Safranski, Μετάφραση Shelley Frisch, W.W. Norton & Company, New York, London, 2002, σελ. 267.

[2] *Uttaradhyayana*, Διάλεξη II, σελ. 11, *The Sacred Books Of The East* Translated By Various Oriental Scholars and Edited By F. Max Müller, Vol. XLV, At The Clarendon Press, Oxford, UK, 1895, σελ. 11.

[3] *Tattvartha Sutra – That Which Is*, Μετάφραση Nathmal Tatia, Πρόλογος L. M. Singhvi και μια εισαγωγή στην πίστη του Τζαϊνισμού από τον Padmanabh S. Jaini, The Institute of Jainology, HarperCollins Publishers, Manchester UK, 1994, σελ. 169.

[4] Sangave, Dr. Vilas A. (2001). *Facets of Jainology: Selected Research Papers on Jain Society, Religion, and Culture*. Mumbai: Popular Prakashan. σελ.123. Βλ. *Tattvartha Sutra*, 5.21 –"Souls render service to one another."

[5] *Should Trees Have Standing? Law, Morality, and the Environment*, Oxford University Press, Oxford and New York, 3rd edition, 2010.

[6] *Biophilia – The human bond with other species*, του Edward O. Wilson, Harvard University Press, Cambridge, Massachusetts, and London, UK, 1984, σελ. 130-131.

[7] ό.π., σελ. 145.

[8] Βλ. *The Theoretical Individual – Imagination, Ethics and the Future of Humanity*, των M. C. Tobias και J. G. Morrison, Springer, New York, σελ. 52. Αρχικά εφαρμόσαμε αυτήν την έννοια στο έργο μας *Anthrozoology: Embracing Co-Existence in the Anthropocene*, των M. C. Tobias και J. G. Morrison, Springer, New York, 2017, https://www.springer.com/gp/book/9783319459639

[9] Βλ. *Gandhi Memorial Center – Selfless Service*, http://www.gandhimemorialcenter.org/selfless-service, Πρόσβαση στις 24 Δεκεμβρίου 2020.

Κεφάλαιο 22

Η ΠΑΡΟΜΟΙΩΣΗ ΤΕΤΑΡΤΗΣ ΤΑΞΗΣ

Τόσο ο χρόνος όσο και η εγγενής δομή, τα παροδικά τρωτά σημεία και το πλήθος άλλων μεταβλητών σε όλα τα βιολογικά καθεστώτα αντιτίθενται επιδέξια στην οικολογική ομοιόσταση. Η προσαρμογή επιφέρει τριβές. Η ένταση διακλαδίζεται μες στη σταθερότητα, έτσι ώστε όλα τα θερμοδυναμικά χαρακτηριστικά στο σύμπαν φαίνεται να εξαρτώνται από αντιθετικές δυνάμεις. Αποδίδουν τις εκτεταμένες (π.χ. εντροπία, μάζα, όγκος), εντατικές (π.χ. πίεση, θερμοκρασία, θερμική αγωγιμότητα) και ειδικές (π.χ. θερμοχωρητικότητα) ιδιότητες που προκαλούν διαρκείς μεταβολές, πριν, κατά τη διάρκεια και μετά την ανθρώπινη αντίληψη.[1] Εντός ενός μαθηματικού πλαισίου, θα μπορούσε κανείς να πολλαπλασιάσει τις σύνθετες πολυπλοκότητες της φύσης αρκετά κατά τον τρόπο του προβλήματος της αίγας, του παλιού αυτού προβλήματος του αιώνα. Αυτή η αρκετά αυθαίρετη, αν και άστατη αναζήτηση λύθηκε πρόσφατα αναγνωρίζοντας "τον πολυδιάστατο χώρο και όχι ένα χωράφι με γρασίδι" ως το μέσο για τον προσδιορισμό του μήκους του λουριού μιας πεινασμένης κατσίκας μέσα σε ένα στρέμμα γρασίδι, όπου η πρόθεση είναι να επιτραπεί στην κατσίκα η πρόσβαση στο μισό στρέμμα.[2] Η πολύπλοκη ανάλυση βασίζεται τόσο σε πραγματικά όσο και σε φανταστικά στοιχεία και αυτό, όπως είναι αναμενόμενο, είναι η μόνη λύση για την ανθρώπινη φύση που φιλοδοξεί να συγχωνευτεί με τη φύση, ενώ παράλληλα υποστηρίζει την ιδανική συμπεριφορά ως προτιμότερη από ένα κατώτερο, ηθικό όριο. Τι σημαίνουν όλα αυτά; Είναι όλα τα μισά στρέμματα, ή οι κατσίκες, το ίδιο; Η γνώση ότι σίγουρα δεν είναι, προσδίδει στη βιολογική αφθονία των μη οριοθετημένων διαστάσεων μια ακόμη παρομοίωση τέταρτης τάξης.

Το βιβλίο του Νίτσε *Πέρα από το Καλό και το Κακό: Πρελούδιο σε μια Φιλοσοφία του Μέλλοντος* (1886) χρησιμοποίησε αρκετές από τις βασικές έννοιές του που αποσκοπούν στην υπέρβαση των εμποδίων τόσο της ανθρώπινης όσο και της εγκόσμιας φύσης. Η "θέληση για δύναμη", ο "φανταστικός αντίκοσμος", η "αλήθεια από τον τάφο" και η "απόλυτη αξία, σε αντίθεση προς τη μικρότητα και τη συμπτωματική θέση του ανθρώπου στη ροή του γίγνεσθαι και του θανάτου".[3] Αυτή ακριβώς η διχοτομική περίσταση, λέει, ωθεί "μόνο" την ανθρωπότητα να έχει "κατασκευασμένους σκοπούς, διαδοχή, αμοιβαιότητα, σχετικότητα, καταναγκασμό, αριθμό, νόμιο, ελευθερία, κίνητρο [και] σκοπό". [4] Αισθάνεται κανείς ναυτία μπροστά σε αυτό το αντιφατικό αποτέλεσμα σε πολιτισμούς όπως οι Μάγια, των οποίων οι επανειλημμένες καταρρεύσεις έχουν αποδοθεί, μεταξύ άλλων, "σε μια χρόνια έμφαση στον πόλεμο και στην ανέγερση μνημείων αντί στην επίλυση των υποκείμενων προβλημάτων...".[5] "Οι Μάγια μάς προειδοποιούν", επισημαίνει ο Jared Diamond, "ότι οι καταρρεύσεις μπορούν να πλήξουν και τις πιο προηγμένες και δημιουργικές κοινωνίες",[6] όπως αναφέρεται και στους πολύ γνωστούς στίχους από το ποίημα "Οζυμανδίας" του ποιητή Πέρσι Σέλλεϋ, (1817): "Το όνομά μου είναι Οζυμανδίας, βασιλιάς των βασιλέων: Κοιτάξτε τα έργα μου, εσείς οι Ισχυροί, και απελπιστείτε!/Τίποτα άλλο δεν απομένει...". Αυτή η διδακτική ιστορία δεν αποδίδεται πουθενά πιο οδυνηρά μέσα στους αιώνες από ότι στο λιθοδομικό προφίλ ενός περήφανα αινιγματικού άνδρα ο οποίος ζει στο "Παλένκε, σε ένα χώρο με γλυπτά, εσωτερική αυλή, κυρίως παλάτι, στο Τσιάπας του Μεξικού", με χείλη αυστηρά σφραγισμένα, ο οποίος δεν μας δίνει καμία καλλιτεχνική πληροφορία για το τι συνέβαινε και τι ήταν πιθανό να πάει στραβά - είτε για τον ίδιο είτε για τον πολιτισμό του - όπως αποτυπώνεται στα λαμπρά φωτογραφικά επιτεύγματα του William Frej.[7] Ο Frej γράφει για τη φωτογραφία του: "Όπως οι αρχαίοι τοίχοι του παλατιού που το περιβάλλουν, το σκαλισμένο πέτρινο πρόσωπο, πιθανότατα ενός αιχμαλώτου, μας θυμίζει τη φευγαλέα φύση των κάποτε ισχυρών πολιτισμών... Ο Πακάλ, ο μεγαλύτερος βασιλιάς του Παλένκε, της αρχαίας πόλης των Μάγια στο Τσιάπας του Μεξικού, ανήγειρε μια υπέροχη πόλη-κράτος που άκμασε τον 7ο αιώνα. Ωστόσο, αυτό το ανάγλυφο, πιθανώς κάποιου αιχμαλώτου, αντιπροσωπεύει για μένα την αρχή του τέλους ενός βασιλείου των Μάγια γεμάτο πόλεμο, που χάθηκε στην τροπική ζούγκλα πριν την εκ νέου ανακάλυψή του χίλια χρόνια αργότερα.". [8]

Γνώριζε ο γλύπτης όλα αυτά που γνωρίζουμε εμείς; Αλλάζει κάτι η γνώση; Είναι η αλλαγή, κατά παρακαταθήκη του *H. sapiens*, ένα θέμα άξιο προς εξέταση; Τι σημαίνει *άξιο*; Μεταφράζεται το νόημα από άτομο σε άτομο, σε όλα τα είδη, μεταξύ ολοτήτων, μέσα στο χρόνο και το χώρο; Τα ερωτήματα είναι ενοχλητικά, οι απαντήσεις, αν υπάρχουν, αναξιόπιστες. Επομένως; Και ποιος είπε ότι η χρήση υποθετικών προτάσεων με τι λέξεις "αν-τότε" θα έπρεπε καν να συνιστάται ως η προτεινόμενη επιλογή; Υπάρχουν και άλλες, όπως αυτές που χρησιμοποιούν τις λέξεις "γιατί-όχι;", "είτε-είτε", "αλήθεια-ή-ψέματα;", "μεγαλύτερο από ή ίσο;".

Υπάρχουν κατηγορήματα "τελεστή εκχώρησης" και "τελεστή ισότητας" που σχηματίζουν μια ιλιγγιώδη σειρά αντιπαραθέσεων, και κάθε μήτρα εξαρτάται από την ίδια λογική ώθηση - τον κώδικα των προτάσεων που χρησιμοποιούν τη λέξη "αν"[9], οι οποίες υποδηλώνουν όλες την ανθρώπινη συναίνεση. Ωστόσο τα ηθολογικά κοινά απαιτούν έναν εντελώς διαφορετικό γνωστικό ή συναισθηματικό τρόπο λειτουργίας.

Σε αυτό το σημείο, καλύτερα να σταματήσουμε για λίγο. Η διανοητική κόπωση μπορεί να γίνει ένας πολύ άνετος καναπές, ιδιαίτερα όταν προσπαθούμε να επιλύσουμε το ανεπίλυτο. Ο Roland Barthes, αναφερόμενος στην "υπερσυνειδητότητα", υποστηρίζει ότι "η σωστή μινιμαλιστική ηθική θα βοηθούσε να επέλθει αρμονία μεταξύ της μέγιστης εσωτερικής έντασης".[10] Ωστόσο, στην πραγματικότητα ο Barthes διερευνά τη λογική του "conatus (θέληση, όρεξη, αόριστη αναγκαιότητα της ύπαρξης, προσπάθεια να επιμείνει κανείς στο είναι του)" του Σπινόζα και το κάνει αυτό ανεβάζοντας το διακύβευμα στο ηθολογικό σημείο εκκίνησης. Γράφει, λοιπόν, ότι "καθώς εξαρτιόμαστε από όλα τα άλλα όντα του κόσμου" υπάρχει μια διακύμανση, ένας διαφορετικός φακός, ένα μεταβαλλόμενο τοπίο, αυτό που αποκαλεί "πλαστικότητα του *conatus*".[11]

Στα ωδικά πουλιά ως νευρική πλαστικότητα έχει αναφερθεί η ικανότητά τους, μεταξύ άλλων, να προσθέτουν συνειδητά νευρώνες σε ένα κύκλωμα που αποκτά ειδικά νέα γλώσσα-τραγούδι ή νευρώνες οι οποίοι προωθούν τη φροντίδα τόσο στα αρσενικά όσο και στα θηλυκά.[12] Η "συναπτική πλαστικότητα" έχει θεωρηθεί ότι συμβάλλει καθοριστικά στη "διαμόρφωση νέας μάθησης και μνήμης" του εγκεφάλου.[13] Η πρώτη περιγραφή έγινε σε μακάκους ήδη από το 1923 από τον συμπεριφοριστή Κάρλ Σπένσερ Λάσλεϊ (1890-1958), ενώ τα στοιχεία υποστηρίζουν πλέον την άποψη ότι διάφορες μορφές νευρικής πλαστικότητας βοηθούν τη νευρογένεση σε ολόκληρο τον εγκέφαλο, ακόμα και στα γεράματα. Πρόκειται για τα σύνολα δεδομένων που υποστηρίζουν το "νευρικό δαρβινισμό", ο οποίος προτάθηκε για πρώτη φορά το 1978 από τον Τζέραλντ Έντελμαν.[14] Πρότυπα είδη, όπως η φρουτόμυγα, έχουν μελετηθεί για να προσδιοριστεί πώς και σε ποιο επίπεδο αριθμητικά η φυσική επιλογή επιδρά στη νευρική αναγέννηση. Λέγεται ότι τα νεύρα σταδιακά ωριμάζουν[15] - γιατί να μην ωριμάζουν; Και δεδομένου του(των) οικοσυστήματος(-ων) των νευρώνων του εγκεφάλου, όπως και σε κάθε άλλο μεγαοικοσύστημα που του έχει δοθεί η επιλογή να διατηρεί κάποιον ζωντανό οργανισμό ως αμοιβαίο παράγοντα των δικών του μεθόδων διαιώνισης, μιλάμε για έναν μηχανισμό επιλογής σε πραγματικό χρόνο προσανατολισμένο προς τη συνεργατική επιτυχία, όπως αυτή μετράται από την επιμονή των δικών του προσπαθειών.[16] Με την πάροδο του χρόνου οι νευρικές συστάδες θα κάνουν την όποια επιλογή τους, εάν όμως το *επιλέξουν,* ή αν το υποκείμενο conatus έχει την απαραίτητη επιμονή για να διατηρήσει τις αρχικές παρορμήσεις και να καθοδηγήσει τη δική του εξέλιξη. Στους βιοστρόβιλους της επιφοίτησης, δεν έχουμε λόγο να αμφισβητούμε μια τέτοια εξέλιξη σε πραγματικό χρόνο, ή επανεξέλιξη. Στο συμβάν του ορίζοντα/ ορίζοντα των γεγονότων μιας παρομοίωσης τέταρτης τάξης, η συνείδηση είναι ικανή για όλα.

Ως εκ τούτου, το μεταβαλλόμενο εσωτερικό βιολογικό τοπίο, καθώς ο νους προσπαθεί να καθορίσει νέες συνθήκες που θα διευκολύνουν την ανταλλαγή θεμελιωδών ελευθεριών, απολαύσεων, τρυφερότητας και αρχέγονης καλοσύνης. Η κλασική περίληψη της οικολογικής ανταλλαγής έχει αντιμετωπίσει το περιεχόμενο των πληροφοριών και των συσχετίσεων ως την ιστορική, "αναδυόμενη ιδιότητα της εξέλιξης του είδους", σύμφωνα με το δόγμα ότι "δεδομένου ενός αυτότροφου, φυτοφάγου και σαρκοφάγου, μπορούμε να προβλέψουμε την οικολογική συσχέτιση που θα προκύψει.". [17] Ωστόσο, η συζήτηση αυτού του είδους προϋποθέτει ότι γνωρίζουμε κάποια πράγματα με βάση, αυστηρά το πώς και τι τρώει κανείς. Το "τρώει" είναι μια αφηγηματική κατηγορία, όπως η "βαρύτητα", δεν μπορεί όμως να υποκαταστήσει την πραγματική ταυτότητα. Ένας Τζαϊνιστής μοναχός στη

Βομβάη έμεινε για πολλούς μήνες νηστικός. Ωστόσο, δεν τον κατηγοριοποιούμε ως προς την ταυτότητά του παρά μόνο ως κάποιον που νηστεύει. Ούτε θα λέγαμε ότι ο Γεωκόκκυξ της Καλιφόρνια *(Geococcyx californianus)* κάνει μόνο αυτό, και ο Λεονάρντο ντα Βίντσι, μόνο αυτό. Ο Μισέλ Φουκώ αντιμετώπισε αποτελεσματικά αυτό το πρόβλημα αναζητώντας "το σύστημα που συνδέει όλα αυτά τα πεδία συνύπαρξης μεταξύ τους" αποδομώντας δηλώσεις που έχουν "επιπτώσεις σε ομάδες ενδείξεων", τα οποία με τη σειρά τους, κατόπιν της έκφρασης ή της κρυφής τους πρόθεσης, βλέπουν την εξέλιξη μιας "θέσης που μπορεί να καλυφθεί υπό ορισμένες συνθήκες από διάφορα άτομα..." [18].

Επιστρέφοντας στη δυάδα που διαφοροποιείται ανάλογα με το κόστος, αλλά και το κέρδος, παρατηρούμε την πλέον θεμελιώδη και συμπαθή εφαπτομένη η οποία κινείται από το ασύμπτωτο στο ολοκληρωτικό/ενοποιητικό είναι. Όπως έχουν γράψει οι Lorraine Daston και Peter Galison: "Για τον Γκαίτε, τα αντικειμενικά και τα υποκειμενικά φαινόμενα ήταν συμπληρωματικά και εξίσου απαραίτητα για την επιστήμη των χρωμάτων".[19] Η συγχώνευση προφανών παραγόντων, μεταξύ άλλων, μια "αμοιβαία εμμονή", όπως αυτό που περιέγραψε ο Κίρκεγκωρ, [20] αντανακλώμενη σ' εκείνη την τάση των ανθρώπων (κατ' επέκταση, όλων των μορφών ζωής) *να ερωτεύονται·* να επιδίδονται σε "εμμενή σκέψη", ώστε "ολόκληρη η πραγματικότητα να μπορεί να είναι πλήρης στην ιδέα...". [21] Ωστόσο, σκεπτόμενοι τόσο, έχοντας βυθιστεί μες σ' αυτήν την απτή πραγματικότητα του έρωτα, όλη αυτή η εμμένεια αναγνωρίζει επίσης από βαθιά μέσα της το άυλο, κάτι πέρα απ' όλα αυτά, κάτι παραπάνω. Αυτό το παραπάνω γεννά μια αξιοσημείωτη περίσταση περιστάσεων, δηλαδή τη μεταγραφή κάθε δυνατής ουσίας σε μια ηθική επιταγή που καθοδηγεί τη ζωή πέρα από τα εμπόδια που γνωρίζουμε πολύ καθαρά σε όλο τον φυσικό κόσμο. Φτάνουμε λοιπόν σε εκείνο το επίπεδο που έχει αποδοθεί στη χάρη και την άνευ όρων αγάπη.

ΣΗΜΕΙΩΣΕΙΣ

[1] *Thermal Engineering*, του Nick Connor, 22 Μαΐου 2019, https://www.thermal-engineering.org/what-is-thermodynamic-property-definition/, Πρόσβαση στις 26 Δεκεμβρίου 2020.

[2] "After Centuries, a Seemingly Simple Math Problem Gets an Exact Solution," των Steve Nadis, *Quanta Magazine*, 9 Δεκεμβρίου 2020, https://www.quantamagazine.org/mathematician-solves-centuries-old-grazing-goat-problem-exactly-20201209/, Πρόσβαση στις 25 Δεκεμβρίου 2020.

[3] *Nietzsche – A Philosophical Biography*, του Rüdiger Safranski, μετάφραση Shelley Frisch, W. W. Norton & Company, New York, London 2002, σελ. 294-297.

[4] *Beyond Good and Evil*, μετάφραση Walter Kaufmann, New York: Random House, 1966; επανατύπωση από την Vintage Books, ως μέρος του έργου *Basic Writings of Nietzsche*, New York: Modern Library, 2000, §21.

[5] Βλ. *Collapse – How Societies Choose To Fail Or Succeed*, του Jared Diamond, Viking Publishers, New York, 2005, σελ. 159-160.

[6] ό.π., *Collapse*, σελ. 159.

[7] Βλ. *Maya Ruins Revisited – In The Footsteps Of Teobert Maler*, του William Frej, με κείμενα των Stephan Merk, Alma Durán-Merk, Tomás Gallareta Negrón, Jeremy A. Sabloff, και Khristaan D. Villela, Peyton Wright Gallery Press, Santa Fe, New Mexico, University of Oklahoma Press Distributer, 2020, σελ. 4-5. Ο Frej μας έδωσε έναν σύνδεσμο μέσω tumblr για τα "Pakal the Great: Architect of Palenque," "A digitized version of an exhibition on K'inich Janab Pakal, ruler of Palenque. This exhibition was a final project in ARTH 321 Maya Art and Architecture at St. Mary's College

of Maryland, Fall 2016," https://pakalthegreatexhibition.tumblr.com/post/154741734763/patio-of-the-captives-late-classic-period-palace

[8] O William Frej μας έκανε τη χάρη να μας στείλει το σχόλιό του στη φωτογραφία του, στις 27 Δεκεμβρίου 2020.

[9] Khan Academy, *Review: Logic and if Statements*, n.a., n.d.., https://www.khanacademy.org/ computing/computer-programming/programming/logic-if-statements/a/review-logic-and-if-statements, Πρόσβαση στις 26 Δεκεμβρίου 2020.

[10] *The Neutral*, Lecture Course at the College de France (1977-1978), του Roland Barthes, μετάφραση των Rosaline E. Krauss και Denis Hollier, Κείμενο που δημιουργήθηκε, σχολιάστηκε και παρουσιάστηκε από τον Thomas Clerc υπό τη διεύθυνση του Eric Marty, Columbia University Press, New York, NY, 2005, σελ. 200, και 257.

[11] ό.π.

[12] Βλ. "Neural Plasticity," του J. H. Kaas, στο *International Encyclopedia of the Social & Behavioral Sciences*, Επιμέλεια N. J. Smelser και P. B. Baltes, Pergamon, Oxford, UK, 2001.

[13] Βλ. "Synaptic Plasticity," των Ruth Heidelberger, John H. Byrne, et al., στο *From Molecules to Networks*, 3rd Edition, Academic Press, Elsevier, Cambridge, Massachusetts, 2014, https://doi.org/10.1016/C2011-0-07251-4

[14] Gerald Edelman, *Neural Darwinism. The Theory of Neuronal Group Selection*, Basic Books, New York, 1987. Βλ. επίσης Yan Sun, Wang Chao, Dai Jiapei (2010). "Biophotons as neural communication signals demonstrated by in situ biophoton autography". *Photochem. Photobiol. Sci.* 9 (3): 315–322. doi:10.1039/b9pp00125e. PMID 20221457.

[15] van Praag, Henriette, Schinder, Alejandro F., Christie, Brian R., Toni, Nicolas, Palmer, Theo D., Gage, Fred H. (Φεβρουάριος 2002). "Functional neurogenesis in the adult hippocampus". *Nature. 415 (6875): 1030–1034.* doi:10.1038/4151030a. ISSN 1476-4687. PMID 11875571. S2CID 4403779.

[16] *Alvarez-Buylla, A; Lim, DA (4 Μαρτίου 2004). "For the long run: maintaining germinal niches in the adult brain". Neuron. 41 (5): 683–6.* doi:10.1016/S0896-6273(04)00111-4. PMID 15003168. S2CID 17319636.

[17] Βλ. *Evolution as Entropy – Toward a Unified Theory of Biology*, των Daniel R. Brooks και E. O. Wiley, The University of Chicago Press, Chicago and London, 1986, σελ. 283.

[18] *The Archaeology of Knowledge*, του Michel Foucault, μετάφραση από τα γαλλικά του A. M. Sheridan Smith, Pantheon Books, Random House, New York, NY, 1972, σελ. 115. Δημοσιεύτηκε για πρώτη φορά με τον τίτλο *L'Archéologie du Savoir*, από τις εκδόσεις Éditions Gallimard, Paris, France, 1969.

[19] *Objectivity*, των Lorraine Daston και Peter Galison, Zone Books, New York, NY, 2007, σελ. 277.

[20] "The First Love – A Comedy in One Act by Scribe, μετάφραση J. L. Heiberg, στο έργο του Søren Aabye Kierkegaard *Either/Or – A One-Volume Abridgment In A New Translation*, επιμέλεια Steven L. Ross, μετάφραση George L. Stengren, Perennial Library, Harper & Row, New York, N.Y., 1986, σελ. 65.

[21] ό.π., σελ. 63.

Tashi Payden Tshering, Διευθύντρια της Βασιλικής Εταιρείας Προστασίας και Φροντίδας Ζώων του Μπουτάν, με μερικούς από τους σκύλους που φιλοξενούν. Φωτογραφία © M. C. Tobias.

Κεφάλαιο 23

ΒΙΟΛΟΓΙΚΗ ΚΛΙΜΑΚΩΣΗ ΓΙΑ ΜΙΑ ΝΕΑ ΦΥΣΗ

"Το επίπεδο της εμμένειας έχει δύο όψεις, όπως η Σκέψη και η Φύση, ο Νους και η Φύσις ως Γένεση, ως Δημιουργία", γράφουν οι Ντελέζ και Γκουαταρί[1]. Η Φύσις, συνδεδεμένη ετυμολογικά με την "ανάπτυξη", αναφέρεται σε κάποια μεταβατική σκέψη μια οντολογική φάση. Αυτό που επιζητούμε δεν είναι η ενημέρωση αλλά η παρώθηση των άλλων, να απηχήσουμε τα όποια όνειρα, κοσμοθεωρίες, ιδεώδη και λογικές έννοιες που μπορεί να μεταφερθούν εκτός του καταστροφικού πλαισίου ενός ιστορικού σολιψισμού ο οποίος έχει κυριαρχήσει στην πρώτη αρμονική της ταχύτητας του "κινητήρα" της ανθρωπότητας, αυτό που ένας μηχανικός αυτοκινήτων θα σκεφτόταν ως παλινδρομικές και εναλλασσόμενες ανισορροπίες.

Πρόκειται, στην πραγματικότητα, για την αναρμονικότητα της ανθρώπινης ιστορίας, η οποία, με τη σειρά της, επικυρώνει τα τρέχοντα θεμελιώδη ερωτήματα. Ποια είναι τα βάσιμα, εύλογα συμπεράσματα στην εποχή του Ανθρωπόκαινου; Είναι η προσωπική και κοινοτική μας αυτοαπορρόφηση επαρκής σε βιολογική κλίμακα ώστε να έχουμε τον υψηλότερο ρυθμό εξάπλωσης, παραφράζοντας το προφανές μοντέλο μετάδοσης; Η φύση είναι παντού. Ωστόσο, είναι τα μεγάλα ερωτήματα απλώς ανθρώπινα ερωτήματα, ή μήπως πρόκειται για ένα είδος ερωτήματος που τίθεται από κάποια φυσική δύναμη *που θέτει* τα ερωτήματα ανεξάρτητα από τις δικές μας πενιχρές φωνές ως είδος, επειδή έχει σημασία σε αυτή τη διάσταση που είναι η ύπαρξή του, ή η λογική του, αλλά αυτά είναι παλιά, πολύ παλιά γνωμικά.

Υπάρχει κάποιο σχετικό μαθηματικό σημείο εισόδου ή σταθερό επίπεδο συνάφειας σε αυτόν τον τομέα; Αν όχι μέσω της επανάληψης, των αριθμών ή οποιουδήποτε προκαθορισμένου συνόλου υποθέσεων, πώς μπορούμε να σκεφτόμαστε χωρίς ανθρώπινη σκέψη; Να υπερβούμε το κλισέ της σιωπής, της ακινησίας; Είναι αρκετή η κραυγή μας για να εξαλείψουμε ρητορικά ψεγάδια, εγωκεντρικές εμμονές και κακά σημάδια των καιρών που συνοδεύουν τα τελευταία χιλιάδες χρόνια την ιστορία μας;

Ας δεχθούμε την υπόθεση ότι υπάρχουν μαθηματικές *ποσότητες* σε αυτή την λεπτή βιόσφαιρα των είκοσι χιλιομέτρων που διαφοροποιούνται σημαντικά από τις βιολογικές *ιδιότητες* της. Οι αριθμοί συναντώνται μόνο στους βιοστρόβιλους – εντός του ξυλώδους παρεγχύματος, του ηθμώδους παρεγχύματος, του σκληρεγχύματος και περιβάλλοντας τις κυτταρικές δέσμες μέσα στα φύλλα, για παράδειγμα· σε πολλαπλασιαστές κορυφής που θα έβρισκε κανείς στα διαγράμματα του Richard Feynman.[2] Θα μπορούσαμε να ρωτήσουμε: οι μαθηματικές ιδιότητες αφορούν σε επίπεδα ατόμου, κυττάρου και μορίου με τρόπους που συμβάλλουν όχι μόνο στη βιολογία, αλλά και στη *βιογραφία* κάθε μιας από αυτές τις μορφές ζωής; Εύκολα μπορεί να πει κανείς πως "ναι", ωστόσο υπάρχουν σχετικές επιπτώσεις σε ένα τέτοιο "ναι";

Μπορούμε να σκεφτούμε το φύλλο ως ένα πολύπλοκο σύστημα επιρρεπές σε στοχαστικές πιθανότητες που επιδεικνύουν μία συμπεριφορά μέσα σε οποιοδήποτε θερμοδυναμικό μοντέλο αλλά χωρίς τον Ευκλείδη, τον Νεύτωνα, τον Όιλερ, τον Δαρβίνο κ.λπ. προκειμένου να αποδεικνύουν τη εκούσια κατάστασή τους, ως σφουγγάρια κάτω από ένα αστέρι, μετατρέποντας όλο αυτό το φως σε ζωή; Δηλαδή, είναι ο φυσικός κόσμος η ίδια "φύση" που παρατηρούμε όταν δεν τον παρατηρούμε; Την ώρα που ένα δέντρο καπόκ συντρίβεται στον Αμαζόνιο, δισεκατομμύρια οργανισμοί, τρισεκατομμύρια κύτταρα, μόρια, νευρώνες καταλαβαίνουν προφανώς ότι έγινε κάποια σύγκρουση. Η επίγνωση αυτή δεν αποτελεί τίποτα λιγότερο από τον δημιουργό της εξίσωσης, τον φακό. Το γεγονός ότι οι άνθρωποι ενδιαφέρονται περισσότερο για τα δέντρα παρά για τις πέτρες, ή περισσότερο για τα μωρά παρά για τα υπολείμματα φυτοπλαγκτόν, δεν εμποδίζει μια νέα αρμονική/θερμοδυναμική: Ο/Οι παρατηρητής/παρατηρητές και το/τα υποκείμενο/υποκείμενα παρατήρησης δεν υπόκεινται σε κρίση. Οι συντελεστές κλιμάκωσης του βάρους του εγκεφάλου, ακόμη και ο ίδιος ο νεοφλοιός, δεν επισκιάζουν το πανίσχυρο *σύμβολο* της ισότητας (=). Πέρα από το άπειρο εύρος του, μπορούν να περάσουν ένας Άγιος Βασίλης, ένας Δούρειος Ίππος, ένας μάντης, μια Αλίκη. Φαίνεται, ωστόσο, ότι μόνο το είδος μας μπορεί να κάνει κάτι τέτοιο με μια διφορούμενη ατζέντα που, αν εξεταστεί σε υποατομικό επίπεδο, είναι προφανώς βιολογικά μπλεγμένη σε ένα σύνολο αντιφάσεων. Αυτό το σύνολο αντιφάσεων αμφισβητεί το προαναφερθέν "ναι" και απαιτεί κάποιου είδους διόρθωση.

Ο θεωρητικός φυσικός Φρανκ Γουίλτσεκ (γεννηθείς το 1951) επινόησε την έννοια της "ασυμπτωτικής ελευθερίας"[3] (για την οποία ο ίδιος, ο Ντέιβιντ Γκρος και ο Χιου Ντέιβιντ Πόλιτζερ έλαβαν το 2004 το Βραβείο Νόμπελ Φυσικής). Πρόκειται για ένα ενδιαφέρον παράδειγμα που θα μπορούσε να εξηγήσει την ενέργεια που ενυπάρχει στη βιοσημειωτική σηματοδότηση μεταξύ των ειδών μέσα σε μια γενική κατανομή του λεγόμενου χώρου Sobolev (το όνομά του προέρχεται από τον Σοβιετικό μαθηματικό, Sergei Lvovich Sobolev, 1908-1989). Η κατανομή την οποία διατύπωσε ως θεωρία (με αριθμούς, όχι σήματα, αλλά εξίσου ισχυρή εφαρμογή στους βιοστρόβιλους) περιλαμβάνει τον σύγχρονο λογισμό με το άπειρο φάσμα διαφοροποιήσεών του (με άλλα λόγια: τη βιοποικιλότητα)[4] σε έναν εξίσου και ακατανόητα ποικιλόμορφο καταιγισμό πινάκων (π.χ., τον πίνακα Toeplitz και τις πολικότητες οργάνων σε όλες τις μορφές ζωής)[5].

Σύμφωνα με τις ευρύτερες συμπεριφορικές ανωμαλίες, καθώς και τους κανόνες της φυσικής της βιολογίας και αντιστρόφως, ανιχνεύουμε νέα μέτρα και νέες λειτουργίες - ως παρατηρητές με θέληση – που, ενώ αναγνωρίζουμε ψευδώς θετικές νεγεντροπίες (π.χ., το σχεδόν ακίνητο μάτι ενός τυφώνα) μπορούν ελεύθερα να ακολουθήσουν πρωτότυπα παραδείγματα. Οι πολιτικοί τα αποκαλούν "δυνατότητες", το κοινό "ελπίδα", οι οικονομολόγοι "ερέθισμα", οι φιλόσοφοι "οντολογία". Ωστόσο η παράγωγη βιολογία μας επιτρέπει να δημιουργήσουμε μερικά πολύ βασικά αιτιώδη συστήματα που πιστεύουμε ότι έχουν αποδειχθεί σε όλη την εξελικτική φυλογένεια. Καμία μορφή ζωής δεν θέλει να υποφέρει. Καμία μορφή ζωής δεν γνωρίζει ότι είναι μια μορφή ζωής παρά έχει μόνο μια φευγαλέα, χρονικά εξαρτώμενη, αίσθηση. Τα συναισθήματα έχουν σημασία, μόνο όμως για άλλα συναισθήματα. Το άθροισμα των συναισθημάτων σε οποιοδήποτε βιολογικό τομέα μπορεί να εκτιμηθεί ως προς την πιθανότητα μιας εξίσου διάχυτης ισοτιμίας σε άλλα αθροίσματα επειδή τα συναισθήματα αντιπροσωπεύουν έναν ομοιόμορφο τελεστή. Παρόλα αυτά, οι κώδικες της ζωής παραμένουν ασαφείς ως προς την προέλευση και τα αίτια. Ως εκ τούτου, η βιοσημειωτική έρευνα πρέπει να αντιμετωπίσει συντελεστές που δεν είναι ούτε ονομαστικοί ούτε τακτικοί, οι οποίοι δεν τηρούν διαστήματα και μπορούν να επινοούν εκ νέου αναλογίες.

Η κλίμακα αυτών των τελεστών φαίνεται να δρα μέσα σε συναθροίσεις τόσο οργανικές όσο και ανόργανες, ακέραιες ή κατατμημένες. Η εποικοδομητική προσαρμογή και ανθεκτικότητά τους συνδυάζει τις πολυπαραγοντικές λειτουργίες τόσο των ανοικτών όσο και των κλειστών συνόλων, των τοπολογιών που λειτουργούν προς οποιαδήποτε κατεύθυνση, αποφεύγοντας έτσι την ταξινομική εξαναγκαστική εξεύρεση ιεραρχιών.

Επιπλέον, όπως και στην περίπτωση της επανεξέλιξης, αυτά τα ιεραρχικά πρότυπα παρέχουν ολοκληρωμένη αναστρεψιμότητα. Βλέπουμε, όπως και το φύλλο και το δέντρο του, προς όλες τις κατευθύνσεις. Φανταστείτε την κυριολεκτική ολοκληρωτική αίσθηση της όρασης που ενυπάρχει στην πολύχρωμη γαρίδα μάντις (*Odontodactylus scyllarus*), ένα οστρακόδερμο 340 εκατομμυρίων ετών, που ανήκει στη σειρά Stomatopoda, στην υποκατηγορία Unipeltata, με την πιο ισχυρή όραση οποιουδήποτε ζώου – και διαθέτει από *12* έως *16* τύπους κυττάρων *φωτοϋποδοχέων*. Οι άνθρωποι έχουμε μόνο 3.

Είναι σημαντικό ότι μπορούμε να καθορίσουμε κάποιο επίπεδο κλίμακας για τα βιολογικά πρότυπα και την περιοδικότητα σε έναν πεπερασμένο πλανήτη όπου οι εμβληματικές συγκυρίες ξεχωρίζουν στις μεγαλύτερες βιολογικές διεπαφές. Για παράδειγμα, τα Ιμαλάια και η ινδική χερσόνησος, οποιοδήποτε από τα 36 χερσαία θερμά σημεία, ή, σε μικρογραφία, το νεκροταφείο της φυτείας Mulberry Grove και το δάσος Tillie Fowler στην περιοχή της Τζάκσονβιλ της Φλόριντα, μια ανθρώπινη "πόλη-φάντασμα" μέσα σε ένα ακμάζον οικοσύστημα. Μπορούμε να εφαρμόσουμε τον μονόπλευρο μετασχηματισμό Λαπλάς-Στάιλτζες[6] στο ζήτημα των "ειδών-φάντασμα"[7] καθώς και των οργανισμών που περιγράφονται στην ενότητα "Το φαινόμενο του Λαζάρου"[8]. Είδη που κάποτε θεωρούνταν αγνοούμενα για πάντα, πού και πού επανεμφανίζονται. Για παράδειγμα το πτηνό Τακάχε (*Porphyrio hochstetteri*) στο South Island της Νέας Ζηλανδίας, το οποίο στο βιβλίο του Λόρδου Γουόλτερ Ρότσιλντ *Extinct Birds ("Εξαφανισμένα Πουλιά")*[9] του 1907 θεωρείτο εξαφανισμένο, και ανακαλύφθηκε ξανά στα Όρη Murchison από τον Geoffrey Orbell το 1948. Ή η επιστροφή της άγριας ζωής στο Τσερνομπίλ, η επανεξέλιξη σε πραγματικό χρόνο που συμβαίνει στην κορυφή πιθανότητας οποιουδήποτε από τους μακράς εκμετάλλευσης πρότυπους οργανισμούς, όπως *το E. coli*, *το D. melanogaster* ή *το P. ubique*, που θεωρείται ως ένα είδος δύναμης ζωής με εξαιρετικές ικανότητες όσον αφορά το μέγεθος, τη κατεύθυνση και την επιτάχυνση.

Αυτό που πραγματευόμαστε, λοιπόν, είναι μια κριτική αισθητηριακή και καθοδηγούμενη από τη συνείδηση *τακτική* για την αποτελεσματική παράκαμψη των ανθρωπογενών ανακολουθιών που πρέπει να προκύψουν από έναν ανενδοίαστο παρατηρητή/ένα άκριτο παρατηρούμενο παράδοξο, με τις σύνθετες προκαταλήψεις του με τις οποίες τίποτα δεν μπορεί να ειπωθεί ότι είναι πραγματικό, συμπεριλαμβανομένης της μαθηματικής κλίμακας που επί του παρόντος εκτιμά ότι υπήρξαν περίπου "4 δισεκατομμύρια είδη" τα οποία έχουν έρθει και παρέλθει στην ιστορία της βιολογίας[10]. Μήπως αυτός ο ανθρώπινος φακός διαστρεβλώνει τις προβλεπόμενες επιπτώσεις, όπως όσον αφορά την "κλιματική αλλαγή και την αλλαγή της χρήσης γης στην παγκόσμια ποικιλομορφία των πτηνών";[11] Ή τις αντιλήψεις του παρελθόντος σχετικά με βιοχημικές διαταράξεις;[12] Τους καθορισμούς σχετικά με τα "κατώτατα όρια" των διαταράξεων των ενδιαιτημάτων,[13] τα άλλα όρια που καθορίζουν την εξαφάνιση ολόκληρων πληθυσμών;[14] Ή ακόμα και τα μέλη του δικού μας είδους που αυτοκτονούν;[15]

Οποιοδήποτε μωσαϊκό διεπιστημονικών ερευνών στην ψυχολογία *του Homo sapiens* και αν χρησιμοποιείται για τη βελτίωση της κατανόησης της ταυτότητάς μας, σε οποιοδήποτε ιστορικό και οικολογικό πλαίσιο, θα παραμένουμε σε μια δίνη ασάφειας, αυτό που σε πολλές ασιατικές ηθικές παραδόσεις λέγεται *καρμικά εμπόδια*, όπως η σκόνη και τα σωματίδια που κρύβουν μια κατά τα άλλα καθαρή επιφάνεια, εμποδίζοντας έτσι την ικανότητά μας να βλέπουμε με ξεκάθαρη, αμερόληπτη σαφήνεια. Δεν υπάρχει κάποια αλχημική σειρά φιλοσοφικών ή επιστημονικών ταχυδακτυλουργιών για την αντιμετώπιση των βιολογικών ορίων μας στο πλαίσιο της προσπάθειας της κατανόησής τους. Η αναγνώριση της ύπαρξης αυτών των παραμέτρων αποτελεί το πρώτο βήμα για την αναγνώριση της ταπεινής μας κατάστασης, ακόμα κι αν δεν έχουμε τρόπο να την τοποθετήσουμε μέσα σε κάποιο εύρος αισθητηριακών ή γνωστικών *δυνάμεων*. Σε καμία κλίμακα, καμία αρμονική, κανένα ενεργειακό παλμό ή μέτρηση που να σχετίζεται με την ηθική διόρθωση.

Μόνο μια ανοιχτά ηθική νομοτέλεια, η οποία γίνεται αποδεκτή χωρίς κλιμάκωση, μπορεί να μας οδηγήσει στην αρχή αυτής της *νέας φύσης*.

Σημειώσεις

[1] Gilles Deleuze and Félix Guattari, *What is Philosophy?*, Columbia University Press, New York, NY, 1994, σελ. 38, που αναφέρεται στη σελ. 143 του "Gilles Deleuze: A Philosophy of Immance," της Fredrika Spindler, στο: *Phenomenology and Religion: New Frontiers*, [ed] Jonna Bornemark, Hans Ruin, Huddinge: Södertörns

högskola, 2010, σελ. 149-163. https://www.diva-portal.org/smash/get/diva2:406664/FULLTEXT01.pdf;Gilles; Βλ. επίσης Daniel. W. Smith, "The doctrine of univocity. Deleuze's ontology of immanence" στο *Deleuze and religion*, ed. Mary Bryden, London/New York: Routledge, 2000, σελ. 160. Βλ. επίσης François Zourabichivili, Deleuze. *Une philosophie de l'événement*, Paris: PUF, 1994.

[2] Βλ. 'Diagramology' Types of Feynman Diagram," του Tim Evans, 2 Ιανουαρίου 2018, https://www.imperial.ac.uk/media/imperial-college/research-centres-and-groups/theoretical-physics/msc/current/qft/handouts/qftdiagramtypes.pdf

[3] Βλ. https://www.nobelprize.org/prizes/physics/2004/summary/

[4] *Sergei Lvovich Sobolev (1908-1989). Bio-Bibliography*, επιμέλεια S. S. Kutateladze, Novosibisrk, Sobolev Institute, 2008.

[5] Ye, Ke; Lim, Lek-Heng (2016), "Every matrix is a product of Toeplitz matrices", *Foundations of Computational Mathematics,* 16 (3): 577–598, arXiv:1307.5132, doi:10.1007/s10208-015-9254-z

[6] Feller, William. *An introduction to probability theory and its applications*. Vol. II. Second edition, John Wiley & Sons, Inc., New York-London-Sydney 1971 σελ. xxiv+669.

[7] Βλ. "Mysterious 'ghost' populations had multiple trysts with human ancestors," της Ann Gibbons, 20 Φεβρουαρίου 2020, *Science*, https://www.sciencemag.org/news/2020/02/ mysterious-ghost-populations-had-multiple-trysts-human-ancestors, Πρόσβαση στις 3 Ιανουαρίου 2021.

[8] Βλ. "The Lazarus effect: can the dynamics of extinct species lists tell us anything about the status of biodiversity? των David A. Keith και Mark A. Burgman, *Biological Conservation*, Volume 117, Issue 1, Μάιος 2004, σελ. 41-48, https://doi.org/10.1016/S0006-3207(03)00261-1, Elsevier, https://www.sciencedirect.com/science/article/abs/pii/S0006320703002611#:~: text=In%20general%2C%20the%20lists%20are,their%20distribution%20in%20remote%20locations.

[9] Hutchinson, London, σελ. 142, 1907.

[10] Βλ. *The Mathematics of Extinction Across Scales: From Populations to the Biosphere*, των Colin J. Carlson, Kevin R. Burgio, Tad A. Dallas, & Wayne M. Getz, PeerJPreprints, σελ. 2, https://doi.org/10.7287/peerj.preprints.3367v2 | CC BY 4.0 Open Access | rec: 11 Sep 2018, publ: 11 Sep 2018, https://peerj.com/preprints/3367.pdf

[11] Βλ. "Projected Impacts of Climate and Land-Use Change on the Global Diversity of Birds," των Walter Jetz, David S. Wilcove, Andrew P. Dobson, *PLOS BIOLOGY*, ημερομηνία δημοσίευσης: 5 Ιουνίου 2007, https://doi.org/10.1371/journal.pbio.0050157

[12] Βλ. "Methanogenic burst in the end-Permian carbon cycle," των Daniel H. Rothmana, Gregory P. Fournierc, Katherine L. Frenchb , Eric J. Almc , Edward A. Boyleb , Changqun Caod , και Roger E. Summons, *PNAS*, 27 Σεπτεμβρίου 2013, https://www.pnas.org/content/ pnas/111/15/5462.full.pdf

[13] Βλ. "Methods and models for identifying thresholds of habitat loss: των Deyi Yin, Shawn J. Leroux, και Fangliang He, Πρώτη δημοσίευση: 10 Νοεμβρίου 2016, Ecography, Wiley Online Library, https://doi.org/10.1111/ecog.02557, https://onlinelibrary.wiley.com/doi/full/ 10.1111/ecog.02557

[14] Βλ. "Recent responses to climate change reveal the drivers of species extinction and survival,"των Cristian Román-Palacios και John J. Wiens, *PNAS* 25 Φεβρουαρίου 2020 117 (8) 4211-4217; Πρώτη δημοσίευση: 10 Φεβρουαρίου 2020; https://doi.org/ 10.1073/pnas.1913007117, PNAS, https://www.pnas.org/content/ 117/8/4211

[15] Βλ. *Firearms and Suicide," Κεφ. 7 του Firearms And Violence – A Critical Review,* Committee to Improve Research Information and Data on Firearms, επιμέλεια Charles F. Wellford, John V. Pepper, και Carol V. Petrie, National Research Council of the National Academies Press, Washington D.C., 2004, σελ. 152, https://www.nap.edu/read/10881/chapter/9 Βλ. επίσης "More and More Americans are dying by suicide. What are we missing?" της Alla E. Dastagir, *USA Today*, 30 Ιανουαρίου 2020, https://www.usatoday.com/story/news/nation/ 2020/01/30/u-s-suicide-rate-rose-again-2018-how-can-suicide-prevention-save-lives/4616479002/

Κεφάλαιο 24

ΠΑΡΑΔΟΞΑ ΑΜΟΙΒΑΙΟΤΗΤΑΣ

Μιλάμε για μια ταπεινή κατάσταση στο πλαίσιο μιας νέας φύσης. Αν πρόκειται απλώς για μια λειτουργία της γραμματικής, ως κάτι συνειδητό, τότε λέμε μόνο ό,τι έχουμε πει, είτε σε έναν λεγόμενο "καθολικό χώρο εργασίας" είτε ως μέρος μιας "ολοκληρωμένης θεωρίας πληροφοριών".[1] Αν συνδυάσουμε τις "δορυφορικές κάμερες με ένα συνελικτικό νευρωνικό δίκτυο", μπορούμε να εντοπίζουμε μεμονωμένους ελέφαντες από το διάστημα.[2] Ένα απλό παράδειγμα των εργαλείων που έχουμε συγκεντρώσει, τα οποία - αν επικεντρωθούν στην καθυστέρηση και την παρεμπόδιση της έκτης περιόδου της μαζικής εξαφάνισης των ειδών η οποία βρίσκεται σε εξέλιξη - αντιπροσωπεύουν την ταπεινότητα και τον κίνδυνο της εποχής μας.[3] Η πιθανότητα να εξαφανιστούν όλα τα έντομα μέχρι το 2119[4] θέτει ένα αποτελεσματικό χρονοδιάγραμμα για τη δημιουργία του τελικού προστατευτικού φραγμού για την περαιτέρω ανθρώπινη ύπαρξη, επειδή χωρίς έντομα η επικονίαση της χλωρίδας και η βιολογική υποδομή για την ύπαρξη των μικροβιοτικών οργανισμών θα καταρρεύσουν. Το είδος μας θα εξαφανιστεί, όντας μάρτυρας της εξαφάνισης όλων των μεγάλων θηλαστικών, της συντριπτικής πλειονότητας των πτηνών και της εξαφάνισης της πλειονότητας των θαλάσσιων σπονδυλωτών. Η επίγνωση των *ανθρώπων* μέσα σε ένα βιολογικό εύρος σχετικά λίγων χιλιετιών θα αποδειχθεί ότι δεν είναι τίποτα περισσότερο από τη μαρτυρία εξόφθαλμων αιτιών και αιτιατών από μια ομάδας καταδικασμένων, οικοκτόνων ναρκισσιστών.

Ούτε οι φυσικοί ούτε οι οικονομολόγοι μπορούν να λύσουν αυτό το πλέον σημαντικό όλων κλιμακούμενο πρόβλημα. Τη διάρκεια των στρεσογόνων παραγόντων. Τους φαντάζονται μόνο ως αιτιώδη παράδοξα που εκδηλώνονται με όχι περισσότερο άμεσο ενδιαφέρον από αυτό της αστάθειας των χρηματιστηρίων. Η κοσμολογική αμοιβαιότητα είναι Νευτώνεια, οι αστεροειδείς συγκρούονται και απωθούνται. Οι οικονομικές συναλλαγές, αν και θεωρητικά λογίζονται ως αγαθοεργία η οποία γεννήθηκε από καλόπιστες διαπραγματεύσεις – όπως οραματίστηκε ο Ιμμάνουελ Καντ στην πραγματεία του Προς την αιώνια ειρήνη: Ένα φιλοσοφικό σχεδίασμα (1795), δεν επιλύουν τα μεγαλύτερα ζητήματα των ανθρώπινων κατώτατων ορίων σε παγκόσμιο επίπεδο για αφειδή αλτρουισμό σε μεγάλη διακύμανση.[5] Τα ειρηνικά μοτίβα αντιπραγματισμού, ή το δημοτικό θέατρο μπορεί να υποδηλώνουν ότι μπορούμε να διατηρούμε ένα δίκτυο εξουσίας, τα συνεκτικά νήματα της δημοκρατίας. Ξέρουμε όμως ότι όλα είναι εύθραυστα. Υπήρξαν πάντα δύο απόψεις για την εξεύρεση μοντέλων λύσεων, η "εξελικτική και λειτουργική οικολογία", στην οποία προσπαθούμε να "την οραματιστούμε ["την αφθονία και κατανομή ενός οργανισμού" ή ακόμα οποιοδήποτε σύμπτωμα της ζωής στη γη] ως το προϊόν μιας ιστορίας προσαρμογής στην κοινότητα στην οποία βρίσκεται".[6] Αυτό, ωστόσο, ελάχιστα εξυπηρετεί τον σκοπό μας. Ακόμα κι αν λάβουμε σοβαρά υπόψη την "προσαρμογή", κάτι που έχουν κάνει ελάχιστοι, εξ όσων γνωρίζουμε. Ίσως οι λίγοι εναπομείναντες μοναχοί Jain Digambara στην Ινδία ή οι κοινοβίτες μοναχοί στο Όρος Άθως, ή στο Μοναστήρι της Αγίας Αικατερίνης στο Σινά. Τα τελευταία εναπομείναντα μέλη των ελάχιστων Προγραμματισμένων Φυλών των Νήσων Άνταμαν, ή των Τόντα της ορεινής επαρχίας Νιλκγίρις στην Ινδία[7,8], αποτελούν το καλύτερο σενάριο.

Κρίσιμο εμπόδιο για την κατανόηση των βιολογικών ιδιοτήτων και προοπτικών για την επίτευξη ανιδιοτελούς (κρίσιμο επακόλουθο) αλτρουισμού είναι η ευρέως διαφοροποιημένη προσέγγιση της αμοιβαιότητας, στην εύφορη ημισέληνο της οποίας βρίσκονται οι ρίζες τόσο της συμπόνιας όσο και της ανοχής. Από αυτές τις δύο ενστικτώδεις και/ή διαισθητικά καλές νόρμες προκύπτει η αναπόφευκτη συνειδητοποίηση ότι κάθε είδος αποτελεί πράγματι μια ξεχωριστή, αν και όχι απαραίτητα αφαιρεθείσα, επανάληψη της αρχής. Για παράδειγμα, μια μελέτη που συγκρίνει "τον αρουραίο της Νορβηγίας, την κοινή νυχτερίδα βαμπίρ και τον καφέ πίθηκο καπουτσίνο" τονίζει αυτό που δύο ερευνητές έχουν περιγράψει ως "πιθανώς μία από τις πιο αμφιλεγόμενες θεωρίες στην εξελικτική έρευνα" ("Αμοιβαιότητα"). [9] Μία ομάδα ερευνητών κατατείνει στο συμπέρασμα ότι η σημειολογία αποτελεί μέρος της πρόκλησης, μια "σύγχυση" σχετικά με τη χρήση των όρων, καθώς και "η ανάγκη διάκρισης μεταξύ των κατά προσέγγιση (του μηχανισμού) και των τελικών (της επιβιωτικής αξίας) εξηγήσεων συμπεριφοράς". [10]

Πρόσφατα "αποδείχθηκε πειραματικά"[11] η συντονισμένη, συνεργατική συμπεριφορά μεταξύ άσχετων κιχλίδων (με Κρητιδική προέλευση, με περισσότερα από 3.000 είδη ψαριών στην τάξη περκόμορφα, συμπεριλαμβανομένης των ειδών του γένους τιλάπια). Αυτού του είδους η διεξαγωγή ad hoc τυχαίων ερευνών διασποράς διεξάγεται εδώ και πολλές χιλιάδες χρόνια στο πλαίσιο της ανθρώπινης έρευνας στο όνομα της φυσικής επιστήμης. Το 1987, ο Ρόμπερτ Τρίβερς δημοσίευσε ένα δοκίμιο με τίτλο "The Evolution of Delayed Return Altruism" ("Η εξέλιξη της καθυστερημένης επιστροφής του αλτρουισμού")[12] στο οποίο διατύπωσε τη συσσώρευση καθαρών οφελών για τον αλτρουιστή έναντι του μη αλτρουιστή, μια σχέση που επινοείται και επιταχύνεται από τη φυσική επιλογή. Θεώρησε τα επίπεδα "αμοιβαίας εξάρτησης", ως χρονικά πλαίσια, ακόμη και θεωρητικά αλληλόμορφα σε διαφορετικούς τόπους στους γονότυπους των ατόμων. Κυριαρχία, θεωρία παιγνίων, οι συμβιώσεις που προκύπτουν όταν "οι λαβρίδες του είδους Labroides dimidiatus καθαρίζουν έναν άλλο οργανισμό (π.χ. τον ροφό *Epinephelus striatus*) από εκτοπαράσιτα (π.χ., καλυκοειδή κωπήποδα)" τα οποία μπαίνουν στο στόμα ή στους θαλάμους των βραγχίων του "ξενιστή".[13]

Ο Τρίβερς θεώρησε τους προειδοποιητικούς ήχων μεταξύ των πουλιών ως μια λειτουργικότητα των διακριτών προτύπων βοήθειας, και, όσον αφορά τους ανθρώπους, υπέδειξε ότι υπάρχουν τουλάχιστον 5 μορφές "αλτρουιστικής συμπεριφοράς" οι οποίες θεωρούνται σίγουρα πλήρεις αρχές. Αυτές περιλαμβάνουν την "παροχή βοήθειας σε περιπτώσεις κινδύνου", την "ανταλλαγή φαγητού", την "παροχή βοήθειας στους αρρώστους", την "ανταλλαγή εργαλείων" και την "ανταλλαγή γνώσης".[14] Αυτές οι "προϋποθέσεις για την εξέλιξη του αμοιβαίου αλτρουισμού", υποστήριξε, ενδέχεται να υπάρχουν ακόμη και πριν από την περίοδο της "Πλειστόκαινου" εποχής,[15] το οποίο σημαίνει, δυνητικά, πριν από 2,6 εκατομμύρια χρόνια. Η πραγματική αποτυχία - πέρα από την κοινή λογική - να ληφθεί υπόψη η ευρεία κατανομή της φιλίας, της συμβίωσης, των αμοιβαία επωφελών συνεργασιών που θα έχουν ως αποτέλεσμα το κέρδος και για τα δύο μέρη της σχέσης, σε ένα πολύ ευρύτερο χρονικό πλαίσιο, ας πούμε δισεκατομμύρια χρόνια, και που περιλαμβάνει μονοκύτταρους οργανισμούς, είναι πιο δύσκολο να προσδιοριστεί.

Μεμονωμένες ομάδες ατόμων (δήμοι) μπορεί να έχουν ελάχιστους λόγους να αλλάξουν αρχετυπικά μοτίβα, εκτός αν μπει στην εξίσωση μια ξαφνική ανταμοιβή από ξένους. Ποια ήταν η αρχική κινητήριος δύναμη που ώθησε δύο οργανισμούς να "κρατηθούν χέρι-χέρι"; Ή να φιληθούν, να τραφούν μαζί, να δώσουν θρεπτικά συστατικά ο ένας στον άλλο; Η σύγκριση της εγωιστικής με την ανιδιοτελή φύση των σχέσεων έχει από καιρό υποβληθεί σε δοκιμές στα μικροσκόπια της φυσικής επιλογής με ελάχιστη ανάλυση.[16] Θα μπορούσε κανείς απλά να ρωτήσει το εξής: "Τι είναι αγάπη"; Και (ακόμα πιο σημαντικό ερώτημα) πώς λειτουργεί η χωρίς ανταπόκριση αγάπη μέσα στο πλαίσιο οποιωνδήποτε εξελικτικών κανόνων;

Ορθώς ο Martin Nowak απέρριψε τον ισχυρισμό ("Κανόνα") του Βρετανού/Ινδού επιστήμονα Τζ. Μπ. Σ. Χαλντέιν (1892-1964) ότι – αποκλειστικά – "η φυσική επιλογή μπορεί να ευνοήσει τη συνεργασία εάν ο δότης και ο αποδέκτης μιας αλτρουιστικής πράξης είναι γενετικοί συγγενείς". Προφανώς, τα "άσχετα" άτομα είναι γενναιόδωρα μεταξύ τους σε αμέτρητες περιπτώσεις και κατηγορίες περιπτώσεων (το "δίλημμα του φυλακισμένου", για παράδειγμα). Οι έννοιες του "Άμεσου", "Έμμεσου", "Δικτυακού", "Ομαδικού" και "Εξελικτικού" αλτρουισμού είναι τόσο αληθινές όσο τα μέλη της οικογένειας που μοιράζονται μεταξύ τους ελάχιστες προμήθειες.[17] Το πρόβλημα, λοιπόν, δεν είναι η διαχειριστική εξουσία ή η ηθική συνειδητοποίηση που είναι η αμοιβαιότητα, αλλά η υποκείμενη συνείδηση που κινείται προς τη συνείδηση, ως υψηλότερη λειτουργικότητα της συνειδητοποίησης γενικά. Στην πραγματικότητα, μια πιο προφανής περίπτωση ήταν η ζωή του ίδιου του Χαλντέιν. Ποια ήταν η αμοιβαία ανταμοιβή που ανέμενε όταν ζήτησε να παραδοθεί το σώμα του στην επιστήμη μόλις πεθάνει από καρκίνο; Θα μπορούσε κανείς να υποστηρίξει ότι μια τέτοιου είδους επίδειξη μετά θάνατον αλτρουισμού ήταν μια θεμελιώδης μορφή συζευκτικής επίδρασης στη φύση του ισομορφισμού. Στις μελέτες των μαθηματικών και των φυσικών επιστημών υπάρχουν σχεδόν είκοσι γνωστές περιπτώσεις τέτοιων συζεύξεων, και όλες περιγράφονται εντός της τροχιάς της αμοιβαιότητας: είτε στον κλασικό ηλεκτρομαγνητισμό (κυκλώματα, διακλαδώσεις και σχετική τάση), μια κεραία που μεταδίδει και λαμβάνει σήματα, ιδιότητες περιορισμού και επαγωγής (αμοιβαιότητα Frobenius), οι αμοιβαίες εντάσεις φωτός και οι αιτίες τους και τα αποτελέσματα της έκθεσής τους στη φωτογραφία (πτυχές της αμοιβαιότητας Helmholtz), αμοιβαίες συνδέσεις στα νευρωνικά δίκτυα, δεξιόστροφο B-DNA και αριστερόστροφο Z-DNA, κ.λπ.

Αλλά από όλες τις συζεύξεις, καμία δεν είναι σαν αυτή που περιγράφεται στο βιβλίο του Σάιμον Σάμα "Rembrandt's Eyes". Στη μεγάλη αυτή βιογραφία του καλλιτέχνη, και τα δυο του μάτια είναι πάντα διαφορετικά σε κάθε ένα από τα πορτρέτα του ζωγράφου. [18] Ομοίως, υπάρχουν πολλές οπτικές μελέτες περί μη συμβατικών καθρεφτών στις οποίες "δύο άνθρωποι [που παραδοσιακά θα ήταν σε θέση να] κοιτάζουν ο ένας τον άλλο μέσω της αντανάκλασής τους" θα έβλεπαν ξαφνικά μια τέτοια "συμμετρία" να διαλύεται.[19] Η διάλυση αντιπροσωπεύει μια καθολική ευαλωτότητα, είτε σε αυτό που ο Φράνσις Σκοτ Φιτζέραλντ ονόμασε "Η ρωγμή",[20] ή στο πλαίσιο των "αέναων ταλαντώσεων μεταξύ συνεργασίας και αποστασίας" - την αποσύνθεση των "στρατηγικών ισορροπίας" όταν η "ένταση μεταξύ ιδιωτικού και κοινού συμφέροντος" υπερβαίνει την αντοχή εφελκυσμού των συμβατικών φυσικών ή πνευματικών δεσμών που διαμορφώνουν τις κοινότητες.[21]

Άλλες εκδηλώσεις αυτής της έντασης υπάρχουν στις διαρκώς αβέβαιες γκρίζες ζώνες που υφίστανται μεταξύ της άμεσης και της έμμεσης αμοιβαιότητας. Πριν από μερικά χρόνια διεξήχθη μια πολύ περίεργη μελέτη του όμορφου μικρού ωδικού πτηνού Καλόγερος της οικογένειας *των Παριδών*. Η ιδέα ήταν να καθοριστεί σε ποιο σημείο αυτά τα πανέμορφα και κοσμοπολίτικα πουλιά συνδυάζουν την άμεση και έμμεση εμπειρία στη δημιουργία εντυπώσεων πιθανών αντιπάλων.[22] Αποδεικνύεται ότι, όπως όλα τα έμβια όντα σε κάποιο επίπεδο, έτσι και αυτά κάνουν το ίδιο. Η ακριβής υπόθεση διατυπώθηκε μερικά χρόνια αργότερα: "Εάν τα πουλιά μπορούν να συνδυάσουν τα δύο… τότε είναι πιθανό ότι και οι άνθρωποι θα είναι σε θέση να το κάνουν".[23] Η υπόθεση είναι ιδιαίτερα σημαντική στο πλαίσιο της φήμης, έτσι ώστε διατυπώθηκε ότι "η ΕΑ [έμμεση αμοιβαιότητα] δεν μπορεί να εξηγήσει την ανησυχία κάποιου σχετικά με τη φήμη και την υπόληψη στις τυπικές κοινωνίες, εκτός εάν η υπόληψη αποτελεί κάποιον εξίσου αξιόπιστο οδηγό για τη συνεργατική συμπεριφορά, όπως η εμπειρία".[24] Εμπλέκεται ένας δημόσιος χώρος, δημόσια αγαθά. Η πιθανότητα μελλοντικών συναντήσεων μεταξύ των συγκεκριμένων ατόμων που συμμετέχουν στην παρατήρηση ή/και τη συναλλαγή. Στο πιο εσωτερικό επίπεδο, αυτή είναι η εμμονή του Εαυτού που είναι καταδικασμένος στον δικό του διάλογο. Ή, *πόσες φορές πρέπει να σου το πω…* Έχει τη δική του μονότονη πολιτική και φειδωλή στάση, μια επιτηδευμένη κοινωνιολογία που απεκδύεται την ποίηση αναζητώντας κάτι αξιόπιστο σε έναν κόσμο επιρρεπή σε παροξυσμούς και στο απροσδόκητο.

Όλα ωστόσο αυτά τα συμπεράσματα υποθέτουν ότι η αντιπαλότητα είναι το παν στον κόσμο, όταν δεν υπάρχει Ειδικός Νόμος Περί Ανταγωνισμού. Πράγματι, στα τέλη του 18^{ου}

αιώνα, η Ακαδημία των Γροθιών του Τσέζαρε Μπεκαρία εμπνεύστηκε από ένα πάθος για τη "διεξαγωγή αδυσώπητου πολέμου κατά της οικονομικής αναταραχής, της γραφειοκρατικής τυραννίας, της θρησκευτικής στενόμυαλης σκέψης και της διανοητικής σχολαστικότητας".[25] Ενώ οι εντάσεις που αναφέρθηκαν παραπάνω μπορεί να ισχύουν στη φυσική και τη μηχανική, από τις συγκρούσεις των μπαλών του μπιλιάρδου μέχρι τη διάτμηση του ανέμου στα αεροσκάφη και σε όλες τις εφαρμογές της θερμοδυναμικής, με πραγματική αμοιβαιότητα εντός της βιοημίσφαιρας, είμαι όντα που *αισθανόμαστε* και παίζουμε με γενικές παρορμήσεις που περιλαμβάνουν την αγάπη και την ευγνωμοσύνη, την περιέργεια και τη φιλία, το πάθος και τη συγχώρεση, τη χαρά της προσφοράς. Αυτά τα qualia (δηλαδή, οι άρρητες υποκειμενικές ποιότητες της εμπειρίας) καταστρέφουν κάθε μαθηματική, ισοπεδωμένη στατιστική, υπονομεύουν την πρόβλεψη και σχεδόν καταστρέφουν την παρέκταση. Πρέπει να ξεκινήσουμε τα πάντα από την αρχή με αληθινή συνείδηση.

ΣΗΜΕΙΩΣΕΙΣ

[1] Βλ. "What is Consciousness? Scientists Compete to Find Out," του Andreas Kluth, 23 Ιανουαρίου 2021, *Bloomberg Opinion*, https://www.bloomberg.com/amp/opinion/ articles/2021-01-23/what-is-consciousness-scientists-around-the-world-compete-to-find-out, Πρόσβαση στις 26 Ιανουαρίου 2021.

[2] *New AI system counts endangered elephants from space*, του Thomas Macaulay, 20 Ιανουαρίου 2021, Neural, https://thenextweb.com/neural/2021/01/20/new-ai-system-counts-endangered-elephants-from-space/amp/

[3] "How the Sixth Mass Extinction Affects the U.S. Economy? του Kimberly Amadeo, Treehugger, 21 Αυγούστου 2020, https://www.treehugger.com/mass-extinction-definition-events-economic-effect-of-6th-extinction-4589742

[4] "Last year, 40% of honey-bee colonies in the US died. But bees aren't the only insects disappearing in unprecedented numbers," by Aylin Woodward, June 21, 2019, *Business Insider*, https://www.businessinsider.com/insects-dying-off-sign-of-6th-mass-extinction-2019-2, Πρόσβαση στις 26 Ιανουαρίου 2021.

[5] *Behavioral Economics.com*," "Reciprocity, https://www.behavioraleconomics.com/resources/ mini-encyclopedia-of-be/reciprocity/, Πρόσβαση στις 26 Ιανουαρίου 2021.

[6] "A Critical Look at Reciprocity in Ecology and Evolution: Introduction to the Symposium," των David N. Reznick, Editor, Judith L. Bronstein, The University of Chicago Press Journals, *The American Naturalist*, Vol. 181, Supplement, May 2013, https://www.journals.uchicago.edu/doi/abs/10.1086/670030?journalCode=an

[7] Βλ. *The Toda Landscape – Explorations In Cultural Ecology*, του Tarun Chhabra, Harvard University Press, Orient BlackSwan, 2015.

[8] Βλ. "Ahimsa – NonViolence," μονόωρο ντοκιμαντέρ του PBS, Σενάριο και σκηνοθεσία του M.C.Tobias, Direct Cinema Ltd., 1987.

[9] "Reciprocity: Different behavioural strategies, cognitive mechanisms and psychological processes," των Martin K. Schweinfurth και Josep Call, *Learning & Behavior* 47, 284-301(2019), https://doi.org/10.3758/s13420-019-00394-5, Δεκέμβριος 2019, https://link.springer.com/article/10.3758/s13420-019-00394-5#citeas

[10] Βλ. "Social semantics: altruism, cooperation, mutualism, strong reciprocity and group selection," των S. A. West, A. S. Griffin και A. Gardner, *Journal Of Evolutionary Biology*, 16 Οκτωβρίου 2006, https://doi.org/10.1111/j.1420-9101.2006.01258.x, https://onlinelibrary.wiley.com/doi/full/10.1111/j.1420-9101.2006.01258.x

[11] "Coaction vs. Reciprocal Cooperation Among Unrelated Individuals in Social Cichlids," των Michael Taborsky και Thomas Riebli, *Frontiers in Ecology and Evolution,* 21 Ιανουαρίου 2020 | https://doi.org/10.3389/fevo.2019.00515, https://www.frontiersin.org/articles/ 10.3389/fevo.2019.00515/full

[12] Βλ. *The Quarterly Review of Biology*, Vol. 46, No.1, Mar. 1971, σελ. 35-37, The University of Chicago Press, http://www.jstor.org/stable/2822435, https://greatergood.berkeley.edu/ images/uploads/Trivers-EvolutionReciprocalAltruism.pdf; Βλ. επίσης "Reciprocal altruism: 30 years later," in Cooperation in Primates and Humans, Επιμέλεια P. M; Kappeler και C. P. van Schaik, Springer, Berlin Heidelberg, https://doi.org/10.1007/3-540-28277-7_4, doi: https://doi.org/10.1007/3-540-28277-7_4, https://link.springer.com/chapter/10.1007%2F3-540-28277-7_4

[13] ό.π. σελ.40.

[14] ό.π. σελ.45.

[15] ιό.π.. σελ.45. Βλ. επίσης A. Gouldner, "The norm of reciprocity: a preliminary statement," *Am. Sociol. Rev.*, 47: σελ.73-80, 1960.

[16] Βλ. *The Evolution of Reciprocal Sharing," Ethology and Sociobiology* 5: 5-14 (1984), https://pages.ucsd.edu/~jmoore/publications/Recip.html; Βλ. επίσης "Effects of multiple gene control on the spread of altruism by group selection," του Tomás Kulich, Jaroslav Flegr, https://arxiv.org/pdf/1002.4204.pdf

[17] "Five rules for the evolution of cooperation," του Martin A. Nowak, Science. 2006 Dec 8; 314(5805): 1560–1563. doi: 10.1126/science.1133755. https://www.ncbi.nlm.nih.gov/ pmc/articles/PMC3279745/, Βλ. επίσης "Upstream reciprocity and the evolution of gratitude," των Martin A. Nowak και Sébastien Roch, Proc Biol Sci. 2007 Mar 7; 274(1610): 605–610. Published online 2006 Dec 19. doi: 10.1098/rspb.2006.0125, *The Royal Society Publishing, Proceedings B,* PMCID: PMC2197219, PMID: 17254983, https://www.ncbi.nlm.nih.gov/pmc/articles/PMC2197219/

[18] *Rembrandt's Eyes*, Simon Schama, Alfred A. Knopf, New York, NY, 1999.

[19] Βλ. "Non-reciprocal Optical Mirrors Based on Spatio-Temporal Modulation," των R. Fleury, D. L. Sounas, και A. Alù, 10.1088/2040-8986/aaaa3e, *Physics, Optics*, Cornell University, https://arxiv.org/ftp/arxiv/papers/ 1709/1709.00702.pdf

[20] *The collection of posthumously published essays* by F. Scott Fitzgerald, επιμέλεια Edmund Wilson, New Directions Publishers, New York, NY 1945.

[21] Βλ. "Direct Reciprocity in structured populations," των Matthijs van Veelen, Julián Garcia, David G. Rand και Martin A Nowak, PNAS, *Proceedings of the National Academy of Sciences of the United States of America, PNAS* 19 Ιουνίου 2012 109 (25) 9929-9934; https://doi.org/10.1073/pnas.1206694109, https://www.pnas.org/ content/109/25/9929

[22] Peake T.M, Terry A.M.R, McGregor P.K, Dabelsteen T. Do great tits assess rivals by combining direct experience with information gathered by eavesdropping? *Proc. R. Soc. B.* 2002;269:1925–1929. doi:10.1098/rspb.2002.2112 [PMC free article] [PubMed] [Google Scholar]

[23] "Evolution of direct and indirect reciprocity," του Gilbert Roberts, The Royal Society Publishing Proceedings B, PMC, *Proc Biol Sci.* 2008 Jan 22; 275(1631): 173–179. Ημεομηνία δημοσίευσης: 31Οκτ 2007. doi: 10.1098/rspb.2007.1134, PMCID: PMC2596181, PMID: 17971326, https://www.ncbi.nlm.nih.gov/pmc/ articles/PMC2596181/#bib17

[24] ό.π., Βλ. επίσης "Evolution of indirect reciprocity," των Martin A. Nowak και Karl Sigmund, *Nature*, 2005/11/01, 1291, 8, Vol. 437, DO – 10.1038/nature04131, https://www.researchgate.net/figure/Direct-and-indirect-reciprocitya-Direct-reciprocity-means-that-A-helps-B-and-B-helps-A_fig2_7514734

[25] Βλ. *Cesare Beccaria (1738-1794), Biography*, 16 Ιουλίου 2020, https://www.biography.com/scholar/cesare-beccaria, Πρόσβαση στις 17 Φεβρουαρίου 2021.

Ο πεμπτουσιακός δεσμός, ζευγάρι αλόγων, Socotra Highlands, Υεμένη. Φωτογραφία © M. C. Tobias.

Κεφάλαιο 25

ΘΕΩΡΙΕΣ ΣΥΖΕΥΞΗΣ ΣΕ ΕΝΑΝ ΚΟΣΜΟ ΓΕΜΑΤΟ ΔΙΑΦΟΡΕΣ

Einfühlungk, empatheria, empathy, ενσυναίσθηση[1]. Θεωρούμε ότι η συγκεκριμένη ιδιότητα αποτελεί το φυσιολογικό κίνητρο για μία διαρκή αμοιβαιότητα, σε αντίθεση με μια ψευδή αποτελεσματικότητα η οποία βασίζεται σε εντελώς άλλα κίνητρα τα οποία δεν είναι διοικολογικά ευεργετικά. Η λατινική και παλαιογαλλική προέλευση της λέξης empathy (ενσυναίσθηση) υποδηλώνουν ει λύπη, τονίζοντας τη Βουδιστική πρώτη ξεχωριστή αλήθεια, Dukkha, εξισώνεται με την ίδια τη ζωή ως ατελής λύπη. Θα μπορούσαμε να ισχυριστούμε ότι η παλιά γαλλική *χαρά*, από το λατινικό *gaudium*, ή *gaudere* στο "δίνω χαρά" θα μπορούσε να ταιριάξει εξίσου σε οποιοδήποτε πραγματικό βιολογικό περιβάλλον, το ακριβώς αντίθετο από αυτό που κήρυξε ο Βούδας Gautama. Υπάρχει ωστόσο και μια τρίτη, δυνητικά ελλειμματική ψυχολογική πραγματικότητα στη δυάδα. Νιώθουμε ένοχοι ή ντροπή όταν η παραμικρή αιτία χαράς ξεσπάσει όταν βρισκόμαστε σε καθεστώς έντονης θλίψης, όμως δεν υποφέρουμε όταν νοιώθουμε τη χαρά ενός παιδιού που γελάει ή ενός λατρεμένου συντρόφου. Και οι δύο εμπειρίες πλαισιώνονται από την οικολογική επίγνωση/συνειδητοποίηση, το γεγονός δηλαδή ότι γνωρίζουμε ότι σε κάθε κοινότητα αρκετές φορές επικρατεί ηρεμία, αλλά και ότι οποιοσδήποτε οργανισμός μπορεί να βιώσει ταυτόχρονα και ακατάπαυστα την απόλυτη κόλαση.

Σε ένα περιβάλλον εγκρατούς ψυχολογίας, η πιθανότητα επέλευσης γεγονότων λύπης που θα έρθουν σε αντιδιαστολή στο εφήμερο τέχνασμα που αποκαλούμε ευτυχία είναι κάτι αναμενόμενο. Δίνοντας υπόσταση σε αυτές τις εμπειρίες σε κάθε έμβιο ον, σε όλους όσους θέλουν τη χαρά και όχι σκληρότητα, υπάρχει το λεγόμενο *idée fixe* δηλ. η προδιατεθειμένη σκέψη (έως και *εμμονή*) που βλέπει έναν κόσμο γεμάτο με καινούργιες πληγές ή επιλέγει να τις αγνοήσει στον δικό του χρόνο και με τον δικό του τρόπο. Τι, λοιπόν μπορούμε να περιμένουμε να εκλογικεύσουμε, ζώντας σε ένα περιβάλλον αμοιβαιότητας, έχοντας κατά νου τον στόχο της συγκαταβατικότητα μεταξύ των ειδών; Μια προσέγγιση στην καλοσύνη που ξεπερνά όλες τις υπάρχουσες θεωρίες;

Η πρόσφατη βιβλιογραφία σχετικά με τη φύση της αμοιβαιότητας βρίθει διαφόρων θεμάτων. "Ποια είναι τα βασικά στοιχεία για την καθιέρωση ενός συστήματος *ηθικών πόρων και εγκάρδιων αγαθών;"*, ρωτάει ο Πατρίτσι Κάλβο.[2] Έχει κάθε είδος τη *δική* του οικονομία, λειτουργεί με γνώμονα το "οφθαλμός αντί οφθαλμού", εκφράζει κοντόφθαλμο προσωπικό συμφέρον που καθημερινά αποκτά μεγαλύτερη βαρύτητα από την ιδέα μιας εξωτερικής αρχής που καλεί σε μεγαλύτερες πράξεις αλτρουισμού που θα βοηθήσουν την κοινότητα πολύ περισσότερο από τον Τόμας Χομπς σε κάθε άτομο; Ας θέσουμε αυτό το ερώτημα ξανά, με διαφορετικό τρόπο όμως αυτή τη φορά: Η καλοσύνη εκδηλώνεται με 100 εκατομμύρια διαφορετικές εκδοχές ανά είδος της ίδιας βούλησης; Είναι αυτή η ομοιότητα κατανεμημένη με παρόμοιο τρόπο στον χάρτη κάθε ατόμου, τρισεκατομμύρια από αυτά; Γιατί να μας ενδιαφέρει αυτή η ερώτηση; Έχει σημασία για όλους; Θα την απλοποιήσουμε με αυθαίρετα συμπεράσματα. Λέγεται ότι όλα τα θηλαστικά ανατρέφονται με κάποια εκδοχή του ίδιου τρόπου ανατροφής. Γνωρίζουμε ότι αυτό δεν συμβαίνει ποτέ σε ποσοστό 100%, τουλάχιστον σίγουρα όχι μεταξύ των ανθρώπων, όπου η ενδοοικογενειακή βία αντιπροσωπεύει το μεγαλύτερο πεδίο επιθετικότητας, ιδιαίτερα η βία προς τις γυναίκες και τα νεαρά κορίτσια. Οι ερευνητές της συμπεριφοράς των ζώων και των φυτών μπορούν να παραθέσουν εύκολα τεράστια βιβλιογραφία με αντιφάσεις έναντι οποιασδήποτε απεικόνισης της βιόσφαιρας ως ειρηνικής, όπως έχουμε συζητήσει νωρίτερα σε πολλές περιπτώσεις. Επομένως, αν επικεντρωθούμε αυστηρά στην καλοσύνη των ανθρώπων, τι μπορούμε να περιμένουμε όσον αφορά τις προβλέψιμες γενικεύσεις; Μόνο ότι η τάση υπάρχει, είναι σαφώς διαδεδομένη, καθολικά εφαρμόσιμη, αλλά ποτέ εγγυημένη. Δεν υπάρχουν οριστικές μελέτες εντός του είδους μας που να ρυθμίζουν τον βαθμό στον οποίο διάφοροι στρεσογόνοι παράγοντες (περιβαλλοντικά επιβλαβείς παράγοντες όπως η πείνα, η αδιάκοπη θερμότητα, η ατμοσφαιρική ρύπανση και η ρύπανση των υδάτων κ.λπ.) μειώνουν τον στατιστικό ρυθμό ή την ίδια την ικανότητα γενναιοδωρίας, συμπόνιας, πόσο μάλλον την ικανότητα ανοχής. Συνδυάστε τις μεταβλητές ακόμα και με ένα μόνο άλλο είδος, και βρισκόμαστε σε μία απέραντη θάλασσα.

Τελικά, υποστηρίζουν οι Όστρομ και Γουόκερ, με την ενεργοποίηση του ανθρώπινου μετωπιαίου λοβού όταν εμπλέκονται η εμπιστοσύνη και η συνεργασία, είναι σαφές ότι "η σημασία των αμοιβαίων σχέσεων στην επεξήγηση της προέλευσης της εμπιστοσύνης και της αξιόπιστης συμπεριφοράς"[3] υπερβαίνει κατά πολύ τα γονίδια εγωισμού, τη θεωρία των παιγνίων και τις μαθηματικές αφαιρέσεις.[4]

Η έκφραση της ανθρώπινης αμοιβαιότητας περί ανταπόδοσης αμφιταλαντεύεται μεταξύ της λέξης RSVP [ακρωνύμιο της γαλλικής φράσης *Répondez s 'il vous plait* ("Παρακαλώ απαντήστε") η οποία έχει καταστεί πλέον παρωχημένη και έχει αντικατασταθεί από την πολύ λιγότερο φιλική *Réponse attendue avant le*.... ("Η απάντησή σας αναμένεται έως τις…"), ή "Με λύπη σας ενημερώνουμε ότι…(*regrets only*)". Οι οικονομικές βάσεις της πηγάζουν από συστήματα δωρεάς και ανταλλαγής, των οποίων η ανακρίβεια ενισχύθηκε από τον χρηματικό υποκειμενισμό: Ποια είναι η αξία ενός ποτηριού με νερό σε σχέση με ένα ποτήρι με χυμό καρύδας; Προφανώς, αυτό εξαρτάται από το χρονικό σημείο της δίψας. Η αμοιβαιότητα των καταστάσεων που δημιουργεί τις αφηγήσεις που κάμπτουν το πλαίσιο ώστε να ταιριάζει σε μια μοναδικότητα χρόνου/χώρου, είναι ζωτικής σημασίας για την αναγνώριση του τρόπου με τον οποίο πρέπει να αυξομειώνεται το ιδιοτελές συμφέρον, προλαμβάνοντας έτσι την ανάγκη γενίκευσης των αρχών της προσφοράς και της λήψης. Η οικονομική αμοιβαιότητα δεν υπόκειται στα χαρακτηριστικά των κύκλων άνθισης και μαρασμού των οικοσυστημάτων, αλλά στην αίσθηση ότι στο επίκεντρο των ανταλλαγών υφίσταται σε κάποιον βαθμό δικαιοσύνη.[5,6] Ωστόσο η ευγένεια δεν αποτελεί ένα από τα μακροχρόνια συστατικά της οικονομικής θεωρίας ή των αγορών.

Με τη δικαιοσύνη ως πρακτικό σημείο εκκίνησης, δύσκολα θα εξηγηθεί πώς, για παράδειγμα, μια υποτιθέμενη φτωχή κοινότητα μπορεί παρόλα αυτά να χαρακτηρίζεται ως εκπληκτικά φιλάνθρωπη[7] ή ποιοι μηχανισμοί έχουν την τάση γενετικά και πολιτιστικά να παρακινούν δυο άτομα ξένα μεταξύ να βοηθούν ο ένας τον άλλο, ακόμη και με μεγάλη ταλαιπωρία, κόστος ή ακόμη και κίνδυνο για τον εαυτό τους. Και επιπλέον, να διατηρεί μια τέτοια συμπεριφορά ως αναγκαίες "ηθικές νόρμες".[8] Αυτό έχει παρατηρηθεί εδώ και αρκετούς αιώνες στους Ιάπωνες, των οποίων οι ζωές έχουν κατακλυστεί τόσο από σεισμούς, ηφαίστεια και – πιο πρόσφατα – από τη φρίκη της πυρηνικής καταστροφής και του τσουνάμι στη Φουκουσίμα. Παρά το όποιο ιστορικό τραγωδιών, οι Ιαπωνικές κοινότητες επιδεικνύουν πάντα αξιοσημείωτο στωικισμό στον τρόπο με τον οποίο οι άνθρωποι αλληλοβοηθούνται, επιδεικνύοντας κάθε είδους ανιδιοτελή ηρωισμό. Μπορεί να ειπωθεί το ίδιο και για τη ναζιστική Γερμανία; Οι διαφορές είναι έντονες, και φέρνουν στο φως σε ορισμένους ένα γονίδιο άμεσης παροχής βοήθειας και σε άλλους το αντίθετο. Η δικαιοσύνη, ως μέσο αθώωσης μίας προδήλως μη αμοιβαίας συμπεριφοράς, υψώνεται με την ευκαιρία των νομικών μέσων και ορισμένων, αλλά όχι όλων των δικαιωμάτων αναγωγής.[9]

Η διαφοροποίηση των αμοιβαίων κανόνων από τις διαθέσεις κατανέμεται οργανικά υπό την έννοια ότι η εξέλιξη παρέχει ανταμοιβές και ποινές, για τις οποίες κάποιος μπορεί να ευελπιστεί ή να τις αγνοεί.[10] Αν δούμε τις επιλογές ως πράξεις εξιλέωσης για τα λάθη του παρελθόντος, θα δούμε μια πολύ διαφορετική στιγμή από εκείνους που δεν μπορούν να συναισθανθούν ή ακόμα και να αναγνωρίσουν τον πόνο των άλλων (το πιο χαρακτηριστικό παράδειγμα είναι ο Καρτέσιος [1596-1650] που περιγράφει τα ζώα ως απλές μηχανές χωρίς αίσθηση του πόνου). Ο τρόπος με τον οποίο προβάλλουμε τα επιχειρήματά μας σχετικά με την επιλογή μας να πράττουμε το καλό παρά να προκαλούμε τη δυστυχία παραμένει ένα άγνωστο πεδίο σχέσεων, προσωπικότητας, συμπεριφοράς και προσωπικής ιστορίας. Η κοινωνιολογική επισκόπηση των κοινοτήτων μπορεί να συλλάβει κάποια ροπή, η οποία ωστόσο καταρρέει στο επίπεδο των ατόμων που είναι ικανά για τα πάντα.[11] Τέτοιου είδους ροπές μπορούν να υποκινούνται από οποιοδήποτε αριθμό ερεθισμάτων. Ένας μεγάλος ηγέτης, μια άψογη φιλία, ο πνιγμός ενός ζώου του οποίου τα μέσα διάσωσης προφανή, η πίστη, η ελπίδα, η τιμή, η αυτοεκτίμηση, η υπεροχή των συνδεδεμένων συναισθημάτων. [12]

Η σύζευξη είναι η επικοινωνία σε κάποια φάση. Δεν μπορούμε να εξακριβώσουμε ποια είναι αυτή η φάση στο κενό, παρά την επίδειξη των ιδανικών εγγυήσεων στο σημείο της συλλογικής, κοινής προσπάθειας. Η Πράσινη Νέα Συμφωνία, για παράδειγμα, διεγείρει τον ενθουσιασμό σχετικά με την αμοιβαιότητα των οικολογικών στόχων, μια αμοιβαία πολιτική που ευνοεί θεμελιωδώς μια συγκεκριμένη κοινή πίστη σε μια εκδοχή της πραγματικότητας έναντι μιας άλλης.[13] Κάθε φορτισμένο *ζήτημα* προκαλεί το ίδιο. Μια κοινότητα σχηματίζεται και υπάρχει μέσα από τους πιο πιεστικούς περιορισμούς και ανωμαλίες και τον τρόπο επίλυσής τους. Εάν δεν υπάρχει κάποια παρόμοια ένταση, δεν υπάρχει κάτι που να μπορεί να συγκρατήσει μια κοινότητα. Ένα πλήρως απελευθερωμένο άτομο ή ομάδα τέτοιων ατόμων θα μπορούσε να εξισωθεί με σχέδια αγροτικών κατοικιών τα οποία είναι προσανατολισμένα προς την αυστηρή ανεξαρτησία, όπως διαμορφώνεται από τον αριθμό των στρεμμάτων ανά νοικοκυριό, δηλαδή την οικονομική κατάσταση. Αντιστρόφως, οι οριακές συνθήκες θα υπαγορεύουν την εμφάνιση και την τοπολογία της εξάρτησης. Και στα δύο άκρα, εξακολουθούν να υπάρχουν υδραυλικές και ηλεκτρικές υποδομές ή έλλειψη μίας αξιόπιστης υποδομής, όπως σε μια πολιτεία όπως το Τέξας, της οποίας το ενεργειακό δίκτυο εξαρτάται σε μεγάλο βαθμό από το φυσικό αέριο, αλλά είναι απροετοίμαστο για μαζικά ψυχρά μέτωπα. Στρατηγικές απόκτησης τροφίμων και φαρμάκων. Τελικά, οι περισσότεροι μοναχικοί τύποι τις δύσκολες νύχτες περιπλανιούνται σε μια ζεστή φωτισμένη παμπ. Κανένα κοινωνικό ζώο ή φυτό δεν θα απεκδυθεί εύκολα κάποια πτυχή της ομαδικότητας ή της επισήμανσης. Αυτό ισχύει και σε μονοκυτταρικό επίπεδο. Πράγματι, από το 2012, οι ερευνητές - αναγνωρίζοντας την αξιοσημείωτη ποσότητα αποθηκευτικής ικανότητας του DNA – προσπαθούν να χειραγωγήσουν μεμονωμένα στελέχη του DNA για να κωδικοποιήσουν ολοένα και περισσότερα δεδομένα.[14] Αλλά ένας κόσμος σημάτων και πληροφοριών προσκαλεί αμοιβαίους κανόνες μετάφρασης που είναι ακατέργαστοι, προ-φιλτραρισμένοι για να προκαλούν οικολογικά αίτια και συνέπειες, τόσο για το άτομο όσο και για ένα μοναστήρι ή μια μεγαλούπολη ατόμων.

Ως εκ τούτου, η βιολογία των συζεύξεων - αυτές οι αμοιβαίες μεταφράσεις - είναι η πεμπτουσία της οικολογικής κοινότητας. Μία θεμελιώδες *βασική* σύζευξη μας δίνει DNA και εγγενώς κβαντικές μηχανικές εκπλήξεις.

Στην περίπτωση της κλιματικής αλλαγής ή της κατανάλωσης ή οποιασδήποτε άλλης περιβαλλοντικής καταστροφής, ο εχθρός *των εμπειρικών δεδομένων* που κωδικοποιούνται φυσικά στο DNA, είναι ο εγωισμός, ο οποίος, φυσικά, έχει τους δικούς του ακλόνητους κώδικες που υπόκεινται σε αλλαγές. Το αντίπαλο δέος είναι η τάση αμοιβαιότητας – που υπόκειται επίσης σε μεγάλη αστάθεια – η οποία μπορεί να είναι κοινή στην πλειοψηφία ή, όταν είναι απαραίτητο, να εμφυτευτεί με κάποια μέθοδο (ή ενστικτώδη νευρική λειτουργία) ισοδύναμη με *εκτελεστική εντολή*, υποθέτοντας ότι η ενστικτώδης νομοθετική ή βιολογική ιεραρχία την υποστηρίζει. Ως εναλλακτική λύση, αναγνωρίζεται πολύ εύκολα ο όχλος, η ομήγυρη. Στην περίπτωση των κοινωνιών με αρετές που έχουν διαμορφωθεί έσω πολλών γενεών, θα ήταν αδιανόητο για ένα άτομο να υπονομεύσει μακροχρόνιες ηθικές συμφωνίες, όπως, για παράδειγμα, στην περίπτωση των Ινδιάνων Salish και Pend d' Oreille που καταφέρνουν να επιβιώσουν κατά μήκος του ποταμού Jocko, καταναλώνοντας την ντόπια πέστροφα, για χιλιάδες χρόνια μέσα στον ινδιάνικο καταυλισμό του Flathead της βορειοδυτικής Μοντάνα.[15] Μια τεράστια εθνογραφική βιβλιογραφία σχετικά με αυτό το θέμα (βιωσιμότητα) υποστηρίζεται από τους μηχανισμούς που περιγράφονται σε ολόκληρη την εκτεταμένη ιστορία της κοινωνικής ψυχολογίας και της έρευνας για την ειρήνη,[16,17] καθώς και από την οικολογική ηθική που περιγράφει οποιονδήποτε αριθμό περιπτώσεων του "ο ρυπαίνων πληρώνει" και σενάρια "τραγωδίας των κοινών".

Αμέσως κάποιος εδώ μπορεί να θυμηθεί τον στίχο του Λευιτικού "Ο ξένος ο κατοικών μεθ' υμών θέλει είσθαι εις εσάς ως εις εκ των πολλών υμών θέλετε αγαπήσει αυτόν ως σεαυτόν, διότι ξένοι εσταθήτε εν τη γη της Αιγύπτου: Εγώ ο ΚΥΡΙΟΣ ειμί ο Θεός υμών".[18] Αντιλαμβανόμενοι τη Γη ως Θεό, όλες οι αδύναμες ανθρώπινες ενορχηστρώσεις και προσανατολισμοί αλλάζουν ριζικά κατεύθυνση ή τουλάχιστον θα ήταν σοφό να το πράξουν, σύμφωνα με τα δόγματα της αμοιβαιότητας που παρακάμπτουν τον αδέξιο ανθρωπικό φακό της ιδιοτέλειας. Στην περίπτωση ενός πρεσβύτερου της φυλής των Αλγκονκίν, ο άνθρωπος προτείνει ότι η "βιωσιμότητα" μοιάζει να μην είναι τίποτα περισσότερο από τη συνεχή διαιώνιση της εκμετάλλευσης, με τον απλουστευτικό υπολογισμό ότι αν εκμεταλλευτείς αρκετά μπορείς επίσης να δώσεις πίσω μέρος του πλεονάσματος, ή κάποιες παρόμοιες ιδεολογίες. Ωστόσο κάτι τέτοιο δεν ισχύει, σωστά; Δηλώνει επίσης ότι "...το μέλημά μας ως λαός, δεν είναι τι μπορούμε να πάρουμε από τη γη, αλλά τι μπορούμε να δώσουμε".[19]

Πρόκειται για ένα εξαιρετικά δυνατό σημείο εστίασης για την εξέταση του τι πρέπει να γίνεται και με ποιον τρόπο, και ανάμεσα σε ποιο σύνολο συνανθρώπων μας, αν θέλουμε να είμαστε ειλικρινείς μεταξύ μας, άνευ όρων και με την πλέον γενναιόδωρη έννοια του όρου.

ΣΗΜΕΙΩΣΕΙΣ

[1] Βλ. *Center for Building a Culture of Empathy*, http://cultureofempathy.com/ References/History.htm

[2] Βλ. *The Cordial Economy – Ethics, Recognition and Reciprocity*, του Patrici Calvo, Springer International Publishing, New York, 2018.

[3] Βλ. *Trust and Reciprocity: Interdisciplinary Lessons for Experimental Research*, επιμέλεια Elinor Ostrom και James Walker, The Russell Sage Foundation Series on Trust, New York, NY, 2003, https://www.jstor.org/stable/10.7758/9781610444347; Βλ. επίσης *Trust and Reciprocity*, των Gary Charness και Valentin Shmidov, Now Publishers, Norwell, Mass., Delft, Ολλανδία 2014.

[4] Βλ. τις συζητήσεις σχετικά με τους νόμους αμοιβαιότητας για τις τετραγωνικές και κυβικές εξισώσεις, τα υπολείμματα και τους ορθολογικούς νόμους αμοιβαιότητας στο *Reciprocity Laws From Euler to Einstein*, των Franz Lemmermeyer, Springer Berlin Heidelberg, 2000.

[5] *The Economics Of Reciprocity, Giving and Altruism*, επιμέλεια J. Ythier, S. Kolm, L. Girard-Varet, Palgrave Macmillan UK, Camden, UK, 2000.

[6] Βλ. *Justice and Reciprocity in Aristotle's Political Philosophy*, του Kazutaka Inamura, Cambridge University Press, Cambridge, UK 2015. Βλ. επίσης *Reciprocity in Ancient Greece*, επιμέλεια Christopher Gill, Norman Postlethwaite και Richard Seaford, Oxford University Press, Oxford UK 1998. Βλ. επίσης *Resilience, Reciprocity and Ecological Economics: Northwest Coast Sustainability*, του Ronald Trosper, Taylor & Francis, Routledge Studies in Ecological Economics, Abingdon-on-Thames, UK, 2009; και *Giving Back: Research and Reciprocity in Indigenous Settings*, επιμέλεια R. D. K. Herman, Oregon State University Press, Corvallis, OR, 2018.

[7] *The Culture of Giving in Myanmar: Buddhist Offerings, Reciprocity and Interdependence*, του Hiroko Kawanami, Bloomsbury Academic, London, UK, 2020.

[8] *A Cooperative Species – Human Reciprocity and Its Evolution*, των Samuel Bowles και Herbert Gintis, Princeton University Press, Princeton, NJ, 2011.

[9] *The Persistence of Reciprocity in International Humanitarian Law*, του Bryan Peeler, Cambridge University Press, Cambridge UK, 2019. Βλ. επίσης *Reciprocity and the Art of Behavioural Public Policy*, του Adam Oliver, Cambridge University Press, Cambridge, UK, 2019; Βλ. επίσης *Reciprocity in International Law: Its impact and function*, του Shahrad Nasrolahi Fard, Taylor & Francis, Abingdon-on-Thames, UK, 2015.

[10] Βλ. *Altruistically Inclined? The Behavioral Sciences, Evolutionary Theory and the Origins of Reciprocity*, του Alexander J. Field, University of Michigan Press, Ann Arbor, MI, 2004. Βλ. επίσης *The Philosophers' Gift: Reexamining Reciprocity*, by Marcel Hénaff, μετάφραση Jean-Louis Morhange, Fordham University Press, New York, NY, 2019.

[11] Βλ. *Reciprocity, Altruism and the Civil Society: In praise of heterogeneity*, του Luigino Bruni, Taylor & Francis, Abingdon-on-Thames, UK, 2008.

[12] Βλ. *Reciprocity in Human Societies: From Ancient Times to the Modern Welfare State*, των Antti Kujala και Mirkka Danielsbacka, Springer International Publishing, New York, 2019, στις οποίες εξετάζονται οι μελέτες της Marcel Mauss για τη "θεωρία δώρων" και τον Barrington Moore σχετικά με τις αμοιβαίες υποχρεώσεις μεταξύ ανθρώπων που βρίσκονται σε πολύ διαφορετικές κοινωνικές καταστάσεις. Βλ. επίσης *Emotion, Identity, and Religion – Hope, Reciprocity, and Otherness*, του Douglas J. Davies, Oxford University Press, Oxford, UK, 2011, με έμφαση σε μια πολυάριθμη πειστική συλλογιστική ανταλλαγής που μπορεί να περιλαμβάνει τον Θεό και τις μυριάδες ερμηνείες του. ή αυτή η πίστη σε μια οντότητα αλλοτρίωσης - του Άλλου - που καθοδηγεί με μια φωνή που χρονολογείται από τους προγόνους μας Βλ.επίσης *Reciprocity and Dependency in Old Age – Indian and UK Perspectives*, της Sue Thompson, Springer New York, 2015, και *Reciprocity*, του Lawrence C. Becker, University of Chicago Press, Chicago, Ill., 1990.

[13] Βλ. *Contemporary Collaborative Consumption: Trust and Reciprocity Revisited*, επιμέλεια Isabel Cruz, Rafaela Ganga, και Stefan Wahlen, Springer Fachmedien Wiesbaden, Γερμανία, 2018.

[14] Βλ. "DNA could store all of the world's data in one room," του Robert F. Service, *Science*, 2 Μαρτίου 2017, https://www.sciencemag.org/news/2017/03/dna-could-store-all-worlds-data-one-room, Πρόσβασης στις 30 Ιανουαρίου 2020.

[15] *Bull Trout's Gift: A Salish Story about the Value of Reciprocity*, Confederated Salish and Kootenai Tribes, Flathead Indian Reservation, Montana, 2011. Βλ. επίσης, "The Natural Model of Reciprocity," της Madronna Holden, *Our Earth/Ourselves*, https://holdenma.wordpress.com/culture-and-environment/reciprocity/2008. Ιδιαίτερου ενδιαφέροντος, δείτε τη συμμετοχή της Holden στη συζήτηση της Winona LaDuke σχετικά με την αμοιβαιότητα από την άποψη της παράδοσης της Anishinabe.

[16] Βλ. "Social Psychology and The Study of Peace: Personal Reflections," του Herbert C. Kelman, στο *The Oxford Handbook of Intergroup Conflict*, επιμέλεια Linda R. Tropp, Oxford Handbooks Online, Nov 2012doi: 10.1093/oxfordhb/9780199747672.013.0021,
https://www.oxfordhandbooks.com/view/10.1093/oxfordhb/9780199747672.001.0001/oxfordhb-9780199747672-e-21

[17] Βλ. *Learning Non-Aggression: The Experience of Non-Literate Societies*, επιμέλεια Ashley Montagu, Oxford University Press, Oxford UK, 1978. Βλ. επίσης *Mutual Service and Cooperation*, του Charles T. Sprading, Libertarian Publishing Company, Los Angeles, CA, 1930.

[18] Λευιτικόν, 19:34

[19] Kimmerer R. (2011) Restoration and Reciprocity: The Contributions of Traditional Ecological Knowledge. στο: Egan D., Hjerpe E.E., Abrams J. (eds) *Human Dimensions of Ecological Restoration. Society for Ecological Restoration.* Island Press, Washington, DC. https://doi.org/10.5822/978-1-61091-039-2_18, https://link.springer.com/chapter/ 10.5822/978-1-61091-039-2_18

Κεφάλαιο 26

ΤΟ ΦΑΙΝΟΜΕΝΟ PROCACCINI

Το έργο του Τζον Μπάρτλετ "Familiar Quotations" ("Γνωστά Γνωμικά/ Αποφθεύγματα") αποτελεί τον βασικό εξοπλισμό της οικολογίας, ιδιαίτερα με την έννοια της *προσφοράς*. "Ωφελεί διπλά τους άπορους αυτός που δίνει γρήγορα", δηλώνει ο Ποπλίλιος Σύρος (1^{ος} αιώνας π.Χ.). 1 "Κερδίζουμε τη ζωή από αυτά που δίνουμε", είπε κάποτε ο Ουίνστον Σ. Τσόρτσιλ, μία "Έξυπνη Φράση"[2] που ομοιοκαταληκτεί με την ιστορία της φιλανθρωπίας, όπως αυτή αρθρώνεται από τόσο πολλούς ισχυρούς ηθικούς αυτουργούς και γαλβανολάστες της ιστορίας. Είτε με τη μορφή της αγάπης, του χρόνου, της υπομονής, του γέλιου, της πιο ανιαρής στοχαστικότητας και της αιθέριας τέχνης. Στον ρυθμό του χρήματος ή κάποιας άλλης φιλανθρωπίας πιστεύουμε, όπως είπε μόνο ο Σαίξπηρ, ότι "η ποιότητα του ελέους δεν εκβιάζεται, πέφτει σαν την απαλή βροχή…".[3] Ο Άλφρεντ Ράσελ Γουάλας, γράφοντας το 1863 για την αξιοσημείωτη ενδημική βιοποικιλότητα στο Σουλαουέσι, επανέλαβε αυτό το συναίσθημα με τη μορφή μιας πρόκλησης: "Οι μελλοντικές γενιές θα μας θεωρούν σίγουρα ανθρώπους που απορροφήθηκαν τόσο στο κυνήγι του πλούτου που αδυνατούσαν να δουν πολυτιμότερα αγαθά", αναφέροντας πρωτίστως μεταξύ αυτών, την αξιοσημείωτη δύναμη ζωής η οποία, με τη σημερινή πορεία των πραγμάτων, θα μπορούσε να "εξαφανιστεί για πάντα από προσώπου γης, χωρίς φροντίδα και ως ένα άγνωστο στοιχείο".[4] Το αριστουργηματικό του έργο με θέμα τη βιολογική φιλοσοφία και ανακάλυψη και τίτλο *Το Αρχιπέλαγος της Μαλαισίας*, του οποίου η έρευνα πραγματοποιήθηκε μεταξύ 1854 και 1862 και δημοσιεύθηκε για πρώτη φορά το 1869, ο Γουάλας αμφισβήτησε πλήρως τον αγγλικό πολιτισμό, τον χριστιανισμό και τη νομολογία, προκαλώντας την ισχυρότερη δυσαρέσκεια, χαρακτηρίζοντας το καθεστώς της βρετανικής ιδιοκτησίας ως την απόλυτη "κατάσταση βαρβαρότητας" στην απόλυτη απαλλοτρίωση και καταστροφή του από τους λίγους "συνανθρώπους του", οι οποίοι έχουν έναν πραγματικό λόγο με βάση της θεωρία της εξέλιξης να βρίσκονται σε κάθε ένα από αυτά τα ιδιοποιημένα εδάφη.[5] Ο Γουάλας έγραψε με την ηθική δύναμη του Τόλστοϊ και του Γκάντι και γνώριζε σε όλη τη διάρκεια της οικολογικής δίνης και πλοκής ότι η φυσική επιλογή δεν συνδέεται με τους ανθρώπινους τρόπους και μέσα. Η δική μας ωμή δύναμη πληροφοριών επιτάσσει τους καλύτερους αγγέλους της και αφήνει ταραχές εκεί όπου ο προηγούμενος συντονισμός και η χάρη είχαν επικρατήσει πολύ πριν μας το πουν τα απολιθώματα.

Αυτό όμως ανήκει στη σφαίρα επιρροής του ελεύθερου χρόνου. Μόνο εκείνοι για τους οποίους η επιλογή υπήρξε καλοπροαίρετη, έχουν τόσο πολύ χρόνο στη διάθεσή τους τους για να προβαίνουν σε μάταιες εικασίες. Για την πλειοψηφία, που απαρτίζεται από ανθρώπους παγιδευμένους σε γενιές φτώχειας, πείνας, ασθένειας και κακουχίας, η φυσική επιλογή είναι απλώς η τελευταία λέξη της σκληρότητας, μετά από τόσες άλλες κακές ιδέες. Πόσο εύκολα και γρήγορα το ρομαντικό τοπίο στρέφεται εναντίον των υποστηρικτών του. Πώς οι άφθονες αναμνήσεις σβήνουν τόσο εύκολα όσο τα παραμελημένα πατώματα, οι στέγες, οι τοίχοι. Μέχρι ο χρόνος της ειδυλλιακής αφήγησης να μεταβεί από τον ενεστώτα, από τις ζεστές εστίες στα κατεστραμμένα από τον καιρό ερείπια. Ανάμεσα σε σμήνη από σκνίπες και εκτυφλωτικές χιονοθύελλες, και πολλαπλές ιστορίες αιματοχυσίας μεταξύ φυλών και πεδία μάχης που ανταγωνίζονται για εθνική υπερηφάνεια, ένα μέρος τόσο αγαπητό όπως τα υψίπεδα της Σκωτίας, για εκείνους που πιάστηκαν, ή κυνηγήθηκαν, δεν είναι τίποτα άλλο από εφιάλτης. Δεν μπορούμε να προσποιούμαστε το αντίθετο. Όπως και στην περίπτωση του πλέον φρικιαστικού θανάτου του Φαραώ Σεκενενρέ Ταό Β, κυβερνήτη της νότιας Αιγύπτου μεταξύ των ετών 1558-1553 π.Χ., του οποίου τα μουμιοποιημένα τραύματα στο πρόσωπο, όπως αναλύθηκαν με υπολογιστική τομογραφία, αποκάλυψαν τις πολυάριθμες μαχαιριές που δέχθηκε σε διάφορα μέρη του προσώπου και του κρανίου του (προφανώς επειδή ο νέος Φαραώ ήταν δυσαρεστημένος με τους ιπποπόταμους του Ταό που έκαναν τόσο πολύ θόρυβο τη νύχτα στην ιερή πισίνα τους). [6]

Αυτό μας κάνει να παγώνουμε. Συμβατική πρόοδος, πλήθος που ψάχνει για προστασία, αδικία με οποιοδήποτε πρόσχημα ή σε οποιαδήποτε εποχή, και μετά; Ποια δύναμη αναφοράς, ποια πηγή παρηγοριάς θα είναι ο οδηγός σου στην ύπαιθρο, όπου το πάχος του τριχώματος ή του φτέρωματος σπανίως κατανέμεται σε ίση έκταση, και καμία οικολογία ή μαθηματικά δεν μπορεί να το επιτύχει;

Καθώς οι θεοί και οι οντολογίες κινητοποιούνται ενάντια σε αυτές τις πολυτάραχες καταστάσεις, αλλοπρόσαλλες γεωγραφικές περιοχές και συμφορές που συγκρούονται, μόνο η ανταπόδοση ενάντια σε όλες αυτές τις πιθανότητες και τη σίγουρη βία μπορεί να προσφέρει κάποια άλλη σανίδα σωτηρίας πάνω στην οποία μπορούμε να ανέβουμε και να εναποθέσουμε τις εσωτερικές μας ελπίδες. Για την υπεράσπιση των μυριάδων αδικημένων. Εδώ βασιλεύει μια εντελώς ξεχωριστή αίσθηση του κόσμου, με την οποία τα μεγαλύτερα ένστικτά μας συντονίζονται μεταξύ τους, από είδος σε είδος, νυχθημερόν. Χωρίς απληστία και χωρίς την επιθυμία του εαυτού μας να προκαλέσει οποιαδήποτε άλλη αντίδραση. Ευδοκιμεί με το ένστικτο και τον υψηλό σεβασμό που δεν κρατά κακία και δεν προάγει κανενός είδους καταστροφή. Κερδίζει από την καθαρή επιθυμία να αισθανθούμε αυτό που κάποιος διαισθάνεται ελαφρά, να δούμε όπως μας βλέπουν, να κρατηθούμε από τον εαυτό μας σαν να μην ήταν ποτέ τίποτα άλλο παρά εκείνο το Άλλο, ζωντανό μέσα στο σώμα της προέλευσης. Το ποίο είμαστε *εμείς*, εμείς με τις ίδιες ανάγκες, ελπίδες και την ίδια απελπισία.

Τέτοια μυστήρια διαχέονται εύκολα μέσα στην έκσταση ή στον αχαλίνωτο ενθουσιασμό της φυσικής ιστορίας. Όμως για να φέρουμε πίσω κάποια μορφή βαθμονομημένης αλήθειας, μη απορρίπτοντας κάποια, αγκαλιάζοντας τις όλες, απαιτείται κάτι περισσότερο από μια απλή θεωρία για να σταματήσουν τα δάκρυα ή οι πολλές προδοσίες της ιδιοτέλειας.

Η αρχέγονη αυτή αμοιβαιότητα αποτελεί έναν στόχο αντάξιο κάθε είδους επιβράβευσης εν ζωή και αλλά και μετά θάνατον. Μια τέλεια αίσθηση αυτού του μεγαλειώδους αυτού αργού περάσματος από το φωτεινό στο σκιερό όπως το εφέ μποκέ και η τεχνική σφουμάτο μπορεί να αντληθεί από μια μελέτη του Camillo Procaccini (1561-1629) που αποδίδεται μοναδικά στο έργο του “The Transfiguration” (“Η Μεταμόρφωση”, 1587-1590).[7] Σε αυτή την απεικόνιση ενός από τα μάλλον πλέον δυσνόητα σημεία σε όλα τα συνοπτικά ευαγγέλια από το δέκατο έβδομο κεφάλαιο του Ευαγγελίου του Ματθαίου, ο Ιησούς εξαλείφει κάθε διαμάχη. Εκεί, πάνω σε ένα βουνό στο οποίο ο Ιησούς, ο Πέτρος, ο Ιάκωβος και ο Ιωάννης έρχονται να προσευχηθούν, συναντούν τον Μωυσή και τον Ηλία. Ένα “φωτεινό σύννεφο τους επισκιάζει” και η “φωνή από το σύννεφο” έρχεται επίσης με ένα μήνυμα μεταφρασμένο προς τον Χριστό: “Μη πείτε σε κανέναν το όραμα, μέχρις ότου ο Υιός τού ανθρώπου αναστηθεί από τους νεκρούς”.[8] Το όρος Θαβώρ, στο Ισραήλ, θεωρείται παραδοσιακά η τοποθεσία της σκηνής. Άλλοι υποστηρίζουν ότι η ακριβής τοποθεσία είναι αυστηρά συμβολική. Οι εικόνες που καλλιτεχνήθηκαν κυρίως μεταξύ του 12ου και 16ου αιώνα, από τον Τζοβάνι Μπελλίνι έως τη ρωσική σχολή του Νόβγκοροντ, αποτελούν τον πυρήνα για την αντίληψη μιας γραφικής, γλωσσικής και φιλοσοφικής συζήτησης. Η φράση “ὁ υιός τοῦ ἀνθρώπου”, και αντίστοιχη Εβραϊκή םדא–ןב- (,πεν αντάμ) επαναλαμβάνονται 132 φορές, κατοχυρώνοντας κάθε πιθανή σχέση μεταξύ της ανθρωπότητας και του Θεού, του ανθρώπου και της φύσης, της φύσης και της φύσης. Κάθε γέφυρα μεταξύ των δυάδων σχηματίζεται ως όραμα, όμως κανένας δεν διασχίζει αυτή τη γέφυρα. Αντ' αυτού, υπάρχει επανάληψη κατ' επανάληψη, υπέρθεση των πεδίων δύναμης, των κυμάτων και των σωματιδίων. Μια μορφή του Χριστού υπερβαίνει μπροστά μας. “Και την ώρα που προσευχόταν, η όψη του προσώπου του έγινε διαφορετική και τα ρούχα του άσπρα κι αστραφτερά”.[9]

Βιβλικοί μελετητές και διάφοροι πιστοί θεωρούν τη Μεταμόρφωση ως απόδειξη της εμμένειας, μιας παγκόσμιας ψυχής που πηγάζει από *το νου* σε κάθε μεμονωμένο θνητό. Τα δόγματα αυτά που συνδέουν τη γη και την αντίληψή μας γι' αυτήν με τους *ουρανούς* αποτελούν τον πυρήνα των περισσότερων θεολογικών αναζητήσεων. Σύμφωνα με την αφήγηση Shussan Shaka, ο Βούδας βίωσε δύο φορές τη μεταμόρφωση, μια κατά τη διάρκεια της φώτισης και μια κατά τη διάρκεια του θανάτου του.[10] Παρομοίως, ο Μαχαβίρα αφήνει το θνητό του σώμα κατά τον θάνατό του, ωστόσο περιγράφει, σαν να διέθετε ακόμα σάρκα και οστά, την υπερβατική εμπειρία, την καθαρή και κλασική κοσμολογία.[11] Σημεία αναφοράς σε αυτό το ανθρωποειδές ταξίδι από τον Εαυτό στο Άλλο. Η μεγάλη διάχυση. Οτιδήποτε σκεφτόμαστε, κάνουμε και ονειρευόμαστε στηρίζεται στις αναμνήσεις τόσο του παρελθόντος όσο και του μέλλοντος. Οι υπερβατικές εξισώσεις ανακατασκευάζονται στα καθημερινά εδάφη και τις καιρικές συνθήκες της εμπειρίας μας. Όποιος και αν είναι ο εκπρόσωπος αυτού του Υιού, αυτής της Κόρης, αυτού του Θεού, σε κάθε αναγέννηση παραμένει το πιο ταιριαστό από όλα τα καθήκοντα, η βελτίωση όλων των δεινών που είναι η μεταμόρφωση. Η εκπλήρωσή του δεν μας προσφέρει κάποια αναγνώριση ή δόξα, αλλά αποτελεί το μοναδικό μας καθήκον, μέσα στο μεγάλο δράμα αυτής της οικολογικής ωρολογιακής βόμβας που έχουμε δημιουργήσει.

ΣΗΜΕΙΩΣΕΙΣ

[1] Απόφθεγμα 6, *Familiar Quotations -A collection of passages, phrases and proverbs traced to their sources in ancient and modern literature*, του John Bartlett, Fourteenth Edition, Emily Morison Beck, Editor, Little, Brown and Company, Boston, Toronto, 1968, σελ. 125.

[2] Βλ. https://www.brainyquote.com/quotes/winston_churchill_131192

[3] Από το έργο "The Merchant of Venice," Act IV, Scene I. Βλ. https://poets.org/poem/merchant-venice-act-iv-scene-i-quality-mercy-not-strained

[4] "He Helped Discover Evolution, And Then Became Extinct," του Anthony Kuhn, NPR, 30 Απριλίου 2013, https://www.npr.org/2013/04/30/177781424/he-helped-discover-evolution-and-then-became-extinct

[5] Alfred Russel Wallace, *The Malay Archipelago – The Land Of The Orang-Utan And The Bird Of Paradise; A Narrative Of Travel With Studies Of Man And Nature*, Macmillan And Co. Limited, London, 1906, σελ.458.

[6] Βλ. "Egyptian pharaoh was executed on the battlefield, mummy reveals," της Stephanie Pappas, 16 Φεβρουαρίου 2021, https://www.livescience.com/egyptian-pharaoh-battlefield-death.html

[7] "The Transfiguration. Medium: etching. Museum: National Gallery of Art, Washington DC. Author: Camillo Procaccini, Image ID: R2R41F, https://www.alamy.com/the-transfiguration-medium-etching-museum-national-gallery-of-art-washington-dc-author-camillo-procaccini-image 225472171. html; Βλ. επίσης RKD/Artists, https://rkd.nl/nl/explore/artists/record? filters%5Bsterfplaats%5D=Milaan&query=&start=40; Βλ. Επίσης *Christ -Life, Death, & Resurrection*, University of San Diego, in Collaboration with the British Museum and the Timken Museum of Art in San Diego, CA, aby Hugo Chapman, Foreword by Hartwig Fischer, Director of the British Museum, 2019, σελ. 64.

[8] *Holy Bible – King James Version, Containing The Old And New Testaments*, The Christian Science Publishing Society, Boston MA, n.d., σελ. 1213-1214.

[9] ό.π., σελ. 1284, Κατά Λουκά 9:29.

[10] E.W. Hopkins, "The Message of Buddhism to Christianity," The Biblical World, Vol. 28, No. 2, Aug., 1906, σελ. 94-107. Βλ. επίσης Robert S. Cohen, *Beyond Enlightenment: Buddhism, Religion, Modernity*, Routledge, Abington, UK, 2006, σελ. 1-3. Βλ. κεφάλαιο 47 σχετικά με τη διπλή φώτιση στο Βουδισμό, στο *On The Nature of Ecological Paradox*, των M.C.Tobias και J.G.Morrison Springer Nature, New York, NY 2021.

[11] *Mahavir - Prince of Peace*, Ranchor Prime, Εικονογράφηση B. G. Sharma, πρόλογος Vinod Kapashi, εισαγωγή Michael Tobias, Mandala Publishing, 2006, σελ.133, https://jainqq.org/explore/020465/133.

Κεφάλαιο 27

ΤΟ ΔΥΝΑΜΙΚΟ BEREITSCHAFTS

Η Χάνα Άρεντ γράφει ότι "Η φιλοσοφική ελευθερία, η ελευθερία της βούλησης, είναι σχετική μόνο για τους ανθρώπους που ζουν εκτός πολιτικών κοινοτήτων, ως μεμονωμένα άτομα… [και] Μιλούσαμε εδώ για τη μεροληψία του ηθοποιού ο οποίος, λόγω του ότι εμπλέκεται, δεν βλέπει ποτέ το νόημα του συνόλου".[1] Θα υποστηρίζαμε ωστόσο το αντίθετο, εξευγενίζοντας τη συνδεσιμότητα με την πλήρη γενναιοδωρία της εσωτερίκευσης ενός μοναχικού ανθρώπου που εκδηλώνεται σε έναν γεμάτο από ζωή κόσμο ο οποίος προκαλεί σύγχυση, και στον οποίο τόσο η θέληση *όσο* και η αναπαράσταση συνδυάζουν την υπέρβαση της ατομικής αγάπης για τη φύση με την ίδια τη φύση.

Στο πλαίσιο του Μεγάλου Θανάτου (της Ανθρωπόκαινου), μπορούμε εύκολα να κάνουμε λίγη οικονομία άμεσα ώστε να αναζωογονήσουμε τις ενσωματωμένες συνδέσεις μας χωρίς να χάσουμε τα πάντα μονομιάς. Μια έκθεση φωτογράφων κήπου του 2005 με τίτλο *PrivatePlaces* μετέφερε επιλογές που είχαν κάνει μοναχικοί άνθρωποι, καθώς και κοινότητες για να διατηρήσουν τις αρετές και τη διαρκή υπόσχεση της βιοποικιλότητας που συνδέεται με τις ψυχές μας.[2] Οι ασπρόμαυρες, και κυρίως έγχρωμες φωτογραφίες από πολλές χώρες θύμισαν τη διάσημη φράση του Σερ Τόμας Μπράουν (1605-1682), "… στα πιο προφανή και παραμελημένα κομμάτια της Φύσης, που χωρίς περαιτέρω ταξίδι μπορώ να κάνω στην Κοσμογραφία του εαυτού μου. Κουβαλάμε μαζί μας τα θαύματα που αναζητούμε χωρίς εμάς: όλη η *Αφρική* και τα θαύματά της υπάρχουν *εντός μας*".[3] Το εξώφυλλο του *PrivatePlaces* κοσμείται από την αξιοσημείωτη φωτογραφία του Peter Baistow που απεικονίζει ένα ζευγάρι στη μικρή, γεμάτη πλίνθους πίσω αυλή τους, σε ένα μικρό στρογγυλό τραπέζι, με τη γυναίκα να μιλά στο κινητό της τηλέφωνο, τον άντρα να κλαδεύει ένα φυτό σε γλάστρα σε εκείνο το τραπέζι στο East Ham του Λονδίνου. Ουσιαστικά όμως απεικονίζει έναν άντρα που νοιάζεται για μια άλλη μορφή ζωής. Πίσω του, εκατοντάδες άλλοι άνθρωποι, που έχουν μεταβεί σε κάποιον άλλο κόσμο, με τις ταφόπλακες τους ολοένα και πιο ακατάστατες. Ωστόσο, μέσα από αυτόν τον "ιδιωτικό κήπο"[4] αυτοί οι δύο άνθρωποι, όντας πραγματικά ζωντανοί, μπορούν να μας υπενθυμίζουν ότι είμαστε όλοι φτιαγμένοι από τις ίδιες παρορμήσεις και λεπτομέρειες. Οι ίδιοι παράγοντες, σε όλες τις πλευρές.

Αυτό που ξεχωρίζει τις ενέργειες και τα κίνητρα των ανθρώπων μπορεί να είναι η κλίμακα των φιλοδοξιών τους ή τα βάρη από τα οποία δεν μπορούν να ξεφύγουν. Ίσως άλλες υπολειμματικές επιδράσεις ενός τόσο σύντομου χρόνου ωρίμανσης, της πλέον βεβιασμένης εξέλιξης οποιουδήποτε από τα πάνω από είκοσι είδη *Homo*. Κάποιοι επιλέγουν να ασχοληθούν με τον κήπο τους, με τον ένα ή τον άλλο τρόπο, το οποίο αποτελεί την πιο αρχαία ιστορία στο χρονολόγιό μας. Άλλοι, όπως ο σπάνιος Τζον Ροκφέλερ δίνουν 100 δις δολάρια σε φιλανθρωπίες. Συγκριτικά, η συνολική φιλανθρωπική προσφορά στις Ηνωμένες Πολιτείες για το 2019 έφτασε σχεδόν τα 450 δισεκατομμύρια δολάρια.[5] Στο έργο του Περὶ *ποιητικῆς* (περίπου το 335 π.Χ.), ο Αριστοτέλης έγραψε: "μεταφορὰ δέ ἐστιν ὀνόματος ἀλλοτρίου ἐπιφορὰ ἢ ἀπὸ τοῦ γένους ἐπὶ εἶδος ἢ ἀπὸ [τοῦ] εἴδους ἐπὶ τὸ γένος ἢ ἀπὸ τοῦ εἴδους ἐπὶ εἶδος ἢ κατὰ τὸ ἀνάλογον" (*Μεταφορά* είναι η απόδοση σε ένα πράγμα ενός ονόματος που ανήκει σε ένα άλλο πράγμα. Συγκεκριμένα: ή από το γένος σε είδος ή από το είδος στο γένος ή από ένα είδος σε ένα άλλο είδος ή κατά την αναλογία)".[6] Αυτός ήταν ο αρχικός ορισμός της μεταφοράς, μιας έννοιας που έχει γίνει αποδεκτή ως γενικό μοτίβο περιγραφικών επιλογών για το τοπίο, τη βιοποικιλότητα και τη βιοπολυπλοκότητα, για να μην αναφέρουμε τη φιλανθρωπία, πολύ πιο μακριά από αυτό που ο Αριστοτέλης οραματίστηκε σε στενά πλαίσια.[7]

Ωστόσο, αποδεχόμενοι τη μεταφορά ως την πιο κρίσιμη κατάσταση της συνείδησης και της εμφάνισης στον *H. sapiens*, όλα τα ευφάνταστα Όντα και Δώρα πηγάζουν από το πρωταρχικό τους θεμέλιο: την αναγνώριση των Άλλων και του Εαυτού ανάμεσά τους, μέσα στο ίδιο τους το κέντρο. Όλα φαίνονται παραδόξως απλοϊκά. Αν ήταν τόσο απλό, σίγουρα θα *το κάναμε όλοι στους άλλους* σύμφωνα με τους ευσεβείς περιορισμούς με τους οποίους μεγαλώσαμε όλοι. Τι όμως διαπερνά το μυαλό *πριν* από τη μεταφορά; Ο Μπέντζαμιν Λίμπετ (1916-2007) μελέτησε το χρονικό χάσμα μεταξύ της πραγματοποίησης μιας δράσης και της σκέψης της, αυτό που οι νευροβιολόγοι ονομάζουν δυναμικό Bereitschaft (Bereitschaftspotential-BP) ή δυναμικό ετοιμότητας.[8] Το κενό των 0,35 δευτερολέπτων είναι εν τέλει μικρής βραχυπρόθεσμης σημασίας όταν εξετάζουμε τους μηχανισμούς που θα μπορούσαν να δημιουργήσουν ανθρώπινα ηθικά κοινά, μια ανεμπόδιστη, διαδεδομένη γενναιοδωρία που δεν απαιτεί νομισματοποίηση του κεφαλαίου της φύσης ή όρους και συνθήκες υλικής αιτίας και συνέπειας για να μας ξυπνήσει από αυτό το λήθαργο που διαφορετικά θα αρνούνταν την εσωτερική πρόγνωση μιας άνευ όρων και αρχέγονης ποιότητας αρετής.

Το σχεδόν 8 εκατοστών γλυπτό του Laugerie Basse το οποίο απεικονίζει ένα ζώο που μοιάζει με αρκούδα που συλλογίζεται, ένα γλυπτό της Μέσης Μαγδαληνίας (17.000-12.000 έτη πριν από την εποχή μας) λαξευμένο από κέρατο ταράνδου[9], αντιπροσωπεύει πολύ περισσότερα από το *chaîne opératoire*, τον όρο με τον οποίο ο Αντρέ Λερουά-Γκουράν (1911-1986) περιέγραψε τους κύκλους ζωής που έχουν περάσει στην κατασκευή ενός τεχνουργήματος. Πρόκειται για μια νοητική αμοιβαιότητα που συνδέει τον βαθύ διαλογισμό με την έκφραση. Αυτή η αρκούδα αναμένει τη διάχυση της γενετικής και πολιτιστικής μνήμης του Τζιακομέτι πίσω στο χρόνο, μέχρι εκείνες τις στιγμές όπου το άγχος της πρόβλεψης ξεπερνιέται με κάτι εντελώς διαφορετικό. Μια αισθητική ευλάβεια που αλλάζει τα πάντα στην ανθρώπινη εμπειρία. Το εάν βρίσκεται εντός του ατόμου, δεν έχει καμία σημασία. Σημασία έχει μόνο ότι η προσωπικότητα *είναι* αυτή η αλλαγή πέρα από όλα τα σύνορα και τα εμπόδια μεταξύ του είδους και του γένους, του γένους και του είδους του Αριστοτέλη. Ως μεταφορές για τη ζωή, τέτοιες ταξινομικές διακρίσεις, από μόνες τους, καθίστανται αμέσως μη συναφείς. Παρωχημένη επιστήμη, σε σχέση με τη σπουδαιότητα των γεγονότων της εποχής μας.

Καμία οικοδυναμική παρομοίωση δεν αρκεί για να δώσει υπόσταση στην εξαίσια υπόσχεση αυτής της αποκαλυπτικής προοπτικής, τη βαθιά λατρεία για τους Άλλους που - μόλις δέκα, είκοσι χιλιάδες χρόνια πριν κυριαρχούσε ακόμα σε ολόκληρο το ανθρώπινο Weltanschauung (κοσμοθεωρία).

Παρόλο που οι ευαισθησίες μας στο δέλεαρ της σπάταλης δύναμης σχεδόν την έχουν αντικαταστήσει, το δυναμικό αμοιβαιότητας παραμένει ριζωμένο μέσα μας. Ο συνεχής σχολιασμός για τόσες δεκάδες χιλιάδες χρόνια προσδίδει αφοσιωμένη πίστη στην εμπλοκή του είδους μας με εκείνους από τους οποίους δεν περιμένουμε τίποτα σε αντάλλαγμα, αλλάζοντας έτσι ολόκληρο το τοπίο της αμοιβαιότητας από τα συμβατικά παραδείγματα του δούναι και λαβείν και της προσφοράς και απολαβής σε κάποιο τρίτο επίπεδο ζωντανής μετεμψύχωσης. Τα εκτεταμένα διαγράμματα του Λερουά Γκουράν που περιγράφουν λεπτομερώς τη "Συχνότητα" των "Θεμάτων", των "Ψυχοσύνθετων Υποκειμένων", των "Ατόμων", των "Συμπληρωματικών Ζώων", των διαφόρων συζεύξεων και των "Συζευχθέντων Σημάτων" καθώς και της αφθονίας των ανθρώπινων άλλων ζωικών μορφών,[10] αποτελούν τα ίδια υποστρώματα μιας τεράστιας ιστορίας της τέχνης του τοπίου και των ανθρώπινων εμμονών με άλλα είδη που κορυφώνονται σήμερα με την καθοριστική ασάφεια της εποχής μας. Πράγματι, το είδος που είμαστε σήμερα διασαφηνίζει συντριπτικά το πρόβλημα του να είσαι άνθρωπος με τρόπο που μόνο σήμερα μπορούμε πλέον να εκτιμήσουμε. Ο κορεσμός της δημογραφικής μας κρίσης – η έμφορτη με πυκνότητα βία και η απαξίωσή της - δεν προωθεί καμία ανώδυνη συμφιλίωση εκτός από την ένταση και τον σκοπό των υπόλοιπων βιοφιλελεύθερων δεσμεύσεών μας[11]. Η αποθέωση σε όλα τα επίπεδα των Άλλων. Βλέπουμε και αισθανόμαστε αυτόν τον προσανατολισμό να ενσωματώνεται στο έργο τέχνης ενός σύγχρονου φωτογράφου όπως ο Σκοπιανός Γκόραν Αναστασόφσκι[12] ή στις σύγχρονες ταινίες του Κάρολ Μπάλαρντ[13] ή, ακόμα πιο πρόσφατα, στο έργο "Penguin Bloom" του Γκλέντιν Άιβιν, που εξιστορεί δύο εντελώς ανταποδοτικές ανακάμψεις υγείας, η μία εκ των οποίων αφορά μια γυναίκα (τη Σαμ Μπλουμ) που έχει παραλύσει από μια άσχημη πτώση και η άλλη μια ορφανή καρακάξα (τα αγόρια της κυρίας Σαμ Μπλουμ πιστεύουν ότι το πουλί μοιάζει με πιγκουίνο) που η γυναίκα αρχίζει να αγαπά.[14] Γοητευόμαστε από τα πάθη και τους κοινούς χώρους εκείνων που εργάζονται απεγνωσμένα για να σώσουν τη βιοποικιλότητα σε όλο τον κόσμο.[15] Το παγκόσμιο όμως βασίλειο στην πραγματικότητα είναι μεγάλο μόνο στο βαθμό της μεταφοράς που σκιάζει το άμεσο περιβάλλον μας, ανεξάρτητα από το πόσο γήινα ή έξυπνα μπορεί να θεωρήσουμε τον πέρα από τα σύνορά μας αλτρουισμό. Παραδόξως, τα μεμονωμένα άτομα μας χρειάζονται πιο άμεσα από όσο οι ολόκληροι πληθυσμοί. Ανεξάρτητα από το τι κρύβουν οι αριθμοί, συσχετιζόμαστε άμεσα με τον ενικό, με μονάδες και όχι με τετράκις εκατομμύρια.

Και έτσι προκύπτει φυσικά το ερώτημα: Σε ποια μοναδικότητα θα μπορούσαμε να αποδώσουμε τη μεταμόρφωση της ανθρώπινης ψυχής, δεδομένου του δυναμικού της ετοιμότητας για προσφορά και του δυναμικού της (ακλόνητης) ετοιμότητας για συσχετισμό με τη φύση; Είμαστε πιο επιρρεπείς στην καλοσύνη από κάτι άλλο; Οι ψυχολόγοι έχουν υποστηρίξει εδώ και καιρό ότι η πλειοψηφία της ανθρωπότητας ζούσε πάντοτε ειρηνικά, τις περισσότερες φορές. Τα στατιστικά στοιχεία σχετικά με την εγκληματικότητα υπογραμμίζουν ασφαλώς αυτό το γεγονός.[16] Όπως και το ρεπορτάζ από τους ψυχιάτρους του Υγειονομικού Σώματος, οι οποίοι, διεξάγοντας έρευνα στο πεδίο της μάχης κατά τη διάρκεια του Δευτέρου Παγκοσμίου Πολέμου, διαπίστωσαν ότι "ο φόβος της αφαίρεσης μιας ζωής και όχι ο φόβος της απώλειας της ζωής [ήταν] η πιο συνηθισμένη αιτία για την αποτυχία στη μάχη".[17] Οι κυνικοί επισημαίνουν ότι στις πρώτες τρεις δεκαετίες μετά τον Δεύτερο Παγκόσμιο Πόλεμο, άλλα 25 εκατομμύρια και περισσότεροι άνθρωποι σφαγιάστηκαν κατά τη διάρκεια 133 μικρότερων πολέμων. Ότι η διχασμένη διανοητική εντροπία της σφαγής του Μι Λάι ή το γεγονός ότι υπήρχε μια στις τρεις πιθανότητες η εισβολή στον Κόλπο των Χοίρων να είχε εύκολα εξελιχθεί σε πυρηνικό πόλεμο, δεν θα πρέπει να μας δώσει κανένα λόγο αισιοδοξίας. Πράγματι, οι απόψεις που έχουν εκφράσει ιστορικοί και ερευνητές συμπεριφοράς, όπως ο Κόνραντ Λόρεντζ, ο Έρικ φον Χολστ, ο Καρλ φον Φρις και ο Νίκο Τίμπεργκεν, όσον αφορά την Χομπεσιανή έννοια του "πολέμου όλων εναντίον όλων" (ή "φάτε και φαγωθείτε" – *bellum omnium contra omnes*) είναι διφορούμενες[18]. Ο Ρέιμοντ Νταρτ υπήρξε ίσως ο πλέον σαφής στην περίφημη διαπίστωσή του ότι "Τα αιματοβαμμένα αρχεία των σφαγών της ανθρώπινης ιστορίας, από τα πρώτα αιγυπτιακά και σουμερικά αρχεία μέχρι τις πιο πρόσφατες θηριωδίες του Δευτέρου Παγκοσμίου Πολέμου [που] συμφωνούν με τον πρώιμο καθολικό κανιβαλισμό… διακηρύσσοντας αυτόν τον κοινό διαφοροποιητή της δίψας για αίμα, αυτή την κυρίαρχη συνήθεια, αυτό το σημάδι του Κάιν που διαχωρίζει τον άνθρωπο διατροφολογικά από τους ανθρωποειδείς συγγενείς του και αντίθετα τον συνδέει με τα πλέον θανατηφόρα σαρκοφάγα".[19]

Είμαστε πρόθυμοι να δημιουργήσουμε σιγά-σιγά μια θεωρία από τον κόσμο άλλων ειδών - όπως οι κάμπιες Hyponomeuta - που αποκτούν τόσο ισχυρούς και θετικούς κοινωνικούς ρόλους. Τα κύτταρα Van Tieghem του μυξομύκητα Dictyostelium που διατηρούν την ατομικότητά τους, ακόμη και όταν συσσωρεύονται σαν τα κοράλλια, ή οι αχινοί, οι γλάροι και κάθε σμήνος πουλιών.[20] Ωστόσο, η πραγματική κάθαρση απέχει παρασάγγας από το σύνολο όλων αυτών των περιπτωσιολογικών μελετών. Για να πειστείς πραγματικά πρέπει να γίνεις μέλος της οικογένειας. Ως εκ τούτου, η δύναμη των αληθινών επιφοιτήσεων,[21] της συμπόνιας που γεννιέται από τον πόνο του καθενός[22] ή εκείνων που έχουν αγωνιστεί να διαπραγματευτούν για την ειρήνη μέσω αυτής ακριβώς της μεταφοράς, αναγκάζοντάς μας για άλλη μια φορά να αλλάξουμε, να μεταφράσουμε, να δούμε τους Άλλους όπως θα βλέπαμε τους εαυτούς μας.[23] Ένα μοτίβο πίστης που κρυφοκοιτάζει έξω, αισθάνεται έντονα ή φαντάζεται τη δική του εξέλιξη καθοδηγούμενο από αυτό που πρέπει να είναι η πλέον υπερλογική σειρά παρατηρήσεων και επιλογών σε συγχρονισμό με τις αλληλοεξαρτώμενες εκκλήσεις τους. Με άλλα λόγια, να συμμετέχουμε σε κάθε ευκαιρία που έχουμε να συναναστραφούμε άλλα είδη με την αγάπη και την υπομονή ενός αγίου ή ενός απλού αγγέλου, που κατεβαίνει στη γη για να χρησιμοποιήσει αυτή τη *ζωντάνια*, κάθε ανάσα στο σώμα μας, προς την κατεύθυνση της έναρξης μιας σχέσης αγάπης, και τίποτα λιγότερο από αυτό.

ΣΗΜΕΙΩΣΕΙΣ

[1] Βλ. *The Life Of The Mind*, Volume Two, "Willing," της Hannah Arendt, Harcxourt Brace Jovanovich, New York and London, 1978, σελ.99, 272.

[2] *PrivatePlaces – International garden and plant photography*, Πρόλογος James Alexander-Sinclair, Εισαγωγή Henrietta Van den Bergh και Helen Fickling, "A Patch of Earth," Chris Young, Editor; Garden Design Journal, Third Millennium Publishing, Ltd., London 2005.

[3] *Religio Medici*, Part 1, Section 15, σελ. 31, 1669.

[4] ό.π.., *PrivatePlaces*, σελ. 95.

[5] Βλ. https://www.forbes.com/sites/jenniferwang/2020/01/20/from-buffett-to-bloomberg-the-25-philanthropists-in-america-making-the-biggest-donations/?sh=6d5c1e3f4e8f; Βλ. "Giving USA 2020," June 16, 2020, https://givingusa.org/giving-usa-2020-charitable-giving-showed-solid-growth-climbing-to-449-64-billion-in-2019-one-of-the-highest-years-for-giving-on-record/. Περισσότερα από τα μισά από αυτά τα χρήματα διατέθηκαν για την παροχή θρησκευτικών υπηρεσιών, υπηρεσιών εκπαίδευσης και άλλων υπηρεσίες, ενώ περισσότερα από τα μισά χρήματα δόθηκαν από ιδιώτες. ενώ λιγότερο από το 3% του συνόλου πήγε σε περιβαλλοντικούς σκοπούς.

[6] Μετάφραση Ingram Bywater, Κεφάλαιο 21, 1457b1-30, στο "Aristotle on Metaphor," του Dr. Jan Garrett, 28 Μαρτίου 2007, http://people.wku.edu/jan.garrett/401s07/arismeta.htm

[7] Βλ. "Ecology, Complexity, and Metaphor," των James D. Proctor και Brendon M. H. Larson, *BioScience*, Volume 55, Issue 12, December 2005, σελίδες 1065–1068, https://doi.org/10.1641/0006-3568(2005)055[1065:ECAM]2.0.CO;2, Δημοσίευση: 01 Δεκεμβρίου 2005, Oxford Academic, https://academic.oup.com/bioscience/article/ 55/12/1065/407139

[8] Βλ. "Biological Consensus Mechanisms: The Future of Coexistence," Κεφάλαιο στο έργο *The Hypothetical Species: Variables of Human Evolution*, των M.C.Tobias και J.G.Morrison, Springer Nature, New York, NY, 2019, σελ. 254.

[9] Βλ. Κεφάλαιορ 4 "Objects of Religious Significance," στο έργο *Treasures of Prehistoric Art*, του André Leroi-Gourhan, μετάφραση από τα γαλλικά του Norbert Guterman, Harry N. Abrams, Inc., Publishers, New York, NY, 1967, σελ. 89.

[10] ό.π., Κεφάλαιο 12, "Charts Supporting The Hypotheses," σελ. 491-523.

[11] Βλ. *The Biophilia Hypothesis*, επιμέλεια Stephen R. Kellert και Edward O. Wilson, Island Press, Shearwater Books, Washington D.C., Covelo, California, 1993. Βλ. επίσης *biophilia -Johanna Pegler*, Sarjeant Gallery Te Whare O Rehua, Whanganui, New Zealand, 2007. Βλ. επίσης *Art & Nature – Three Centuries of Natural History Art from Around the World*, της Judith Magee, Greystone Books, Natural History Museum, London, UK 2009, και *The Exultant Ark – A Pictorial Tour of Animal Pleasure*, του Jonathan Balcombe, University of California Press,

Berkeley Los Angeles London, 2011, και *Birds & People*, του Mark Cocker, Φωτογραφία David Tipling, Με εξειδικευμένη έρευνα από τον Jonathan Elphick και τον John Fanshawe, Jonathan Cape London, UK 2013.

[12] "Wildlife Photographer Captures What Love Looks Like in the Animal Kingdom," της Margherita Cole, 3 Φεβρουαρίου 2021, *My Modern Met*, https://mymodernmet.com/goran-anastasovski-affectionate-animals-photography/See https://www.gorananastasovski.com/ Other/ABOUT-ME/

[13] Carroll Ballard, σκηνοθέτις των: "The Black Stallion"(1979), "Never Cry Wolf"(1983), "Fly Away Home"(1996) και "Duma"(2005).

[14] Βλ. https://www.rogerebert.com/reviews/penguin-bloom-movie-review-2021

[15] Βλ. *Saving Endangered Species – Lessons in Wildlife Conservation from Indianapolis Prize Winners*, επιμέλεια Robert W. Shumaker, πρόλογος Harrison Ford, Johns Hopkins University Press, Baltimore, MD, 2020, Βλ. επίσης *Sanctuary: Global Oases of Innocence*, των M.C.Tobias και J.G.Morrison, A Dancing Star Foundation Book, Council Oak Publishers, Tulsa OK and San Francisco, CA, 2008. Βλ. επίσης "Ten conservation success stories when species came back from the brink," by Adam Vaughan, *The New Scientist*, 17 Φεβρουαρίου 2021, https://www.newscientist.com/article/mg24933223-400-ten-conservation-success-stories-when-species-came-back-from-the-brink/, Πρόσβαση στις 18 Φεβρουαρίου 2021.

[16] https://www.unodc.org/unodc/en/data-and-analysis/statistics/reports-on-world-crime-trends. html

[17] Βλ. *Men Against Fire*, του S. L. A. Marshall, William Morrow, New York, 1947.

[18] Βλ. *After Eden: History, Ecology & Conscience*, του M.C.Tobias, Avant Books, San Diego, CA, 1985, σελ. 245.

[19] Βλ. "The Predatory Transition from Ape to Man," του Raymond Dart, *International Anthropological and Linguistic Review*, 1, 1953.

[20] Βλ. "A Descriptive Study of the Development of the Slime Mold Dictyostelium discoideum," του John T. Bonner, *American Journal of Botany*, Vol. 31, No. 3 (Mar., 1944), σελ. 175-182, Έκδοση από: Wiley, https://doi.org/10.2307/2437641, https://www.jstor.org/stable/2437641, https://www.jstor.org/stable/ 2437641?seq=1

[21] Βλ. McDonald, M. G. (2008). The nature of epiphanic experience. *Journal of Humanistic Psychology*, 48(1), 89–115. https://doi.org/10.1177/0022167807311878, https://psycnet.apa. org/record/2008-00641-006

[22] Βλ. "Transformation Through Suffering: A Study of Individuals Who Have Experienced Positive Psychological Transformation Following Periods of Intense Turmoil," του Steve Taylor, *Journal of Humanistic Psychology* 52(1): 30-52, doi: 10.1177/0022167811404944, https://www.researchgate.net/publication/241653920_ Transformation_Through_Suffering_A_Study_of_Individuals_Who_Have_Experienced_Positive_Psychologi cal_Transformation_Following_Periods_of_Intense_Turmoil

[23] F. Aquilar, F., and M. Galluccio, (Eds.). (2011). *Psychological and political strategies for peace negotiation: A cognitive approach.* Springer Science + Business Media. https://doi.org/10.1007/978-1-4419-7430-3, https://psycnet.apa.org/record/2010-25371-000

Πρώτη επαφή. Φωτογραφία © M. C. Tobias.

Κεφάλαιο 28

ΖΩΟΣΗΜΕΙΟΛΟΓΙΚΕΣ ΜΕΤΑΦΡΑΣΕΙΣ

Η συνεχιζόμενη απώλεια αυτού που οι οικονομολόγοι προτιμούν να χαρακτηρίζουν ως "φυσικό κεφάλαιο" έχει πρόσφατα πέσει κατακόρυφα σε ποσοστό περίπου 27% ανά άτομο σε ολόκληρο τον πλανήτη[1]. Τι σημαίνει αυτό; Οι σχεδόν οκτώ δισεκατομμύρια ανθρώπινες διασυνδέσεις έχουν δημιουργήσει ένα επιθετικό δίκτυο τοπολογιών πολύ πέρα από τη γεωμετρική τους διαμόρφωση, επιβάλλοντας αυτό που ο Ράντολφ Τ. Χέστερ αποκάλεσε "πλαίσιο προτεραιότητας" στην κορυφή μίας "αμετάβλητης δομής". Το αποτέλεσμα σε πλήρως ηθικούς όρους, είναι αυτό που είπε ο Έστερ, παραθέτοντας τα λόγια του Ιάπωνα μελετητή τοπίου Ναγκασάσι Ταμεσούκε: "Ένας καλός άνθρωπος [ο οποίος,] με κακή πληροφόρηση είναι ένας κακός άνθρωπος".[2] Επιπλέον, πολύ πριν από τη 15η Διάσκεψη των Συμβαλλομένων Μερών στη Σύμβαση για τη Βιοποικιλότητα (COP15) και την 26η Διάσκεψη των Συμβαλλομένων Μερών στη Σύμβαση Πλαίσιο των Ηνωμένων Εθνών για την Αλλαγή του Κλίματος (COP26) η *Έκθεση Dasgupta* ανέφερε ότι επιδοτήσεις της τάξεως μεταξύ των 4 και 6 τρισεκατομμυρίων δολαρίων ΗΠΑ - από τα περίπου 84 τρισεκατομμύρια δολάρια της παγκόσμιας οικονομίας - υποστήριζαν την υπονόμευση του φυσικού κόσμου.[3] Αν και ανατριχιαστικά, οι αριθμοί και τα ποσοστά αυτά είναι επίσης παραπλανητικά, ακόμη και με την πιο στοιχειώδη έννοια. Μέχρι το 2005, προβλεπόταν, για παράδειγμα, ότι υπήρχαν 1 δισεκατομμύριο είδη βακτηρίων παγκοσμίως.[4] Μέχρι το 2016, οι εκτιμήσεις για τον αριθμό των μικροβιακών ειδών (βακτήρια, μύκητες, αρχαία, πρώτιστα, όχι όμως ιοί και πρωτεΐνες πρίον)[5] είχαν αυξηθεί σε 1 τρισεκατομμύριο,[6] με τον πληθυσμό τους να εμφανίζει αυξητική πορεία. Επειδή όμως δεν υπάρχει απολύτως καμία σχέση μεταξύ 1 δολαρίου και 1 είδους, η πρόκληση για την εκδήλωση ενός λογικού ίσου πρόσημου (=) απαιτεί κάτι περισσότερο από απλή μεταφορά. Όπως γράφει ο Έστερ στον Επίλογο της διατριβής του σχετικά με το σχεδιασμό και την οικολογική δημοκρατία, "Ενώ είμαι συγκλονισμένος από την απόγνωση γύρω μας και την κακία που αντιμετωπίζουμε στη κατοίκησή μας, η οικολογική δημοκρατία ριζώνει, βλαστάνει και ανθίζει εδώ και εκεί. Έχω αναφέρει ορισμένες από τις περιπτώσεις".[7]

Θα πρέπει να εφαρμόσουμε μια παρόμοια προσδοκία για στατιστική μετατόπιση και ασάφεια στον πλήρη βιολογικό χάρτη των αμοιβαιοτήτων.

Το περιθώριο καθυστέρησης για την εξέταση των "περιπτώσεων" αυτού του είδους είναι ελάχιστο αν όχι μηδενικό. Κάθε μέρα πεθαίνουν περίπου 150.000 άνθρωποι. Ο αριθμός στροβιλίζεται γύρω μας σε ποιητική απόσταση. Όπως και η ταφή του διάσημου ορνιθολόγου (και "συλλέκτη") Χένρι Σίμπομ (1832-1895). Ο πρώην μεγιστάνας του χάλυβα τελικά παθιάστηκε με το δέρμα και τα αυγά των πουλιών, τα περισσότερα από τα οποία τελικά δωρίστηκαν στο Βρετανικό Μουσείο. Οι περιπέτειές του τον ταξίδεψαν σε ολόκληρη τη Σιβηρία, την Ιαπωνία, τη Νότια Αφρική και όλη την Ευρώπη, καθώς αφιέρωσε τη ζωή του στην καταγραφή των πτηνών, ενστερνιζόμενος την τριωνυμική έλευση των κατηγοριών υποειδών. Καθώς το φέρετρό του κατέβαινε στη γη στο Χίτσιν, βόρεια του Λονδίνου, οι παρευρισκόμενοι προφανώς παρακολούθησαν ένα πουλί (μία τσίχλα) να τραγουδά σε ένα κοντινό δέντρο και το θεώρησαν αυτό ως σημάδι επιβεβαίωσης, πιθανώς μιας ισόβιας δέσμευσης προς τα πουλιά που τώρα επιβραβευόταν, η οποία αναγνωρίστηκε από έναν πρεσβευτή από τη Γη της Πτηνοπανίδας.[8] Στον πρόλογο της Έκδοσης του 1855 του έργου του "Leaves of Grass" ("Φύλλα Χλόης"), ο ίδιος ο Γουόλτ Γουίτμαν παρέπεμψε στην επεξηγηματική πρόζα: "Η γη και η θάλασσα, τα ζώα τα ψάρια και τα πουλιά, ο ουρανός των συννέφων και οι ουράνιες σφαίρες, τα δάση τα βουνά και τα ποτάμια, δεν είναι μικρές ιδέες… όμως οι άνθρωποι περιμένουν από τον ποιητή να δείξει κάτι περισσότερο από την ομορφιά και την αξιοπρέπεια που πάντα προσδίδουν στα χαζά πραγματικά αντικείμενα… περιμένουν από αυτόν να δείξει το μονοπάτι μεταξύ της πραγματικότητας και της ψυχής τους".[9]

Αυτός ακριβώς ο δρόμος έχει χαραχθεί σχεδόν σε κάθε *γνήσια αρχική* ανθρώπινη κοινότητα, σε ορισμένες με πιο μακροχρόνια επίδραση από άλλες. Κοντά στο Tassajara, τα μέλη της φυλής Excelen της περιοχής Έσσελεν που βρίσκεται κατά μήκος της ακτής Μπιγκ Σουρ της Καλιφόρνια άφησαν περίπου 250 κομψά ζωγραφισμένα αποτυπώματα χεριών σε ένα καταφύγιο βράχων ("CA-MNT-44")[10] χρησιμοποιώντας κάποιο λευκό ορυκτό, ίσως ασβεστόλιθο, και μια κόλλα που μπορεί να ήταν φτιαγμένη από "θερμαινόμενο λίπος αρκούδας" ή ενδεχομένως κάποιο είδος χυμού δέντρου. Τα ακροδάχτυλά τους αποτυπώθηκαν απαλά. Γράφει ο Πέτερ Φρίντμπεργκ: "Το όνομα Esselen πιθανότατα προέρχεται από μια τοποθεσία της φυλής γνωστή ως Ex 'selen ("ο βράχος"), η οποία με τη σειρά της προέρχεται από τη φράση "Xue elo xonia eune", η οποία σημαίνει "Προέρχομαι από τον βράχο". Όπως διακηρύσσουν σήμερα την αρχαία ύπαρξή τους τα εικονογράμματα στους τοίχους των βράχων…".[11] Έχοντας ζήσει στο Μπιγκ Σουρ για μερικά χρόνια και εξερευνώντας πολλούς από τους πιο απομακρυσμένους αρχαιολογικούς χώρους κατά τη διάρκεια πολλών δεκαετιών, μπορούμε να επιβεβαιώσουμε την απόλυτη κομψότητα του νου, την ομορφιά της κοινότητας, που διαπέρασε κάθε πτυχή της βιοκληρονομιάς του Έσσελεν. Δεν έχει φύγει πέρα από τα βαθιά παράκτια φαράγγια του Ρέντγουντ. Καθώς κάποιος κινείται όλο και βαθύτερα στο παρελθόν του, μέσα από τις σύγχρονες ζωντανές συλλογές της φυσικής ιστορίας, είναι συντριπτικά προφανές ότι αυτοί οι λαοί εξακολουθούν να υφίστανται. Πρέπει απλώς να συντονιστούμε. Και να προσπαθήσουμε να αγνοήσουμε τον Αυτοκινητόδρομο 1 που όχι μόνο διχοτομεί το ευάλωτο οικοσύστημα, αλλά αποτελεί και τον πλέον κραυγαλέο πρεσβευτή της κατάκτησης.

Δύο ισπανικές εκστρατείες, οι οποίες διέπονταν από την επεμβατική τους επίδραση, ξεκίνησαν στις αρχές του 17ου αιώνα με σκοπό να κατακτήσουν ένα σύνολο από πολυάριθμους αυτόχθονες πληθυσμούς κατά μήκος της Μπιγκ Σουρ με ιστορία 6.000 έως 10.000 ετών. Υπήρχαν περισσότερα από 1.940 περίπου στρέμματα γης των περιοχών αυτοχθόνων, με πολιτισμούς όπως οι Ohlone/Costanoan, Esselen, Salinan και Yokut ανακατεμένους μεταξύ τους, των οποίων ο συνολικός αριθμός ήταν πιθανώς λιγότερα από 1.500 άτομα μέσα στα τραχιά παράκτια φαράγγια. Οι ισπανικές αποστολές εγκαταστάθηκαν νότια και βόρεια του Μπιγκ Σουρ στα τέλη του 18ου αιώνα, και το Μοντερέι Πρεσίντιο (ένα από τα τέσσερα στρατιωτικά φρούρια στην Καλιφόρνια εκείνη την εποχή) έγινε η πρωτεύουσα της Άλτα Καλιφόρνια.[12] Δεν είναι περίεργο ότι, παρά τη σχετικά απροσπέλαστη ακτογραμμή της Μπιγκ Σουρ, αυτοί οι ντόπιοι αφομοιώθηκαν γρήγορα, ενώ πολλοί βαπτίστηκαν με τη βία. Μέχρι τη δεκαετία του 1880, ο Αμερικανός εθνολόγος Χιούμπερτ Χόουι Μπάνκροφτ (1832-1918) από τον οποίο πήρε το όνομά του η κεντρική βιβλιοθήκη του Πανεπιστημίου Καλιφόρνια στο Μπέρκλεϋ, κατέβαλε μεγάλες προσπάθειες για να επισημάνει την κακομεταχείριση των Ινδιάνων από τους Φραγκισκανούς, γράφοντας τα εξής: "Δεν πέρασε ποτέ από το μυαλό του [Χουνιπέρο Σέρα] να αμφισβητήσει το απόλυτο δικαίωμά του να μαστιγώνει τους νεοφώτιστούς του για οποιαδήποτε ελαφρά αμέλεια σε θέματα πίστης…". Ενώ άλλοι επικριτές κάνουν συζήτηση για ένα σύστημα που επιδοκίμασε "τη γενοκτονία, τη δουλεία και ένα ευρύ φάσμα θηριωδιών".[13] Ωστόσο, η παλαιότερη ευρωπαϊκή αποστολή παρατήρησης το 1602 με ηγέτη τον Ισπανό καπετάνιο Σεμπάστιαν Βισκάινο, ο οποίος είχε πλεύσει στον Κόλπο του Μοντερέι, περιέγραφε τους Έσσελεν ως "ευγενικούς και ειρηνικούς ανθρώπους…" και "σύμφωνα με τον Φρέι Αντόνιο ντε λα Ασενσιόν, ο οποίος συνόδευε τον Βισκάινο, "το λιμάνι είναι περικυκλωμένο από... φιλικούς Ινδιάνους, καλούς ιθαγενείς και καλοπροαίρετους, που τους αρέσει να δίνουν ό,τι έχουν... είναι όλοι γυμνοί σε αυτό το λιμάνι."". [14]

Η βιοπολιτιστική ιστορία της Καλιφόρνια είναι πλέον εύκολα ορατή από το διάστημα: η αυξανόμενη λεκάνη απορρύπανσης που είναι πλέον η Κεντρική Κοιλάδα και ο Αυτοκινητόδρομος 1 που διασχίζει με αγένεια την Ακτή Μπιγκ Σουρ, και από τις μεγαλουπόλεις πάνω και κάτω από την Πολιτεία, το γεγονός ότι τη νύχτα στη Νότια Καλιφόρνια υπάρχει μόνο μια σκοτεινή λωρίδα έκτασης (περίπου 28 χιλιομέτρων), η Βάση Πεζοναυτών του Πέντλετον, η μοναδική περιοχή που δεν φωταγωγείται μέσα σε όλο αυτό το συνονθύλευμα, από την Τιχουάνα μέχρι τη Σάντα Μπάρμπαρα. Στην οποία οι πράσινες και λαδί θαλάσσιες χελώνες, ο ποντικοί του Ειρηνικού, οι τσικνιάδες και οι χρυσαετοί μπορούν να κοιμούνται χωρίς τη λάμψη των φώτων γύρω τους. Το τελευταίο τους καταφύγιο.

Οι οικολόγοι έχουν δανειστεί από την πειραματική οικονομία φράσεις όπως "μη συνεργάσιμες συνθήκες ισορροπίας" και "ενοχλητικές μεταβλητές".[15] Από τη βιβλιογραφία της ηθικής δανείστηκαν έννοιες όπως η "συνολική ευημερία" και η "γενική ευημερία".[16] Οι εκτάσεις σκότους όπως αυτή της Βάσης του Πέντλετον τη νύχτα, έντονα παραμορφωμένες σε όλες τις πλευρές από έντονα φώτα, αποτελούν, φυσικά, ένα πρόσφατο φαινόμενο σε γεωλογικό χρόνο. Το 1855, όταν οι πρώτοι δυτικοί απεσταλμένοι έφτασαν στην Ιαπωνία, τα αθόρυβα φώτα των φαναριών δεν είχαν ακόμη αντικατασταθεί από τη δύναμη της ηλεκτρικής ενέργειας, η προέλευση της οποίας θα προέκυπτε, στην πραγματικότητα, από μια ρεματιά με μπαμπού που επέλεξε ο Τόμας Έντισον η οποία συνόρευε με ένα ναό Shinto όχι πάρα πολλά μίλια νότια του σύγχρονου Κιότο. Σε εκείνη την επιδρομή του 1855 σε εκείνο το τόσο μυστηριώδες έθνος, ο αξιωματικός του ναυτικού Πλωτάρχης Τζον Ρότζερς έμεινε έκπληκτος όταν έμαθε ότι δεν υπήρχαν στη χώρα πυροβόλα. Τα είχαν λιώσει όλα αιώνες πριν. Σε αντίθεση με τις δημογραφικές συνθήκες που επικρατούσαν την εποχή εκείνη στις ΗΠΑ, οι Ιάπωνες διατηρούσαν ένα χαμηλό ποσοστό γονιμότητας και διαβίωναν για εκατοντάδες χρόνια σε μια "φούσκα" σταθερής κατάστασης. Ο Τάουνσεντ Χάρις, ο πρώτος πρέσβης των ΗΠΑ σε αυτό το έθνος είχε δηλώσει σχετικά ότι "Μοιάζει περισσότερο με τη χρυσή εποχή της απλότητας και της ειλικρίνειας από ό,τι έχω δει ποτέ σε οποιαδήποτε άλλη χώρα". Και ο βοηθός του Χάρις, Χένερι Χέουσκεν, έγραφε στο ημερολόγιό του, περιγράφοντας την πρώτη επίσκεψή του στο Ιαπωνικό Αυτοκρατορικό Παλάτι: "Η απλότητα της Αυλής του Έντο, η ευγενής και αξιοπρεπής συμπεριφορά των αυλικών, οι εξευγενισμένοι τρόποι τους που τιμούν την πιο λαμπρή Αυλή, αποκαλύπτουν ένα μεγαλείο πιο εκθαμβωτικό κι από όλα τα διαμάντια των Ινδιών... Ούτε μια ξιφολόγχη δεν τους ακολουθεί... Φοβάμαι, ω, Θεέ μου, ότι αυτή η σκηνή ευτυχίας φτάνει στο τέλος της και ότι οι δυτικοί θα φέρουν εδώ τα θανατηφόρα ελαττώματά τους".[17]

Κατά ειρωνικό τρόπο, ο Χέουσκεν σκοτώθηκε γύρω στην ηλικία των 30 από έναν σαμουράι, την εποχή που ο νέος αυτοκράτορας Μουτσουχίτο διέταζε τους παρωχημένους στρατούς του να παραδώσουν τα σπαθιά τους. Ο Ιάπωνας ιστορικός τέχνης που έγινε επιμελητής του ανατολίτικου τμήματος τέχνης του Μουσείου Καλών Τεχνών της Βοστώνης, Οκάκουρα Κακούζο, περιέγραψε το χάος των πολιτιστικών παρεξηγήσεων (το οποίο θυμίζει τον θάνατο του Κυβερνήτη Τζέιμς Κουκ το 1779 κατά τη διάρκεια της τρίτης επίσκεψής του στην ομάδα των νησιών του Ειρηνικού της Χαβάης): "Πότε θα καταλάβει η Δύση ή έστω θα προσπαθήσει να καταλάβει, την Ανατολή;", έγραψε ο Κακούζο. "Είναι αρκετά παράξενο το ότι η ανθρωπότητα έχει μέχρι τώρα συναντηθεί σε ένα φλιτζάνι τσάι. Είναι το μόνο ασιατικό τελετουργικό που έχει κερδίσει τον παγκόσμιο σεβασμό. Ας σταματήσουμε τις ηπείρους από το να εκτοξεύουν απειλές η μία στην άλλη και ας είμαστε πιο ευχαριστημένοι, αν όχι πιο σοφοί, από το αμοιβαίο κέρδος του μισού ημισφαιρίου".[18]

Μπορούμε να εξάγουμε ορισμένα συμπεράσματα από τα διάφορα αυτά παραδείγματα αρχικών επαφών και της επακόλουθης αλλαγής και επιθετικότητας. Όπως έγραψε ο Άλντους Χάξλεϊ, "Ακόμη και τα καλύτερα συντάγματα και οι προληπτικοί νόμοι θα είναι ανίσχυρα απέναντι στις σταθερά αυξανόμενες πιέσεις του υπερπληθυσμού και της υπεροργάνωσης που επιβάλλουν η αύξηση των αριθμών και η εξέλιξη της τεχνολογίας".[19] Οι στρατηγικές διατήρησης ερευνούσαν πάντοτε τα "εστιακά είδη" των οποίων η "κατανομή και η αφθονία... προσφέρουν συμπεράσματα για την κατάσταση των άλλων ειδών".[20] Στην περίπτωση των λεγόμενων "ειδών ομπρέλας", είμαστε πλέον περιορισμένοι σε έναν πλανήτη ενός τέτοιου πλάσματος, παρά τα εκατομμύρια ακάρεα ωοθυλακίων που ξεχειλίζουν τις βλεφαρίδες μας ή βακτήρια στα χέρια μας και στα έντερά μας. Εάν οι στρατηγικές μας για τον περιορισμό των επιπτώσεων από έναν οργανισμό ομπρέλα (εμάς) σε όλα τα άλλα εστιακά είδη - εκείνα των οποίων οι σχέσεις φέρουν όλες τις ομοιότητες των αλληλεπιδράσεων αμετάβλητες για εκατοντάδες εκατομμύρια χρόνια – η βία μας πρέπει να περιοριστεί στο βαθμό που ο χαρακτήρας της δεν έχει διαμορφώσει τις παρατηρήσεις μας. Αυτό, από μόνο του, αποδεικνύεται εξαιρετικά δύσκολο. Μόνο η αμείωτη αμοιβαιότητα μεταξύ γειτόνων, αγνώστων, φίλων και συμμάχων, εχθρών και εκείνων με τους οποίους θα προτιμούσαμε να μην έχουμε καμία σχέση, σε καθέναν από αυτούς, καθώς και σε όλα αυτά τα πλήθη, άγνωστα, σε όλα τα βιολογικά βασίλεια για τα οποία είμαστε οι ξένοι. Για κάθε έναν από αυτούς, η μοναδική απάντηση είναι η επίδειξη αμοιβαιότητας, καλοσύνης, περιέργειας και αγάπης αυτού του είδους.

Επειδή όλοι οι ανθρωπικοί φακοί θολώνουν από εκείνους τους πολύ ανθρώπινους παρατηρητές που αναπόφευκτα συνδέονται με το είδος τους, αντιμετωπίζουμε τον πιο αρχόμενο από όλους τους θανάτους: οτιδήποτε σκεφτόμαστε, αισθανόμαστε, αντιλαμβανόμαστε και κάνουμε παραμένει παγιδευμένο μέσα στη δική του τροχιά. Ωστόσο, γνωρίζουμε επίσης ότι στο παρελθόν, αλλά και στο παρόν, είναι πολλοί οι άνθρωποι του είδους μας έχουν δείξει το δρόμο προς την αγάπη. Προς τους Άλλους. Ο Σάμιουελ Μπέκετ χαρακτήρισε μια ομάδα ανθρώπων καταδικασμένων από τη μοίρα, την ανθρώπινη κοινότητα που αναφέρει στο σύντομο έργο του, *Αυτοί που έχουν χαθεί*, για τους οποίους έγραψε: "Έτσι, επ' άπειρον μέχρι το αδιανόητο τέλος, αν αυτή η έννοια διατηρείται, ένα τελευταίο σώμα όλων από αδύναμες κρίσεις και αρχίζει να ψάχνει ακόμα".[21] Μπορούμε να ενστερνιστούμε αυτή την άποψη είτε με μοιρολατρία είτε με μια αναζωπυρωμένη μαγεία με την οικολογία και όλα τα εθνογραφικά, μυστικιστικά μέρη της. Ακλόνητη, σε ισορροπία, και με μια καθολικά μη βίαιη αποφασιστικότητα.

ΣΗΜΕΙΩΣΕΙΣ

[1] "Estimates show that between 1992 and 2014, produced capital per person doubled, and human capital per person increased by about 13% globally; but the stock of natural capital per person declined by nearly 40%...At their core, the problems we face today are no different from those our ancestors faced: how to find a balance between what humanity takes from Nature and what we leave behind for our descendants. While our ancestors were incapable of affecting the Earth system as a whole, we are doing just that." Estimates show that between 1992 and 2014, produced capital per person doubled, and human capital per person increased by about 13% globally; but the stock of natural capital per person declined by nearly 40%." Από το *The Economics of Biodiversity: The*

Dasgupta Review, Headline Messages, February 2021, Crown copyright, www.gov.uk/official-documents, https://assets.publishing.service.gov.uk/ government/uploads/system/uploads/attachment_data/file/957629/Dasgupta_Review_-_Headline_Messages.pdf, σελ. 4.

[2] Βλ. *Design for Ecological Democracy*, του Randolph T. Hester, The MIT Press, Cambridge, MA, 2006, σελ. 264 και 325.

[3] ό.π., Dasgupta, σελ. 2.

[4] "Species Numbers in Bacteria," του Daniel Dykhuizen, Proc Calif Acad Sci. Author manuscript; available in PMC 2011 Aug 24. Published in final edited form as: *Proc Calif Acad Sci.* 2005 Jun 3; 56(6 Suppl 1): 62–71, PMCID: PMC3160642, NIHMSID: NIHMS222630, PMID: 21874075, https://www.ncbi.nlm.nih.gov/pmc/articles/PMC3160642/

[5] "What are microorganisms?" Centre for Geobiology, University of Bergen, https://www.uib.no/en/geobio/56846/what-are-microorganisms#:~:text=Microorganisms %20can%20be%20bacteria%2C%20fungi,generally%20classified%20as%20non%2Dliving

[6] Βλ. "Earth May Be Home to a Trillion Species of Microbes," του Nicholas Bakalar, May 23 Μαΐου 2016, *The New York Times*, https://www.nytimes.com/2016/05/24/science/one-trillion-microbes-on-earth.html

[7] ό.π., Hester, σελ. 419.

[8] Βλ. *The Travelling Naturalists*, της Clare Lloyd, Croom Helm, London, 1985, σελ. 119.

[9] *Leaves Of Grass* – Selected and with an Introduction by Christopher Morley, Illustrations by Lewis C. Daniel, Doubleday, Doran & Co., Inc., New York, 1940, σελ. 299.

[10] Βλ. *The Esselen Indians of the Big Sur Country – The Land and the People*, των Gary S. Breschini και Trudy Haversat, Coyoto Press, Salinas CA, 2004, σελ. 140-167.

[11] "The Esselen Hands," του Peter Friedberg, Wayback Machine, Esalen Institute, https://web.archive.org/web/20111115115734/http://www.esalen.org/air/essays/esselenhands.html

[12] ibid., Breschini and Haversat, p. 173.

[13] ό.π., σελ. 172.

[14] ό.π., Friedberg.

[15] *Experimental Economics*, των Douglas D. Davis και Charles A. Holt, Princeton University Press, Princeton New Jersey, 1993, σελ. 524.

[16] *Moral Differences – Truth, Justice and Conscience in a World of Conflict*, του Richard W. Miller, Princeton University Press, 1992, σελ. 292.

[17] *Japan Journal 1855-1861*, του Henry Heusken, Μετάφραση και επιμέλεια των Jeannette C. van der Corput και Robert A. Wilson, Rutgers University Press, Brunswick, New Jersey, 1964, αναφέρεται στη σελίδα 84 του *After Eden: History, Ecology & Conscience*, του M.C.Tobias, Avant Books, San Diego, CA 1985.

[18] *The Book of Tea*, του Kakuzo Okakura, Dover Publishers, New York, NY, 1964, σελ. 3.

[19] *Brave New World Revisited*, του Aldous Huxley, Harper & Brothers, 1958, New York, NY, σελ. 137.

[20] "Species-Level Strategies for Conserving Rare or Little-Known Species," στο *Conservation of Rare or Little-Known Species – Biological, Social, and Economic Considerations*, Επιμέλεια Martin G, Raphael και Randy Molina, των Bruce G. Marcot και Curtis H. Flather, Πρόλογος της Nancy Molina, Island Press, Washington, Covelo, London, 2007, Κεφάλαιο 6, σελ. 135.

[21] *The Lost Ones*, του Samuel Beckett, Evergreen Press, New York, NY, 1972; αρχική έκδοση από τον οίκο Les Editions de Minuit in Paris το 1971 με τον τίτλο Le Depeupleur, σελ. 60.

Κεφάλαιο 29

ΣΥΝΕΙΔΗΤΗ ΕΙΔΟΓΕΝΕΣΗ

Κάθε εβδομάδα περιγράφονται νέα είδη. Τουλάχιστον 270 ανακαλύφθηκαν το 2018, τουλάχιστον 71 το 2019, 503 το 2020[1]. Τα περισσότερα, αν όχι όλα αυτά τα νέα είδη *είναι νεοανακαλυφθέντα*, όχι *νέα*. Αν και η ηλικία του γένους μας είναι άνω των 2,8 εκατομμυρίων ετών, η ηλικία του είδους μας χρονολογείται πλέον μεταξύ των 250.000 και 334.000 ετών, όπως έδειξαν και τα ευρήματα στην Τζεμπέλ Ιρχούντ στο Μαρόκο[2] και το κρανίο του Florisbad στη Νότιο Αφρική[3]. Κατατασσόμαστε, εξ' όσων μπορούμε να προσδιορίσουμε, μεταξύ των νεότερων ειδών στη γη. Το σκουλήκι *Rhagoletis pomonella* είναι ακόμα νεότερο, περίπου 150 ετών. Πρόκειται στην πραγματικότητα για μια φρουτόμυγα που μετατόπισε τους ξενιστές από τον κράταιγο στο *Malus domestica*, μια μηλιά που βρίσκεται στις βορειοανατολικές Ηνωμένες Πολιτείες. Η διαδικασία ονομάζεται συμπατρική μετατόπιση. Ο Βρετανός βιολόγος Σερ Έντουαρντ Μπάγκναλ Πόουλτον (1856-1943) περιέγραψε για πρώτη φορά αυτή τη διαδικασία με την οποία τα αδελφά είδη μπορούν να εξελιχθούν από κοινές προγονικές γενεαλογίες εντός επικαλυπτόμενων περιοχών. Στα βακτήρια, οι γρήγορες οριζόντιες γονιδιακές μεταναστεύσεις αιτιολογούν εύκολα την ύπαρξη τέτοιων νέων ειδών καθώς απαιτούνται πολύ λιγότερες μεταλλάξεις πριν από την προέλευσή τους σε ξεχωριστές νέες θέσεις. Η ομοζυγωτική κυριαρχία στα αντίγραφα των αλληλόμορφων γονιδίων που αποδίδεται ειδικά στα πλεονεκτικά χαρακτηριστικά σε έναν γονότυπο, είναι πιο πιθανό να επιταχύνει τη φυσική επιλογή προς όφελός τους.

Η εμφάνιση νεοσύστατων ειδών μπορεί να προσδιοριστεί από διάφορες κατανομές αλληλόμορφων γονιδίων εντός ενός συνόλου παραμέτρων του γονιδιώματος κάποιου πληθυσμού. Για παράδειγμα, η συμπατρική ειδογένεση μπορεί εύκολα να συμβεί στα γενετικά περιθώρια των πανμεικτικών συνθηκών που καθιστούν δυνατό το τυχαίο ζευγάρωμα εντός μακρο-συμπατρικών γεωγραφικών περιοχών, επικαλύπτοντας οικολογικούς θώκους. Υπάρχουν και άλλοι καθοριστικοί παράγοντες. Για παράδειγμα, οι μυριάδες μη χωρικές διαστάσεις που μπορούν να επιταχύνουν έναν νέο οργανισμό, όπως τα ημερήσια ή νυκτερινά χαρακτηριστικά που μεταβάλλουν το ρυθμό και την κλίμακα των αλληλεξαρτήσεων. Ή ο βαθμός στον οποίο το ενδιαίτημα έχει γίνει "ανομοιογενές" - ένας συντελεστής γονιδιακής διάχυσης που δίνεται σε υψηλότερη πιθανότητα ετεροπαθητικής ειδογένεσης ή, με τη σειρά της, η συνύπαρξη ειδών κατά την οποία η γονιδιακή ροή μπορεί να επιταχυνθεί με το συνδυαστικό ζευγάρωμα ή κάποια άλλη αλλοιωμένη δύναμη σαγήνευσης ή/και διάσπασης. Οι συγκεκριμένες καινοτομίες έχουν παρατηρηθεί σε σαλιγκάρια, φρύνους, όρκες, κιχλίδες της Ανατολικής Αφρικής, διάφορα σκαθάρια φύλλων και σε μια μεγάλη ποικιλία ειδών ωδικών πτηνών, μεταξύ πολλών άλλων[4].

Το σωρευτικό βάρος αυτών των νέων μορφών ζωής, κυρίως υβριδικών, αλλά και βιοσυνθετικών (συχνά κατοχυρωμένων με δίπλωμα ευρεσιτεχνίας) πλασμάτων[5] δεν παρουσιάζει καμία καθοδήγηση όσον αφορά την ανθρωπότητα. Τα "βιο-τούβλα" ή αλλιώς οι βιολογικές πρότυπες μονάδες του συνθετικού εμπορικού κόσμου της βιοτεχνολογίας αποτελούν εκδηλώσεις σχετικά με κάτι πέρα από τα γονίδια του Φρανκενστάιν. Κατά τη γνώμη μας, αντιπροσωπεύουν ένα απροσδόκητα νέο ύψος αλαζονείας όσον αφορά τη ρύπανση, μια υπαρξιακή ύβρι των μηχανικών που παριστάνουν τον Θεό, με οποιοδήποτε αριθμό ιδιοτελών στάσεων, κινήτρων και στόχων.[6] Η βιαστική διασφάλιση βιολογικών διπλωμάτων ευρεσιτεχνίας - συνήθως κωδικοποιημένων για 20 χρόνια - συμπεριέλαβε διπλώματα για εμβρυϊκά βλαστοκύτταρα, χιλιάδες γενετικά τροποποιημένους οργανισμούς, γονίδια από ρύζι, χοίρους, διαγονιδιακά ποντίκια και ολόκληρες αλληλουχίες DNA· ίσως περισσότερα από 6.000 διπλώματα ευρεσιτεχνίας για ανθρώπινα γονίδια (τουλάχιστον το 20% όλων των 30 γνωστών ανθρώπινων γονιδίων) και πάνω από 50.000 άλλα διπλώματα ευρεσιτεχνίας που σχετίζονται με τη γενετική.[7] Αρχικά, στην υπόθεση *Parke-Davis εναντίον Mulford* του 1906, ο δικαστής έκρινε ότι η ανθρώπινη τροποποιημένη αδρεναλίνη ήταν καλύτερη από την αδρεναλίνη στη φυσική της κατάσταση και, ως εκ τούτου, τα χημικά πρόσθετα της μπορούσαν να κατοχυρωθούν με δίπλωμα ευρεσιτεχνίας.[8]

Έτσι, όχι μόνο σφετεριστήκαμε τη λεπτή βιόσφαιρα του πλανήτη και τα πρωτογενή προϊόντα της φωτοσύνθεσης, αλλά και επιδιώξαμε επιθετικά να διακηρύξουμε την ιδιοκτησία του ίδιου του γενετικού υλικού, των βιολογικών οδών και μηχανισμών με τους οποίους η ζωή, μέχρι την παρουσία μας, διαιωνίστηκε και διαφοροποιήθηκε με ασφάλεια και με τους δικούς της κανόνες.

Τα ερωτήματα που ελπίζαμε να θέσουμε στην αρχή του παρόντος κεφαλαίου παραμένουν αναπάντητα. Τι μπορούμε να μάθουμε εμείς, ένα νέο είδος, από άλλα νέα είδη; Πώς το κάνουν; Ποια σοφία έχουν αντλήσει από τη δοκιμή και το λάθος; Οι προσπάθειες αυτού του είδους για τη σκιαγράφηση των συγκρίσιμων εμπειριών μεταξύ ειδών έχουν αγνοηθεί, έχουν αποκρυφτεί και έχουν αποξενωθεί από την αιτία και την έννοιά τους. Τα μόνα ερωτήματα που έχουμε θέσει σχετικά με τους Άλλους είναι τι μπορούμε να τους πάρουμε δια της βίας ή μέσω της ομήγυρης για τους φυσικούς πόρους που θα αξιοποιηθούν όσο το δυνατόν πληρέστερα, με την Αγία Γραφή στο χέρι. Ωστόσο, είναι εκεί, στον Ιώβ, Κεφάλαιο 12, 7-8: "Αλλά, ρώτησε τώρα τα ζώα, και θα σε διδάξουν· και τα πουλιά τού ουρανού, και θα σου αναγγείλουν· ή, μίλησε στη γη, και θα σε διδάξει· και τα ψάρια τής θάλασσας θα σου διηγηθούν".[9]

Επιστρέφουμε με άδεια χέρια σε μια βιολογική πρώτη βάση. Ούτε κατανοούμε την εμμονή ορισμένων γονιδίων πάνω σε άλλα - άσχετα με αναπαραγωγικούς φραγμούς ή όχι,[10] ούτε έχουμε κατανοήσει αυτό που είναι γνωστό ως "γενομική νησίδα" σε σχέση με τις χρωμοσωμικές κατανομές και τις πολυεπιστημονικές μελέτες in situ και ex situ που είναι απαραίτητες για την απόδοση μεγαλύτερης σαφήνειας στα πρότυπα επιβίωσης.[11]

Ωστόσο, τουλάχιστον μία μελέτη, πάνω από μία δεκαετία νωρίτερα, παρείχε στοιχεία σχετικά με το γεγονός ότι η χρωμοσωμική εξέλιξη μπορεί να συμβεί σε πολλαπλές γεωγραφικές περιοχές κοντά η μία στην άλλη, σχηματίζοντας βιώσιμα αναπαραγωγικά συμπλέγματα, μια εξελικτική διαδικασία που είναι συνήθως αργή και σταθερή. Σε αυτή την περίπτωση, το στατιστικώς προβλεπόμενο αυτό χρονικό πλαίσιο έχει αλλάξει εξ ολοκλήρου λόγω των συμπεριφορικών κινήσεων ενός συνυπάρχοντος είδους: άνθρωποι που ζουν σε απομονωμένες κοιλάδες σε όλο το πορτογαλικό αρχιπέλαγος της Μαδέρας, όπου ένα είδος ποντικιού ηλικίας 500 ετών έχει εξελιχθεί σε "τουλάχιστον έξι διαφορετικά είδη", καθώς "ορισμένα χρωμοσώματα έχουν συγχωνευθεί" θέτοντας αναπαραγωγικούς περιορισμούς σε μια απομονωμένη ομάδα ιδρυτών.[12] Αντί να υποκύψει σε ένα γενετικό cul de sac, συνέβη το ακριβώς αντίθετο, προκαλώντας ταχεία ειδογένεση των θηλαστικών.[13] Τα άγρια ποντίκια των σπιτιών ευδοκιμούν, όχι ως υβρίδια αλλά ως διακριτά, βιώσιμα είδη, ένα κλασικό παράδειγμα αυτού που χαρακτηρίζεται ως "Ρομπερτσόνια (Rb) μετατόπιση, μια χρωμοσωμική αναδιάταξη που περιλαμβάνει κεντρική σύντηξη δύο ακρο- (τελο)κεντρικών χρωμοσωμάτων ώστε να σχηματίσουν ένα μόνο μετακεντρικό…"[14]

Τα άγρια και πιο εξημερωμένα ποντίκια στη Μαδέρα (*Mus musculus domesticus* και *M. sprelus* Lataste) έχουν καταστεί επιστημονικές διασημότητες διεθνώς, καθώς έχουν μάθει γρήγορα πως να εναρμονίζονται με τις τοποθεσίες τους, τους συγκατοίκους τους, τις επιλογές φαγητού και τα καταλύματά τους. Σε κανένα άλλο τεταρτημόριο του πλανήτη η έκφραση "ποντίκια και άνθρωποι" δεν είναι πιο κατάλληλη για την κοινότητα.[15]

Το 2006 ταυτοποιήθηκε στην Κύπρο ένα νέο υποείδος ποντικού.[16] Παγκοσμίως, υπάρχουν επί του παρόντος 42 γνωστά είδη *Μυών*, μιας φυλής γνωστή ως Μυίδες[17] της οικογένειας των Κρικετιδών (της υποκατηγορίας των Μυόμορφων και της υπεροικογένειας των Μυοειδών, που περιλαμβάνει τα χάμστερ, τους αρουραίους, και τα ποντίκια). Έχει γραφτεί ότι "η οικογένεια των Κρικετιδών αποτελεί ένα από τα μεγαλύτερα και πιο επιτυχημένα πειράματα της μεγάλης τάξης των θηλαστικών τρωκτικών" με "142 γένη, 765 είδη, τουλάχιστον 1505 υπάρχουσες τάξεις". Περίπου 3 είδη Κρικετιδών ανά αιώνα έχουν εξαφανιστεί από το 1600, ενώ 32 παραμένουν απειλούμενα με εξαφάνιση και 40 είδη χαρακτηρίζονται ως ευπαθή.[18] Αυτή η εξαιρετικά έμπειρη ομάδα θηλαστικών, αν και προέρχεται από την Ηώκαινο εποχή, μεταξύ 33,9 και 56 εκατομμύρια πριν, έχει διευκολυνθεί σημαντικά από την έκρηξη του ανθρώπινου πληθυσμού. Αξίζει να αναφερθεί σε κάποιο βαθμό το σχόλιο του Θορώ για ένα συγκεκριμένο ποντίκι, για το οποίο έγραψε στο *Walden* (1854): "Κατά πάσα πιθανότητα δεν είχε δει ποτέ ξανά άνθρωπο. Σύντομα όμως απέκτησε αρκετή εξοικείωση και έτρεχε πάνω από τα παπούτσια μου και πάνω στα ρούχα μου… Αργότερα, όπως έγειρα με τον αγκώνα μου στον πάγκο μια μέρα, έτρεξε πάνω-κάτω στα ρούχα μου και κατά μήκος του μανικιού μου και γύρω-γύρω από το χαρτί που είχα τυλίξει το δείπνο μου, που το κρατούσα κοντά μου, και έκανα στο πλάι και έπαιξα μαζί του κουκουτσά. Και όταν τελικά κράτησα ακόμα ένα κομμάτι τυρί ανάμεσα στον αντίχειρα και τον δείκτη μου, ήρθε και τσίμπησε λίγο, καθισμένο στο χέρι μου και στη συνέχεια καθάρισε το πρόσωπο και τα πόδια του, σαν μύγα, και απομακρύνθηκε".[19]

Η βιβλιοθήκη New York Society Library έχει δημοσιεύσει το "The Great Mice of Literature".[20] Σίγουρα η *ιστορία των δύο κακών ποντικιών*[21] της Beatrix Potter προωθεί το πάνθεον εκείνο του πειστικού συναισθήματος, όπως έκανε και ο Γιαν Μπρίγκελ ο πρεσβύτερος στην ελαιογραφία του με τίτλο "Ποντίκι, Τριαντάφυλλο και Πεταλούδα" το 1618, η οποία του ανατέθηκε από τον Αρχιεπίσκοπο του Μιλάνου, Φεντερίκο Μπορομέο.[22] Ο Πλίνιος ο Πρεσβύτερος, στο έργο του *Φυσική Ιστορία*, βρήκε ότι τα ποντίκια του Νείλου μπορούν να προμηνύουν το μέλλον, θαύματα της δημιουργίας, εραλδικά και εφ' όλης της ύλης.[23] Ο Τζον Κόνσταμπλ ζωγράφισε το κατοικίδιο ποντίκι του, τον Τζακ, σε ένα πορτρέτο του 1824, "Ένα ποντίκι με ένα κομμάτι τυρί", ενώ ο Ιερώνυμος Μπος μετέφερε μια πιο αξιοσημείωτη εικόνα ενός ποντικιού και ενός ανθρώπου που κοιτούν ο ένας τον άλλον σε ένα οριζόντιο κυλινδρικό σωλήνα στο κέντρο του πίνακά του "The Garden of Delrthly Delights" (1510-1490). Στο θαυμάσιο αφιέρωμά της στα ποντίκια, η Όρνα Όουεν επισημαίνει μερικές από τις πιο δυναμικές συμπεριφορικές ιδιότητες αυτής της πανέξυπνης γλυκειάς συμμορίας, συμπεριλαμβανομένου του γεγονότος ότι τραγουδούν μελωδικούς υπέρηχους,[24] καθώς και πληθώρα άλλων ιδιοτήτων όπως τις αντιλαμβάνονται ο Αντρέ Κερτέζ, ο Πικάσο, ο Κλάες Όλτενμπουργκ, ο Φρανς Μαρκ, ο Γκούσταβ Κλιμτ και ο Χοκουσάι και ο Γουώλτ Ντίσνεϊ. Όσον αφορά τον Ντίσνεϊ, ο χαρακτήρας του Μίκυ Μάους εμφανίστηκε για πρώτη φορά το 1928 και είναι σαφές ότι χωρίς τον Μίκυ δεν θα υπήρχε ποτέ Ντίσνεϊλαντ.[25]

Το ποντίκι, πολιτισμικά κατοχυρωμένο στο δικό μας είδος, παρέχει ένα "πρόσθετο σημείο υπεροχής"[26] από το οποίο μπορεί κάποιος να ατενίσει την εξέλιξη σε πραγματικό χρόνο και τη γενετική που φροντίζει για την ταχεία ειδογένεση σε είδη σε ένα νησί όπως η Μαδέρα ή στον εγκέφαλο ενός κινουμένου σχεδίου που ζει στο Χόλιγουντ. Ωστόσο στις μέρες μας υπάρχει η πεποίθηση ότι είναι πιο πιθανό να εμφανιστούν νέα είδη σε περιοχές που έχουν χάσει την αφθονία των ειδών, όπου η αφθονία έχει μετατραπεί σε έλλειψη, ιδιαίτερα μεταξύ των πτηνών.[27] Εξ ου και οι όροι "θερμά σημεία" και "ψυχρά σημεία", με την ισοδυναμία τους από την άποψη των δυνάμεων της ειδογένεσης.

Γνωρίζουμε ότι αυτές οι δυνάμεις ενεργοποιούν δύο τύπους αιτίων και συνεπειών: κινητήριες δυνάμεις/στρεσογόνους παράγοντες που υπονομεύουν τα υφιστάμενα γένη και πληθυσμούς, κατακερματίζουν τα ενδιαιτήματα, ομογενοποιούν τη βιοποικιλότητα, και πιέσεις από κοινού αξιολόγησης που αφορούν πληθυσμούς σε άνθηση, όπως αυτός του ιού Covid στις παραλλαγές τους, των μυρμηγκιών, και σε μικρότερο βαθμό των κριλ και του *H. sapiens*. Αυτό σημαίνει ότι ελάχιστα από αυτά, αν όχι κανένα, παγιδεύονται στη μήτρα μεταξύ αυτών των δύο απαιτητικών μηχανισμών γονιδιακής ροής. Υπάρχουν σαφείς ενδείξεις για διάφορες προσαρμογές που επέρχονται σε νέα (ανθρωπογενή) οικοσυστήματα. Τα κουνούπια, για παράδειγμα, έλκονται από ορισμένες οσμές που εκλύονται από τους ανθρώπους όταν είναι φορείς μίας ασθένειας.[28/29] Μέσα σε δεκαπέντε χρόνια μετά την κατασκευή ενός φράγματος στην κεντρική Βραζιλία, οι σαύρες γκέκο (*Gymnodactylus amarali)* που κατοικούσαν κοντά, μεγάλωσαν τα κεφάλια τους προκειμένου να καταναλώνουν περισσότερα από τα πολλά και διάφορα έντομα που προήλθαν από το φράγμα. Ωστόσο, όπως έχουν γράψει οι J.W. Bull και M. Maron: "Ενώ έχει αποδειχθεί ότι οι άνθρωποι μπορούν να οδηγήσουν τη σύγχρονη εξέλιξη σε βαθμό σημαντικά υψηλότερο από αυτόν που οφείλεται στα φυσικά αίτια, ο εκτιμώμενος αριθμός περιπτώσεων ειδογένεσης που αποδίδεται σε ανθρώπινες δραστηριότητες δεν είναι διαθέσιμος για τους περισσότερους οργανισμούς".[30] Αυτό ισχύει ιδιαίτερα για τους μικροβιοτικούς οργανισμούς. Αυτό που έχει συμβεί στο μεταξύ είναι ότι μεταξύ της ανθρώπινης γεωργίας, του κυνηγιού/λαθροθηρίας, της θανάτωσης γενικά, η ανθρωπότητα έχει "ξεπεράσει" (ίσως περισσότερο από 300%) όλα τα άλλα ισοδύναμα κληρονομικά χαρακτηριστικά που υπήρχαν στη Γη πριν από την παρουσία μας. Πρόκειται για μια συντηρητική μεν, παρόλα αυτά εκπληκτική συνειδητοποίηση.[31]

Παρά την τρέχουσα αναγέννηση των εμπειρικών φυσικών ιστοριών και του μέλλοντος, οι περισσότερες στρατηγικές διατήρησης στοχεύουν σε εγγύς, ανθρωπίνως καθορισμένα αποτελέσματα. Παρόλο που η έρευνα της ζωοσημειολογικής μετάφρασης βρίσκεται σχεδόν στην αρχή της χρονικογράφησης των δεδομένων επικοινωνίας μεταξύ των ειδών του φυσικού κόσμου, όσον αφορά την ενεργό συμμετοχή μας στη βιοσημειόσφαιρα, αυτή παραμένει ακόμα σε αρχικό στάδιο, αν όχι καταστροφική, παρά τις χίλιες καλές προθέσεις.

Αυτό αφήνει μόνο μία πρακτική ιδεολογία στα οπλοστάσια της βιολογίας της απελευθέρωσης και της διατήρησης των ζώων: να συνεχίσουμε να κάνουμε ακριβώς αυτό που κάνουν τα ταπεινά και ενάρετα άτομα, παρά το είδος τους.

Σημειώσεις

1 University of Turku. "New species described in 2020." *ScienceDaily. ScienceDaily*, 1 July 2020. www.sciencedaily.com/releases/2020/07/200701100030.htm, https://www.sciencedaily.com/releases/2020/07/200701100030.htm; Sources from: Zamani A. and Marusik Yu.M. A new and easternmost species of Loureedia (Aranei: Eresidae) from Iran. Arthropoda Selecta, 2020 doi: 10.15298/arthsel.29.2.09; Diego Galvão de Pádua, Ilari Eerikki Sääksjärvi, Ricardo Ferreira Monteiro, Marcio Luiz de Oliveira. Seven new species of spider-attacking Hymenoepimecis Viereck (Hymenoptera, Ichneumonidae, Pimplinae) from Ecuador, French Guiana, and Peru, with an identification key to the world species. ZooKeys, 2020; 935: 57 doi: 10.3897/zookeys.935.50492; Valter Weijola, Varpu Vahtera, André Koch, Andreas Schmitz, Fred Kraus. Taxonomy of Micronesian monitors (Reptilia: Squamata: Varanus): endemic status of new species argues for caution in pursuing eradication plans. *Royal Society Open Science,* 2020; 7 (5): 200092 doi: 10.1098/rsos.200092

[2] Hublin, JJ., Ben-Ncer, A., Bailey, S. et al. New fossils from Jebel Irhoud, Morocco and the pan-African origin of *Homo sapiens*. *Nature* 546, 289–292 (2017). https://doi.org/10.1038/nature22336, June 8 2017, https://www.nature.com/articles/ nature22336#citeas

[3] J Anthropol Sci, 2020 Dec 31;98. doi: 10.4436/JASS.98014. "The skull from Florisbad: a paleoneurological report," Emiliano Bruner, Marlize Lombard· *PMID:* 33130632, doi: 10.4436/JASS.98014, National Library of Medicine, PubMed, https://pubmed.ncbi.nlm.nih.gov/33130632/

[4] Smith, J. Maynard (1966). "Sympatric speciation". The American Naturalist. 110 (916): 637–650. doi:10.1086/282457. See also, M. Malinsky et al. "Genomic islands of speciation separate cichlid ecomorphs in an East African crater lake," *Science (*2015). doi: 10.1126/science.aac9927

[5] See "Synthetic Life: Made from Scratch," by Niko McCarty, *synbiobeta*, November 22, 2019, https://synbiobeta.com/synthetic-life-made-from-scratch/; on the legality, see, "BRIA 23 4 c Patenting Life," Constitutional Rights Foundation, Winter 2008 (Volume 23, No. 4), https://www.crf-usa.org/bill-of-rights-in-action/bria-23-4-c-patenting-life; See also, "Intellectual Property in Higher Life Forms: The Patent System and Controversial Technologies," by Robert P. Merges, https://cyber.harvard.edu/IPCoop/88merg1.html. See also, "The Supreme Court Holds Genes Are Patent-Ineligible Products of Nature," by Teige P. Sheehan, NYSBA Bright Ideas, Fall 2013, Vol. 22, No. 2, New York State Bar Association, https://www.hrfmlaw.com/wp-content/uploads/2018/04/The-Supreme-Court-Holds-Genes-Are-Patent-Ineligible-Products-of-Nature-TPS-Fall-2013-H1209113xD16FB.pdf

[6] See *Mendel's Ark – Biotechnology and the Future of Extinction*, by Amy Lynn Fletcher, Springer New York, NY 2014, p. 96.

[7] "Gene Patents," by Robert Cook-Deegan, MD, The Hastings Center September 23, 2015, https://www.thehastingscenter.org/briefingbook/gene-patents/#:~:text=There%20are%203% 2C000%E2%80% 935%2C000%20U.S.,and%20genetic%20therapies%20unacceptably%20high. See also, "Gene Patenting," *AMA*, https://www.ama-assn.org/delivering-care/precision-medicine/gene-patenting#:~:text=Gene%20 Patenting%20background&text=The%20total %20number%20of%20human,the%20government%20and%20 even%20individuals.

[8] Parke-Davis & Co. v. H.K. Mulford Co., 189 F. 95 (C.C.S.D.N.Y. 1911), Learned Hand, J. See also, Dutfield, Graham. "DNA Patenting: Implications for Public Health Research." *Bulletin of the World Health Organization*, May 2006, Volume 85, Issue 5. See also, Eisenberg, Rebecca S (December 2002). "Why the Gene Patenting Controversy Persists: Academic Medicine". Academic Medicine. 77 (12): 1381–1387.

[9] "Job," 11: 7-8, *Holy Bible, King James Version*, The Christian Science Publishing Society, Boston, MA, n.d. p. 709.

[10] See "Macroevolutionary speciation rates are decoupled from the evolution of intrinsic reproductive isolation in *Drosophila* and birds," by Daniel L. Rabosky and Daniel R. Matute, *PNAS* September 17, 2013 110 (38) 15354-15359; https://doi.org/10.1073/pnas.1305529110, https://www.pnas.org/content/110/38/15354

[11] "Speciation: The Origin of New Species," by Rebecca J. Safran, and P. Nosil, Knowledge Project, *Nature Education Knowledge* 3(10): 17, 2012, https://www.nature.com/scitable/ knowledge/library/speciation-the-origin-of-new-species-26230527/; See also, "Ecotypes and the controversy over states in the formation of new species," by David B. Lowry, *Linnean Society, Biological Journal*, 21 December 2011, https://davidbryantlowry.files. wordpress.com/2013/01/lowry_david_3.pdf

[12] "Are new species still evolving?" by Abbie Thomas, an interview with Michael Schwarz and Arthur Georges, *ABC Science*, March 10, 2010, https://www.abc.net.au/science/ articles/2010/03/10/2820949.htm, Accessed February 9, 2021.

[13] See Britton-Davidian, J., Catalan, J., da Graça Ramalhinho, M. et al. Rapid chromosomal evolution in island mice. *Nature* 403, 158 (2000). https://doi.org/10.1038/35003116, https://www.nature.com/articles/35003116 #rightslink

[14] See "Telomeres and mechanisms of Robertsonian fusion," by P. Slijepcevic, *Cromosoma*, 1998 May;107(2):136-40. doi: 10.1007/s004120050289. https://pubmed.ncbi.nlm.nih.gov/ 9601982/

[15] "On the origin and colonization of house mice in the Madeira Islands," Maria Da Luz Mathias and António Mira, *Biological Journal of the Linnean Society,* Volume 46, Issue 1-2, May 1992, Pages 13–14, https://doi.org/10.1111/j.1095-8312.1992.tb00846.x, 14 January 2008, https://academic.oup.com/biolinnean/article-abstract/46/1-2/13/2654404, Accessed February 9, 2021.

[16] T. Cucchi, A. Orth, J.-C. Auffray, S. Renaud, L. Fabre, J. Catalan, E. Hadjisterkotis, F. Bonhomme, J.-D. Vigne (23 June 2006). "A new endemic species of the subgenus Mus (Rodentia, Mammalia) on the Island of Cyprus". *Zootaxa*. Magnolia Press. 1241 (1).

[17] See *Handbook Of The Mammals Of The World*, 7, II, Rodents, Chief Editors, Don E. Wilson, Thomas E. Lacher, Jr., and Russell A. Mittermeier, Published by Lynx Edicions in association with Conservation International and IUCN, Barcelona, Spain, 2017, pp. 788-798.

[18] ibid., p. 204.

[19] See "Henry and the Mouse," by Corinne H. Smith, *Thoreau Farm*, October 8, 2015, https://thoreaufarm.org/2015/10/henry-and-the-mouse/Accessed February 9, 2021.

[20] https://www.nysoclib.org/collection/great-mice-literature

[21] Frederick Warne & Co., London, UK, September 1904.

[22] See *Nature and Its Symbols*, by Lucia Impelluso, Translated by Stephen Sartarelli, The J. Paul Getty Museum, Los Angeles, 2004, p. 231. See also, "Jan Brueghel Complete Catalogue," https://www.janbrueghel.net/object/mouse-rose-and-butterfly-paris

[23] See "Pliny the Elder, The Natural History," John Bostock and H.T. Riley, Chapter 84 (58.) –"The Mice Of The Nile," http://www.perseus.tufts.edu/hopper/text?doc= Perseus% 3Atext%3A1999.02.0137%3Abook%3D9%3Achapter%3D84.

[24] *Mouse Muse: The Mouse In Art*, by Lorna Owen, Monacelli Press, New York, NY, 2014.

[25] Holliss, Richard; Brian Sibley, *Walt Disney's Mickey Mouse: His Life and Times*. New York: Harper & Row, New York, NY, 1986, pp. 40–45. See also, Inge, M. Thomas (September 30, 2014). "Mickey Mouse". In Apgar, Garry (ed.). *A Mickey Mouse Reader*. University Press of Mississippi. pp. 341–342.

[26] See "Natural selection in action during speciation," by Sara Via, *Proc Natl Acad Sci U S A.* 2009 Jun 16; 106(Suppl 1): 9939–9946. Published online 2009 Jun 15. doi: 10.1073/pnas.0901397106, PMCID: PMC2702801, PMID: 19528641, https://www.ncbi.nlm.nih.gov/pmc/articles/PMC2702801/

[27] "Diversity hotspots: Coldspots of speciation?" by Hélène Morlon, *Science* 11 Dec 2020:Vol. 370, Issue 6522, pp. 1268-1269, doi: 10.1126/science.abf0830, https://science.sciencemag.org/content/370/6522/1268

[28] See "Evolution of mosquito preference for humans linked to an odorant receptor," by Carolyn S. McBride, et al., *Nature*. Author manuscript; available in PMC 2015 May 13, Nature. 2014 Nov 13; 515(7526): 222–227. doi: 10.1038/nature13964, PMCID: PMC4286346, NIHMSID: NIHMS635307PMID: 25391959, https://www.ncbi.nlm.nih.gov/pmc/articles/PMC4286346/

[29] See "Geckos evolve rapidly in Brazil after new dam constructed," by Katie Langin, *Science*, July 31, 2017, https://www.sciencemag.org/news/2017/07/geckos-evolve-rapidly-brazil-after-new-dam-constructed, Accessed February 9, 2021.

[30] See "How humans drive speciation as well as extinction," by J. W. Bull and M. Maron, The Royal Society, *Proceedings Of The Royal Society B, Biological Sciences*, 29 June 2016, https://doi.org/10.1098/rspb.2016.0600, https://royalsocietypublishing.org/doi/full/10.1098/rspb.2016.0600

[31] See "Human predators outpace other agents of trait change in the wild," by Chris T. Darimont, Stephanie M. Carlson, Michael T. Kinnison, Paul C. Paquet, Thomas E. Reimchen, and Christopher C. Wilmers. *PNAS* January 20, 2009 106 (3) 952-954; https://doi.org/10.1073/pnas.0809235106, *PNAS, Proceedings of the National Academy of Sciences of the United States of America.* https://www.pnas.org/content/106/3/952. For an excellent commentary on these issues, see "Are Humans Creating More Species Than We're Killing?" by Emily Anthes, The Atlantic, October 11, 2016, https://amp.theatlantic.com/amp/article/503597/

Νεαρός ουρακοτάγκος, Εθνικό Πάρκο Tanjung Puting, Κεντρικό Μπόρνεο, Ινδονησία. Φωτογραφία © M. C. Tobias.

Κεφάλαιο 30

ΚΑΤΑΚΛΕΙΔΑ

Όλα τα γεγονότα του έτους 1802 στην Αγγλία φαίνονται σήμερα τόσο καθοριστικά όσο και οικεία. Ο απολυταρχικός Γεώργιος Γ' διένυε τα ώριμα χρόνια της μοναρχίας του, ενώ η Δυτική Ευρώπη παρακολουθούσε τον Ναπολέοντα να υπονομεύει συστηματικά τα ιδανικά που είχαν καθοδηγήσει τη Γαλλική Επανάσταση. Η λαϊκή κουλτούρα είχε μαγευτεί από το πρώτο μουσείο κέρινων ομοιωμάτων, την πρώτη αναφορά αστεροειδή, τα φωτογραφικά πειράματα, τη στήλη της Ροζέτας και μια πτήση ρεκόρ με αερόστατο με δύο άνδρες μέσα στο καλάθι. Ο παππούς του Κάρολου Δαρβίνου, ποιητής/φιλόσοφος/βοτανολόγος Έρασμος Δαρβίνος πέθανε, και ο μεγάλος Σκωτσέζος γεωλόγος Χιου Μίλερ αυτοκτόνησε αμέσως μετά τη συγγραφή ενός ποιήματος με τίτλο "Strange But True".[1] Ενώ ο Ζαν-Μπατίστ Λαμάρκ και ο Γκότφριντ Ρέινχολντ Τρεβιράνους πρότειναν ριζικά πρωτότυπες αναθεωρήσεις του δέντρου της ζωής και της εξέλιξης των ειδών, στο Παρίσι ο Φρανσουά Αντουάν Ροχ προειδοποιούσε τον κόσμο να σταματήσει την αποψίλωση των δασών. Ο Τσαρλς Μπελ δημοσίευσε το πρώτο ολοκληρωμένο του έργο με τίτλο "Anatomy of the Brain" (*Ανατομία του Εγκεφάλου*) και εντυπωσιακά χαρακτικά, και στο Έξετερ, ο Τζορτζ Μπόντλεϊ κατοχύρωσε με δίπλωμα ευρεσιτεχνίας τον πρώτο φούρνο κουζίνας.

Είναι η χρονιά που λέγεται ότι ξεκίνησε η *εκλαϊκευμένη επιστήμη*, κατά τη διάρκεια της οποίας το δημόσιο αίσθημα προκλήθηκε το ίδιο έντονα τόσο από την είδηση για την εύρεση απολιθωμάτων - τα πρώτα αποτυπώματα δεινοσαύρων που βρέθηκαν από ένα αγόρι στη Μασαχουσέτη – όσο και για τη νεοεφευρεθείσα "μηχανή υψηλής πίεσης" του Ρίτσανρτ Τρέβιτικ. Παρ 'όλα αυτά, ήταν η στιγμή που ο ποιητής Γουίλιαμ Γουόρντσγουορθ (1770-1850) ένιωσε διχασμένος ανάμεσα στην αγάπη του για τους ασφόδελους, και την ειλικρινή, κυνική του περιφρόνηση για την αυξανόμενη ηθική κατάπτωση και οικολογική υποβάθμιση στο Λονδίνο και τη βρετανική κουλτούρα γενικότερα, στο έργο του "Λονδίνο, 1802", στο οποίο απηύθυνε κάλεσμα προς την ιδιοφυΐα του παρελθόντος, της Χρυσής Εποχής, τον Τζον Μίλτον, και την εποχή που εκείνος έγραψε την απαισιόδοξη εκτίμησή του για τη ζωή, σε ένα απελπισμένο σονέτο 14 στίχων, με τίτλο "Ο κόσμος δεν μπορεί να μας ανεχτεί άλλο": "Με το να παίρνουμε και να ξοδεύουμε σπαταλάμε τις δυνάμεις μας … Έχουμε χαρίσει τις καρδιές μας αλλού, μια άσχημη ευλογία!... Γι 'αυτό, για τα πάντα, είμαστε εκτός σκοπού".[2]

Μόλις δύο χρόνια πριν, η τελευταία εναπομείνασα Γαλάζια Αντιλόπη (*Hippotragus leucophaeus*) είχε πυροβοληθεί μέχρι θανάτου από Ευρωπαίους αποίκους, καθιστώντας την έτσι την πρώτη γνωστή εξάλειψη είδους στην αφρικανική ήπειρο.

Μέχρι το 1802 εμφανίστηκαν άφθονοι μεταφορικοί προάγγελοι των γεγονότων του Ανθρωπόκαινου, ενώ οι οικο-ψυχολογικές επιπτώσεις τους εμφανίστηκαν σε πολλούς τομείς, της επιστήμης, της πολιτικής και των καλών τεχνών. Ήταν θέμα χρόνου ότι θα αναζητούσαμε απαντήσεις σχετικά με τη βιοχημεία της φιλανθρωπίας, τους φυσιολογικούς μηχανισμούς που εμπλέκονται στην αγάπη και τους συγκεκριμένους νευροδιαβιβαστές που σηματοδοτούν τη συμπεριφορά που ονομάζουμε καλοσύνη προς τους ξένους. Πιθανά αντίδοτα για την εποχή μας, αμυδρές αναλαμπές ελπίδας που προέρχονταν από μέσα μας.

Το 1936 απονεμήθηκε βραβείο Νόμπελ στους Ότο Λόουι και Χένρι Χάλετ Ντέιλ για την έρευνά τους σχετικά με τη χημική νευροδιαβίβαση, τη λεγόμενη και "αρχή της συνύπαρξης" σύμφωνα με την οποία οι νευρικοί παλμοί και κάθε νευρώνας περιέχουν πολλαπλά χημικά και ορισμένα ηλεκτρικά συναπτικά σήματα.[3] Το 1906 ο Ντέιλ είχε ανακαλύψει την πρωτεϊνική ορμόνη/νευροπεπτίδιο γνωστή ως ωκυτοκίνη. Έως τη δεκαετία του 1950 έγινε κατανοητό ότι αυτή συμμετέχει σε ένα παράξενο και θαυμαστό ταξίδι – που παρατηρείται σε σχεδόν όλα τα σπονδυλωτά – από τον υποθάλαμο και την οπίσθια υπόφυση προς το κυκλοφοριακό σύστημα, κατά τη διάρκεια, μεταξύ άλλων, του τοκετού στα θηλαστικά, και αποτελεί κατά κάποιον τρόπο βασική λειτουργία για την κοινωνικοποίηση, την ανάπτυξη δεσμών, το σεξ και τον έρωτα.[4] Στο πνεύμα αυτής της έρευνας, η Πατρίσια Σ. Τσέρτσλαντ ανέλυσε πρόσφατα διάφορα στοιχεία της χημείας του εγκεφάλου ως υποκείμενο πρότυπο για αυτό που θεωρούμε έκφραση της ανθρώπινης συνείδησης. Η έρευνά της την οδήγησε στη σφαίρα του "μονογαμικού δεσμού ζευγαριών" "και της πυκνότητας των υποδοχέων για συγκεκριμένες νευροχημικές ουσίες", ιδιαίτερα τις ορμόνες "ωκυτοκίνη και βασοπρεσίνη", όπως φαίνεται στις διαφοροποιημένες πυκνότητες των "υποδοχέων ωκυτοκίνης" στους αρουραίους των χωραφιών συγκριτικά με τους αρουραίους του βουνού, εκ των οποίων οι πρώτοι είναι πολύ πιο μονογαμικοί από τους δεύτερους.[5] "…η νέα γνώση των ιδιοτήτων της ωκυτοκίνης έχει αποδειχθεί χρήσιμη για την επεξήγηση αρκετών αινιγματικών χαρακτηριστικών της αγάπης", γράφουν η Σου Κάρτερ και ο Στήβεν Πόρτζες.[6] Η ίδια η καλοσύνη εξακολουθεί να μην αποτελεί θέμα ανάλυσης κανενός επιστημονικού πεδίου και στην περιφέρεια όλων αυτών[7]: από τις ανθρωπολογικές γνώσεις που αντλούνται από την ίδια τη φύση της προσφοράς δώρων,[8] τη "μοιρασιά του φαγητού" και τις "τελετουργίες φιλοξενίας"[9] στη βαθιά ριζωμένη ψυχολογία *όλων* των ανταλλαγών [10].

Σε γενικές γραμμές, η προσφορά - ο απόλυτος εκπρόσωπος της καλοσύνης, της ενσυναίσθησης, της ανοχής, της ενσυναίσθησης και της αγάπης – αποτελεί τον τελικό πρόδρομο αυτού που ο Γουόρντσγουορθ βρήκε ότι έλειπε στην εποχή του: της οικολογικής απελευθέρωσης. Εκτιμάται ότι η για την εκπλήρωση των 17 Στόχων Βιώσιμης Ανάπτυξης των Ηνωμένων Εθνών[11] απαιτείται ετήσια επένδυση της τάξης των περίπου 5 με 7 τρισεκατομμυρίων δολαρίων ΗΠΑ,[12] αν και οι αριθμητικές μεταφράσεις αυτού του είδους είναι μεταθετικοί υπολογισμοί, άγνωστοι σε οποιαδήποτε πραγματικά βιωματικά και μικροαναλυτικά εργαλεία από τη γεωγραφική περιοχή στον πολιτισμό.[13] Ακόμα όμως και αν μπορούμε να τους αξιοποιήσουμε, τότε μπορούμε να υιοθετήσουμε ένα εύκολα επιτεύξιμο σκεπτικό για την κατανομή των χρηματικών πόρων, λέγοντας ότι όλα τα χρήματα μπορούν να διατίθενται για χρήση κατά περίπτωση σε οποιοδήποτε παζλ οικοσυστημάτων και να καλύπτουν εξελικτικές ανάγκες και προορατικά μέτρα τα οποία είναι όλα εξαιρετικά σαφή σε αυτό το προχωρημένο στάδιο και στην ταχεία επιδείνωση της βιόσφαιρας. Απλά τυπώνεις χρήματα όποτε κινδυνεύουν ανθρώπινες ζωές και οι ζωές όλων των άλλων ειδών και πληθυσμών. Ας αφήσουμε τις μελλοντικές γενιές να ασχοληθούν με το χρέος, και ας επικεντρωθούμε στη διάσωση των ζωών σε αυτή τη γενιά, όσο υπάρχει ακόμα η ευκαιρία να το πράξουμε.

Ακριβώς όπως δεν μπορούμε να συσχετίσουμε λογικά οποιαδήποτε συγκεκριμένη αξία με τη φύση, δεν πρέπει να υποκύπτουμε στη χρήση των αλγόριθμων αντί της κοινής λογικής ή της υπερβολικής συμπόνιας. Έχει προταθεί ότι "τα υψηλά επίπεδα αβεβαιότητας παράγουν υψηλά ποσοστά αμοιβαιότητας".[14] Ωστόσο αυτή η έννοια υποδηλώνει επίσης ότι η *προσφορά* βασίζεται σε κάποιο επίπεδο ανασφάλειας, επιθυμίας για κάποια αξιωματική ισότητα ή αντασφάλιση, ενώ η αληθινή προσφορά, όπως και η αληθινή συνείδηση, είναι ξεκάθαρη: δεν υπάρχουν μισές αλήθειες και μισές αρετές, ούτε φιλοσοφίες μισής καλοσύνης οποιασδήποτε λογικής έλξης. Δεν επιδιώκουμε να συνδυάσουμε την αμοιβαιότητα με οποιαδήποτε υπόνοια ποιητικής ή εκδικητικής δικαιοσύνης, ούτε· να προσφέρουμε μόνο μέσω της υπόσχεσης μιας κάποιας ανταμοιβής. Δεν είναι αυτό το αγνό ιδανικό που προαναγγέλλουμε εδώ. Πράγματι, αυτού του είδους οι συναλλακτικές νοοτροπίες - δώσε για να πάρεις - έχουν σταθεί τεράστιο εμπόδιο για τη λογική βάση της προσφοράς και έχουν μπλοκάρει τη ροή οποιασδήποτε άνευ όρων επίδειξης ενδιαφέροντος και γενναιοδωρίας. Σπασμένο από δεσμούς που δένουν χωρίς ελεύθερη συναίνεση, δεν έχει υπάρξει καμία διαγενεαλογική ιεροτελεστία διέλευσης που να παρέχει καλοσύνη στο σύνολό της. Καλοσύνη που ζητά και δεν χρειάζεται τίποτα άλλο παρά τον ανιδιοτελή πολλαπλασιασμό της μεταξύ όλων εκείνων με τους οποίους έχουμε την τιμή και την τύχη να μοιραζόμαστε την ευλογία μιας δύναμης ζωής.

Αυτή η έλλειψη αιτημάτων μεταξύ των ειδών είναι που αποτελεί τη βάση για την άγρια συντροφικότητα και την ειλικρινή καταγραφή των πράξεών μας. Η πραγματική πρόγνωση και η εκτίμηση των γεγονότων του μέλλοντος.

Σημειωσεις

[1] Robert Farquharson Sharp, *A Dictionary of English Authors: Biographical and Bibliographical*, K. Paul, Trench, Trübner & Company, Limited, London, UK, 1904, σελ. 198.

[2] Tianyu Ma, "Boons, authority, and imagination: A reading of 'The World Is Too Much with Us'". ANQ: *A Quarterly Journal of Short Articles, Notes and Rev*iews, 2017; doi:10.1080/0895769X.2017.1385377.

[3] R.M.E. Sabbatini, "Neurons and synapses. The history of its discovery. IV. Chemical transmission," *Brain & Mind*, 2004, https://cerebromente.org.br/n17/history/neurons5_i.htm

[4] Βλ. D.D. Chiras, *Human Biology* (7th ed.). Jones & Bartlett Learning, Sudbury, Massachusetts, 2012, σελ. 262.

[5] *Conscience – The Origins of Moral Intuition*, by Patricia S. Churchland, W.W. Norton & Company, Inc., New York, NY 2019, σελ. 46.

[6] "Biochemistry of Love," by Sue Carter and Stephen Porges, NOBA, https://nobaproject.com/modules/biochemistry-of-lovenew%20knowledge%20of%20 the%20properties%20of%20oxytocin%20has%20proven%20useful%20in%20explaining%20several%20enigmatic%20features%20of%20love, Accessed February 11, 2021. See also, "The biochemistry of love: an oxytocin hypothesis," της Sue Carter και του Stephen Porges, *EMBO Rep*, v.14(1); 2013 Jan, PMC3537144, Journal List

[7] Βλ. "On the Biology of Kindness: is Kindness the Antidote of our Time?" by Nadia Waber, June 15, 2020, *Association of Women in Business,* http://awib.org.et/newsite/on-the-biology-of-kindness-is-kindness-the-antidote-of-our-time/Accessed February 11, 2021. See also, Zeki, S (June 2007). "The neurobiology of love". *FEBS Lett.* 581: 2575–9. doi:10.1016/j.febslet.2007.03.094. PMID 17531984. See also, McGregor IS, Callaghan PD, Hunt GE (May 2008). "From ultrasocial to antisocial: a role for oxytocin in the acute reinforcing effects and long-term adverse consequences of drug use?". Br. J. Pharmacol. 154 (2): 358–368. doi:10.1038/bjp.2008.132. PMC 2442436. PMID 18475254.

[8] "Gift Giving in Anthropological Perspective," by John F. Sherry, Jr., *Journal of Consumer Research*, Volume 10, Issue 2, September 1, 1983, σελ. 157-168, Oxford Academic, https://academic.oup.com/jcr/article-abstract/10/2/157/1801227?redirectedFrom=PDF

[9] Chris Hann, Hann, Chris, "The Gift and Reciprocity: Perspectives from Economic Anthropology," *Handbook on the Economics of Giving, Reciprocity and Altruism*, στο: S. Kolm & Jean Mercier Ythier (ed.), 2006, edition 1, volume 1, chapter 4, σελ. 207-223, Elsevier, https://ideas.repec.org/h/eee/givchp/1-04.html

[10] "The Psychology of Gift Exchange," των C. Mayet και Karen J. Pine, University of Hertfordshire, 2010, https://karenpine.com

[11] "Take Action for the Sustinable Development Goals," https://www.un.org/ sustainabledevelopment/sustainable-development-goals/, Πρόσβαση στις 11 Φεβρουαρίου 2021.

[12] "Shaping Philanthropy," του Josef Stadler, UBS Wealth Management, UBS Philanthropy, https://www.ubs.com/global/en/ubs-society/philanthropy/experiences/philanthropy-reports/shaping-philanthropy.html

[13] Βλ. για παράδειγμα: Colin Robson, *Real World Research. A Resource for Social Scientists and Practitioner-Researchers* (Second ed.), σελ. 624, Malden: Blackwell, Malden MA, 2002.

[14] Βλ. Richard L. West and Lynn H. Turner, *Introducing communication theory: analysis and application*, pp.147–165, McGraw-Hill Publishers, New York, NY, 2014

Σχετικα με τους Συγγραφεις / Μεταφραστρια

Συγγραφεις

Ο Μάικλ Κάρολος Τομπάιας, ο επί σειρά ετών πρόεδρος του Ιδρύματος Dancing Star Foundation (dancingstarfoundation.org), έκανε το διδακτορικό του στην ιστορία της συνείδησης από το Πανεπιστήμιο της Καλιφόρνιας – Σάντα Κρουζ, που ειδικεύεται στην παγκόσμια οικολογική ηθική και τα βιολογικά συστήματα, καθώς και στις διεπιστημονικές ανθρωπιστικές επιστήμες. Το ευρύ φάσμα του έργου του εκτείνεται στις οικολογικές επιστήμες της υφηλίου, στην Τέχνη, στη συγκριτική λογοτεχνία, την ιστορία των Ιδεών, στη φιλοσοφία και τη φυσική ιστορία. Ο Δρ Τομπάιας είναι συγγραφέας περισσότερων από 60 βιβλίων – μυθιστορήματα και επιστημονικά ή φιλοσοφικά βιβλία - και έχει γράψει, σκηνοθετήσει ή/και παράγει πάνω από 50 ταινίες (κυρίως ντοκιμαντέρ, δραματοποιημένα ντοκιμαντέρ και τηλεοπτικές σειρές). Τα έργα του Τομπάιας έχουν διαβαστεί, μεταφραστεί και μεταδοθεί σε ολόκληρο τον κόσμο και, επιπλέον, έχει πραγματοποιήσει αυτοπροσώπως οικολογικές έρευνες σε σχεδόν 100 χώρες. Πιο πρόσφατα, το βιβλίο του, "The Earth In Fragments: A Memoir" εκδόθηκε από τον εκδοτικό οίκο: Nova Science Publishers (2021).

Η Τζέιν Γκρέι Μόρρισον είναι οικολόγος που έχει ταξιδέψει σε περισσότερες από 30 χώρες. Η κα Μόρρισον ήταν παραγωγός πολυάριθμων ταινιών για τηλεοπτικά δίκτυα όπως το Discovery, PBS (όπου συν-σκηνοθέτησε το "A Day in the Life of Ireland" για την ιρλανδική τηλεόραση (Irish Television) και το WNET της Νέας Υόρκης), "Hotspots" (www.hotspots-thefilm.com) και την Turner Broadcasting, για την οποία εργάστηκε ως παραγωγός για το "Voice of the Planet", μια δραματική σειρά 10 ωρών, βασισμένη στην ιστορία της ζωής στη Γη. Ως Πρεσβευτής Καλής Θελήσεως στο Εθνικό Πάρκο Yasuni του Εκουαδόρ, δημιούργησε την ταινία μικρού μήκους "Yasuni - Ένας Διαλογισμός για τη Ζωή", της οποίας η πρεμιέρα παρουσιάστηκε στη Διάσκεψη Κορυφής των Ηνωμένων

Εθνών Ρίο+20 (UNCSD) το 2012. Τα βιβλία της περιλαμβάνουν: "Sanctuary: Global Oases of Innocence" (www.sanctuary-thebook.com), "Donkey: The Mystique of Equus Asinus", "God's Country: The New Zealand Factor" και "No Vacancy" (το οποίο συν-επιμελήθηκε). Έχει συγγράψει επτά βιβλία που εκδόθηκαν από τον εκδοτικό οίκο Springer Nature. Από το 1999, η Τζέιν Μόρρισον έχει υπηρετήσει ως Διευθύνουσα Αντιπρόεδρος του "Dancing Star Foundation", ενός μη κερδοσκοπικού οργανισμού που εστιάζει στα δικαιώματα των ζώων, στις διεπιστημονικές ανθρωπιστικές σπουδές και σε κινήματα κοινωνικού και οικολογικού δικαίου, καθώς αφορούν τη σχέση της ανθρωπότητας με τον κόσμο της Φύσης.

ΜΕΤΑΦΡΑΣΤΡΙΑ

Η Νίκη Σταύρου είναι Διευθύνουσα Σύμβουλος των Εκδόσεων Καζαντζάκης (1964), του εκδοτικού οίκου με έδρα την Αθήνα, η οποία εκδίδει τα βιβλία του Νίκου Καζαντζάκη. Ως πνευματική δικαιούχος του έργου του συγγραφέα, η Νίκη Σταύρου ίδρυσε και διαχειρίζεται το τμήμα δημοσίων σχέσεων και ξένων δικαιωμάτων των Εκδόσεων Καζαντζάκη, και συνεργάζεται με εκδοτικούς οίκους στην Ευρώπη, την Ασία και τη Βόρεια και Νότια Αμερική, όπως ο Simon & Schuster στις ΗΠΑ και ο Faber & Faber στο Ηνωμένο Βασίλειο. Ο πρόλογός της εμφανίζεται στην τρέχουσα έκδοση του Zorba the Greek, το οποίο εκδίδεται από τον Simon και Schuster. Η Νίκη διοργανώνει συνέδρια και εκδηλώσεις παγκοσμίως, δίνοντας διαλέξεις για τον Νίκο Καζαντζάκη, στην Ελλάδα και διεθνώς. Έχει δώσει πάμπολλες συνεντεύξεις σε ντοκιμαντέρ, τηλεοπτικές και ραδιοφωνικές εκπομπές και έχει διασκευάσει έργα του Νίκου Καζαντζάκη για το θέατρο, που ανέβηκαν και παρουσιάστηκαν σε διάφορα ιδρύματα, σχολεία και το Μέγαρο Μουσικής Αθηνών. Ήταν Επίκουρη Καθηγήτρια Αγγλικής Λογοτεχνίας στο Webster University, Athens, και Λέκτορας στο University of Indianapolis, Athens Campus.